Polar Oceans from Space

ATMOSPHERIC AND OCEANOGRAPHIC SCIENCES LIBRARY

VOLUME 41

For other titles published in this series, go to
www.springer.com/series/5669

Josefino Comiso

Polar Oceans from Space

Springer

Josefino Comiso
Cryospheric Sciences Branch, Code 614.1
NASA Goddard Space Flight Center (GSFC)
Greenbelt, MD 20771
USA
josefino.c.comiso@nasa.gov

ISBN 978-1-4939-3887-2 ISBN 978-0-387-68300-3 (eBook)
DOI 10.1007/978-0-387-68300-3
Springer New York Dordrecht Heidelberg London

Published by Springer Science+Business Media, LLC 2010
Softcover re-print of the Hardcover 1st edition 2010

Cover illustration: Wave@2009 JupiterImages Corporation

Printed on acid-free paper

Springer is part of Springer Science+Business Media (www.springer.com)

To my wife Diana who taught
me the warmth and joy of living

Preface

Only a few centuries ago, we knew very little about our planet Earth. The Earth was considered flat by many although it was postulated by a few like Aristotle that it is spherical based on observations that included the study of lunar eclipses. Much later, Christopher Columbus successfully sailed to the West to discover the New World and Ferdinand Magellan's ship circumnavigated the globe to prove once and for all that the Earth is indeed a sphere. Worldwide navigation and explorations that followed made it clear that the Earth is huge and rather impossible to study solely by foot or by water. The advent of air travel made it a lot easier to do exploratory studies and enabled the mapping of the boundaries of continents and the oceans. But aircraft coverage was limited and it was not until the satellite era that full coverage of the Earth's surface became available.

Many of the early satellites were research satellites and that meant in part the development of engineering measurement systems with no definite applications in mind. The Nimbus-5 Electrically Scanning Microwave Radiometer (ESMR) was a classic case in point. The sensor was built with the idea that it may be useful for meteorological research and especially rainfall studies over the oceans, but success in this area of study was very limited. It turned out that one of the most intriguing applications of this sensor was in the study of the sea ice cover since the emissivities of frozen and liquid sea water turned out to be very different. For centuries, the sea ice cover was just regarded as a part of the frozen North (or South) and nobody bothered to study it except the few curious scientists and explorers who felt that it has to be important since it is so vast. During the cold war, the military did a lot of research on the subject but such studies were mainly classified because the ice covered Arctic Ocean was part of a strategic warfare and was an ideal place for a well armed nuclear submarine to maneuver without being detected. It was soon realized that the sea ice cover and the polar oceans are important entities to study, not just for military but also for economic, environmental, and climatic reasons.

ESMR was among the first sensors to image and characterize the entire polar regions on a daily basis. It was the first sensor to reveal the true extent and distribution of the global sea ice cover as well as its seasonal variability. Data from ESMR led to many unexpected discoveries, including the existence of an unusually large Winter Polynya (unfrozen surface water within the ice pack) in the Weddell Sea in the 1970s, the size and persistence of which is counter intuitive and still a subject of scientific inquiry up to this day. We have come a long way since ESMR. Many sensors much more advanced than ESMR have been launched providing us with enormous capabilities for studying the complicated processes occurring in the polar regions and the variability of many of the parameters that are being measured by satellite sensors. Satellite data are now capable of providing a detailed history of the evolution of our planet during the last three decades including that of the polar regions.

This book provides introductory material about the polar oceans, more specifically the ice covered areas and peripheral seas as observed by satellite sensors. I owe a lot of materials for this book from many experts in the field some of whom I worked with as co-investigators and collaborators. I specially want to express my gratitude to my NASA/Goddard colleagues including, Drs. Jay Zwally, Per Gloersen, Claire Parkinson, Chuck McClain, Frank Carsey, Bob Bindschadler, and Don Cavalieri for many years of collaborative studies. My first love was particle physics and it was Jay who made me realize that the study of the polar regions and climate is an important and a very exciting continuation of physics research. The encouragements and support of Drs. Stan Wilson, Bob Thomas and more recently, Waleed Abdalati, Seelye Martin, and Tom Wagner who managed or manages the Cryospheric Sciences Program at NASA headquarters made it possible for me to do polar research at GSFC over the years. I am especially grateful to Dr. Norbert Untersteiner and Dr. Willy Weeks for many in-depth discussions about the polar climate system and the intricate processes associated with the sea ice cover. I have benefited from close collaborations with Drs. Arnold Gordon, Peter Wadhams, Stan Jacob, Steve Ackley, Manfred Lange, Neal Sullivan, Walker Smith, Koni Steffen, Rob Massom, and Ron Kwok. The writing of this book was started in 2006 while I was a fellow of the Japanese Society for the Promotion of Science and working as a Guest Professor at Chiba University, and I am grateful to Prof. Fumihiko Nishio, who was my host, for his encouragements and kind hospitality. Larry Stock of SGT and Robert Gersten of Adnet/RSIS have provided me excellent programming and analysis support at GSFC during the last several years, and I appreciate very much their dedicated efforts. Last but not least, I wish to thank my wife, Diana, my sons Glen and David and my daughter Melissa for encouragements and patience not just during the period of book writing but throughout my entire career as a polar researcher and adventurer. Many useful comments and suggestions were provided by Drs. Arthur Cracknell of the University of Dundee, UK, Markus Thorston, Head of the NASA Cryospheric Sciences Branch, Lawrence Mysak of McGill University and Edward Hanna of the University of Sheffield, UK. My daughter Melissa, daughter-in-law Cindy, and my wife Diana also provided some editorial help.

While writing this book, my mother, who had been a great source of inspiration all my life, died at the age of 100 years. It was an enormous personal loss for me, my brothers, and sisters but it is wonderful to realize that we can live to be centenarians and are able to experience a significant window of our changing planet and its changing climate. Our current collection of continuous satellite data is only a few decades long but we have undoubtedly learned a lot from them about the Earth's environmental and climate system and personally observed the subtle but important changes since the launch of Sputnik in the late 1950s. With all the satellite data we now have and with the advent of international ventures like the Global Earth Observation System of all Systems (GEOSS), we will find the answers of many of our current questions, but at the same time, we will undoubtedly find many more questions that need to be answered. The advances in our understanding of the Earth's climate engine, including the impact of the polar oceans have been tremendous and we look forward to being able to forecast effectively the state of our planet in the immediate and foreseeable future.

Greenbelt, MD Dr. Josefino Comiso

Contents

Chapter 1
Background

Abstract The polar oceans are important to study because they are key components of the Earth's climate engine. We have learned that many processes in the polar regions are highly linked to the thermohaline circulation of the World's oceans as well as their physical, chemical and biological characteristics. The region is vast, and it is only through satellite remote sensing that we have been able to examine and investigate the mesoscale and large scale properties of many parameters in the region, including surface temperature, albedo, sea ice cover, plankton concentration and primary productivity. A brief history of the polar regions and satellite remote sensing is provided, and the role of the polar regions in the climate system is discussed.

Keywords Polar regions • Satellite observing system

1.1 Introduction

What makes the Earth so unique in the planetary system and, perhaps, in the universe is the abundance of water. Water is the result of the bonding of one atom of oxygen and two atoms of hydrogen. In solid form, it covers about 12% of the oceans in the Northern Hemisphere and 9% of that of the Southern Hemisphere during midwinter period. In liquid form, it occupies more than 70% of the surface of the Earth. In gaseous form, water makes up more than 60% of all greenhouse gases in the Earth's atmosphere.

Water is an indispensable commodity, but its importance could easily be undermined because it is so readily available. Without water, the Earth would be a lot warmer during daytime than it is today, and it would be impossible for life as we know it to exist. The lush vegetation and the very diverse forms of life that inhabit the Earth came about only because of the presence and abundance of water. Without water, the Earth, when viewed from space, wouldn't have the color and radiance associated with cloud formations, remarkable land geometries, and beautiful sceneries.

J. Comiso, *Polar Oceans from Space*, Atmospheric and Oceanographic Sciences Library 41,
DOI 10.1007/978-0-387-68300-3_1, © 2010 United States Government as represented
by the Administrator of the National Aeronautics and Space Administration

Why do we need to study the oceans? The oceans are the largest bodies of water and are integral components of the Earth's climate engine. Averaging about 4 km in depth and spread out over almost three quarters of the Earth's surface, the oceans are big reservoirs of heat and serve as a moderating factor that keeps fluctuations of the surface temperature of the Earth from getting into extremes. They also serve as habitats of many forms of life at all trophic levels from microorganisms to mammals. And for the unsuspecting observer, their physical properties are constantly changing. Through evaporation and precipitation, through growth and decay of sea ice, through wind forcing and iceberg calving, and through river runoff, the physical characteristics of the oceans are always evolving. The chemical compositions of both the ocean and the atmosphere are also being altered by exchanges between these two systems through the emission and/or absorption of atoms and molecules. The oceans are also very dynamic as they are occasionally being battered by strong winds causing the formation of fronts, eddies, and a more organized movement of mass, called ocean currents. Underneath the surface, its character is slowly evolving as large masses of water are transported around the globe through the so called thermohaline circulation. Why and how the oceans are changing must be studied and analyzed to understand why and how the climate is changing.

Why do we study the Polar Oceans? By polar oceans, we usually refer to the relatively cold portions of the World's oceans located at high latitudes and mainly covered by sea ice in winter. In the Northern Hemisphere, they correspond to the Arctic Ocean and peripheral seas including the northern parts of the Pacific and Atlantic Oceans. In the Southern Hemisphere, they correspond to the southern portions of the Pacific, Atlantic, and Indian Oceans, where all these big oceans are united and intermixed together as huge water masses transported around the vicinity of the Antarctic continent through the Antarctic Circumpolar current. These regions have been combined into a single entity, sometimes referred to as the Southern Ocean. While generally ignored because of its remoteness and inaccessibility, many important processes that affect the global oceans, climate, weather and sea level occur in these polar oceans. Even just by virtue of its vast extent, the sea ice cover, which encompasses about 6% of the world's oceans, is an important climate parameter. More specifically, it serves as an effective insulator that keeps the warmth of the ocean from leaking to the atmosphere during the winter while it reflects solar radiation efficiently and keeps solar heat from being absorbed by the surface during the summer. It also causes spatial redistribution of salt during ice formation and decay. Furthermore, it causes the initiation of deep ocean convection that enables the ocean to ventilate itself. And during growth stages, it causes the formation of cold and high salinity bottom water which helps drive the aforementioned thermohaline circulation of the world's oceans.

The shelf regions of the polar oceans and peripheral seas are among the most productive regions of the World's oceans, as observed from space. The regions are sites of convergence zones where the cold and denser surface water sinks while the warmer and nutrient rich water upwells to take their place. At the same time, the melt of sea ice causes the formation of a stable layer of low density sea water at the surface where planktons can readily grow and multiply rapidly because of the abundance of sunlight needed for photosynthesis. In particular, sea ice causes the water to be conditioned for optimum productivity. There are other factors that affect the distribution of planktons, such as wind that induces upwelling, iron, clouds, and grazing, but these are not strictly polar parameters. Furthermore, the polar oceans are the coldest regions of the oceans where the oxygen content is relatively high. Such conditions are favorable to animal life, and with plankton being among the primary source of food, it is not a surprise that the most productive region and richest fishing grounds on Earth have been found at high latitudes. At the same time, there are large regions that are unproductive especially in the deep ocean regions of the Arctic and the southern part of the Indian Ocean where iron and/or nutrients are likely limiting factors.

A very contentious issue that human society has been dealing with in recent years is the potential change in the Earth's climate associated with anthropogenic greenhouse gases released to the atmosphere since the start of the industrial revolution. At the center of the issue is the net impact of atmospheric carbon dioxide concentration which has increased by 10% alone in the last 30 years and is now at 385 ppm (parts per million), which is significantly higher than the average range of 180–290 ppm during the last 650,000 years, as inferred from Antarctic ice cores. Concurrently, other greenhouse gases like methane and nitrous oxide have also been increasing. Although a significant warming signal has been documented in several studies, it is not clear what fraction of the signal can be attributed to greenhouse gas and what fraction is due to natural climate variability. An early resolution to the problem is likely to be found in polar regions because of a much greater sensitivity of the latter to global change. Modeling studies have shown that global warming signals, such as those associated with the observed increases in greenhouse gases in the atmosphere, could be amplified by 2–5 times in the polar regions because of ice-albedo feedback associated with the snow and ice that covers much of the regions. We now have a few decades of satellite data and our preliminary analysis of that data has already revealed that large changes on the surface of the Earth are occurring, especially in polar regions. Whether or not the reasons for these changes can be resolved to the satisfaction of every concerned segment of society is still not obvious in part because of the complexity of the Earth's climate system and the very demanding requirements in data and computer capabilities of physical numerical models that are needed for simulating such a system and for making sensitivity studies and forecasts for the future. Previous comparisons of results predicted by several of these models show big discrepancies if not inconsistencies. This indicates that we are still not capable of simulating the physics of such a complex system very well. But things have improved a lot in the last decade through many dedicated programs to gain insights into some of these complex processes. And as more and more global data accumulates, as our understanding of the important processes within the system improves, and as computer systems get more versatile and powerful, our assessment of the impacts of anthropogenic activities will become more and more accurate. The most recent IPCC report (2007) has indicated a 90% confidence that the observed warming is indeed mainly associated with the increasing greenhouse gases. This is already a big improvement to the previous report indicating a 50% confidence.

Why do we use satellite remote sensing? Satellite techniques currently provide the only practical way of obtaining synoptic and comprehensive data over the oceans and sea ice regions because the latter are so vast, generally inaccessible, and the site of hostile and harsh environment. Some of the important polar phenomena are, in fact, only possible to study using satellite remote sensing. Also, it is fortuitous that satellite data over ice-free and ice-covered oceans are relatively easier to interpret than those of land surfaces where the presence of vegetation, forests, human communities and mountains complicates retrieval of some parameters. We also have continuous data sets from the 1970s to the present which coincides with a time period when significant changes are occurring in the polar regions. Sensors observing the oceans from space, however, are not without limitations, since they are mainly sensitive only to the top layer of relatively deep oceans as discussed in Chapters 3 and 4. Nevertheless, some of the parameters can be inferred indirectly and the sensor data, in conjunction with in situ measurements and models, provide a very powerful tool for studying the processes that govern the Earth's climate system, and for evaluating on a global scale how this system has been changing.

This manuscript is meant primarily for those interested in polar oceanographic processes, sea ice and how the polar oceans have been changing as observed by satellite data. The basic remote sensing techniques and the algorithms used to transform digital data to geophysical parameters are also discussed in enough detail to help the uninitiated assess the amount of work needed to process and analyze satellite data, evaluate the limitations of the data and appreciate the enormous potential in the use of the data. The main thrust, however, is the presentation of results from analyses of the time evolution of polar surfaces as observed by satellite sensors and their relevance to the study of cryospheric and oceanic processes and our changing climate. Although the historical record of about 3 decades of continuous satellite observation is relatively short, these measurements have provided useful insights into the current state of the polar environment and serve as a powerful tool for studying our changing planet. The key satellite parameters that have been put together for polar studies are surface temperature, ice concentration, clouds, albedo, and ocean color. Satellite research data of the oceans and sea ice are not meant to take the place of conventional data derived from scientific cruises, moorings, buoys, submarines, AUVs and other oceanographic platforms. Neither can it replace acoustic measurement, which is another form of remote sensing, and are known to be extremely valuable for process and climate change studies. However, we should also realize that there are limitations of in situ and other data sets primarily in coverage in space and time. The strategy is to use both data sets and concurrently with models to expand our knowledge and extend our understanding of the Earth's climate system.

1.2 Early Perspective of the Polar Regions

While record shows that human colonization of the Arctic region may have started as early as 25,000 years ago, coordinated explorations of the Arctic Ocean and peripheral seas and also of the Southern Ocean did not start until only a few hundred years ago. A location map of the Arctic and the Antarctic regions as we now know it is depicted in Figs. 1.1 and 1.2. The first recorded encounter with sea ice

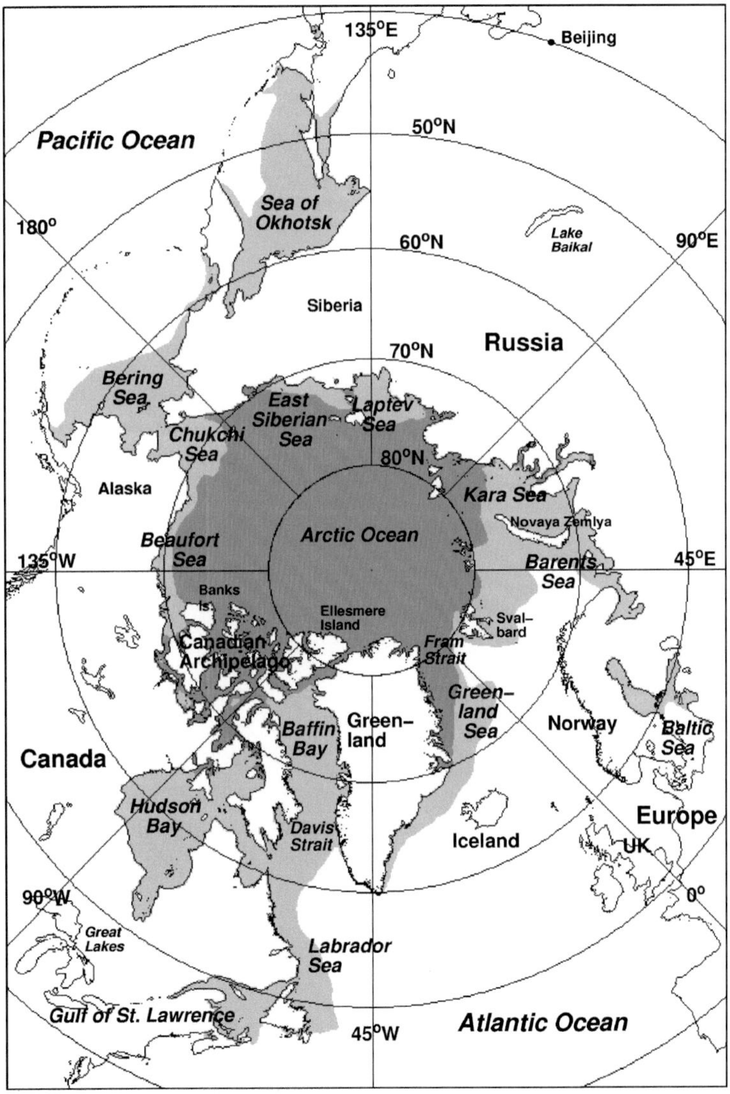

Fig. 1.1 Location map of the Northern Hemisphere at high latitudes including the Arctic Ocean and peripheral seas. Light gray area represents average location of the sea ice cover in winter while the dark gray area represent the average location of the sea ice cover during summer minimum extent

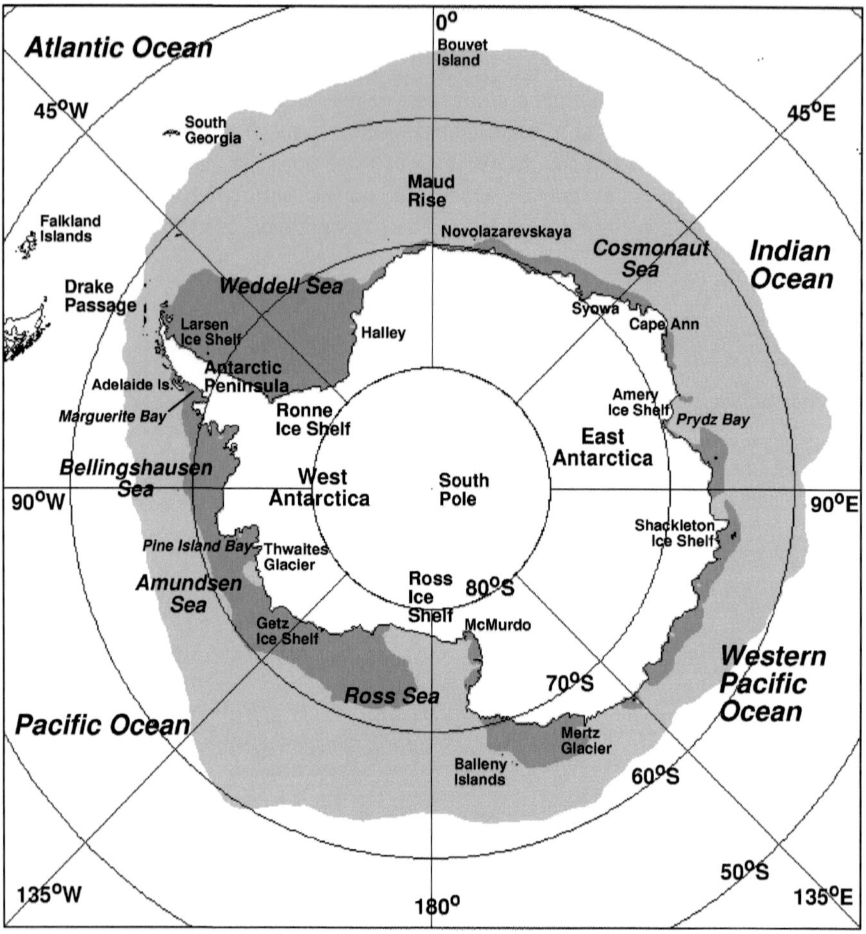

Fig. 1.2 Location Map of the Southern Hemisphere at high latitudes including the Southern Ocean and Antarctica. Light gray area represents average location of the sea ice cover in winter while the dark gray area represents the average location of the sea ice cover during summer minimum extent

by humans was that of a Greek explorer, Pytheas of Massilia, who in 325 BC sailed north and reached a place called Ultima Thule near Jan Mayen Island. Pancake ice was reportedly observed and it was possible that what they saw was part of what is now known as the Odden ice tongue, one of the distinct oceanographic features in the Greenland Sea. At about 550 AD, an Irish monk, Saint Brendan, undertook a voyage to the north, duplicating the feats of Pytheas of Massilia. It was possible that Saint Brendan went beyond Pytheas' farthest reach northward and if so, he set the record of having been to the farthest north and closest to the North Pole during his time. The next Arctic explorers were the Vikings who had previously lived exclusively in Southern Scandinavia (i.e., Norway and Denmark) for many centuries until the pressure of increasing population around 780 AD encouraged many to seek their fortune elsewhere. The Vikings were notorious as sea raiders and known to have terrorized all the coast of Western Europe, including most of Ireland and Scotland. They ventured north primarily to cover more territory to loot but, perhaps, they were accidentally blown to the northwest by stormy weather and encountered Iceland and Greenland. In the process, they settled in these regions and established communities which existed for a few centuries until the climate changed, and in the fourteenth century the settlements were abandoned because of extremely cold temperatures.

The Arctic basin was basically untouched until it was realized that alternative routes through the northern regions would drastically shorten the shipping lanes between the East and the West. The tales of Marco Polo prompted a strong interest in reaching the East in the fifteenth Century. At that time, the ship routes from Europe to Eastern Asia (or vice versa) were through the so called "Southeast Passage" and "Southwest Passage," which were dominated by the Portuguese and the Spaniards, respectively. The Southeast Passage involves a trip towards the southern part of Africa, through the Indian Ocean, then the Indonesian Seas and through the South China Sea. The Southwest passage consisted of a long journey across the Atlantic Ocean to the tip of South America, through Cape Horn, followed by a crossing of the vast Pacific Ocean. A new route through the Arctic from England to Japan for example, would cut these distances by almost a factor of three. The early explorations were done primarily by the English and the Dutch, who were then trying to compete with the Spaniards and the Portuguese, the World's naval powers of that era. The French, the Germans and the Russians were also very much involved in the explorations.

Two possible options for alternative routes were proposed: one called the Northeast Passage along the coastal regions of Russia and the other called the Northwest Passage along the North American side. Theoretically, such routes would provide a substantial savings in time and money, but these routes were conceived without prior knowledge of the predicaments in the region, and it took many years before explorers realized that it involved going through adversely cold and hostile conditions and what was apparently an almost impenetrable sea ice cover. The growing demand for fur in the fifteenth Century led to early explorations of the northern part of Russia, including Siberia, and it became obvious that much of the coastal regions were located north of the Arctic Circle and locked with sea ice, except for a few weeks in the summer. To get to the Pacific Ocean from the East, one needs to go through these coastal regions all the way to the end of the continent in Siberia and then through the Bering Strait which turned out to be the only opening from the Arctic to the Pacific Ocean. This made the Northeast Passage a very unattractive option. Nevertheless, early explorations for the Northeast Passage option were actually useful since they led to the discovery of Novaya Zemlya, Franz Josef Land, and other Arctic islands and much was learned about the Arctic during the period.

Although the North America region was already inhabited by Native Americans (Eskimos or Inuits) for quite some time, the western world knew very little about the region in the fifteenth century and progress on the exploration of the Northwest passage was very slow and incremental. Among the pioneers in the search for a direct passage to the Pacific Ocean were the French, who in 1524, outfitted an expedition under an Italian navigator, Giovanni da Verrazano, and reached Cape

Fear in North Carolina in April 17th of that year. He then navigated up north with the hope of finding a connection to the Pacific Ocean until he reached Newfoundland, but found nothing but land. Ten years later, a French navigator, Jacques Cartier investigated Newfoundland in greater detail and found the Gulf of St. Lawrence which he first thought would lead to the Pacific Ocean, but instead led to the Great Lakes. The first major attempt to move further north in search for a Northwest Passage was made by the English, headed by Martin Frobisher, who started at the coast of Labrador, crossing a strait and reaching the Baffin Island and as far north as 63° N. In 1587, John Davis followed where Frobisher left off and moved up the western coastline of Greenland reaching as far as 72.7° N, but gave up when he did not find any connection to the west. It was Henry Hudson, an Englishman, who found the strait that both Frobisher and Davis missed, which, to his surprise, led to an enormous inlet which we now call Hudson Bay. He thought that this would lead to China and headed to the southern part of the region reaching James Bay where his ship was frozen for 6 months. When the ice started retreating, he wanted to go further west but his crew refused and abandoned him adrift together with his son and seven loyal followers. Among the ironies of polar exploration, Hudson and his party probably died of hunger and cold. Further explorations of Hudson Bay were not so promising and in 1612, William Baffin decided to try the Davis Strait again and went as far north as the Ellesmere Island going through the narrow stretches of water in Smith Sound, Jones Sound and Lancaster Sound and concluded incorrectly that they were merely inlets. It turned out that these Sounds are openings to the Arctic but are blocked by sea ice, and, in fact, are not practical pathways to the Pacific Ocean.

When Baffin went back to England and asserted the nonexistence of a Northwest Passage, interest subsided in part, because by this time, the focus has shifted to the colonization of North America, and for a few explorers, the driving force was to be the first to get to the North Pole. It was not until the nineteenth century that new expeditions to explore the feasibility of a Northwest passage came about. One of the most famous of these expeditions was that of John Franklin, who, with two ships and a crew of 128, had laid out a well planned and well equipped expedition in 1845. They went north of Baffin Island and entered the Lancaster Sound but then disappeared and were not heard of again. The failed expedition captivated the English public, and the desire to find out what went wrong attracted at least 40 subsequent expeditions in the next 14 years. Eventually, in 1903–1906, the first transit of the entire Northwest Passage was completed by Roal Amundsen of Norway. Also, in 1932, a Russian icebreaker Sybiryakov completed the Northeast Passage in one season.

These aforementioned expeditions were usually used concurrently as platforms for scientific studies to gain an improved understanding about the physical characteristics of the Arctic environment. The collection of meteorological data during the First International Polar Year from 1882 to 1883 provided the framework for a more systematic scientific study of the weather and climate of the Arctic. A US ship, Jeannette, was crushed by sea ice in the Laptev Sea in 1881 and part of the wreckage turned up in the Greenland Sea, more than 4,800 km away 5 years later. This was the first evidence of the existence of a Transpolar current from Laptev Sea to

Greenland Sea. This information was used by Fridtjof Nansen who took a bold step and had his ship, FRAM, frozen in the Siberian Sea and let the ship and its crew drift around the Arctic from 1893 to 1896. During this period, Nansen studied many aspects of the Arctic environment and came up with the first quantitative report about the seasonal and interannual variability of many geophysical parameters in this region. Among the highlights of his study were the discovery of the existence of a deep Arctic Ocean and the confirmation of the presence of the Transpolar Current. Nansen's discoveries led to enhanced interest in the Arctic region and to many international projects dedicated to the study of the region. The ensuing studies led to a recognition that changes in the oceanographic conditions of the Arctic basin would have a big influence in the ocean circulation and therefore on the Earth's climate. More than a century later and with the help of numerical models and satellite data, the Arctic is now being viewed as among the most important regions for studying global climate change.

Observations by Harald Sverdrup on board the Maud from 1922 to 1924 provided much needed data to supplement Nansen's data. Climate records were routinely maintained at trading posts of the Hudson's Bay Company in Canada and along some Russian Arctic coastline harbors, but data over the Arctic Ocean were hard to come by until the first Soviet North Pole drifting station started in 1937 and continued until 1991. Also, the US ice island stations Arlis II, T-3 and others operated in the period 1952–1974 collecting surface and weather information. Arctic weather stations were established in the Canadian Archipelago in 1950 and in the 1960s, the Distant Early Warning (DEW) line radar sites at 65–70° N were established across North America. The cold war also led to extensive exploration of the Arctic Ocean by nuclear submarines but much of the information was generally classified. Nevertheless, submarine upward looking sonar data provided the first assessments of large scale thickness distribution of Arctic sea ice. In the 1970s, the drifting buoy program put together by the University of Washington got started to understand sea ice dynamics and continues to the present. Valuable data needed to study atmospheric and oceanic processes were acquired during the Arctic Dynamics Joint Experiment (ADJEX) project in 1975–1976, while ice edge processes were studied during several Marginal Ice Zone Experiment (MIZEX) projects in the late 1980s and early 1990s. The Surface Heat Budget of the Arctic Ocean (SHEBA) experiment was conducted from September 1997 to October 1998 as a follow-on to the ADJEX project and to better understand the physics of vertical energy exchange and help resolve some of the large discrepancies between global climate models. During this time, a ship was frozen in the study region and was used as the platform by several scientists for making many of the much needed atmospheric and surface measurements. After the SHEBA experiment, there was a realization that the Arctic is changing and this led to the launch of yet another project called SEARCH. Many other dedicated experiments by many countries were also performed during the last few decades.

Failure to find alternate routes in the Arctic was, in some sense, fortuitous because it meant that more expeditions to the south and around the Americas were needed. Some of these expeditions went much farther south than usual, leading to the discovery of the Antarctic continent surrounded by a very expansive and seasonal sea ice cover. A modern map of the Antarctic region is shown in Fig. 1.2. The discovery of Antarctica came about more because of a curiosity than a need. The lack of balance in the extent of land areas in the Southern Hemisphere when compared with that in the Northern Hemisphere had made many to wonder if there exists some undiscovered land area in the south. In 322 B.C., Aristotle was among the first to postulate the existence of a continent in the Southern Hemisphere. The first who tried to explore this possibility was the French naval officer, Pierre Bouvet who, in 1738, went as far south as 55° S, reaching an island which was named after him. Many other attempts were made after him but it was not until January 1773 that James Cook of England made it beyond the Antarctic Circle. This was the first time ever that a human being crossed this circle (i.e., 66.56° S). Cook went as far south as 71.17° S and could not go beyond that because of inability to penetrate sea ice. However, his discovery that the Antarctic waters are rich with seals and whales became one of the big incentives for future expeditions to the south. Furthermore, his observations of large icebergs made him to surmise the existence of a frozen continent to the south.

The Antarctic region became an important fishing ground, especially for sealers and whalers. In 1820, the Antarctic Peninsula was discovered separately by Edward Bransfield, a British naval officer, Nathaniel Palmer, a 21 year-old American sealer, and Fabian Bellingshausen, a Russian explorer. The first who really set foot on the Peninsula was an American sealer, John Davis, who in February 7, 1821 correctly conjectured in his log that the peninsula is part of an Antarctic continent. Twenty years later, an American explorer, Charles Wilkes followed the long stretches of coastline between Enderby Land and Adelie Land in the east, and upon combining his findings with the discoveries from the previous two decades, he was the first to proclaim that the land mass in the south that had already been called Antarctica is of continental size. The incentive to establish claiming rights and to be the first to reach the South Pole led to many expeditions that provided early assessments of the extent of the sea ice cover and the size of the continent. One of the most intriguing voyages was that of Ernest Shackleton, whose ship was beset by sea ice in the Weddell Sea on 14 December 1914. He had to abandon the ship, but he and his crew managed to drift with the ice for several months and miraculously got back to civilization without any fatality. Despite the big challenges of staying alive during the ordeal, some crew members of the Shackleton expedition managed to keep records and provided the first detailed account of the seasonal characteristics of the environmental conditions in the Southern Ocean, including those of sea ice and the polar atmosphere.

Many scientific cruises followed that of Shackleton, in part, to understand the oceanography of the region and the variability and significance of the sea ice cover. Scientific studies were carried out by many countries including those on board the German ship called Deutschland and the English vessels called Discovery II and William Scoresby. Contributions from the United States were mainly through the U.S. Navy Deep Freeze operations and the circumpolar study of the Antarctic Ocean from 1962 to 1972 by the ship USNS Eltanin operated by the National Science Foundation. A mysterious open water area within the ice pack, later called the "Weddell Polynya" during the middle of winter, was observed by a satellite sensor in 1974 and also became a subject of scientific inquiry and strong interest. Among the specific cruises that were meant to explore this phenomenon were the Weddell Polynya Expedition (WEPOLEX) in 1981, the Winter Weddell Experiment in 1986, and the Weddell Gyre Experiment in 1989. Many insights into the physics of polynya formation came out from these studies, but since the polynya did not recur during these expeditions, a definitive explanation is still missing.

Being the coldest, windiest, iciest and harshest environment on the surface of the Earth, the Arctic and the Antarctic regions have been difficult to explore. Nevertheless, over the years, our knowledge about the polar regions, and in particular, the polar oceans, have been tremendously enhanced with the advent of the satellite era. The polar regions are regarded as the heat sinks of the world, and are important components of the climate system. They have been also the source of ice cores and sediment information that have provided hundreds of thousands of years of environmental and climate records that are currently being exploited to enhance our knowledge about the history of the Earth's climate system. Undoubtedly, surface exploration will continue to be important, and knowledge acquired on the surface will strongly complement data from new satellite technologies.

1.3 Development of Satellite Earth Observation Systems

Remote sensing has been defined simply as the act of making measurements from a distance. The human eye has been used as a remote sensing tool for centuries and has been effective for studying the general characteristics of the cosmos leading to important applications in navigation and the construction of calendars. But as a tool for making quantitative assessments, there are obvious limitations. With the introduction of high powered optical telescopes in the sixteenth century, our knowledge about the physical characteristics of these foreign objects vastly improved. The studies by Galileo during this period and subsequently by Newton and others culminated in concepts that became the cornerstone of physical sciences. These studies were the catalyst for orderly and organized remote sensing studies of the universe. The development of modern optical and other imaging devices, especially those launched on board satellites like the Hubble Telescope, coupled with the use of computer visualization, led to major breakthroughs in our understanding of the planets, the galaxies, and the universe. These were mainly Earth based studies of space objects and it was unthinkable only a few decades ago to do something similar for the Earth until the advent of satellite technologies. It is interesting to know that the size and spatial distribution of the polar ice cap in Mars were quantified (by Hershel and others in the seventeenth century) before we were able to do the same for the ice caps on Earth.

The launch of Sputnik in 1958 marked the realization that the Earth can be studied effectively from space. Photographs of the Earth about 36 years apart are shown in Fig. 1.3a and b. Figure 1.3a was taken by an astronaut during the Lunar missions in 1972 and Fig. 1.3b was taken by a high definition television camera on board a Japanese lunar explorer (KAGUYA) in 2008. Both images make it obvious that from a perspective away from the Earth, the Earth can be monitored as the other

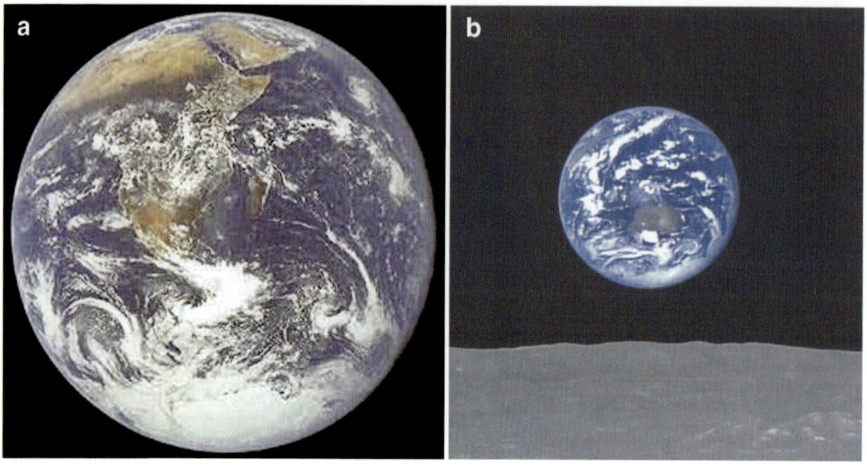

Fig. 1.3 (**a**) Image of the Earth as taken by Apollo 17 lunar module pilot and scientist astronaut Harrison A Smitt during the flight from the Earth to the Moon on 7 December 1972. Antarctica is located at the bottom of the image. Photo courtesy of NASA. (**b**) Image of the Earth as observed on 30 September 2008 by the Japanese lunar explorer "Kaguya" which is currently flying in a lunar orbit at an altitude of 100 km. Photo courtesy of JAXA (Japan)

planets. Among the earliest research satellite systems intended for such purpose is the Television Infrared Observation Satellite (TIROS) which was launched by NASA in April 1960. A scanning vidicon system (similar to those used by television cameras) on board the satellite provided one of the first imaging data of the planet Earth. Important practical applications were quickly recognized, such as improvements in weather prediction, assessments of natural resources, and studies of land and atmospheric pollution.

For optimal applications, sensors exploiting′ the visible, infrared and microwave part of the electromagnetic spectrums were developed. Initially, visible data were easiest to interpret since they relate closely to human experience, but other frequencies provided unique data that provided new information. For example, the infrared spectrum makes it possible to obtain data even in the dark and to get complementary information about the atmosphere and atmospheric effects needed to retrieve surface parameters. But surface observations are possible to obtain with visible and infrared data only under clear skies conditions. Fortuitously, microwave sensors provide the means to obtain surface information even during cloudy and adverse weather conditions. Resolution is sometimes a factor, since it is inversely proportional to wavelength, and while the visible sensors usually provides very good resolution, the resolution from microwave sensors has not been as good. However, microwave sensors provide global coverage at a good temporal resolution. The information inferred from the different types of sensors is usually different but complementary and sometimes the concurrent use of all three provides the best information.

The first US infrared sensor was launched on board the second TIROS satellite, called TIROS-2, in March 1961, while 8 other TIROS satellites using a combination of

IR and visible sensors were launched in subsequent years. This set of satellites clearly showed that space-based data can be extremely useful. However, the orbits of these earlier satellites have relatively low inclination (48–58°) and therefore, the coverage of the polar regions was poor. It was not until the launch of TIROS-9 and 10, which were put on a quasi-polar sun-synchronous orbit that near global coverage and synoptic observations of the polar regions became possible. With the success of the TIROS satellite series came the need of more data, especially after the realization that they will lead to more accurate weather forecasts. The system was followed by the launch of the TIROS Observational Satellite (TOS) series that began in February 1966 with the idea of making them operational, which means providing daily global coverages without interruptions. As an operational system that guarantees continuity, it was renamed the Environmental Science Services Administration (ESSA) series. Launched at altitudes between 1,500 and 1,700 km at an inclination of 102° and with onboard tape recorders when receiving stations were out of sight, the data found many applications in the polar regions including sea ice data for ship navigation in the region. The system was further enhanced with the Improved TIROS Operational Satellite (ITOS) series started in January 1970 that became the first system that relied on scanning radiometers instead of optical cameras. Further refinements led to the Advanced Very High Resolution Radiometer (AVHRR) onboard a series of NOAA satellites, which began operation in 1978 and has continued to provide data up to this day. The characteristics of this sensor will be discussed in detail in the satellite section.

Concurrent with the operational satellites were research satellites like the NASA Nimbus satellite series, which allowed for the development of improved capabilities through the launch of new and advanced satellite sensors. This series led to the introduction of sophisticated sensors including the forerunner of the Landsat series, namely: the Electrically Scanning Microwave Radiometer (ESMR), the Scanning Multichannel Microwave Radiometer (SMMR), Coastal Zone Color Scanner (CZCS), Thermal Humidity Infrared Radiometer (THIR), Earth Radiation Budget Experiment (ERBE) and Total Ozone Mapping Spectrometer (TOMS). As research for advanced systems continued, data acquired from sensors in the Nimbus series ignited the interest of many scientists who marveled at the unexpected phenomena like the occurrence of the large polynyas in the Weddell Sea as revealed by the ESMR data, the elevated phytoplankton blooms at high latitudes seen in CZCS images, and the big ozone hole in the Antarctic region revealed by TOMS.

The first satellite with the primary mission of monitoring the global oceans was the SeaSat (Sea Satellite), which was launched in July 1978. The sensors on board this satellite were the SMMR, Synthetic Aperture Radar (SAR), Scatterometer (SCAT), and Radar Altimeter (RALT). Details about these or similar instruments will be discussed in Chap. 3. Except for SMMR, which was also on board the Nimbus-7 satellite, these sensors were active microwave systems. Unfortunately, the satellite stopped providing data after only 3 months of operation. The 3 months of data, however, contained a wealth of unique information at that time about the ocean and clearly demonstrated that from space, one can infer wave heights, wind velocity, surface current, dynamic topography, ocean eddies, and sea ice distributions. These alone should have provided the justification for the launch of more satellites carrying these sensors. But cost consideration was a big factor, and it was not until the 1990s that satellites carrying these sensors were launched and they were not US satellites. The first was the European Space Agency (ESA)/ERS-1 satellite that carried SAR, NSCAT, and RALT, launched in 1991. This was followed by a second system called ERS-2, the Japanese NASDA/JERS-1 and JERS-2 satellites that carried SAR, and NSCAT, and a SAR system on board a Canadian satellite called RadarSat which was launched in 1995.

While all these global observing systems were being developed, there was a growing concern about impacts of anthropogenic activities and the changing habitability of the planet. Modeling studies suggested significant warming associated with the increasing greenhouse gases in the atmosphere observed by land based systems, but the accuracy and reliability of these models have been hampered by the lack of global data. A major program that is meant to address this need was started in the 1980s by NASA and has been referred to as the Mission to Planet Earth. The primary argument for this project was that the Earth may be the only entity in the universe with intelligent life and we should do whatever it takes to ensure that our advanced form of life continues to exist. The project was quite ambitious with objectives to understand the Earth's climate and how it is impacted by greenhouse gases through the combined use of satellite data and numerical models of the climate system. It involves the launch of major satellite systems equipped with sophisticated sensors that take advantage of the most current technology to obtain as comprehensive and accurate data of the Earth as technology would allow. After several iterations, the project included the launch of three major US satellite systems called Terra, Aqua, and Aura in 1999, 2002 and 2005, respectively, which are the key component of what is now known as the NASA/Earth Observation System. Among the sensors of interests for polar oceans applications are the Moderate Resolution Imaging Spectroradiometer (MODIS), the Advanced Spaceborne Thermal Emission and Reflection Radiometer (ASTER), and Multi-angle Imaging Spectro Radiometer (MISR) in Terra and MODIS and the Advanced Microwave Scanning Radiometer (AMSR-E) in Aqua. These sensors have provided a wealth of high quality information about the Earth System. Complementary information have been provided by ESA satellite systems like the ENVISAT launched in 2002, which is equipped with an Advanced Synthetic Aperture Radar (ASAR), Radar Altimeter 2 and Microwave Radiometer, Advanced Along-Track Scanning Radiometer (AATSR) and Medium Resolution Imaging Spectrometer (MERIS), and are complemented by similar satellite projects from other countries. Equally ambitious systems like the JAXA/Adeos-2 satellite, launched by JAXA (Japan) in December 2002 with Global Imager (GLI), Advanced Microwave Scanning Radiometer (AMSR), Polarization and Directionality of Earth's Reflectances (POLDER) and SeaWinds sensors, but unfortunately the satellite system malfunctioned after 9 months of operation. High resolution imaging systems were also launched aboard the JAXA/Advanced Land Observing Satellite (ALOS) including the Phased Array type L-band Synthetic Aperture Radar (PALSAR) and the Advanced Visible and Near Infrared Radiometer type 2 (AVNIR-2). In recent years, many satellite sensors have also been launched by other countries like Korea, India, China, Taiwan and other countries.

With so many launches by so many countries, it is difficult to keep track of all the sensors that have been launched and the discussion of all of them is not within the scope of this book. But many of these sensors provide similar capabilities and some countries want to help fulfill the overall mission by providing replacement systems if or when the current ones cease to operate. A comprehensive summary of the various systems and their applications is provided by Kramers (2002), Martin (2004), and Lubin and Massom (2006). With so much data, there is obviously a need to get all satellite data sets organized, consolidated and made available to users. There is currently a major attempt to fulfill such consolidation of all such satellite data under an organizational umbrella called the Global Earth Observing System of all Systems (GEOSS). The long term goal is eventually to involve all countries of the World in the monitoring of the state of planet Earth and make everybody realize that we are all on board the same ship and as individuals or groups or countries and we need to do our part in ensuring that the planet is provided with the attention and care needed to make it habitable not just for the present but also for future inhabitants.

1.4 The Role of the Polar Regions in the Climate System

Many climate studies have been focused on the tropical and subtropical regions, because compared to the polar regions, these regions have a much greater concentration of energy and have much larger effect on the state of the troposphere through latent heat contributions. Also, the tropics control the variability of some of the major climate phenomena like the El Nino Southern Oscillation. It is thus believed that the important processes occur somewhere else other than the polar regions and the latter is being studied only because it is part of the climate system and the role of each component needs to be understood. However, it has been argued by some (e.g., Alley 1985) that it is really the polar regions, and the Arctic in particular, that control global climate change. The major argument in this regard has been that much of the global response relative to changes in the seasonality and strength of insolation reaching the Earth is controlled by conditions at high latitudes. The high frequency as well as the low frequency oscillations in the climate systems are strongly influenced by constants that originate in the Arctic region. For example, it is the ice sheets and the North Atlantic Oceans (through deep water formation) that has been hypothesized to be the leading controls for these oscillations. The Southern Ocean has also been considered the primary source of high salinity bottom water that is a key driver of the global thermohaline circulation (Broecker 1997). Thus, even minor changes in the characteristics of the sea ice cover in the Southern Ocean could cause alterations in the global circulation patterns of the oceans. Such changes could have a considerable effect on the climate.

The role of the ice-albedo feedback in the amplification of the climate signal in the polar regions has also been cited as one of the key elements of the climate system (Kellogg 1983). By feedback, we mean the interaction of related parameters in a circular loop that in effect cause the amplification (or reinforcement) of the magnitude of the change. The variables in the ice-albedo feedback are ice and snow cover, albedo, and surface temperature. An example of such feedback is the reduction of albedo due to a decline in the ice cover that in turn would cause more heat from the sun to be absorbed by the surface, thereby increasing the surface temperature. The increase in surface temperature would cause further decline in the ice cover and the cycle repeats itself in the loop. Thus, through its albedo and thermal properties, the sea ice and snow cover exert tremendous control on the energy budgets at both hemispheres. With the Arctic perennial sea ice cover declining at a relatively fast rate as observed previously (Comiso 2002; Stroeve et al. 2007; Comiso et al. 2008) and discussed in this book, we may get to witness in our lifetime a possible major climate phenomenon. While the impact of greenhouse gases is being debated, this perturbation in the ice cap that is at least in part caused by warming during the last few decades may, in fact, become the center of scientific research and public attention. The advent of a blue ocean in the Arctic during summer is no longer a remote possibility on account of ice-albedo feedback effects and such phenomenon could have profound consequences that include a change in the climate system not just in the region but worldwide.

There are many lingering issues that needs to be unraveled in both polar regions. In particular, our knowledge about accepted high and low frequency phenomena, including the Arctic Oscillation and the Southern Annular Mode, needs to be improved especially since their links with in situ and satellite observations have been relatively weak (Overland and Wang 2005; Kwok and Comiso 2002). The role of the atmosphere on the changing polar ice cover has to be better understood in order to be able to assess the true impact of anthropogenic greenhouse warming as well as ice-albedo feedback effects on the climate system. The changes that are being observed must also be analyzed in greater detail so we get to know the physics behind these changes and minimize misinterpretation of the data or unnecessary speculations.

The availability of satellite data has enabled significant advances in our knowledge of the polar regions and provided the opportunity to better understand the system. In particular, with almost 3 decades of concurrent data on many parameters, we can study phenomenologically how these regions have been changing during this time period and at the same time use the data as input for numerical models that simulate the climate system. Not surprisingly, and because of their global coverage and unique capabilities, space-based data have led to many scientific breakthroughs in polar oceanography. Among these are the discovery or observation of some major phenomena such as the large Weddell Polynyas in the 1970s, the Odden ice tongue, and the Antarctic Circumpolar Wave.

In this book, we discuss the various satellite sensors and their capabilities, the techniques for transforming the data to geophysical parameters and the results of analyses of time series satellite data sets that tells us how the polar oceans have been changing. The key global data sets available for long term time series studies are the passive microwave time series from 1978 to 2007 the visible and infrared time series from 1981 to 2007, and the ocean color series that started in 1978.

However, only ocean color data from 1997 to 2007 are useful for time series analysis because of gaps and the paucity of coverage from the CZCS sensor. Other satellite data sets are presented in the context of providing an improved and consistent interpretation of the long term satellite data sets. A time series of about three decades may be too brief for climate change studies but they provide a unique perspective in terms of global coverage during a period of significant changes in the polar oceans. In addition to the observation of the drastic change in the Arctic perennial ice cover, the series enables the discovery or detection of other phenomena that could profoundly affect our climate, our environment and the inhabitants of our planet.

References

Alley RB (1985) Resolved: The Arctic controls global climate change. In: Smith WO, Grebmeier JM (Eds) Actic Oceanography: Marginal ice zones and continental shelves, American Geophysical Union, Washington, DC

Broecker WS (1997) Thermohaline circulation, the Achilles heel of our climate system: Will man-made CO2 upter the current balance? Science 278(5343):1682–1588

Comiso JC (2002) A rapidly declining Arctic perennial ice cover. Geophys Res Letts 29(20), 1956, doi: 10.1029/2002GL015650

Comiso JC, Parkinson CL, Gersten R, Stock L (2008) Accelerated decline in the Arctic sea ice cover. Geophy Res Lett 35, L01703, doi: 10.1029/2007GL031972

Kellogg WW (1983) Feeback mechanisms in the climate system affecting future levels of carbon dioxide. J Geophys Res 88(C2):1263–1270

Kramer HJ (2002) Observation of the Earth and its environment: Survey of missions and sensors. Springer, Berlin, Germany

Kwok R, Comiso JC (2002) Spatial patterns of variability in Antarctic surface temperature: Connections to the Southern Hemisphere Annular Mode and the Southern Oscillation. Geophys Res Lett 29, doi:10.1029/2002GL015415

Lubin D, Massom R (2006) Polar remote sensing: volume 1: atmosphere and oceans, Springer, Chichlester, UK

Martin S (2004) An introduction to ocean remote sensing, Cambridge University Press, Cambridge, UK

Overland JE, Wang M (2005) The Arctic climate paradox: The recent decrease of the Arctic Oscillation. Geophys Res Letts 32, L06701, doi: 10.1029/2004GL021752

Stroeve J, Holland MM, Meier W, Scambos T, Serreze MC (2007) Arctic sea ice decline: Faster than forecast. Geophys Res Lett 34, L09501, doi: 10.1029/2007GL029703

Chapter 2
Fundamental Characteristics of the Polar Oceans and Their Sea Ice Cover

Abstract The physical, chemical, and biological characteristics of the oceans and their sea ice cover at high latitudes in both the northern and southern hemispheres are discussed in conjunction with the bathymetry, atmospheric and circulation patterns. The exchanges of intermediate to deep water with the World's oceans are quite limited in the Arctic region with the relatively narrow Fram Strait as the only possibility while the World's main oceans circulate freely from one ocean to another at all depths in the Southern Ocean through the Antarctic Circumpolar Current. The perennial sea ice is also located in deep water regions in the Arctic, while the perennial ice in the Antarctic is usually confined in shallow shelf areas. These and other environmental differences make the characteristics of Arctic Ocean very different from those of the Southern Ocean.

Keywords Peripheral seas • Ice drift • Ocean bathymetry • Dynamic topography

2.1 Introduction

The polar regions are unique in that they are practically in total darkness during the winter and in almost continuous daylight in the summer. The associated seasonal changes in solar insolation cause large seasonal variations in surface air temperatures that in turn make both the solid and liquid components of the polar oceans very seasonal. Because of the high heat capacity of water, this also involves a large seasonality in the release of heat fluxes. In the Arctic, the extent of the sea ice cover increases from about $7 \times 10^6 km^2$ in the summer to about $16 \times 10^6 km^2$ in the winter while in the Antarctic, the corresponding values are $2 \times 10^6 km^2$ in the summer to $19 \times 10^6 km^2$ in the winter. The average thickness of sea ice is not as well known, but has been estimated to be between 2 and 3 m in the Northern Hemisphere and less than 1 m in the Southern Hemisphere (Wadhams and Comiso 1992). The process associated with the transformation of such a massive amount of water from liquid to solid and back to liquid within an annual cycle is what makes the physical,

J. Comiso, *Polar Oceans from Space*, Atmospheric and Oceanographic Sciences Library 41, DOI 10.1007/978-0-387-68300-3_2,

chemical and biological characteristics of polar oceans very different from those of other regions.

The geographical setting of the two polar regions and the general location of the sea ice cover during winter and summer based on climatology are shown in Fig. 1.1. and Fig. 1.2. The sea ice cover is generally classified as either perennial ice or seasonal ice. By perennial ice, we refer to ice that survives the summer melt and consists mainly of multiyear ice floes which are the mainstay of the ice cover. By seasonal ice, we refer to ice that forms and decays during the same ice season. In Fig. 1.1, the average location of the seasonal sea ice is represented in gray, while that of the perennial ice is represented in light gray. It is apparent that geographical settings for sea ice are very different in the two hemispheres. In the Northern Hemisphere (NH), a large fraction of the sea ice cover is surrounded by land with the edge of the latter automatically becoming the outer limit of the ice cover in winter. During maximum extent in winter, the ice cover is shown to be fragmented with the largest fraction located in the Arctic basin and the rest in the Canadian Archipelago, bays and peripheral seas located as far south as 40° N (Fig. 1.1). In the summer, the ice cover is basically contiguous and confined mainly to the Arctic Basin (Fig. 1.1). On the other hand, the sea ice cover in the Southern Hemisphere (SH) surrounds the Antarctic continent with the highest latitude being in the Ross Sea at about 80° N but no apparent limit at lower latitudes. The SH sea ice cover is shown as a contiguous ring surrounding Antarctica in winter (Fig. 1.2) and covering a significant part of the Indian, Pacific, and Atlantic Oceans. During minimum extent at the end of the summer (Fig. 1.2, gray), the ice cover is fragmented and located mainly in the Western Weddell Sea, Bellingshausen Sea, and Amundsen Sea and parts of the Ross Sea and the Indian Ocean.

Geographical considerations play an important role in defining the distribution of the sea ice cover and the associated characteristics of the polar oceans. The huge difference in geographical distribution of land and environmental conditions (Fig. 1.1 and Fig. 1.2) are the key factors that make the sea ice cover and the oceans in the Northern and Southern Hemispheres very different. When compared with the other oceans, the Arctic Ocean is relatively small, covering only about $14.1 \times 10^6\,km^2$ or an area almost the same size as the Mediteranean Sea. Some parts of the ocean are relatively deep but because of extensive shallow shelf areas, the average depth is only about 1,300 m and the total volume is about $18 \times 10^6\,km^3$. The borders of the Arctic Ocean are not very well defined, but usually include the Central Arctic, Baffin Bay, Barents Sea, Beaufort Sea, Chukchi Sea, East Siberian Sea, Greenland Sea, Hudson Bay, Hudson Strait, Kara Sea, Laptev Sea and the Canadian Archipelago. In addition to the Arctic Ocean, we include the Bering Sea, the Okhotsk Sea, Japan Sea, Labrador Sea, the Gulf of Bosthnia and the Baltic Sea as parts of the polar ocean in the Northern Hemisphere.

In the Southern Hemisphere, there is no counterpart to the Arctic Ocean. At the same general location, we have the Antarctic continent about the same size as the Arctic Ocean that is surrounded by what has been referred to as the Southern Ocean. The Southern Ocean has been defined as the region bounded by Antarctica to the south and the southern polar front to the north (i.e., >45° S). It has not been officially considered an ocean and is regarded instead as the southern part of the

Pacific, Atlantic and Indian Oceans. The area south of the polar front is about 38×10^6 km^2 and the volume is about 140×10^6 km^3.When compared with the Arctic, it has about 3 times greater area and about 8 times greater volume. About 50% of the region is usually covered by sea ice during winter maximum ice extent. In this book, we will define the boundaries of the Southern Ocean in the same manner. It should be pointed out, however, that in 2000, the International Hydrographic Organization (IHO), decided to declare the Southern Ocean as one of the world's oceans with the boundary to the south being still the Antarctic continent, but to the north it is the 60° S latitude. The total area of this revised boundary, which includes the Antarctic Circumpolar Current (ACC) and the Weddell Sea, Cosmonaut Sea, Ross Sea, Amundsen Sea and Bellingshausen Sea, is 20.33×10^6 km^2. As defined by IHO, the area of the Southern Ocean is almost the size of the sea ice cover during winter maximum extent and does not encompass all of the sea ice covered regions since the ice edge sometimes goes as far north as 55° S. Many countries want the region to be extended further north (mostly to 50° S), while some including many scientific societies do not accept the official designation and continue to assume that the Atlantic, Pacific and Indian Oceans extends to the shores of Antarctica.

2.2 The Polar Oceans

2.2.1 Ocean Bathymetry, Circulation and Dynamic Topography

The depth of the oceans, called bathymetry, is a parameter that does not change much over thousands of years, but its short and long term influences on the physical, chemical and biological characteristics of the oceans can be profound. The realization of the importance of detailed knowledge of bathymetry has led to huge efforts by many institutions to cover at least the critical areas of the World's oceans. Almost continuous echo-soundings by ships to cover as much territory as possible have been undertaken to help meet this goal. While the coverage is still not complete, the efforts led to what is now called the General Bathymetric Chart of the Oceans (GEBCO). Coverage of the polar regions had been relatively poor because of very restricted accessibility in areas covered by thick first year and multiyear ice. However, things have improved considerably with the availability of other platforms, such as submarines and submersibles, and through dedicated efforts of many individuals and institutions (e.g., Vanney and Johnson 1985; Jakobsson et al. 2000; IOC et al. 2003). The technology has also improved with the development of multibeam sonar, sidescan sonar, seismic reflection and refraction, and satellite altimetry. As the name implies, the multibeam sonar enables measurements over a wide swath of ocean bottom and therefore covers more area than the echo sounders. The sidescan sonars are usually installed at the side of a ship and can provide extra details about the relief and irregularities in the sea bottom. The seismic reflection and refraction systems allow for analysis of structure of the bottom surface and can reveal layers, thickness and sediment types. Satellite altimetry is not as accurate as the other techniques, but is useful in areas where there is no sonar data. In the Arctic, much of the available data were derived from submarine measurements. During the period 1957–1997, about 480,000 km of submarine data were incorporated in the record. In addition, 37,800 km of multibeam data were acquired between 1998 and 1999. Similar efforts but from ship based measurements have been going on in the Antarctic region (e.g., Schenke et al. 1998) which is almost ice free in the summer. Although there are still gaps in the observations and the quantity and quality of data in the polar oceans are not as good as those in temperate waters, the bathymetry maps in the polar regions that are currently available are quite impressive in scope and already considered very useful.

Also of special interest are the circulation patterns, dynamic topography and the drift of sea ice in the polar oceans. It is useful to divide the circulation of the ocean into wind driven currents and thermohaline circulation. The first may appear relatively short term since as the name implies, it is driven primarily by the wind. However, in depth studies indicated that the circulation in the upper layers of the oceans are the result mainly of surface wind stresses (Munk 1950). The second is the average motion over a long period of time and driven by many factors including temperature, pressure, and sea ice formation. Ultimately, however, the sun is the primary source of energy since for both types, wind energy is also derived from the sun. Our current knowledge of ocean circulation, ice drift and dynamic topography has been based mainly from ship, mooring and buoy measurements, the spatial coverage of which is limited in many places. Data from satellite sensors have been used to fill in the gaps indirectly through the analysis of wind effects and the movements of surface features like those of sea ice and ocean color.

2.2.1.1 The Arctic Ocean and Peripheral Seas

It was originally thought that the entire Arctic Ocean is relatively shallow since the first impressions came from the readily accessible and shallow shelf regions adjacent to land. During the 3-year expedition on board a ship frozen into the ice pack, Nansen discovered that the central part of the ocean is actually relatively deep with depths of more than 3 km in some regions. Much has been learned about the Arctic since and a modern version of the bathymetric map as presented in Fig. 2.1 reveals a relatively complex bathymetry for the entire Arctic. Shallow shelf regions are shown to be vast in the Arctic basin with the narrowest located north of Canada and the widest in the Barents Sea. The regions along the Barents, Kara, Laptev, East Siberian and Chukchi Seas with depths less than 200 m are quite extensive with the shelves typically about 600–800 km wide. These shelves represent about 30% of the Arctic Ocean. The deep ocean is split into two by the Lomonosov/Trans–Euranic ridge with the Western part being bigger with a maximum depth of about 3,800 m, while the Eastern part being smaller with a maximum depth of about 5,400 m. The Lomonosov Ridge is about 1,700 km long extending from the continental shelf off Ellesmere Islands toward the North Pole, where it continues along about 140° E meridian to the continental shelf off the New Siberian Islands. The Western part is

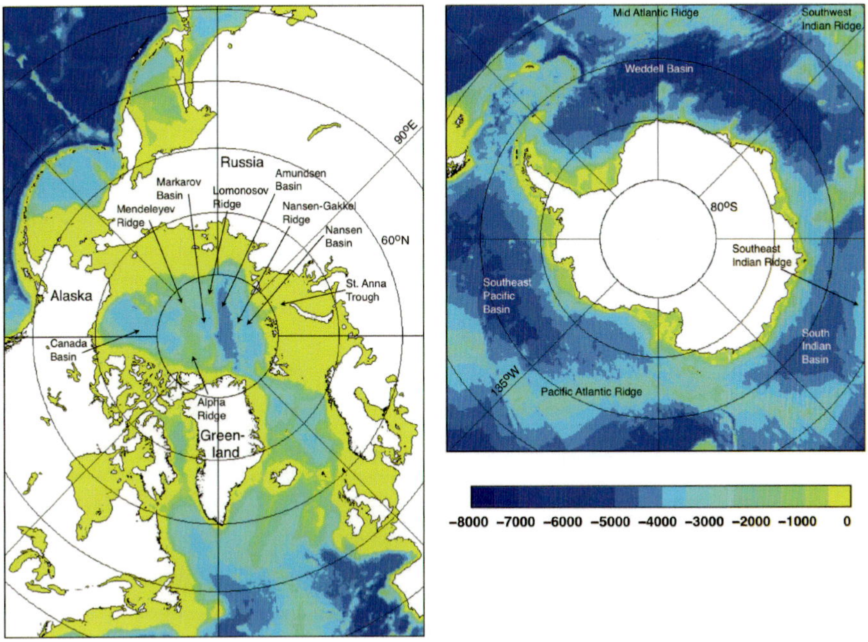

Fig. 2.1 Bathymetry of (a) the Arctic Ocean and surrounding regions and (b) the Southern Oceans. Reproduced from the GEBCO One Minute Grid, version 2.0, http://www.gebco.net

further divided into the Canada and Makarov basins by the Alpha–Mendeleyev Ridge, while the Eastern part is divided into the Nansen and Amundsen basins by the Nansen–Gakkel Ridge. The Nansen–Gakkel Ridge is a continuation of the mid-oceanic ridge system that spreads into the Arctic basin. The deep Arctic water is confined to the Central Arctic and it is apparent that the only possible connection of the deep water to the rest of the World's ocean is to the Atlantic through the Fram Strait, which is 450 km wide and 3,000 m deep. The connection to the Pacific Ocean is limited and possible only through the narrow and shallow Bering Strait, which is about 85 km wide and 45 m deep. Overall, the average depth of the Arctic Ocean is about 1,800 m which is the smallest of all the oceans.

Our ability to assess the circulation of water and the mixing processes in the Arctic Ocean has been limited by the presence of the sea ice cover. When compared with other oceans, the Arctic Ocean is basically isolated being surrounded mainly by land and there are only a few areas where exchanges with other oceans are possible. The observations of the drift of sea ice, initially using buoy data and more recently using satellite data, have been very useful in assessing large scale surface circulation patterns within the pack (Aagard 1988; Thorndike and Colony 1982; Kwok et al. 1998; Liu et al. 1998).

Current knowledge about the large scale circulation pattern of surface water in the Arctic is summarized in Fig. 2.2. As illustrated, water masses exit from or enter the Arctic through the Bering Sea, Fram Strait, the Barents Sea, and the Chukchi

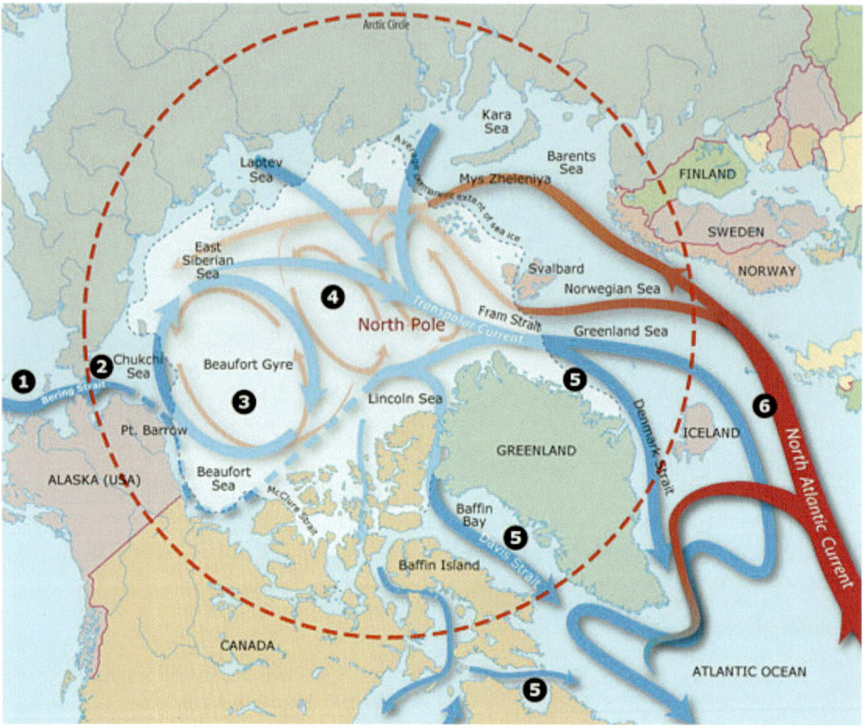

Fig. 2.2 Surface circulation patterns of the Arctic Ocean. The most dominant are the North Atlantic Current (in red), the Transpolar Current (in blue) and the Beaufort Gyre (in blue). Other gyres including a reversal of the Beaufort Gyre are shown in pink. Reproduced with permission of the Wood Hole Institute of Oceanography

Sea and to a more limited extent, the Canadian Archipelago. The analysis of available surface data had led to the discovery of two primary circulation patterns in the Arctic basin, the first of which is the Beaufort Gyre (labeled #3 in the diagram) which is located at the Canadian Basin and has a mean surface flow in the clockwise (anti-cyclonic) direction. Buoy drift data indicate that on the average, it takes about 4 years for the cycle to be completed. The second circulation pattern has been called the Transpolar Drift (#4) and is the flow out in the Eastern/Barents Sea side toward the Fram Strait. The average speed has been estimated to be about 1–3 cm/s and mooring and satellite measurements have yielded approximately 1 million km^2 of ice cover per year drifting out of the region (Vinje and Finnekasa 1986; Kwok and Rothrock 1999). The Beaufort Gyre is primarily wind driven, while the Transpolar Drift is caused by a combination of westerly wind and inflows from Siberian rivers. As indicated, water masses get into the Arctic primarily through the North Atlantic Current (#6). The North Atlantic Current contains replacement water consisting mainly of warm and salty water from the Atlantic and as shown in Fig. 2.2, the current is split into two near Norway, one going directly to the Arctic through the Barents Sea and the other going through Fram Strait. The latter which has been called the West Spitsbergen Current (WSC) has been studied in detail by several investigators (e.g., Gascard et al. 1995) and it turns out that although a main component goes north, as indicated in Fig. 2.2, part of it goes westward in three topographic fracture zones near Svalbard and eventually merge with the East Greenland Current (EGC) going south.

The circulation of surface water in the peripheral seas of the Arctic is similarly complex. An example is the circulation near the North Atlantic Subpolar gyre, adjacent to Greenland and Eastern Canada as illustrated in Fig. 2.2 (lower right) and as described by Lazier and Wright (1993) and by Schmiz (1993). In this case, warm water from subtropical regions (in red) gets split into two, one to the north as part of the North Atlantic Current and the other to the west toward Greenland. The latter gets colder as part of the Iminger Current and becomes very cold as it combines with the East Greenland Current along Denmark Strait. This in turn gets split into two parts at the southern tip of Greenland, one going north toward Baffin Bay, while the other joins the Baffin Island/Davis Strait Current that becomes the Labrador Current to the south. The Labrador Current hits a topographic feature at 46° N and part of it follows the continental shelf going south, while another part goes to the North Atlantic.

In the Bering Sea/Pacific Ocean region, the circulation of surface water is similarly complex as illustrated in Fig. 2.3. Studies using buoy data as tracked by satellites (Stabeno and Reed 1994) indicate a generally cyclonic gyre with the western boundary located along the western side of the basin (called Kamchatka Current). At the eastern side is the Alaska Stream which is the northern boundary current of the Pacific subarctic gyre. This current goes along the southern side of the Aleutians and turn northward near the Amchitka Pass (180°) and follows the bathymetry and becomes part of the Bering Slope Current. The tracks shown in Fig. 2.3 were derived through a completely different technique that combines tracks of 29,000 toys that fell from a cargo ship in 1992 and a computer simulator called the Ocean Surface Current

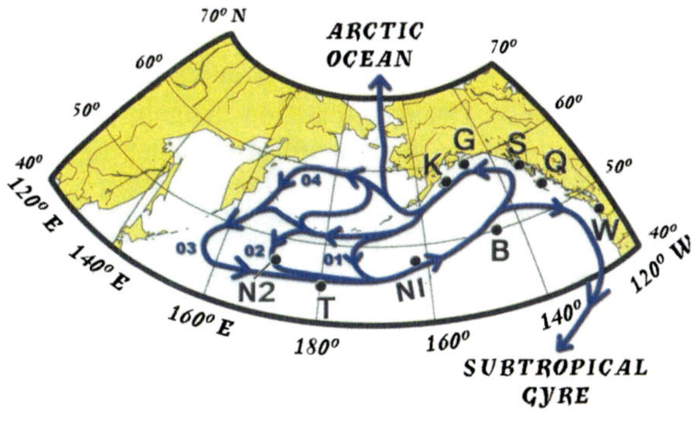

Fig. 2.3 Circulation patterns of subtropical gyre in the Bering Sea as derived from toy drifters and an ocean surface-current simulator program. From Ebbesmeyer et al (2007) with permission from AGU

Simulator, a more detailed ocean current circulation in the region has been recreated. The tracks show a much more complex circulation system in the region than those derived from buoy data and includes intrusions into the Arctic Ocean and entry into the subtropical gyre. The northward flow into the Arctic Ocean has also been studied previously by Overland et al. (1996) using Argos drifters which are tracked by satellites and the results show drift velocities that are relatively low at about 5–40 cm/s.

The circulation of subsurface water in the Arctic region is generally restricted by the bathymetry of the region. The relatively shallow Bering Sea makes it practically impossible for intermediate to deep water to flow from the Pacific Ocean to the Arctic. The distinction of being the only region where the direct flow of intermediate and deep water between the Arctic Ocean and other oceans can occur belongs to the Fram Strait, which is only 450 km wide. The exchange in this case is with the North Atlantic Ocean. The transport of ice and water in this region has been recognized as an important factor affecting the salt distribution and circulation of North Atlantic water. Moreover, heat, salt and humidity exchanges between the Arctic Ocean and the North Atlantic Ocean are intense in this region. Surface water (including sea ice) that leaves the Arctic through Fram Strait is replaced by warmer Atlantic water at depth and becomes a major part of the Atlantic Intermediate Water (AIW) in the Arctic Basin as we know it today.

Although the exchange is primarily through Fram Straits, studies have shown that the intermediate water in the Arctic is also influenced by water from the Atlantic Oceans through the Barents Sea. (e.g., Rudels et al. 1994). The growth of sea ice at the MIZ in autumn and winter at the Barents Sea and the associated brine rejection is an important factor in this regard (Aagaard et al. 1981) because it leads to the formation of high salinity cold water that accumulates at the underside and becomes bottom water. As this water gets circulated into the Arctic Basin, it can penetrate into and have significant influence on the intermediate and deep water.

The general circulation pattern of intermediate water in the Arctic Basin has been inferred from tracer studies (e.g., Rudels 1987) and is unexpectedly complex as shown in Fig. 2.4. The figure shows water from both Fram Strait and Barents Sea progressing initially eastward until they meet north of the Kara Sea where they follow the continental slope until Laptev Sea. At this point, a fraction of the water splits off and returns westward on a tight loop and bring the water back to the Fram Strait along the Gakkel Ridge. The rest of the water continues on along the continental slope until it approaches the Lomonosov Ridge where again a fraction of the water splits off and returns along this ridge back to the Fram Strait. The water that is left goes into the Canadian basin. Two types of outflowing water from the Canadian basin have been observed suggesting that the circulation pattern in the region goes through loops as shown in Fig. 2.4 (Rudels et al. 1994). It has been observed that the contributions to the intermediate water from the Fram Strait and the Barents Sea are about of equal strengths. The contributions from the warmer, more saline Fram Strait are of relatively lower densities while those of the fresher, colder Barents Sea are of higher densities.

Ocean circulation is driven mainly by changes in water mass density, wind, current, and topography. The density of water changes with location and season and is influenced mainly by salinity and temperature. Of these two parameters, salinity

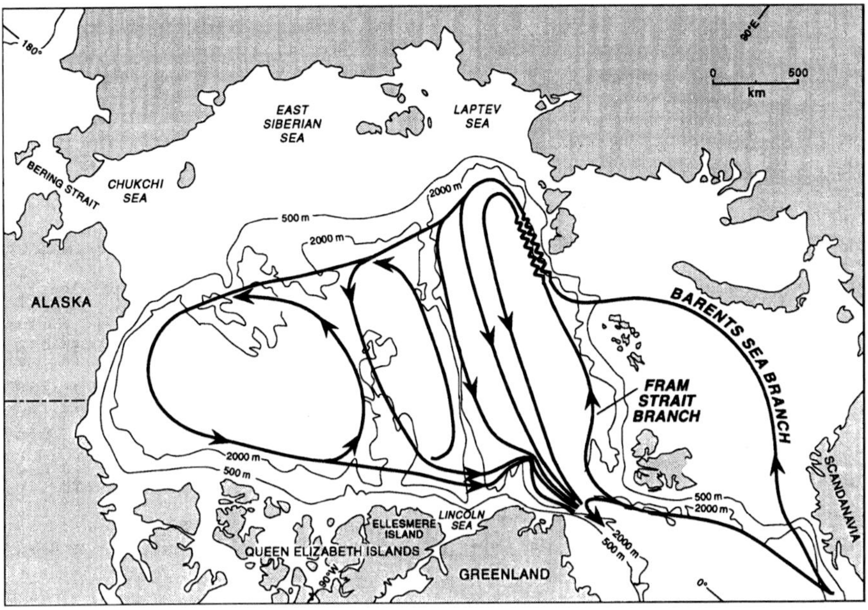

Fig. 2.4 Circulation Patterns of the Atlantic Layer and intermediate depth waters (between 200 m and 1700 m) in the Arctic inferred from tracer studies. From Rudels et al. (1994) with permission of the American Geophysical Union

appears to be more important since the densiest water observed has been the most saline and not the coldest. The key factors that affect water salinity are evaporation and sea ice formation. The latter is associated with the rejection of brine during ice formation especially in polynya regions as will be discussed in chapter 8. The polar regions are known to be among the key sites of deep ocean convection in which dense water near the surface sinks and is replaced by lighter water mass from the deeper regions. This vertical exchange of water mass is an important process because it serves to ventilate the oceans and causes nutrients and chemicals to be redistributed. The exchange can also be initiated by the upward deflection of water masses as can be caused by bottom topography.

Among the most discussed deep convection areas in the Northern Hemisphere is the region called Nordbukta (North Bay), adjacent to where the Odden ice tongue normally forms in the Greenland Sea as discussed in Chap. 8. The Greenland Sea and in particular, Nordbukta, has been cited as one of the main sources of bottom water and its importance has been recognized earlier by Nansen and others. The influence of sea ice on the modification of water masses accompanied by weak stratification and cyclonic oceanic circulation are believed to be the key factors in the initiation of convection in these regions (Carsey and Roach 1994; Toudal and Coon 2001). However, other studies indicate that the presence of the Odden may not be a prerequisite to the formation of deep convection since the latter had happened during non-Odden years (Lherminier et al. 1999).

Not too far from Nordbukta is another site of deep ocean convection. In fact, some of the deepest ocean convections had actually been observed in the Labrador Sea (Lazier 1980; Clarke and Gascaard 1983). Deep convection in the region has been observed to be thermally driven and causes the transport of low salinity upper waters downward to a depth of 2 km where the water mix with higher salinity Labrador Sea water and the even higher salinity (and temperature) Northeastern Atlantic Deep Water beneath (2.5 km). Because of the relatively small size of the Labrador Sea, its contribution to the global heat flux going poleward is moderate, but the region is one of the dominant sources of low salinity water for the mid-depth (1.5 m) World Ocean (Lily et al. 1999).

The water mass circulation in the Arctic region is part of a global phenomenon that is now referred to as the thermohaline circulation (Broecker 1997). It is called "thermohaline" because of the important role of both temperature and salinity in the circulation. It is driven primarily by differences in density which as discussed earlier is controlled by temperature and salinity. The compilation of many years of oceanographic studies has led to a relatively crude but intriguing map of global ocean circulation shown in Fig. 2.5, as adapted from Broecker (1987). This map has also been popularly called the "conveyor belt" of the World's oceans. The North Atlantic and North Pacific are critical turning points of this circulation with apparently reverse roles in that water mass sinks to the bottom in the North Atlantic while they upwell to the surface in the North Pacific. The study of changes in the physical characteristics of the ocean at the northern high latitudes is therefore very important because such changes could cause an alteration in the circulation. For example, a millinneum of very cold conditions, referred to as the Younger Dryas, has been postulated as the result of the temporary shutdown of the conveyer. The other critical turning point is located in the Southern Ocean in which a similar sinking process occurs. This is

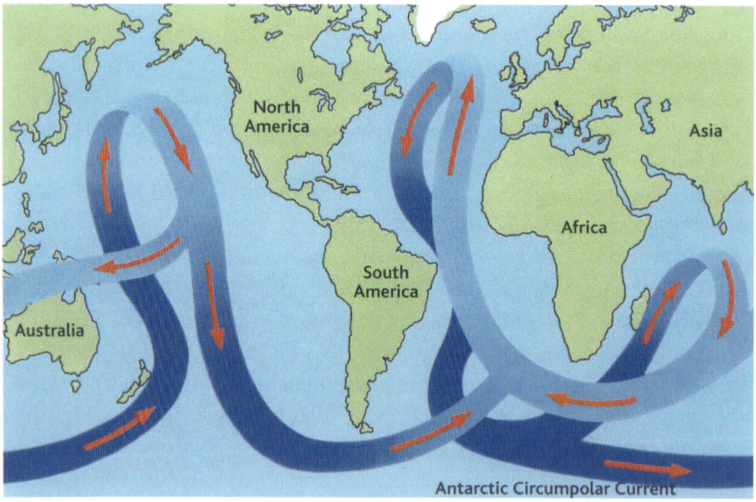

Fig. 2.5 Global Thermohaline Conveyor Belt. Dark blue represents cool bottom water ocean current while light blue correspond to warm surface currents. From Thomas (2004) with permission of the Natural History Museum

aided by the formation of sea ice which rejects much of the salt during the change in phase from liquid to solid. The process leads to the formation of the Antarctic Bottom Water (AABW) which sinks to the bottom and underflows the North Atlantic Deep Water and extends to the north toward the equator in the Atlantic.

2.2.1.2 The Southern Ocean

The bathymetry of the Southern Ocean (Fig. 2.1b) is generally defined by three deep regions (i.e., >4,000 m) namely: the Weddell Basin, the South Indian Basin, and the Southeast Pacific Basin which are separated by the Kerguelen Gaussberg plateau, Mid-Ocean ridge and the Scotia Ridge, respectively. Unlike in the Arctic, the flow of intermediate and deep water from one ocean to another is not so limited by bathymetry. Except for the portion between South America and the Antarctic Peninsula which is known as the Drake Passage, the region between 40° and 60° S is basically wide open. But, even at the Drake Passage, the flow of water is not so restricted since the width of the passage is almost twice that of the Fram Strait at about 800 km and the depth is more than 3,000 m. The Drake Passage is actually the only means by which deep water can be interchanged between the Pacific and Atlantic Oceans because of the aforementioned limitations in the north caused by shallow waters at the Bering Strait between Alaska and Siberia.

A large fraction of the deep ocean regions in the Southern Ocean is covered by sea ice in winter. Within the region are topographic features that have at the same time been the location of some of unique spatial features in the ice distribution. For example, the presence of a topographical mount, called Maud Rise located near 65° S and 3° E in the Weddell Sea, has been shown to be strongly correlated with the formation of the Maud Rise Polynya (de Veaux et al. 1993) usually located directly above or adjacent to the mount. It was postulated that the Maud Rise, which protrudes about 3 km above the ocean floor, and a relatively strong current (i.e., the Weddell Gyre) causes the upwelling of warm water that in turn would melt the underside of the sea ice cover in the region. The presence of these mounts and small islands also causes the diversion of the direction of ocean circulation and the drift of sea ice causing the unique pattern of the current and the sea ice cover. Also interesting are some of the processes associated with bathymetry that occur closer to the Antarctic continent. The continent is surrounded by continental shelf regions that are generally narrow except those off the Ronne/Filchner Ice Shelves in the Weddell Sea and the Ross Ice Shelf in the Ross Sea. On average, these continental shelves are about 400 km wide and are relatively deep (500–900 m) in part because of depression associated with the formation of the massive Antarctic continental ice sheet. As will be discussed later, the presence of such shelves is critical in the formation of dense and highly saline deep water called AABW.

The circulation of the Southern Ocean is dominated by the ACC which is regarded as the largest of the World's ocean currents. The circumpolar nature is obviously made possible by the presence of the Drake Passage. It has been estimated that the ACC carries about 135–150 times more water around the continent than that carried by all the World's rivers combined. The ACC goes on a clockwise direction

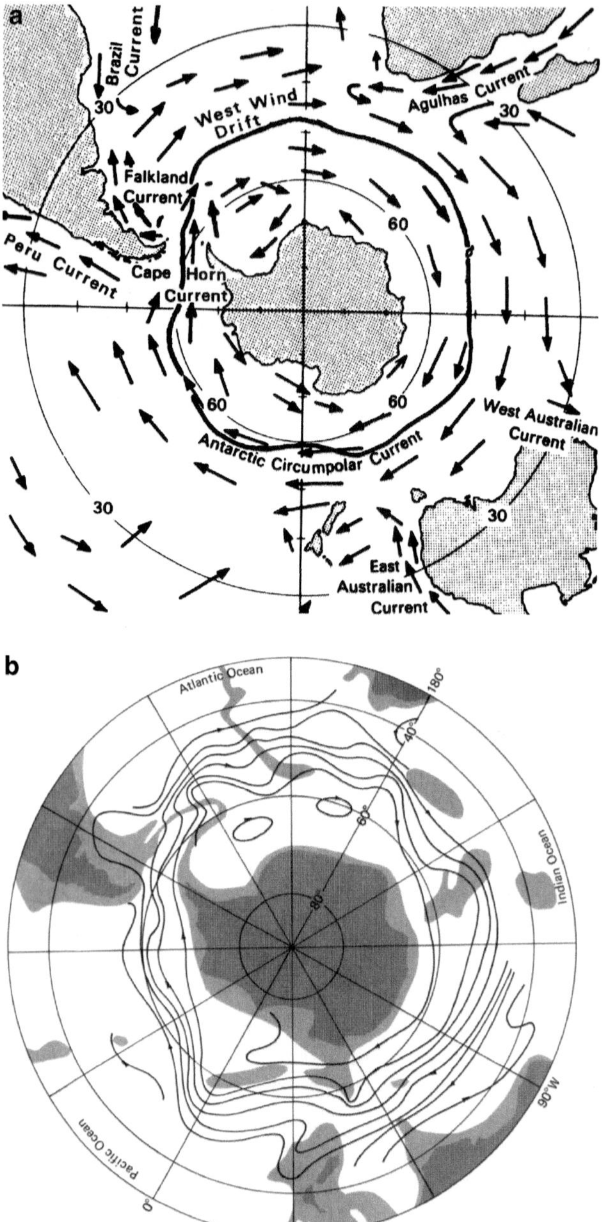

Fig. 2.6 (**a**) Patterns of surface currents in the Southern Ocean. Average location of the Antarctic Circumpolar Current and other currents in the region are indicated. From Wadhams (2000) with permission of Gordon and Breach Science Publishers). (**b**) Contours of the Antarctic Circumpolar Current around Antarctica as it gets affected by bathymetry. The direction is altered by the presence of topographic features which in this case is represented by the light shaded areas (for water depths less than 3000 m). From Krauss (1997) with permission from Prentice Hall, Inc

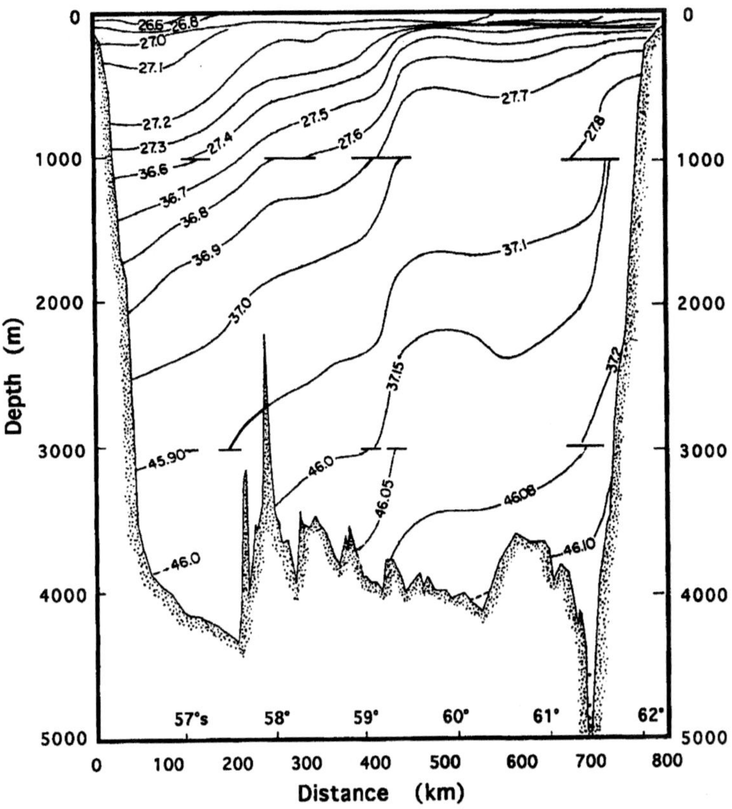

Fig. 2.7 Salinity profiles across the Drake passage. The slopes of the isopycnals in units of σ_t, σ_2, σ_4 (kg m^{-3}) are indicative of surface features going all the way to the bottom. From Nowlin and Klink (1986) with permission from the American Geophysical Union

and is driven primarily by prevailing westerly winds. The general direction of the ACC surface current, as observed from buoys and other systems, is depicted by both a vector map (Fig. 2.6a) and a contour map (Fig. 2.6b). There is ample evidence that the circulation extends all the way to the bottom. For example, the slope of the isotherms and isohalines between the Antarctic continent and 47° S indicates an eastward geostrophic flow increasing upward (Fig. 2.7). A typical eastward speed at about 300 m is 3–9 cm/s while bottom speed of 2 cm/s has been inferred from float data (Gille 2003). The curvature of the current flow in response to bottom topography is also a manifestation that the current extends all the way to the bottom.

The large scale characteristics of the ACC have been studied by many investigators using all historical hydrographic data that were available. The flow of the ACC has been characterized by Orsi et al. (1995) as a zonal band with a relatively large geostrophic shear when compared with subtropical and subpolar regimes. It was noted by Nowlin and Klinck (1986) that the ACC appears to exist as a banded structure, with multiple narrow jets associated with strong lateral density gradients located near the

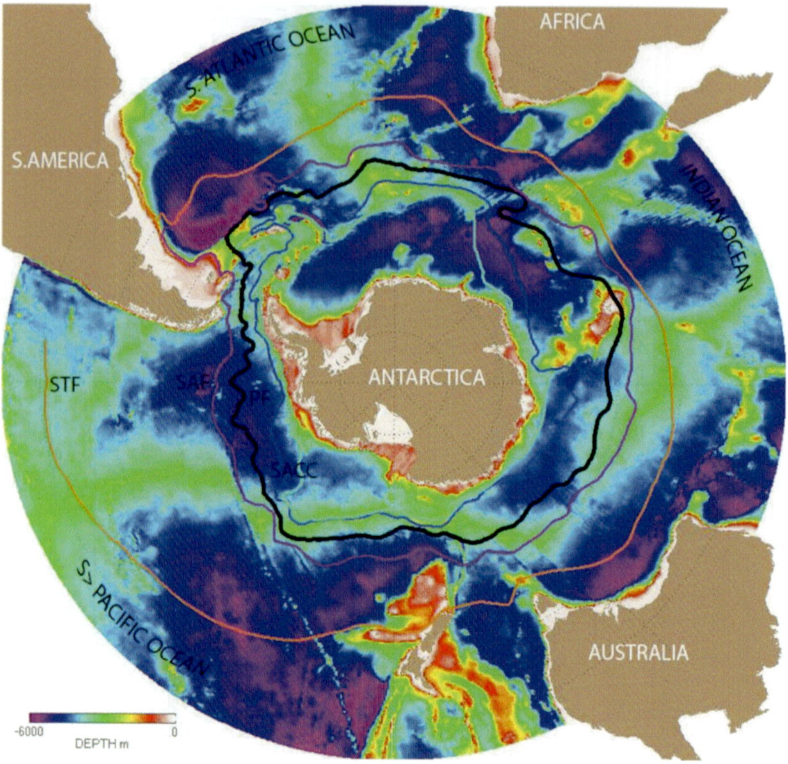

Fig. 2.8 Antarctic circumpolar current and seawater density fronts as derived from Orsi et al (1995) over the Southern Ocean bathymetry. Graphics courtesy of NASA/JPL

polar and subantarctic fronts. The location of these fronts has been observed to vary by as much as 100 km during a 10 day period of observation. Year-long measurements of the transport of the ACC through the Drake Passage by Bryden and Pillsbury (1977) also indicated that about 134 ± 13 Sv is transported annually with the instantaneous flow varying by as much as 20% from average values (Nowlin and Klinck 1986).

Along the continental margin, narrow coastal currents generally going in the counter-clockwise direction have also been observed (Deacon 1937; Sverdrup 1953). The average speed is roughly 0.1 m/s, but they are not circumpolar and instead are part of clockwise gyres that are mainly driven by persistent atmospheric patterns in the region. Among the most visible of such gyres are those in the Weddell and Ross Seas. The Weddell gyre is actually the largest cyclonic gyre south of the circumpolar current extending from a western limit defined by the Antarctic peninsula to as far as 30° E to the east. It is also the most vigorous and best formed of the subpolar gyres and believed to be responsible for much of the abyssal ocean ventilation. An early estimate of the strength of the Weddell Gyre was 97 Sv (by Carmack and Foster 1975), but this value has gone down to 76 Sv (Gordon et al. 1981) and more recently to 60 Sv (Beckmann et al. 1999; Schroder and Fahrbach 1999). It is not known to what extent the strength of the gyre has been changing, but accurate estimates are required to assess and understand the role of the gyre in the global thermohaline circulation.

Within the ACC is the region of Antarctic convergence which is where the Antarctic surface water moving northwards sinks below subantarctic waters (Deacon 1937). The region is also referred to as the Antarctic Polar Front associated with elevated speeds of the current and enhanced gradients in density, temperature, salinity and other oceanographic properties. The front is an important climate boundary noted for intense exchanges between the surface and atmosphere and where strong salt and heat fluxes have been observed. The location of the front has been linked to topography, but significant seasonal and interannual variations have been observed (Moore et al. 1999). The meridional transport across a latitude is also of interest since the southward flow brings relatively warm water into the south and to areas depleted of oxygen but rich in nutrients. As the warm water reaches the surface water, it gets its oxygen content renewed through air–sea interaction. At the same time, phytoplankton growth and therefore marine life in the euphotic zone is enhanced due to the abundance of nutrients. Surface cooling causes the release of heat fluxes to the sub-Antarctic atmosphere causing the latter to be relatively warm.

A number of distinct fronts along the ACC based primarily from studies by Orsi et al. (1995) have been identified as illustrated in Fig. 2.8. One is the subtropical front indicated by the northernmost contour line. This is the boundary between the subtropical waters that are warm and salty and the subpolar waters that are relatively cooler. The next contour toward the south is the subantarctic front which is located at a latitude at which subsurface salinity minimum first appears. The third countour to the south is the point of transition to relatively fresh and very cold water and is referred to as the Polar Front. The water mass that is transported along the ACC is carried mainly by the second and third fronts. The farthest contour to the south is the Southern Boundary Front which is the location where very dense abyssal waters upwell to at least above the 100 m depth.

The complex processes occurring in the polar regions make the surface deviate significantly from average geopotential heights. Such deviations are often referred to as relative dynamic topography. There are some semipermanent features such as the ACC and the Antarctic Fronts that stand out in relative dynamic topography maps. There are also transient features like internal waves and mesoscale eddies that perturb the ocean surface and cause the surface height to significantly deviate from average climatic conditions. Current data on dynamic topography have been based mainly from wind and sea level pressure data and may include ship measurements (Gordon et al. 1978), but surface data have limited coverage in time and space. Major improvements are forthcoming such as the use of satellite radar altimeters data for this purpose has become feasible as will be discussed later (Gille 1994).

2.2.2 *Physical Properties*

The physical property of an ocean mass is typically defined by its temperature, salinity and density. Based on the vertical profiles of these variables, an ocean can be characterized in terms of three well defined layers. The first is the near surface layer, usually referred to as the mixed layer, about tens of meters thick and mainly isothermal primarily because it is kept well mixed by wind and other factors. Immediately below the mixed layer is a layer of rapid change in temperature called the thermocline. The thermocline is an important layer since it controls the degree at which vertical convection could take place. Below the thermocline is the third layer called the deep layer which is usually relatively cold and nearly isothermal. In the polar regions, the temperature of the mixed layer is highly seasonal because of the large seasonal fluctuations in surface air temperature. Also, the upper surface is an effective absorber of solar energy which is abundant during summer but very limited during winter because of long darkness and the presence of the sea ice cover. The deep layer is relatively cold in part because a large fraction of the water comes from coastal polynya regions in the polar regions as discussed in Chap. 8.

Salinity is usually defined as the total quantity of dissolved materials in sea water. The unit for salinity had been "parts-per-thousand" for many years, but the name has been changed and is now almost universally referred to as "practical salinity units" (psu). The fraction of dissolved materials, which consist generally of about 87% sodium chloride and 13% of other types of particulates, changes depending on history, environment and location of the water mass. The temporal changes in salinity in a particular location are generally very small (e.g., from about 34.5 to 35 psu), but even small changes can make a big difference on the physical characteristics of the ocean. The rejection of brine during the growth of sea ice in the polar oceans contributes much to these small changes in salinity in that it causes the formation of high salinity cold water. This in turn can trigger vertical convection that enables the ventilation of the ocean, which is a process in which surface water acquires nutrients from the bottom, while the deep ocean is replenished with oxygen. On the other hand, the melt of sea ice causes the introduction of low salinity water to the surface thereby causing a reduction in the salinity of the mixed layer.

The density of sea water controls the distribution of water masses and is a function of temperature, salinity, depth and in effect the hydrostatic pressure applied to it. For every 10 m increase in depth from the surface, the hydrostatic pressure increases by about 1 atmosphere. Even after the compressibility is removed, the density varies with depth and although the change is only for a few parts per thousand, such variation (i.e., 1,025–1,028 kg/m^3 from surface to deep ocean) influences the physical characteristics of the ocean and also many oceanographic processes.

2.2.2.1 The Arctic Ocean and Peripheral Seas

Since the Arctic is mainly ice covered, it would be intuitive to assume that the temperature of the underlying water column would be close to the freezing point. This is indeed the case in the upper layer adjacent to the underside of the ice, but what was not expected is the presence of a layer of relatively warm water, about 3°C above freezing, at a depth of about 300 m. This phenomenon is illustrated by a typical vertical profile of the potential temperature (T), salinity (S), and potential density

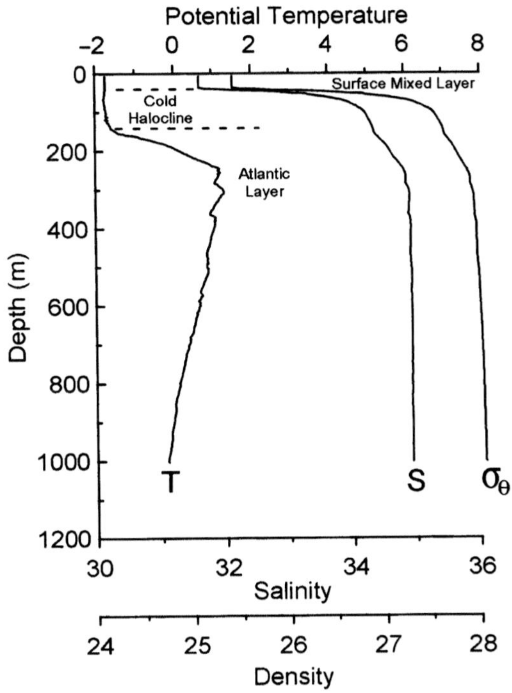

Fig. 2.9 Typical temperature (T), salinity (S) and potential density (σ_0) profiles in the upper layer in the Arctic. From a cast at (84.485°N, 17.212°E) during EUBEX (Perkin and Lewis, 1984). The mixed layer, cold halocline and Atlantic layer are indicated. From Padman (1995) with permission of the American Geophysical Union

(σ_θ) in the Arctic as presented in Fig. 2.9. The warm water layer apparent in the profile has been called the AIW since the source is mainly the Atlantic Ocean advected to the Arctic through the WSC as discussed earlier. The AIW is present in much of the Arctic Ocean where the water is deeper than 1,000 m. The sea ice cover would melt if this warm water reaches the surface, but it is protected by a stratified layer of relatively cold and low density water called halocline, which is about 100 m thick (Aagaard et al. 1981). The halocline layer has a temperature of about −1°C and salinities between about 30.4 and 34.4 psu and has been regarded as one of the most important features of the Arctic Ocean. A good understanding of the formation process and maintenance of the halocline is thus important since any alteration could mean dramatic changes in the physical characteristics of the Arctic Ocean, including its sea ice cover.

There are at least three possible mechanisms that may have caused the formation of the halocline. The first is the salinization of cold surface water during ice formation. The second is the mixing of upwelled AIW with cold and fresh summer shelf water as discussed in Aagaard et al. (1981). The third is the melting of ice in the MIZ as the Atlantic Water first enters the ice covered Arctic basin as a surface current (Steele et al. 1995). The relative rates at which halocline water is formed through these various mechanisms, which are all possible, are not known. All of them have sea ice components which are highly variable in the vast continental shelf region which occupies about 30% of the entire Arctic basin.

The surface water has low salinity at the end of summer because of a combination of melt from the sea ice of the previous winter, river runoff, and infiltration of low salinity water from the Bering Sea through the Bering Straits. In the autumn, surface air temperature drops considerably causing rapid freezing and the re-salinization of the surface through brine rejection during the formation of sea ice. The rate of such process drop as the ice thickens through the winter period. The halocline is cold because of the presence of the ice cover and has intermediate values in salinity and density because the re-salinization is not strong enough to overcome the desalination effects. Upwelled AIW has been observed on several occasions (Garrison and Becker 1976; Aagaard 1977; Hufford 1974) making the second mechanism a very viable possibility. In the third mechanism, near-freezing halocline water can be produced by the incorporation of a relatively small amount of melted ice into the Atlantic Water as it first enters the ice covered Arctic at the MIZ. While the shelf region is mainly ice covered in winter, the location and size of open water areas in the summer are highly variable and depend on wind circulation, surface air temperature, ocean current, and temperature of the mixed layer. Regional variations in the depth and characteristics of the halocline have been observed by Rudels et al. (1996) and postulated to be associated not only with sea ice formation but also with bathymetry and proximity of river run-offs.

The unique distribution of potential temperature and salinity as a function of depth in various ocean areas provides the ability to classify the various water masses in the oceans and the means to evaluate their magnitude and their characteristics. The water masses in the Arctic Ocean (sometimes referred to as Arctic Mediteranean) and adjacent seas have been described in terms of the temperature/salinity (TS) classification schemes by Carmack (1986) as illustrated in Fig. 2.10. Different authors provide different terms for the same type of mass and in our case, we follow the one used by Swift and Aagaard (1981) and by Aagaard et al. (1985). There are three main layers in a water column: a surface layer, an intermediate layer, and a deep layer. In the region, there are in turn three types of surface water: Atlantic Water (AW), which is located in the North Atlantic and is a branch of the Norwegian Atlantic Current; Polar Water (PW), which occupies the upper 200 m layers of the Arctic and includes the surface outflow within the East Greenland Current and the Canadian Archipelago; and Arctic Surface Water (ASW), which is found mainly in the central gyres of Greenland and Iceland seas. AW has temperatures above 3°C and salinities above 34.9 psu; PW has temperatures below 0°C and has salinities below 34.4 psu, while ASW has temperatures above 2°C for salinities between 34.7 and 34.9 psu and above 0°C for salinities between 34.4 and 34.7. ASW is warmer and fresher than the PW and cooler than AW, but it is notably denser than either PW or AW indicating that it is not simply a mixture of the latter two but is significantly influenced locally by air–sea interaction.

The top 30–50 m of Polar Water represents the seasonal mixed layer which in winter tends to be uniform vertically in temperature and salinity, but in summer it exhibits a strong salt stratification due to ice melting. The lower part of PW is the halocline which is a relatively cold saline layer described earlier. In the Eurasian basin, the salinity increases rapidly with depth values of about 35 psu at around 200 m while the temperature stays cold at close to −1.5°C. The halocline is deeper in the Canadian Basin and the salinity increases more slowly with depth, while the temperature shows minima at 31.6 and again at 33.1.

Within the Arctic Ocean, the intermediate layer is the AIW as discussed earlier and usually resides at depths between 200 and 800 m. This is a relatively warm and saline water originating from the Atlantic Ocean, as the name implies, and entering the Arctic Basin primarily through Fram Strait. At Fram Strait, the initially high temperature and salinity (over 3°C and 35 psu) are altered by the atmosphere and by mixing with local water and acquire the T/S properties associated with AIW. It was observed by Swift and Aagaard (1981) that ASW usually lies above a temperature minimum at 75 and 150 m depth, a temperature maximum at 250 m, and a salinity maximum at about 400 m. Carmack (1990) provided the following classification to account for these distinct features of Arctic water masses: lower Arctic Intermediate Water (lAIW) for water lying immediately above deep water; upper Arctic Intermediate Water (uAIW) for water lying between the temperature minimum and the temperature maximum; and Polar Intermediate Water (PIW) for water which has somewhat lower salinities and is distinguished from the other two by its association with overlying PW. Such classification is illustrated in Fig. 2.10.

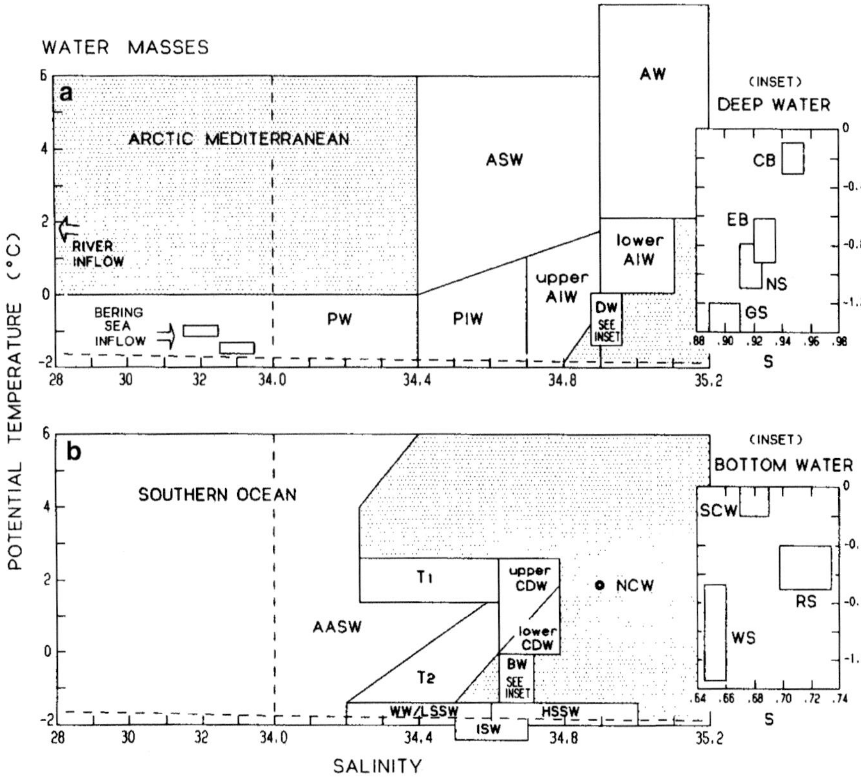

Fig. 2.10 Water mass classification based on TS diagram for the (**a**) Arctic Ocean (referred to as Arctic Mediterranean) and (**b**) Southern Ocean. From Carmack (1990) with permission of the American Geophysical Union

The deep layer of the Arctic Ocean is occupied by cold (<0°C) water masses. The origin of water in this region has been studied and described by many investigators including Aaagard et al. (1985), Rudels (1987) and Swift and Koltermann (1988). The distinct water masses that had been identified are:

Greenland Sea Deep Water (GSDW), which is the coldest (about −1.2°C) and the freshest (<34.90 psu); Canadian Basin Deep Water (CBDW), which is about the warmest (about –0.5°C) and the most saline (>34.95); Norwegian Sea Deep Water (NSDW), with temperatures typically at –0.9°C and salinity of 34.92 psu; and Eurasian Basin Deep Water (EBDW) with temperature and salinity being about –0.7°C and 34.94 psu.

2.2.2.2 The Southern Ocean

The pioneering studies of Brennecke (1921), Mosby (1934) and Deacon (1937) contributed much to our current understanding of the waters in the Southern Ocean. The physical characteristics of the water masses are best described in terms of the zonation with regard to water mass distribution and the water mass structure at various locations around the continent as discussed by Gordon et al. (1977) and Deacon (1982). Comprehensive reviews of the water mass structure around the continent have been provided by Carmack (1977), Reid et al. (1977), Jacobs et al. (1985), and Whitworth and Nowlin (1987) while the structural variability has been discussed in detail by Gordon (1989).

Water masses in the Southern Ocean have been classified by Carmack (1990) according to the T/S classification schemes and presented in Fig. 2.10b. The coldest and freshest of these water masses is the Antarctic Surface Water (AASW) which usually resides in the upper 50–200 m layer north of the continental slope. The AASW originates primarily from deep water that has upwelled and has been modified by exchange processes in the surface. Underneath the AASW is the Winter Water (WW) formed from the AASW during freeze-up and when brine is being rejected during ice formation. This layer is noted as the temperature minimum layer with salinities below about 34.4 psu. The WW is the coldest and deepest being located immediately north of the continental slope (Jacobs et al. 1985) while its salinity has its upper limits set by cabbeling instability (Foster and Carmack 1976). When two masses of sea water with the same density but different temperatures and salinities are mixed together the resulting water mass may have a density greater than the constituent water masses because of nonlinear nature of the equation of state of sea water. This phenomenon is known as cabbeling and an instability cause by the formation of sea ice could lead to relatively dense surface water that is transported vertically and contribute to the formation of bottom water.

The most abundant of all the water masses in the Southern Ocean is the Circumpolar Deep Water (CDW) which is a nutrient rich water layer characterized by a temperature maximum and oxygen minimum at intermediate depth and a salinity maximum at the deeper regions. It has also been divided into upper CDW (uCDW) and lower CDW (lCDW) by a diagonal line passing midway through the CDW bivariate class. Its origin has been traced to as far as the North Atlantic Deep Water (NADW) and ultimately the Arctic Ocean, but its mixing history is complex as it moves with the ACC around the continent. In some regions, as in the Weddell Sea, the cooler and fresher variety of this water gets modified by mixing with continental shelf water and becomes what is known as the Weddell Deep Water (Gordon 1982).

The lower part of the Southern Ocean is occupied by the AABW or simply (BW) which is a key component of the global thermohaline circulation. The characteristics of the AABW varies slightly with region with the extreme varieties being the Weddell Sea Bottom Water (WSBW, or WW), with potential temperature below –0.7°C and salinities of about 34.64 psu and the Ross Sea Bottom Water (RSBW or RW in Fig. 2.10b), with potential temperatures of above –0.5°C and a higher salinity of above 34.7 psu. The AABW has been associated with the formation of sea ice in coastal polynya areas as discussed in Chap. 8.

Typical salinity, potential temperature and oxygen profiles as observed during the Alfred Wegener Institute (AWI)/Polarstern V/2 winter cruise in 1986 (Gordon and Huber 1990) are presented in Fig. 2.11. The cruise was to our knowledge the first scientific expedition in the inner pack of the Antarctic ice cover during winter and lasted for almost 3 months. The region of observation is the Maud Rise and vicinity with the objective of gaining insights into the oceanographic conditions of the region where the large Weddell winter polynya was formed. The average thickness of the mixed layer was typically 111 m with a standard deviation of 33 m. The measurements were taken in three general areas in the Maud Rise region of the Weddell Sea: (a) the cold regime; (b) Maud Rise; and (c) the warm regime. In the cold regime, the depth of the mixed layer was about 125 m and the pycnocline was very well defined. At the Maud Rise, the pycnocline was much weaker and ranged from about 125 m to about 240 m. At the warm regime, the pycnocline was again well defined but at a much shallower depth of about 60 m. The weak stratification of water above the Maud Rise makes the region vulnerable to an upwelling of warm water that can be caused by a spin up of the Weddell Gyre.

Historical hydrographic data have been originally put together by Gordon (1982) and later revised by Olbers et al. (1992) with the purpose of providing a relatively consistent but nonsynoptic picture of the ACC frontal structure. The dynamic topography of the Southern Ocean as revised by Olbers et al. (1992) is presented in Fig. 2.12a. The ACC has been associated with sharp temperature and density fronts, each separating two distinct water masses and supporting a jetlike geostrophic velocity structure (Nowlin and Klinck 1986). More detailed mean sea surface heights have become possible with the advent of radar altimeter data and Fig. 2.12b shows an example of sea surface height as derived from satellite data by Gille (1994) Analysis of this data shows that the fronts are substantially steered by topography and that the jets have an average width of about 44 km in the meridional direction and about 75 km to either side of their mean locations. Also, it was inferred that the average height difference is 0.7 m across the Subantarctic Front (SAF) and 0.6 m across the Polar Front (PF).

2.2.3 Chemical and Biological Properties of Polar Waters

The oceans have been referred to as the origin of life. Although not so apparent at the surface, the ocean has an ecology the diversity of which is unmatched even by

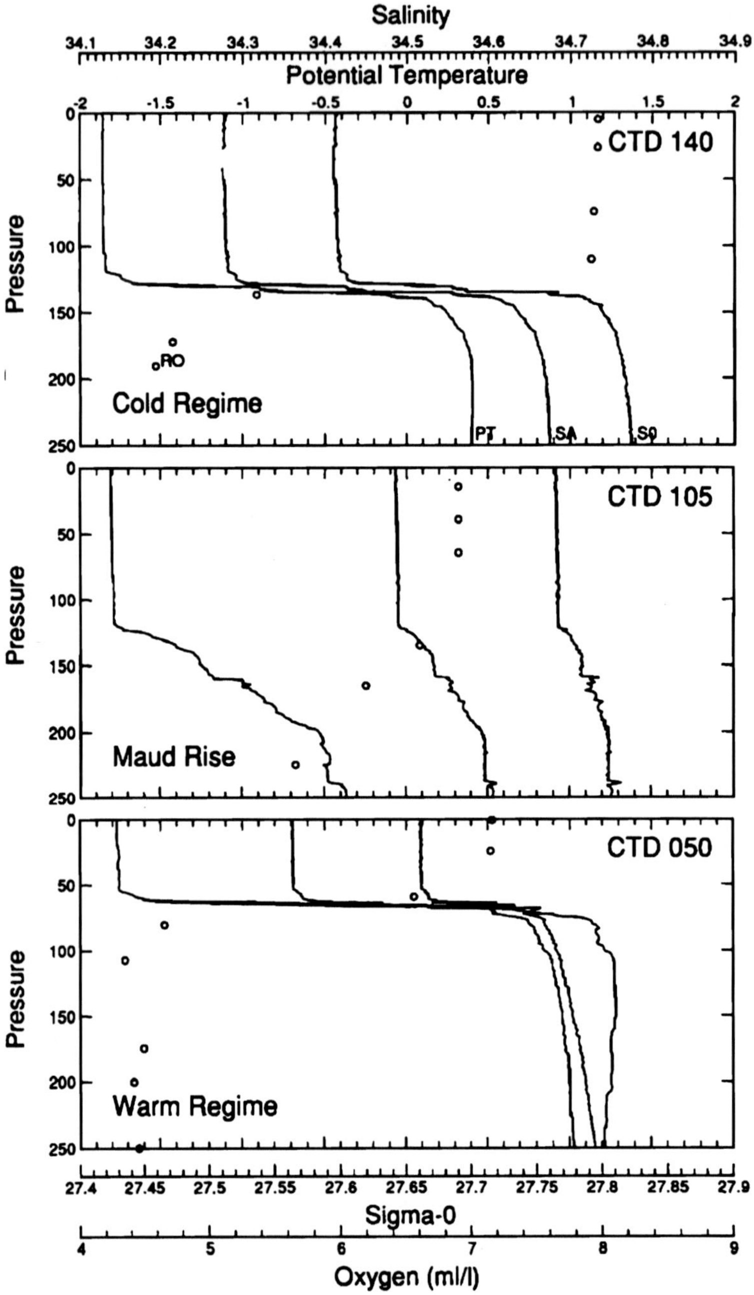

Fig. 2.11 Typical temperature, salinity and potential density profiles in the Antarctic from measurements during the Polarstern Antarctic V/2 winter experiment. From Gordon and Huber (1990) with permission of the American Geophysical Union

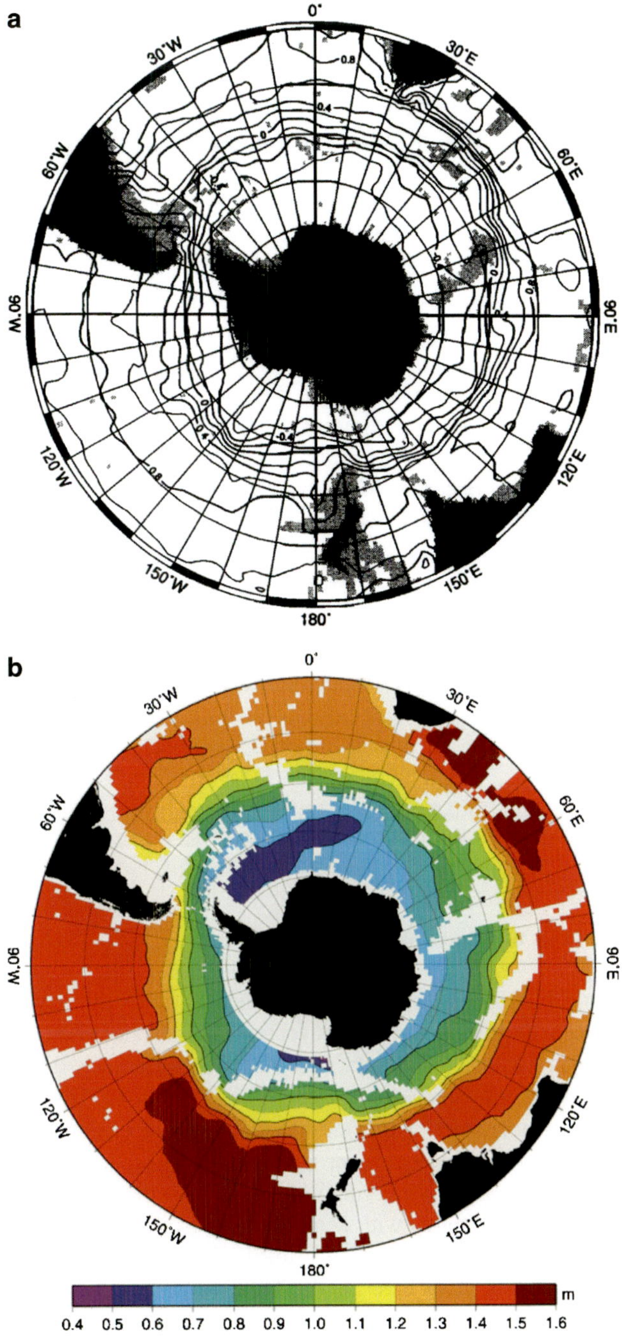

Fig. 2.12 (**a**) Mean dynamic height (in meters) from historical data as put together by Olbers et al. (1992). Dynamic topography is referenced to 2500 m and has been normalized by an arbitrary constant to simplify comparison with reconstructed SSH; (**b**) Mean sea surface height reconstructed from radar altimeter data and contoured at 0.2 m intervals. The 0 m height contour separates the SAF (positive heights) from the PF (negative heights). Typical errors are between 0.04 and 0.05 m. Areas with no data or with errors greater than 0.5 m are in white. From Gille (1994) with permission of the American Geophysical Union

its terrestrial counterpart. Studies of the chemistry and biology of the oceans have revealed many interesting insights into its history, its spatial variability and its evolution over the years. The chemistry and biology of the oceans are closely linked to each other and will be discussed as a unified field in this section.

In the polar regions, the long periods of daylight during spring and summer followed by long periods of darkness during autumn and winter make the biology and chemistry of the regions very highly pulsed. The associated processes occur concurrently with strong seasonal variations in irradiance, temperature and sea ice cover. The chemical characteristics of the ocean change during each season as nutrients and carbon dioxide are consumed by algae/diatoms while the latter release oxygen and introduce organic matter when they decay.

Among the most important parameter associated with the biology of global oceans is what is referred to as its primary productivity, which is loosely defined as the amount of algal, bacterial, or plant biomass built over time through the process of photosynthesis. This parameter has been quantified in terms of the amount of carbon fixed by photosynthesis, per unit area of space or volume, per unit time, but most estimates are referred to as net primary productivity which takes into account the respiration. It has been estimated that about 6.5 Pg (pentagram) of carbon per year are produced within the polar oceans, representing about 6.5% of the total primary production of the Earth including terrestrial contributions (Thomas 2004). The latest technique for estimating the primary productivity in the oceans is through the use of satellite ocean color sensors as will be described in Chap. 3. The primary productivity in Polar Oceans is influenced largely by the chemical and biological characteristics in the upper layers of the ocean as discussed in the following sections.

2.2.3.1 The Arctic and Peripheral Seas

The chemical and biological characteristics of the Arctic Ocean have been studied extensively in recent decades but much has yet to be learned because of limited accessibility and extreme weather conditions that make measurements difficult. Reviews of the current state of knowledge are given by Anderson (1995), Grebmeier et al. (1995), Jones and Anderson (1990), Smith and Sakshaug (1990). The key factors affecting the chemical and biological properties of the Arctic Ocean include processes in the vast shelf regions, river-runoffs and the growth and melt of sea ice. The shelf regions are shallow regions with depth varying from about 50 m to 200 m, as described earlier, and occupy more than a third of the Arctic basin. Except those going through Fram Strait, water entering the Arctic flows usually through these shallow regions. The main rivers that are connected to the Arctic are the Mackenzie and Yukon in the North American side and Lena, Yenisey, Ob, Yukon, and Pechora in the Eurasian side, providing a total yearly runoff discharge of about 68,100 $m^3 s^{-1}$ or about 70% of the total discharges. For comparison, the total volume of sea ice that melts or gets advected out of the region during the summer is about $4 \times 10^{12} m^3$ which amounts to 126,840 $m^3 s^{-1}$ year round. An important consequence of the large river input to the shelf seas is a high supply of nutrients. This supply is further enhanced by high nutrient water input from the Pacific Ocean through the Bering Sea and relatively moderate nutrient water input from the Atlantic Ocean. When light conditions are favorable, the primary production in these regions can be high. In the Siberian Sea where the average water depth is about 50 m, light limitation is normally not a problem because the stratification is strong. However, effects of shading can be pronounced if the water is turbid as can be caused by river-run-off or if the plankton concentration near the surface is high. On the other hand, water depth at the Kara and Barents Seas exceeds 200 m in large areas and vertical stratification is relatively weak. In these regions, the weak stratification is in part compensated by the presence of highly stratified melt water adjacent to the retreating ice edge. In general, high productivity is expected in most shelf regions of the Arctic during summer that may result in extensive biological production at all tropic levels. In winter, however, the productivity is extremely low since the region is in darkness and covered by sea ice.

The intensity of production in parts of the Arctic is limited by processes occurring during the off season as illustrated schematically in Fig. 2.13. For example, during sea ice growth in autumn and winter, the release of brine causes the formation of high salinity bottom water. Decaying organic matter and associated chemical constituents that are formed during this period get trapped in this high density water and dragged to the bottom where they are not available for primary production. The high density material flows along the bottom and eventually ends up in the deep interior. Such process is one of the means by which water masses are modified within the deep central Arctic Ocean but keeps these chemical constituents from being available in the upper part of the ocean. Because of largely varying topography in the Arctic, the impact of this process on productivity varies from region to region.

When conditions are right, enhanced primary production usually follows the retreating ice edge. The primary production extends into the pack where the ice concentration is relatively low. Primary production is associated with the consumption of nutrients and the release of oxygen thereby altering the chemical characteristics of the surface water. When the surface gets super-saturated with oxygen, a degassing to the atmosphere occurs resulting in a decrease in nitrate (NO) content. This happens only near the surface and not in the deeper layers where organic matter decays and the decrease in oxygen gets compensated by the release of nitrate for as long as the water is well oxygenated.

2.2.3.2 The Southern Ocean

One of the unresolved and puzzling issues in the studies of global carbon budget is the role of the Southern Ocean as a carbon sink. The region is regarded as a high-nutrient, low-chlorophyll (HNLC) region (Nelson et al. 2002) and modeling studies have indicated that the primary production should be very high. However, the annual photosynthetic uptake of CO_2 by phytoplankton and the resulting export of particulate organic carbon (POC) to the deep ocean have been much more modest values than those which can be sustained by the available nutrients (Sarmiento et al. 1998). Interpretation of in

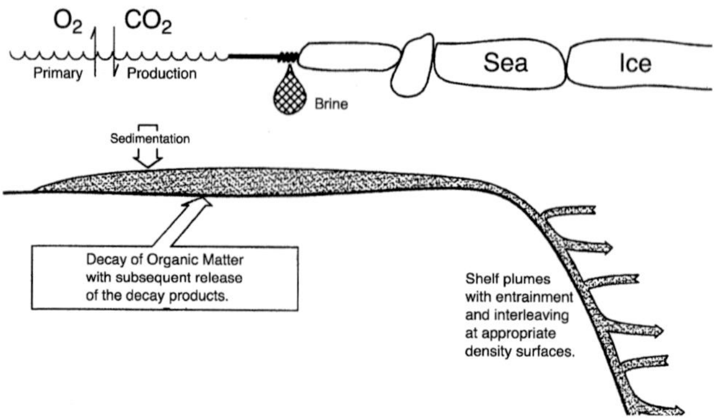

Fig. 2.13 Schematic illustration of different processes taking place over the year in the shallow shelf seas. During the summer season primary production fixate nutrients and CO^2 and release oxygen. At the end of the productive season, organic matter sinks to the bottom where it decays. During ice growth brine is released causing the formation of high salinity bottom water that flows out into the deep interior. From Anderson (1995) with permission of the American Geophysical Union

situ data can be biased by the lack of spatial and temporal coverage, but it is also necessary to fully understand the biogeochemical controls and feedbacks on primary production and the way oceanic biota respond to climate variations. The results from WOCE, JGOFS and other in situ studies and the advent of ocean color satellite data provide enormous opportunity for gaining insights into this problem.

Because there is no river run-off from the continent, the role of sea ice in controlling the salinity balance (and the associated biology) of surface waters in the Southern Ocean is quite important. Unlike the Arctic Ocean, the water masses in the region have relatively weak vertical stratification with a coincident large nutrient reservoir (Smith 1990). The Antarctic's continental shelves are not extensive but productive in some regions (Smith and Nelson 1985; Smith and Comiso 2008). The productivity is in part associated with almost unlimited nutrients. Figure 2.14 shows the distribution of silicate, phosphate and nitrate at the surface and at depths of 100 m, illustrating abundance of nutrients in the region. The data were originally put together at a resolution of 1° latitude by 2° longitude for the Southern Ocean Atlas (Gordon et al. 1982) using 6,313 casts from the region south of 30°C. It is apparent that while surface water in the Southern Ocean already shows abundance of nutrients, the concentration of nutrients at 100 m is even higher. Thus, upwelling and wind forcing would serve to make nutrients even more abundant for phytoplankton growth.

It was postulated by Martin and Fitzwater (1988) and later confirmed by them and other groups that one of the controlling factors that affects phytoplankton growth is iron. The availability of iron is very regional and could set limits to the magnitude of the phytoplankton stocks. In the Southern Ocean, there are regions where the plankton concentrations have been consistently low and have been regarded as part of the High Nutrient Low Chlorophyll (HNLC) system. Recent measurements have confirmed a limitation in iron in some of these regions

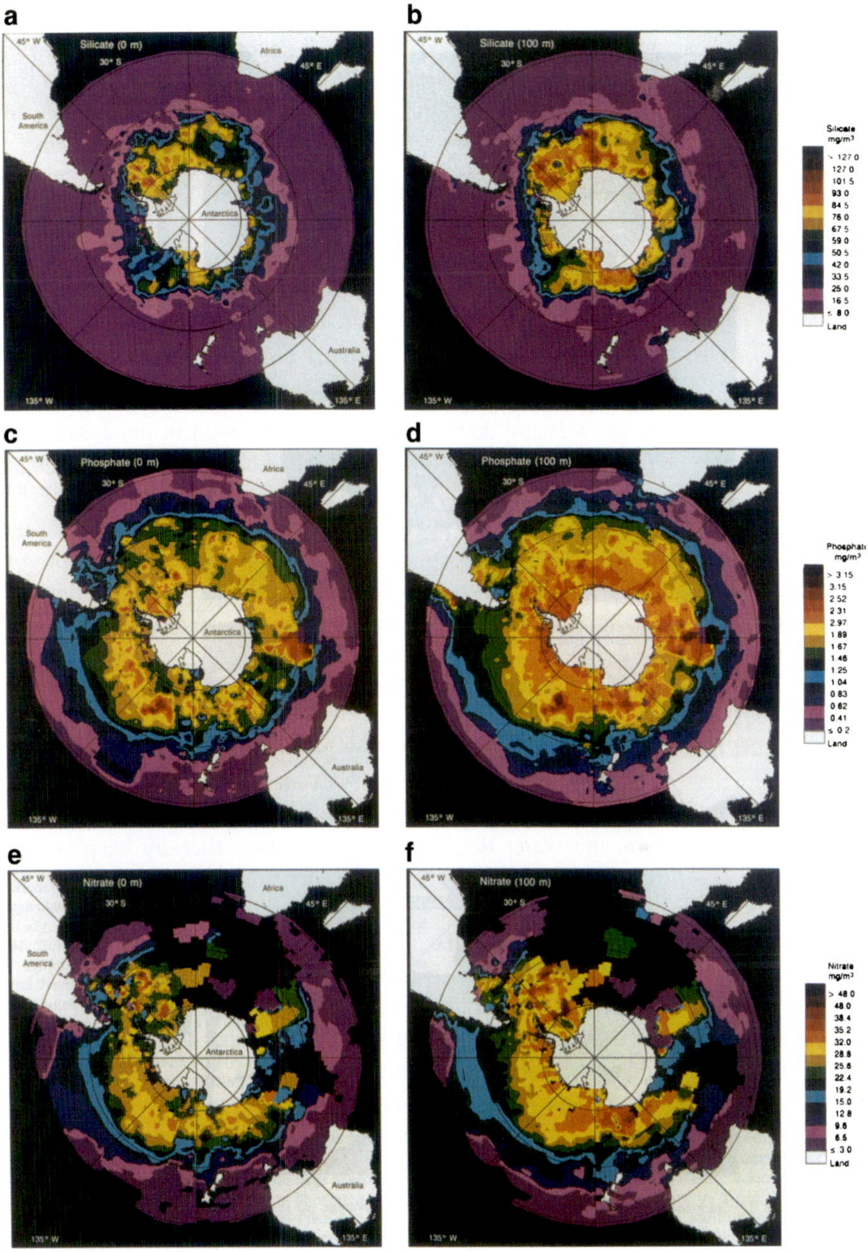

Fig. 2.14 Antarctic nutrients at the surface and 100m. From Comiso et al. (1998) with permission of the American Geophysical Union)

(Arrigo et al. 2003; Tremblay and Smith 2007). Aerosol iron is believed to be a primary source of iron in the Southern Ocean since enhanced productivity has been observed in large areas downwind of dry continental areas (Cassar et al. 2007). Upwelling, especially in coastal areas can also cause an increase in the supply of iron near the surface.

2.3 The Sea Ice Cover

2.3.1 Thermodynamics of Ice Growth and Decay Processes

The physics of the ice growth and decay processes is discussed in great detail in many publications (e.g., Weeks and Mellor 1984; Gow and Tucker 1990; Maykut 1985; Wadhams 2000). The formation and melt of sea ice are the natural consequences of large seasonal fluctuations in surface temperature and associated changes in fluxes between the surface ocean and the atmosphere. When compared with the depth of the oceans, sea ice is just a thin sheet that undergoes large changes in response to wind, waves, tides, and small changes in oceanic heat transport. The heat balance between the surface and the atmosphere in the polar regions also undergoes large changes from periods of long darkness in fall and winter through extended daylight in spring and summer.

One of the most unusual characteristics of fresh water is that as the water cools, its density increases but only down to temperatures of about 4°C. This temperature is sometimes referred to as the temperature of maximum density of fresh water. Below this temperature, the water does not get any denser thereby keeping the surface water layer from sinking before it gets frozen when it reaches 0°C temperature. It is for this reason that ice forms in lakes and rivers even while the deeper part of the water is still relatively warm. The ice floats because the solid version is lighter than the liquid version of water. This is unlike most other solids and only because the packing density of water molecules in ice is lower than in liquid water (Eicken 2003). With sea water, the temperature of maximum density is reduced depending on salinity and reaches the freezing temperature of −1.9°C when the salinity is about 24.7 psu (Wadhams 2000). Thus, the sea surface water continues to sink with lower temperature, but the water mass does not go all the way to the bottom because at the pycnocline, the density of polar water is relatively high. The density of the mixed layer is also usually significantly lower than the deeper part of the ocean because of river runoff and melt of sea ice in the Arctic and because of melt of sea ice, icebergs, and ice shelves in the Antarctic. Since the layer has to be cooled before freezing can commence, the onset of ice growth is in part dictated by the depth of the mixed layer. It has been observed for example that the initial date of ice formation can be delayed by as long as two months in regions where the mixed layer depth is 50 m when compared with those where it is 10 m (Doronin Yu and Kheisin 1975).

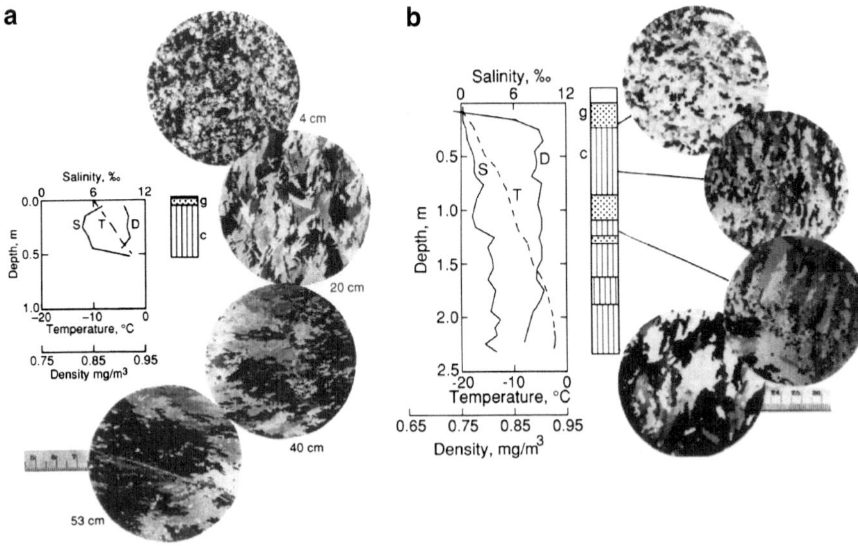

Fig. 2.15 Thin film images salinity profiles and vertical characterization of ice structure of (**a**) lead ice 53 days after freeze-up and (**b**) multiyear ice. From Tucker et al. (1992) with permission of the American Geophysical Union

Ice can form in one of two ways: one as granular ice which consists mainly of frazil ice and the other as columnar ice which is primarily made up of congelation ice (or dendrites). The difference in the structure of frazil and columnar ice is shown in Fig. 2.15. Generally, the formation of sea ice commence with the supercooling in the ocean which occurs when the water is cooled below its freezing point without nucleating ice crystals or changing its composition. Frazil ice crystals are formed at this time at various depths in turbulent water, but all end up floating at the surface to form the foundation of the initial sea ice cover. Frazil ice takes the shape of needles, spicules or platelets the typical size of which is a few tens of millimeters across and less than a millimeter in thickness (Weeks and Ackley 1986). Because of the rapid growth of ice and the relatively pure nature of ice crystals, spaces containing liquids called brine are trapped during the growth process. Within the newly formed ice, the net salinity including the brine is around 12 psu. Usually, the salt concentration of the brines increases as the volume decreases because of colder temperatures and vice-versa. In a sense, this phenomenon keeps the fluctuation of the ice temperature more moderate than would be without the brine. The brines do not get totally frozen because of much higher concentration of salt than the regular sea water. After the formation and consolidation of the ice, the salinity starts to decline through various processes. The main process is through gravitation in which the brine migrates vertically toward the underside of the ice. The rate at which the process occurs varies with season and is optimal during summer. Other processes include brine expulsion or brine rejection which leads, under the right conditions, to the formation of salt flowers at the ice surface. The salt flowers come in various sizes and spatial density and could change the physical and radiative characteristics of the ice surface.

The accumulation of frazil ice at the surface leads to the formation of a surface slush layer called grease ice which then consolidates into circular patterns called "pancake" ice that gets thicker through the accumulation of more frazil ice in the underside or through rafting. A few of the smaller pancakes usually stick to each other to form larger pancakes. Collisions with other pancakes also lead to the raised edges which is the attribute of most pancakes. Frazil ice would normally accumulate between the pancakes and when frozen, they serve as the glue that leads to larger pancakes and eventually, very large ice floes. In the inner zone, as the waves get attenuated by the ice cover and the ice cover becomes a vast and almost continuous ice sheet as individual pancakes are no longer distinct. The ensuing snow precipitation would engender the appearance of a uniform and almost homogeneous surface making it difficult for an ice observer to discriminate the individual components.

Congelation ice (or dentrites) is usually formed under stable water conditions and through the normal thermodynamic process called quiescent growth. Congelation (or dentritic) growth is the most common form of ice growth in consolidated ice regions, especially in the Arctic. They may be formed in areas of leads and polynyas initially as grease ice, consolidating into nilas that appears lighter gray with thickness. The ice sheet gets thick thermodynamically through the formation of congelation ice at the underside and would grow as long as the temperature of the water and ice is at or below the freezing point. To understand the growth of congelation ice, it is useful to discuss the ice crystal structure and associated characteristics. There are at least nine types of ice crystals formed under different conditions of temperature and pressure, but the only structure found under normal conditons, including that of sea ice, is what is called Ih. The "*h*" indicates the crystal symmetry which is hexagonal in form with the oxygen atoms at the apices and the hydrogen atoms along the sides as discussed in the literature (Hobbs 1972; Weeks and Ackley 1984). The water molecules (H_2O) are arranged tetrahedrally around each other with sixfold rotational symmetry. The crystals are normally characterized by one principal hexagonal axis of symmetry called the "*c*-axis" which is the optic axis of the crystal. Perpendicular to this axis is a series of parallel planes called "basal planes" each of which is defined by three crystallographic axes (called the "*a*-axes") which are equal in length and 120° of each other. These basal planes are most favorable for the ice growth of sea ice.

Sea salt ions cannot be accommodated in the crystal structure of ice. Ice is saline because brine is incorporated into the ice as illustrated in Fig. 2.16. In the figure, the most favorable direction of growth is perpendicular to the *c*-axis leading to the creation of skeletal ice structures, as indicated, about 1–3 cm long and 0.5–1 mm apart and consisting of platelet ice. Brine pockets are entrapped between these parallel columns of skeletal structures during the process of ice growth. The salinity of the ice is thus dependent on the efficiency of brine entrapment. The vertical alignment of the brine pockets is also a factor in the desalization process after the formation of the ice sheet. The formation of vertical layers of skeletal ice structures is actually unique

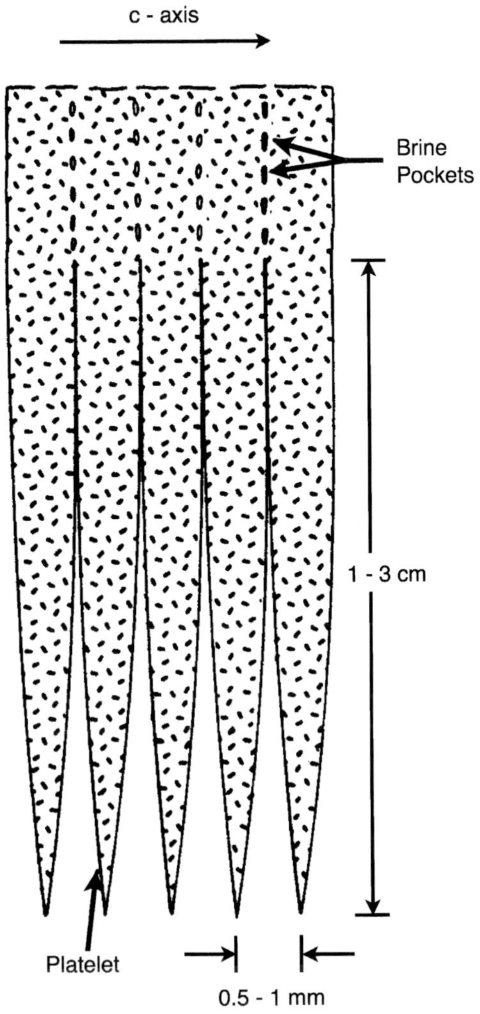

Fig. 2.16 Schematic of dentritic growth of sea ice. From Maykut (1985) with permission from CRC Inc

with sea water since such structures are not found in fresh water. The rejection of brine during ice growth causes the formation of a thin boundary layer of salty water underneath the advancing ice/water interface. Because this thin layer is much more salty than the underlying layer, a downward flux of salt accompanied by an upward flux of heat occurs in this layer. In the process, the actual temperature will be below the freezing point (because of the salinity) and supercooling occurs as described by Weeks and Ackley (1984). Laboratory observations by Harrison and Tiller (1963) also reveal that the length and spacing of the platelets (see Fig. 2.16) are such that the degree at which supercooling occurs at the ice/water interface is minimized.

The stratigraphy of the ice cover as observed from ice cores reveals sequences of granular ice and columnar ice. This indicates that rafting is a significant part of the growth process. The fraction of granular compared to columnar ice depends on environment with the former more dominant in turbulent environment during growth, as in the Antarctic, while the latter is dominant in the more stable water as in the Arctic. The growth process thus involves a combination of the thermodynamics growth as well as the dynamic growth.

The thermodynamics of the growth of sea ice has been studied through empirical, phenomenological and modeling studies. Finding direct relationships between sea ice growth and surface temperatures have been the object of many field studies (Zubob 1945; Bilello 1961). These studies were useful, but the growth process is complex and can be understood only through theoretical and modeling studies. One of the first serious modeling studies was done by Budyko (1966) who reported the simulation of thickness and horizontal extent that compares favorably with observations. The model had some shortcomings, in that it ignored the effects of heat conduction on surface ablation and it did not account for snow cover or the effects of ice salinity on the heat transport through the ice slab. Such shortcomings were taken into account in the model developed by Untersteiner (1964) which was subsequently enhanced and tested by Maykut and Untersteiner (1971). These early models assumed a horizontally infinite sheet of ice and studied the one dimensional growth and decay processes.

The basic equation that controls the growth and decay processes at the upper surface of sea ice after the initial ice sheet has been formed is given by

$$(1-\alpha)\, S_{\mathrm{i}} - S_{\mathrm{t}} + L_{\mathrm{i}} - L_{\mathrm{o}} + \mathrm{TS} + \mathrm{TL} + \mathrm{Cu} - M = 0 \tag{2.1}$$

where S_{i} is the incoming shortwave (solar) radiation, α is the albedo of the ice surface, S_{t} the flux of shortwave radiation which passes through the thin surface layer into the ice interior, L_{i} and L_{o} the incoming and outgoing longwave radiation at the ice surface, TS and TL the sensible and latent heat fluxes, Cu the conducted heat flux at the upper surface, and M the amount of heat loss due to melting or sublimation at the surface. The energy balance at the bottom surface is on the other hand given by

$$Q_{\mathrm{f}} + C_{\mathrm{L}} + F_{\mathrm{W}} = 0 \tag{2.2}$$

where Q_{f} is the heat gain or loss due to freezing or melting on the bottom of the ice, C_{L} the conduction of heat away from the bottom toward the ice surface, and F_{W} the turbulent heat transfer between the ice and the ocean. The two equations are not independent but are coupled together through the conducting terms given by Cu and C_{L}. These terms are usually not trivial to estimate because of complexity associated with the nonlinear nature of the vertical profile of the physical characteristics of the ice, specially its temperature and salinity. In actual modeling studies, linear approximations are made to reduce the complexity.

At the upper surface, the most seasonal parameter is the net solar radiation (i.e., S) which varies from zero in winter when the sun is below the horizon and the region is in darkness to about 300 W m^{-2} during the summer solstice when the sun is above the horizon. The net radiation ($S[1-\alpha]$) is more moderate with its peak lagging that of S because of the reduction in albedo caused by melt and meltponding. The latter is actually caused by the shortwave radiation. The incoming longwave radiation is primarily from clouds and atmospheric water vapor, while the outgoing longwave radiation is that from the ice surface and is proportional to the 4th power of the surface temperature as expressed by the Stefan–Boltzmann law. The net longwave radiation (i.e., L_i–L_o) is relatively seasonally uniform and is slightly negative at about 30 W m^{-2} representing a small loss of energy from the surface. Combining the contributions from the shortwave and longwave radiation, the net radiation is slightly negative in all seasons except in summer, where it goes up to about 80 W m^{-2}. When compared with the contribution of the radiation fluxes, those from the sensible and latent heat fluxes are relatively small. The sensible heat flux is dependent on temperature differences between the ice surface and the atmosphere, while the latent heat flux depends primarily on humidity differences. The sensible heat flux at the surface is positive in winter when a strong temperature inversion usually exists in the lower atmosphere and dips slightly during summer. The latent heat flux is neglible year round and is slightly negative in the summer when wet ice surfaces allow vapor transfer into the atmosphere. While the sensible and latent heat fluxes are relatively small on a monthly basis, daily values can be large. Model simulations by Maykut and Untersteiner indicate that without the heat losses associated with these fluxes, the surface ablation during summer would be 20–100% larger and the equilibrium ice thickness would be reduced by about 50%. The term S_t is generally a small percentage of the incident short wave radiation that is transmitted through the ice and allows the solar radiation to be directly absorbed by brine pockets in the ice interior causing the temperature profile to be altered and the brine volume to increase. This in turn changes the conductivity of the material and hence the complications in the estimate of Cu. The last term is the loss of ice due to melting or the acquisition of ice through sublimation.

The second equation indicates that the ice will grow if the sum of the oceanic flux, F_W, and the lower boundary conduction term, C_L, is negative and melt otherwise. A negative value will cause the release of latent heat that is conducted through the ice to the surface, while a positive value will cause the conduction of heat in the opposite direction as in the summer. The term Cu in the first equation is actually equal to the sum of C_L and the latent heat associated with mass changes in the ice surrounding the brine pockets. In particular, the conductivity flux at the surface is coupled to the lower and upper surface energy balances determined by the heat input from the ocean, the net annual latent heat released by ice growth and stored solar energy and is estimated to have a value of about 9 W m^{-2} in perennial ice and 2 or 3 times more in the thinner ice types (Maykut 1985).

It is important to note that while the same energy fluxes govern both the Arctic and Antarctic regions, there are significant differences in the environmental factors that make the growth and decay processes different in the two regions. The Arctic ice cover is generally located at higher latitudes than those of the Antarctic ice and therefore, the average shortwave components are slightly different. Since the Arctic ice is less divergent and thicker than the Antarctic ice, the oceanic heat fluxes are especially different and are approximately 2 and 12 $W m^{-2}$, respectively. Core samples also indicate that frazil ice production is more dominant in the Antarctic than in the Arctic reflecting the more dynamic environment in the former.

2.3.2 *Sea Ice Types*

The sea ice cover has been classified into two general types, namely, the seasonal ice cover that grows at the beginning of the freeze-up season and melts completely at the end of the summer, and the perennial ice cover or ice that survives the summer melt. The perennial ice cover consists mainly of thick multiyear ice floes that are as old as seven years and are the mainstay of the Arctic sea ice cover. They include second year ice floes which are those that formed in the previous winter and survived the summer melt. There are also some regional classifications such as the shear zone which refers to areas of heavily deformed ice, marginal ice zone which includes the ice edge and adjacent ice cover that generally consists of loose ice floes that can move freely with waves and currents, and fast ice zone, which refers to areas where the ice cover is locked together with the coastline by a grounded iceberg or thick ice. For convenience in the interpretation of satellite data, we also divide the ice cover into the outer zone, which is basically the marginal ice zone, and the inner zone, which is the vast and almost continuous sheet of ice cover following the outer zone. Sea ice has been classified conventionally as one of the distinct types described below.

2.3.2.1 New Ice

This ice type corresponds to those formed at the earliest stages of ice growth. Grease ice is basically the first layer of ice that forms and has been referred to as such because they look like oil slick and have thick soup-like consistency. Depending on surface temperature, they form within an hour and grow to a few centimeters in thickness although wind and wave effects can cause a file up of this material behind obstacles (e.g., edge of an ice floe) to as thick as a meter.

In turbulent water, grease ice eventually form loosely aggregated discs known as pancakes. The size of newly formed pancakes is nominally a few centimeters in diameter. These small pancakes tend to stick together to form larger pancakes and the process is repeated until they grow to giant pancakes that are several meters or kilometers in diameter. Pancakes are usually noted as having raised (or rough) edges that result from constant collisions with neighboring pancakes due to wind or wave action. During growth stages, the open water areas between pancakes are usually filled by frazil ice which serves to eventually glue the pancakes together.

In calmer water, grease ice quickly solidifies into thin sheets called nilas. As the nilas grows in thickness, the color changes and those less than 5 cm are usually referred to as black nilas, while the thicker ones are called gray ice and the thickest are called gray-white nilas. Because they are thin, nilas sheets tend to cut through each

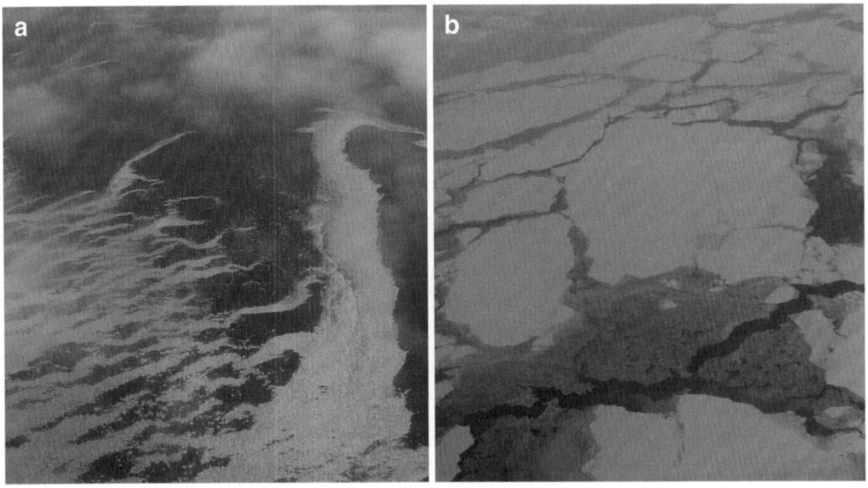

Fig. 2.17 Photos of (**a**) pancakes at the ice edge and (**b**) nilas, leads and first year ice in the Bellingshausean Sea from a P3 Aircraft during the AASI 2003 mission. (Courtesy of Fumihiko Nishio 2004)

other during windy conditions to form finger rafted ice. Also, when the conditions are right, brine rejection goes in the upward direction through a process called capillarity and the ice surface gets covered by salt flowers. Such phenomenon sometimes makes it difficult to interpret satellite data as will be illustrated later.

Both pancakes and nilas grow in thickness in various ways as illustrated in Fig. 2.17. Pancakes can grow through the accumulation of more frazil ice from the underside while nilas would grow through the thermodynamic process and form congelation ice as described earlier. The fastest, most efficient and effective way for the sea ice to grow, however, is through rafting and/or ridging.

2.3.2.2 Young and First Year Ice

When the thin sheet of sea ice reaches a thickness of about 15–30 cm, it becomes young ice. It is distinct from nilas in that the natural color has changed from the gray–white color to lighter gray or white. As the young ice acquires a snow cover and gets thicker, it becomes first year ice which is the dominant ice type in the seasonal region. The length of transition from young ice to first year ice depends on temperature, wind, and location. At some stages, these two types are difficult to discriminate especially when the ice sheet is undeformed and has only a few centimeters of snow cover. During the ice season, undeformed first year ice grows thermodynamically to about 2 m in the Arctic and 1 m in the Antarctic. Because of the desalination process after the initial sheet of ice has been formed, young ice is typically more saline than first year ice.

2.3.2.3 Second Year, Multiyear and Old Ice

Most of the first year ice floes melt completely in the summer, but some are thick enough to survive the melt period and become second year ice. The fraction of first year ice that becomes second year ice is basically unknown, but it depends on how cold the temperature gets in winter and how much rafting and ridging occurred before the summer melt period. The cycle repeats itself during the next winter period and the second year and first year ice that survive become third and second year ice, respectively. Typical ice surface types found in the Central Arctic are shown in Fig. 2.18. While multiyear ice has been defined generally by some as second year ice and older, WMO defines ice that survives at least one summer as old ice while ice that has survived at least two summers is referred to as multiyear ice. In this book, we will adopt the latter definition for multiyear ice. The thickness of multiyear ice has been estimated to average around 3 m but in some places like north of Greenland, they can be as thick as 10 m (Wadhams and Comiso 1992).

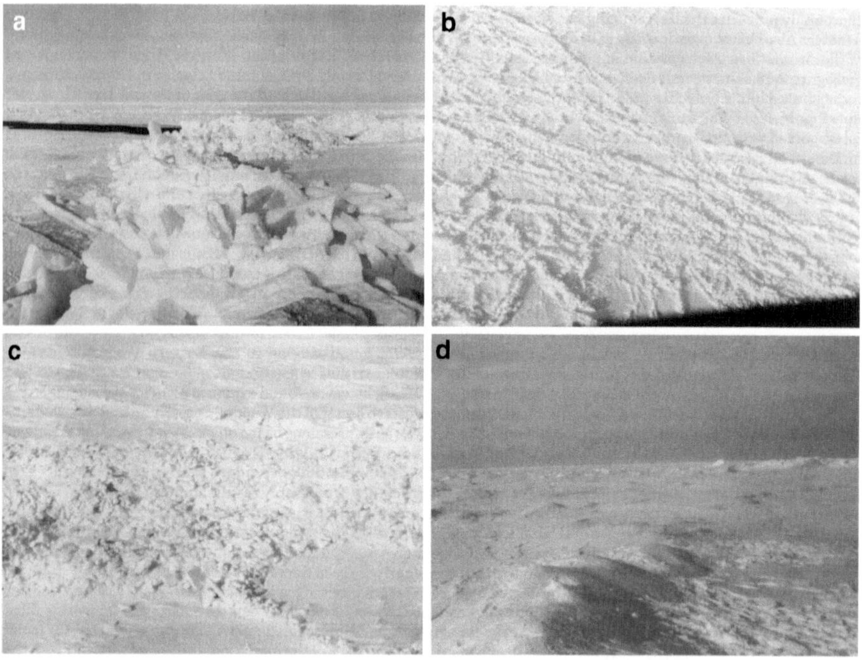

Fig. 2.18 Photos of deformed ice: (**a**) fire-year ridge about 2 m in height; (**b**) grounded first year ridge over 6 m high; (**c**) aerial view of a 400 m-wide rubble in Beaufort Sea near the coast; and (**d**) multiyear ridge about 3m in height. From Tucker et al (1992) with permission of the American Geophysical Union

2.3.2.4 Meltponded Ice

The snow over sea ice melts in the spring and summer and in the process creates pools of water over the ice floes as depicted in Fig. 2.19. This happens over any ice type that can hold the water but mainly over first year, second year and multiyear ice. Although it is not a distinct type of ice, its radiometric signature is very different. Over seasonal ice, the meltponds do not have a lasting impact since the sea ice melts completely, but over the perennial ice areas, the meltponds normally serve as flushing mechanism that further desalinates the sea ice cover. The size of meltponds varies with location and regional temperatures and can be from meters to kilometers in diameter. In some places in summer, the areal coverage has been estimated to exceed 20% (Perovich et al. 2002).

2.3.2.5 Deformed and Undeformed Ice

The sea ice cover has been classified as either undeformed or deformed. When the ice sheet is formed under calm conditions, it is basically undeformed. Strictly speaking, the sea ice cover is undeformed only during its early stages. Because of winds, tides, and ocean currents, the ice cover goes through almost continuous motion and some form of ridging, rafting and deformations at the edges are inevitable. In general, ice sheets that are basically horizontally homogeneous and have uniform thickness are regarded as undeformed ice. This category has been loosely defined to include parts of the vast ice sheet with relatively uniform thicknesses after consolidation of various ice types, including pancake ice, young ice and first year ice. The classification is meant to discriminate such types of ice cover from

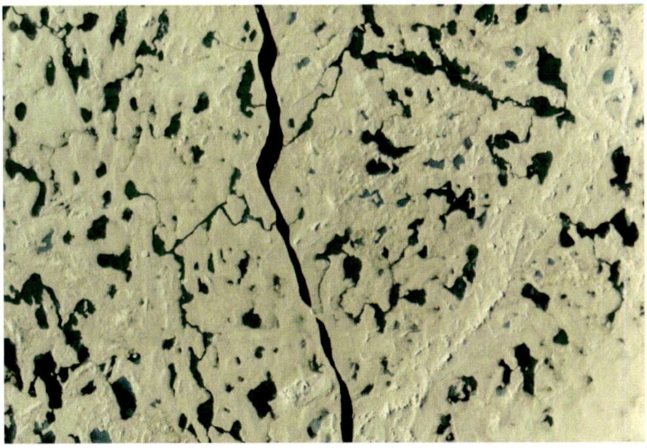

Fig. 2.19 Arctic meltponds as observed from an aircraft during the summer. Photo courtesy of D. Perovich, 2003

deformed ice such as those shown in Fig. 2.18 which refers to those that have been subject to substantial rafting and ridging events caused by storms and powerful winds. These types of deformed ice can have thicknesses of 10 m or even greater and normally survive the summer melt. They are readily identified by uneven distribution of sails at the surface or keels at the underside of the ice.

2.3.2.6 Land-fast Ice

Another ice type is the so called "land-fast ice" which is encountered in most of the coastal regions in the Arctic and the Antarctic. Land-fast ice (sometimes called shore fast ice or near shore ice) is sea ice that is attached to the coastline and is basically motionless, but sometimes they may move tens of meters in response to thermal and mechanical stresses (Tucker et al. 1980). It is usually formed in shallow water early in winter where the coastal geometry allows the water column to cool rapidly to freezing point. At times, they are anchored by grounded pressure ridges as thick as 30–50 m or icebergs. Such thick ridges have been called "stamukhas" (Russian word for "grounded hummock"). In the spring and summer, they are usually detached from the shore and melt but the thick anchor may survive the melt period and enables easy recovery of the land fast ice during the subsequent winter. Land-fast ice is present along Arctic and Antarctic coastlines for as long as 10 months in a year. The extent of the fast ice varies from one location to another and is determined mainly by ocean depth and topography. For example, they are about 10 km wide in some places like Point Barrow (Alaska), but can extend as far as 60 km offshore at Harrison Bay (Alaska) which is only several kilometer away to the west. They are most extensive in regions, where the shelf depth is nearly constant over relatively large areas as in Laptev Sea.

Fast ice can grow in thickness from less than a meter to several meters depending on whether they are seasonal or multiyear ice or whether they are undeformed or deformed. In parts of the Antarctic along the Indian and Western Pacific Ocean, the relative ratio of the volume of fast ice to that of the total sea ice cover has been estimated to be as high as 0.28 (Giles et al. 2008).

From an oceanographic point of view, the presence of land-fast ice is potentially very important for the dynamics of the polar ocean because it creates an extension of the coastline toward the deeper regions as illustrated in Fig. 2.20. Thus, coastal polynyas formed by strong winds would be located at the farther edge of the fast ice closer to the shelf break region, where the dense water mass formed during the ice growth process feeds more readily into the deeper parts of the ocean. In recent years, they have also been used as a tool for climate change studies by monitoring the potential impact of warming on the length of the residence period of the ice. A shortening of the land-fast ice season is likely to have an important impact as well to coastal dynamics since land-fast ice usually protects coastlines from ocean surges, associated wave damage, and ice pile-up. From the practical viewpoint, land-fast ice also offers a good platform for whaling and seal hunting to local communities and animals.

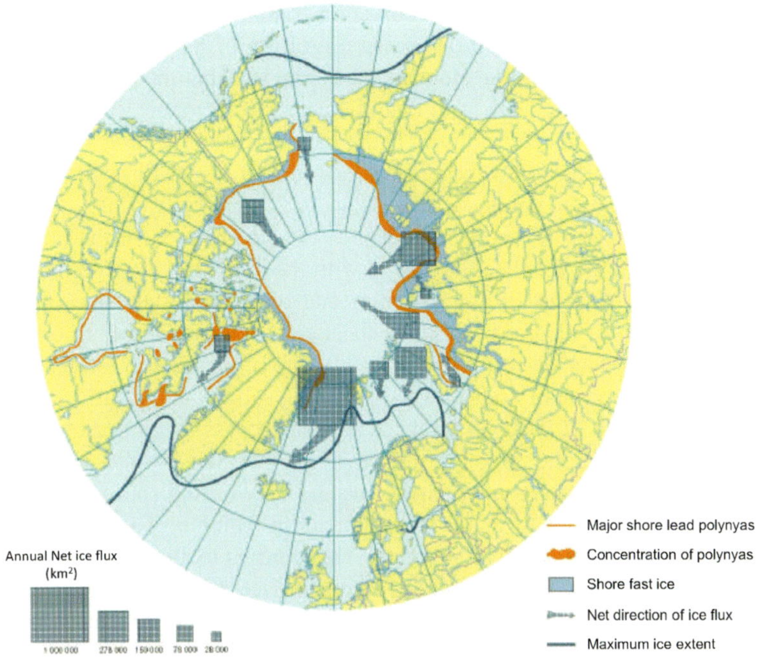

Fig. 2.20 Role of land-fast ice in sea ice exchange and ice production and bottom water formation through recurrences of coastal polynyas. Annual sea ice fluxes in different regions are indicated by the size of the boxes and the directional flow of sea ice is also shown. Reproduced with permission of the Arctic Monitoring Assessment Program (AMAP), 1998, Oslo Norway

2.3.3 Temperature, Salinity, Density and Conductivity

The temperature of sea ice depends on stage of formation, type, thickness, season and age and is influenced by many environmental factors. The temperature of sea ice during the earliest stage of its formation (i.e., frazil or grease ice) is approximately equal to that of freezing temperature. As the ice thickens, the underside of the ice remains at the freezing temperature, while the surface temperature changes depending on thickness and air temperature. For young and thicker ice types, the surface temperature is influenced mainly by surface air temperature the degree of which depends on snow cover. For bare ice, the ice temperature fluctuates almost as rapidly as surface air temperature. As the snow cover gets reasonably thick, the impact is more moderate because of the insulating effect of snow. The temperature profile with depth in autumn and winter indicates the coldest temperature at the surface warming up to freezing temperatures at the underside of the ice. In spring and summer, the temperature profile is more uniform with the average temperature closer to melt temperatures.

The salinity of sea ice is highest during formation stages when the brine pockets are still in place and the ice has not undergone a desalination process.

The desalination process goes through one or a combination of a few possibilities. The primary mechanism is the downward migration of brines often referred to as gravity brine drainage. This is enhanced by the fact that the brines are formed in vertical strings (see Fig. 2.16) and tend to become interconnected through internal stress or changes in temperature. Thus, they would form long vertical tubes called brine channels that allow the vertical movement of brine through the ice. The colder denser brine at the upper parts of the ice would normally sink downward to replace the warmer less denser brine. Also, as the ice grows from the bottom and the ice gets thicker, the ice sheet moves up with the brine, but hydrostatic pressure forces the brine downward in the channels and out to the underside of the ice. During early stages of ice growth, the ice sheet can also lose salinity through brine expulsion. This mechanism results from the buildup of pressure in refreezing brine pockets. The observation of "salt flowers" and high salinity at the surface of young ice is a manifestation of such occurrences but the process is not a major factor in the ice desalination process. Another mechanism is "flushing," which in this case is the dilution of brine and this time usually occurs only during spring and summer. As the surface warms up to melt temperatures, the volume of the brine pockets increases and the ice becomes very porous. This enables meltwater from snow and ice to percolate into the ice further decreasing the ice salinity. This process may have a significant role in the large change in salinity when first year ice becomes second year ice at the end of the summer.

The dependence of ice salinity with depth varies with ice type as illustrated in Fig. 2.21. The profiles in Fig. 2.21 are basically summaries of data collected from many observations (Weeks and Lee 1958; Schwarzacher 1959; Cox and Weeks 1974; Nakawo and Sinha 1981; Weeks and Ackley 1982). For young and first year ice (a-d), the salinity goes down initially but bounces back at the bottom causing the characteristic c-shape profile. The c-shape is an indication that the brine does not completely drain at the bottom until perhaps during the summer. The thicker ice types also show less salinity and less curvature than the thinner and younger ice types. The data show large scatter and the profiles can change significantly depending on ice growth rate and location. For thick multiyear ice, the salinity profile depends on location of the freeboard and Fig. 2.21f shows that elevated freeboards as in hummocks have lower salinity near the surface than those with the freeboard close to sea level.

The salt content of the ice affects not just the physical property but also the radiative characteristics. For completeness, we show in Fig. 2.22 the distribution of data compiled by Cox and Weeks (1974) that shows the way the salinity of the ice varies as a function of thickness both during cold and warm conditions. The figure shows a big drop in salinity (from 14 to 5.5 psu) as the ice grows from near zero thickness to about 40 cm. The drop in the salinity continues after that but at a much more gradual rate. This may be an indication of a shift in the salinization process (i.e., from brine expulsion to brine drainage). With the case of warm ice (open circles), the salinity actually increases slightly with thickness. This may in part be caused by greater uncertainties in the measurement since the latter is complicated by brine leakages from the warm ice cores.

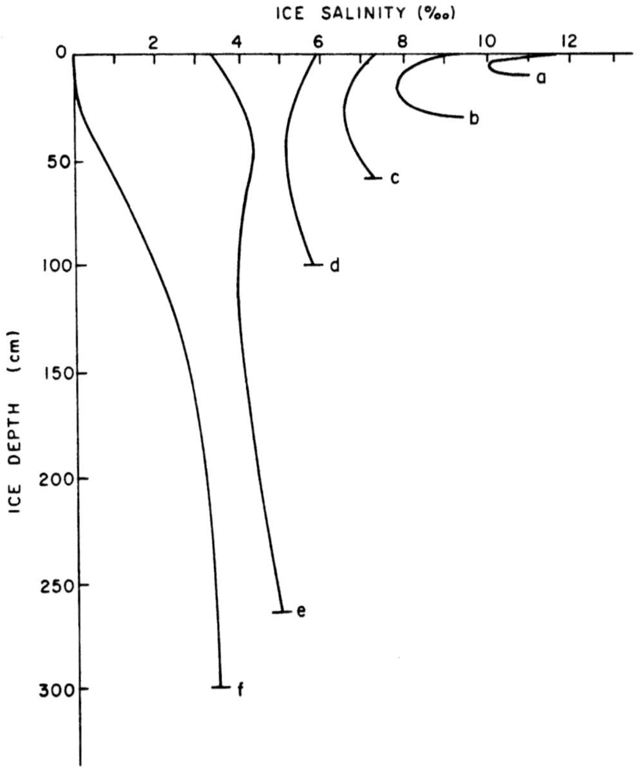

Fig. 2.21 Salinitity profiles of different ice types based on field data. Curves a to d are for young and first year ice while curves e and f are for multiyear ice. Curve e is for freeboards relatively close to sea level while curve f is for more elevated freeboards like hammocks. From Maykut (1985) with permission of the CRC Press, Inc

The density of pure ice is 0.9178 Mg/m^3 and would also be the density of sea ice were it not for the impurities that include brine and air bubbles. Other contaminants would include soot, microorganisms and others. The fractional volume of brine (in %) can be written as

$$v_b = \frac{\rho_i S_i}{\rho_b S_b} \tag{2.3}$$

where ρ_i and ρ_b are the density while S_i and S_b are salinities of the ice and brine, respectively. The effect of changing temperatures on S_i and ρ_b provides some complications but v_b (in units of %) can be derived at reasonable accuracy using the following simple equation developed by Frankenstein and Garner (1967):

$$v_b = S_i\left(0.0532 - 4.919 / T_i\right), -22.9 \le T_i \le -0.5°\text{C}. \tag{2.4}$$

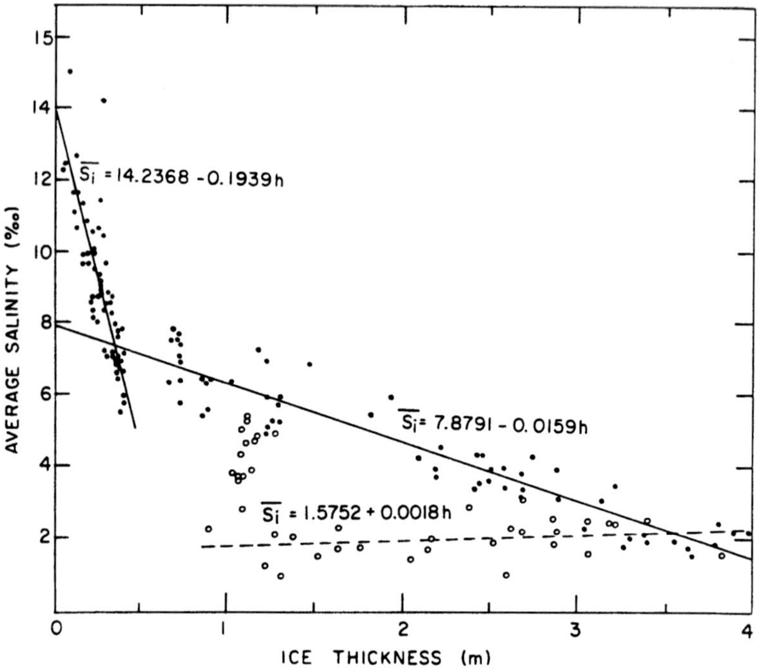

Fig. 2.22 Functional relationships between average or bulk salinity of Arctic sea ice and its thickness during the growth season (depicted by data points in the top two solid lines) and the melt season (data along the dash line). From Cox and Weeks (1974) with permission from the International Glaciological Society

Typical values of v_b ranges from 8% to 40% for young ice, while for multiyear ice, it is usually about 10%. Following the work of Schwerdtfeger, the fraction of air bubbles could also be estimated using the equation:

$$v_a = 1 - \rho_i \left[\left(1 - S_i'\right) / \rho_o + 4.98 S_i' / T_i \right], T_i > -8.2^\circ\text{C}, \tag{2.5}$$

where ρ_o is density of pure ice, ρ_i is the bulk density of sea ice and $S_i' = S_i / 1{,}000$ is the fractional salt content of the ice. This equation can be rewritten and expressed in terms of ρ_i as

$$\rho_i = \left(1 - v_a\right)\left[1 - 0.00456\, S_i / T_i\right] \rho_o. \tag{2.6}$$

The density of ice is thus primarily determined by v_a during cold conditions and by S_i during relatively warm conditions. Knowledge of the magnitude of these parameters is important since they affect the bulk thermal, mechanical, optical, and electrical properties of sea ice.

The heat flux through sea ice has been shown experimentally to be proportional to the change in ice temperature per unit length. Heat is transferred through the ice only if there is a thermal gradient between the surface and the underside.

The constant of proportionality is called the thermal conductivity, k_i. The thermal conductivity of pure ice, k_p, is dependent on the ice temperature and can be estimated from the following empirical equation as determined by Yen (1981):

$$k_p = 1.16\left(1.91 - 8.66\times10^{-3}\,\mathrm{T} + 2.97\times10^{-5}\,\mathrm{T}^2\right) \tag{2.7}$$

where T_i is ice temperature in Celsius. At 0°C, k_p is approximately equal to 2.0 W/m/K. The thermal conductivity of brine, k_b, on the other hand is about a factor of four lower and is given according to Yen (1981) by

$$k_b = 0.4184\left(1.25 + 0.30\,\mathrm{T} + 0.00014\,\mathrm{T}^2\right) \tag{2.8}$$

The thermal conductivity of sea ice is thus a function of these two variables and has been estimated through the use of a model that makes assumptions about the microstructure of sea ice that takes into account such factors as brine pocket distribution, air bubbles and precipitation (Schwedtfeger 1963; Ono 1968). A simple but reasonably good approximation of the thermal conductivity is provided by Untersteiner (1961),

$$k_i = k_p + \beta S_i / T_i, \tag{2.9}$$

where T_i is in °C, β is a constant and is equal to 0.13 W/m. This equation provides a simple functional relationship between k_i, S_i and T_i and has been used in modeling studies. From the equation, one can infer that the dependence on temperature is small at low temperatures but at greater than −4°C, the variation is much larger. Also, for perennial ice, k_i varies by 10–20% while for young ice, the variation is about 50%.

2.3.4 Thickness of Ice and Snow Cover

The thickness of sea ice is a very important attribute because it plays a central role in its interaction with the atmosphere, the underlying ocean and its environment. For example, heat flux from thin ice is known to be an order of magnitude greater than those of thick ice (Maykut 1978). Also, thick ice is less vulnerable to deformation and total melt in summer than the relatively thinner ice. However, as discussed in the previous section, ice is always moving and its thickness distribution is affected by the degree of convergence as well as divergence. In convergent regions, the ice cover is generally thick, but the variability of the thickness is high. In divergence regions, the ice cover is generally thinner because of the abundance of the thinner ice types.

Ice thickness can be estimated from first principles in a stationary environment knowing the air temperature history, ocean heat flux, and the amount of snow precipitation in a region. The ice surface is generally wet during the early stages of the ice cover and the surface temperature is close to the freezing point. Once the ice is consolidated, free water is no longer available and the ice surface temperature approaches that of atmospheric temperature. At this stage, ice growth occurs at the underside of the ice sheet at a rate depending on how rapidly heat can be conducted from the ice water interface to the surface. The rate is influenced primarily by air temperature and empirical studies by Anderson (1961) in Baffin Bay actually reveal that a good approximation of ice thickness can be obtained from the cumulative number of freezing days (f_n) using the equation

$$6.7f_n = 5.1h + h^2 \tag{2.10}$$

where h represents thickness and

$$f_n = \int (T_f - T_a)\,dt \tag{2.11}$$

where T_f and T_a are freezing and air temperatures, respectively while t is time. The relationship is generally true for thicknesses between 10 and 80 cm and with minimal snow cover. As the ice gets thicker, the relationship breaks down because of increasing thermal mass of the ice, the decreasing magnitude of heat flux in the ice relative to the ocean, and increasing snow cover. It also does not work everywhere as other investigators find different regression formula for h and therefore may not be suitable for other regions like the central Arctic or the Antarctic regions.

A more general relationship was developed by Maykut (1985) using a model that accounts for the energy balance at the top and bottom of the ice and makes use of the rate of heat transport between the two boundaries. For young ice, conductive flux is directly proportional to the temperature difference between the two boundaries and he showed that the Anderson regression equation follows from the assumption of a linear temperature gradient in the ice and negligible heat flux in the water. The theoretical formula for thickness as derived by Maykut (1985) is given by

$$h^2 - h_o^2 + \left(2k_i h_s / k_s + 2k_i / C_t\right)\left(h - h_o\right) = 2k_i f_n / \rho_i L \tag{2.12}$$

where h_0 is the initial thickness of the ice, k_i is the average thermal conductivity of the ice, k_s is the thermal conductivity of the snow, ρ_i is the density of the ice, L is the latent heat of fusion, C_t is a bulk transfer coefficient parameterizing turbulent heat transfer between the ice and atmosphere, and h_s is the snow depth which is assumed to remain constant over the period of integration.

The thickness of the snow cover of sea ice depends on the amount of precipitation which is not much within the pack because of relatively low humidity. In the Arctic basin, the estimates range from a few cms to a few tens of cms. In the Antarctic, the amount of precipitation over sea ice is in tens of cms and more than in the Arctic basin because of proximity to open ocean. Because the sea ice is not so thick in the Antarctic, the freeboard is only a few cms above sea level making it much more vulnerable to become negative as snow accumulates. Negative freeboard would allow sea water to get into the snow ice interface and would convert some of the snow into snow-ice which becomes an integral part of the sea ice cover. The fraction of surfaces with negative freeboard has been found to be substantial and estimated by some to be as much as 30% of the Antarctic ice cover.

2.3.5 Dynamics, Ice Drift Characteristics and Pressure Ridges

Except for land fast ice, sea ice is a dynamic entity that is always kept in motion by wind, ocean current, tides and the Earth's rotation. Ice motion enhances the transfer of momentum from the atmosphere to the ice, and hence the ocean, and causes a more vigorous thermodynamic interaction between the ice, ocean and atmosphere. The creation of pressure ridges makes the ice cover generally thicker and to have a higher chance to survive the summer, while the breakup and advection of ice to warmer water can cause an early disappearance of the ice cover in the spring and summer period. Ice motion can also cause the transport of large amount of sea ice to other regions where they melt and cause the freshening of surface water in these regions.

The movement and deformation of sea ice are governed by atmospheric and oceanic stresses and can be quantified using the force balance equation given by

$$m\,\mathrm{d}v/\mathrm{d}t = S_{\mathrm{w}} + S_{\mathrm{a}} + C + F_{\mathrm{i}} + G_{\mathrm{c}} + T_{\mathrm{o}} \tag{2.13}$$

where v is the ice velocity, m the ice mass per unit area, C the Coriolis force, S_{w} the water stress, S_{a} the air stress, F_{i} the force due to the variation of internal stress, G_{c} the force due to geostrophic currents, and T_{o} the force attributed to the tilt of the ocean surface. Intuitively, the major contribution to ice motion is expected to come from the factors associated with the wind. This has been confirmed by modeling studies (Hibler 1979), with the air and water stresses estimated to be about 0.1 $\mathrm{N\,m^{-2}}$, whereas the Coriolis force is about half this value. The internal stress term, F_{i}, is associated with the ability of the ice pack to transmit stresses over very large distances. It is especially relevant in enclosed areas such as bays normally occupied by consolidated ice, the motion of which is not affected by high wind or current forcing. The term is usually calculated as a residual of the force balance. The contributions from geostrophic current and ocean tilt terms are relatively small but long term effects of these factors are important. For example, the ice drift in Fram Strait and in the Bering Sea has been observed to be dominated by these currents (Moritz and Colony 1988; Pritchard 1988).

The drift of sea ice has been studied by monitoring the changing location of buoys that have been anchored on top of them (Colony and Thorndike 1984; Rigor et al. 2002). With satellite data, we now have much better spatial sampling and the means to cover the entire ice pack and get to know the drift characteristics practically everywhere and at all times, except for a few exceptions such as in late spring and summer when the surface signature becomes highly variable and unpredictable. The drift patterns are essentially similar to those of the circulation patterns of the surface water as described earlier. In the Arctic basin, there are two main drift features, previously identified by Gordienko (1958), namely: the transpolar drift stream that transports ice through the Fram Strait into the Greenland Sea and the Beaufort Gyre, which moves ice in the western region in a generally clockwise direction. Interannual changes in wind circulation could alter these patterns as described by Proshutinsky and Johnson (1996). The impact of wind and other external forcing may be described in two ways: convergent, in which the ice floes tend to get closer to each other, or divergent, in which the floes get further away from each other.

The impact of dynamics in the two hemispheres is quite different. Because the Arctic Ocean is surrounded by land, the ice cover is generally convergent and ice could easily file up in a variety of deformation features. A good example is the formation of pressure ridges which are linear features that can extend up to several kilometers in length. This leads to the creation of "keels" beneath the ice and "sails" on the surface of the ice. It has been observed that the keels and sails are primarily made up of ice that is typically 1 m thick (Tucker and Govani 1981). This means that they are made of first year ice sheets (or younger ice) that formed in leads in perennial ice regions. The ridging was thus likely a consequence of the collision of the thicker multiyear ice floes, with the thinner first year ice sheet sandwiched in between, during adverse weather conditions. The process can occur in different places and multiple times each year and it has been estimated that about 40% of the Arctic ice pack has some deformed ice on it (Wadhams 1980).

Conversely, in the Southern Ocean, there is no northern boundary and ice is generally divergent, when compared with that of the Arctic. While rafting and ridging still happen, large pressure ridges are not prevalent except near land mass regions as near the Antarctic Peninsula. The general patterns of the ice drift generally follow the surface circulation patterns as described earlier with ice floes going in the clockwise direction at the outer ice edges and in the counter clockwise direction closer to the coastlines. The gyres, such as the Weddell gyre and the Ross Sea gyre, also cause the ice cover in certain regions to stay within the ice pack. (possible Figure to add-buoy path over an entire 1 year period).

2.3.6 *Chemical and Biological Characteristics of Sea Ice*

Record shows that in 1820, Bellingshausen noticed the discoloration in the sea ice and interpreted it as caused by dust from land or by droppings of seabirds. During his voyage to the Ross Sea in 1839–1843, James Clark Ross was initially inclined by such interpretation, but his surgeon, Joseph Hooker took samples of the material and found out that they consisted mainly of remains of microorganisms which were later identified as diatoms. During the Fram voyage from 1893 to 1896, Nansen was quite excited to see these "unicellular pieces of slime that live by the millions in pools at nearly every ice floe all over." Diatoms are microalgae, or aquatic unicellular organisms that photosynthesize like plants and require nutrients like nitrogen, and phosphorus to grow. They are usually found in lakes, rivers and oceans, as

Fig. 2.23 Brown ice as observed during a ship-based validation campaign in the Bellingshausen Sea in August 2003. Photo courtesy of Koni Steffen, 2003

indicated in Section 2.2.3, but it was totally unexpected that they would be found in sea ice.

We now know that during ice formation algae, bacteria and other particles get attached to frazil ice formed at different depths as they rise to the surface (Garrison et al. 1989). Eventually, the concentration of these particles is considerably higher (by as much as 50 times) than that of underlying sea water. During subsequent ice growth, these particles get to be incorporated in the ice, but it is in the regions of the ice floe in which there is immediate access to nutrients (e.g., sea water at the underside of the ice) that the sea ice algae gets to flourish. Light is usually sufficient for photosynthesis during spring and summer except in ice floes with extremely thick snow cover. Algal communities thus usually reside at the bottom of the ice (Fig. 2.23), forming an active layer that extends upward by as far as 0.2 m from the bottom, being limited only by the availability of nutrients (Arrigo and Sullivan 1992). However, under certain conditions, they can be found in internal frazil ice layers that are porous enough to have accessibility of nutrients from sea water or are formed through rafting. Algal communities can also survive at the snow ice interface if nutrients are provided by sea water during flooding conditions when the snow cover gets to be very thick or during submersion of the ice floe during stormy conditions.

In the polar regions where extended periods of darkness during winter are a norm, it is also reasonable to ask how the communities survive. It has been postulated that those that cannot form stress-resistant spores or cysts, enter into a stage that is similar to "hybernation," during which cell metabolism is reduced to minimal levels but enough to keep the cells alive (Thomas 2004). Some diatom species (from both polar and temperate regions) have been observed to survive periods of darkness for several months or even several years. They also survive despite freezing or below freezing temperatures and also salinities that are much higher than sea water salinities.

Generally, the Antarctic sea ice cover is more conducive for algal growth than that of the Arctic sea ice, primarily because of the availability of more sunlight in the former on account of geographical location. It has been estimated that on average, the algal biomass of chlorophyll *a* in the Arctic ice is 88 mg m^{-2} ,whereas in the Antarctic, it is about 133 mg m^{-2}. *The* highest values are 300 and 400 mg m^{-2} for the Arctic and Antarctic regions respectively (Thomas 2004). The estimates are useful not just for the study of the biology of sea ice, but also how sea ice impacts the open ocean productivity as the meltwater and algal biomass are released to the open ocean in the spring and summer.

References

Aagarrd K (1977) STD measurements in possible dispersal regions of the Beaufort Sea. Annual Report, Contract 03-5-022-67, TO1. Department of Oceanography, University of Washington

Aagaard K (1981) On the deep circulation in the Arctic Ocean. Deep-Sea Res 28:251–268

Aagaard K (1988) Some thoughts on the large-scale circulation of the Arctic Ocean. Proceedings of the Second Conference on Polar Meteorology and Oceanography, American Meteorological Society, March 29–31, Madison, Wisconsin

Aagaard K, Coachman LK, Carmack E (1981) On the halocline of the Arctic Ocean 28A(6):529–545

Aagarrd K, Swift JH, Carmack EC (1985) Thermohaline circulation in the Arctic Mediterranean seas. J Geophys Res 90:4833–4846

Anderson LG (1995) Chemical oceanography of the Arctic and its shelf seas, In: Smith Jr W, Grebmeier JM (eds) Arctic Oceanography: Marginal ice zones and continental shelves, American Geophysical Union, Washington DC, pp 183–202

Arrigo KR, Sullivan CW (1992) The influence of salinity and temperature covariations on the photphysiological characteristics of Antarctic sea ice microalgae. J Phycology 28(6):746–756

Arrigo KR, Worthen DL, Robinson DH (2003) A coupled ocean-ecosystem model of the Ross Sea: 2. Iron regulation of phytoplankton taxonomic variability and primary production. J Geophys Res 108: doi:10.1029/2001JC000856

Beckmann A, Hellmer HH, Timmermann R (1999) A numerical model of the Weddell Sea: Large-scale circulation and water mass distribution. J Geophys Res 104(C10): 23,375–391

Bilello MA (1961) Formation, growth and decay of sea ice in the Canadian Arctic Archipelago. Arctic 14(1):2–24

Brennecke W (1921) Die ozeanographischen Arbeitentder deutschen antarktischen Expedition 1911–12. Arch Dsch Seewarte 39:1–214

Broecker WS (1987) The biggest chill, Natural History Magazine, 74–82

Broecker WS (1997) Thermohaline circulation, the Achilles heel of our climate system: Will man-made CO_2 upset the current balance? Science 278(5343):1682–1588

Bryden H, Pillsbury RD (1977) Variability of deep flow in the Drake Passage from year-long current measurements. J Phys Oceanogr 7:803–810

Budyko M (1966) Polar ice and climate. Proceedings of the Symposium on the Arctic heat budget and atmospheric circulation, Memo RM-5233-NSF, The Rand Corporation, Santa Monica, CA

Carmack EC (1977) Water characteristics of the Southern Ocean south of the polar front. In: Angel M (ed) A Voyage of Discovery, George Deacon 70th Anniversary Volume, Pergamon Press, Oxford

Carmack E C (1986) Circulation and mixing in ice-covered waters, In: Untersteiner (Ed) The Geophysics of Sea Ice, Plenum Press, New York
Carmack EC (1990) Large scale physical oceanography of polar oceans, in polar oceanography, part A: Physical science, Academic Press Inc., San Diego, CA
Carmack EC, Foster TD (1975) On the flow of water out of the Weddell Sea. Deep-Sea Research 22:711–724
Carsey FD, Roach AT (1994) Oceanic convection in the Greenland Sea Odden region as interpreted in satellite data. In: Johannessen OM, Muench RD, Overland JE (eds) The Polar Oceans and their Role in Shaping the Global Environment, Geophysical Monograph 85:211–222
Cassar N, Bender ML, Barnett BA, Fan S, Moxim WJ, Levy H II, Tilbrook B (2007) The Southern Ocean biological response to Aeolian iron deposition. Science 317:1067–1070
Clarke RA, Gascaard JC (1983) The formation of Labrador Sea water, Part I: Large-scale processes. J Phys Oceanogr 13:1764–1788
Colony R, Thorndike AS (1984) An estimate of the mean field of Arctic sea ice motion. J Geophys Res 89:10623–10629
Cox GFN, Weeks WF (1974) Salinity variations in sea ice. J Glaciol 13:109–120
De Veaux R, Gordon AL, Comiso JC, Chase NE (1993) Modeling of topographical effects on Antarctic sea ice using multivariate adaptive regression splines. J Geophys Res 98(C11):20207–20319
Deacon GER (1937) The hydrology of the Southern Ocean. Discov Rep 15:3–122
Deacon (1982) Physical and biological zonation in the Southern Ocean. Deep-Sea Res 29:1–16
Deacon GER, Foster TD (1977) The boundary region between the Weddell Sea and Drake Passage currents. Deep-Sea Res 26:981–995
Doronin Yu P, Kheisin DE (1975) Sea ice, Gidrometeoizdat Publishers, Leningrad, p 323 (Transl. by NSF, TT75-52088, 1977)
Ebbesmeyer CC, Ingraham WJ Jr, Royer TC (2007) Tub toys orbit the Pacific Subarctic Gyre, EOS, Trans 88(1):1, 4
Eicken H (2003) From the microscopic, to the macroscopic, to the regional scale: Growth, microstructure and properties of sea ice. In DN Thomas and GS Dieckmann (eds) Sea Ice: An introduction to its physics, chemistry, biology and geology. Blackwell Publishing, Oxford, UK
Foster TD, Carmack EC (1976) Temperature and salinity structure in the Weddell Sea. J Phys Oceanogr 6:36–44
Frankenstein G, Garner R (1967) Equations for determining the brine volume of sea ice from -0.5oC to -22.9oC. J Glaciology 6(48):943–944
Garrison GR, Becker P (1976) The Barrow submarine canyon: A drain for the Chukchi Sea. J Geophys Res 81:4445–4453
Garrison DL, Close AR, Reimnitz E (1989) Algae concentrated by frazil ice: evidence from laboratory experiments and field measurements. Antarctic Science 1:313–316
Gascard JC, Richez C, Rouault C (1995) New insights on large scale oceanography in Fram Strait: the West Spitsbergen current. In: Arctic Oceanography: Marginal ice zone and continental shelves coastal and estruarine studies, Volume 49, American Geophysical Union, Washington, D.C.
Gille ST (1994) Mean sea surface height of the Antarctic Circumpolar Current from Geosat data: Method and Application. J Geophy Res 99(C9):18255–18273
Gille ST (2003) Float observations of the Southern Ocean, Part I: Estimating mean fields, bottom velocities and topographic steering. J Phys Oceanogr 33:1167–1181
Giles AB, Massom RA, Lytle VI (2008) Fast ice distribution in East Antarctica during 1997 and 1999 determined using Radarsat data. J Geophys Res 113:C02S14: doi:10.1029/2007/JC004139
Gordon AL, Huber BA (1990) Southern ocean winter mixed layer. J Geophy Res 95(C7): 11,655–11,672
Gordon AL (1982) Weddell deep water variability. J Mar Sys 40:199–217

Gordon (1989) Spatial and temporal variability within the Southern Ocean. In: Sahrhge D (ed) Antarctic Variability and Krill Distribution, Springer, Berlin, pp 41–56
Gordon AL, Molinelli EM (1982) Southern Ocean Atlas: Thermohaline-chemical distributions and the Atlas Data Set. Columbia University Press, NY
Gordon AL, Georgi DT, Taylor HW (1977) Antarctic polar front zone in the western Scotia Sea-summer. J Phys Oceanogr 7:309–328
Gordon AL, Martinson DG, Taylor HW (1981) The wind-driven circulation in the Weddell-Enderby Basin, Deep-Sea Res 28A:151–163
Gordon AL, Molinelli E, Baker T (1978) Large-scale relative dynamic topography of the Southern Ocean. J Geophys Res 83(C6):3023–3032
Gordienko P (1958) Arctic ice drift. In: Proceedings of conference on Arctic Sea Ice, National Academy of Sciences Publ. 598, Washington, DC, pp 210–222
Gow AJ, Tucker III WB (1990) Sea ice in the polar regions. In: Smith Jr W, Grebmeier JM (eds) Arctic Oceanography: Marginal ice zones and continental shelves, American Geophysical Union, Washington DC, pp 47–122
Grebmeier JM, Smith Jr WO, Conover RJ (1995) Biological processes on Arctic continental shelves: Ice-ocean-biotic interactions. In: Smith Jr W, Grebmeier JM (eds) Arctic Oceanography: Marginal ice zones and continental shelves, American Geophysical Union, Washington DC, pp 183–202
Harrison JD, Tiller WA (1963) Controlled freezing of water. In: Kingery WD (ed) Ice and Snow, MIT Press, Cambridge
Hibler WD III (1979) A dynamic thermodynamic sea ice model. J Phys Oceanog 9:815-846 (1986)
Hobbs PV (1974) Ice Physics. Claredon Press, Oxford, UK
Hufford GL (1974) An apparent upwelling in the Southern Beaufort Sea. J Geophys Res 79(9):1305–1306
Jacobs SS, Fairbanks RG, Horibe Y (1985) Origin and evolution of water masses near the Antarctic continental margin: Evidence from H218O/H216O ratios of sea water. In: Jacobs S (ed) Oceanology of the Antarctic Continental Shelf, Antarctic Research Series, Vol 43, AGU, Washington, DC
Jakobsson M, Cherkis N, Woodward J, Coakley B, and Macnab R (2000) A new grid of Arctic bathymetry: A significant resource for scientists and mapmakers, EOS Trans Am Geophys Union, 81(9):89
Jones EP, Anderson LG (1990) On the origin of the chemical properties of the Arctic Ocean halocline. J Geophy Res 91:10,759–10,767
Kraus JA (1997) Introduction to physical oceanography. Prentice-Hall, Inc. New Jersey
Kwok R, Schweiger A, Rothrock DA, Pang S, Kottmeier C (1998) Sea ice motion from satellite passive microwave imagery assessed with ERS SAR and buoy motions. J Geophys Res 103(C4):8191–8214
Kwok R, Rothrock DA (1999) Variability of Fram Strait ice flux and North Atlantic Oscillation. J Geophys Res 104(C3):5177–5189
Lazier JRN (1973) The renewal of Labrador Sea water. Deep-Sea Res 20:341–353
Lazier JRN (1980) Oceanographic conditions at Ocean Weather Ship Bravo. Atmos Ocean 18:227–238
Lazier JRN, Wright DG (1993) Annual velocity variations in the Labrador Current. J Phys Oceanogr 23:659–678
Lily JM, Rhines PB, Visbeck M, Davis R, Lazier JRN, Schott F, Farmer D (1999) Observing deep convection in the Labrador Sea during winter 1994/1995. J Phys Ocean 29:2065–2098
Liu AK, Zhao Y, Liu WT (1998) Sea ice motion derived from satellite agrees with buoy observations. EOS Trans 79:353 & 359
Lherminier P, Gascard JC, Quadfasel D (1999) The Greeland Sea in winter 1993 and 1994: Preconditioning for deep convection. Deep-Sea Res II 46:1199–1235
Martin JH (1992) Iron as a limiting factor. In: Falkoski PG, Woodhead AD (eds) Primary productivity and biochemical cycles in the sea, Brookhaven Symposium on Biology, Plenum, NY, pp 123–138

Martin JH, Fitzwater SE (1988) Iron deficientcy limits phytoplankton growth in the north-east Pacific subarctic. Nature 331:341–343

Maykut GA (1978) Energy heat exchange over young sea ice in the Central Arctic. J Geophys Res 83:3646–3658

Maykut GA (1985) The ice environment. In: Horner RA (ed) Sea Ice Biota, CRC Press, Boca Raton, FL, pp 21–82

Maykut GA, Unterstiener N (1971) Some results from a time dependent thermodynamic model of sea ice. J Geophys Res 76:1550–1575

Moore JK, Abbott MR, Richman JG (1999) Location and dynamics of the Antarctoc polar front from satellite sea surface temperature data. J Geophys Res 104(C2):3059–3073

Moritz RE, Colony R (1988) Statistics of sea ice motion, Fram Strait to North Pole. In: Sodhi DS, Luk CH, Sinha NK (eds) Proceedings of the seventh international conference on offshore mechanics and Arctic engineering, Vol 4, Am. Soc. Mech. Eng., New York, pp 75–82

Mosby H (1934) The waters of the Atlantic Antarctic Ocean. Sci Results Norw Antarct Exped 1927–1928 1:1–131

Munk WH (1950) On the wind-driven ocean circulation. J Meteor 7(2):79–93

Nakawo M, Sinha NK (1981) Growth and salinity profile of first year ice in the high Arctic. J Glaciol 27:315–330

Nelson DM, Anderson RF, Barber RT, Brzezinski MA, Buyesseler KO, Chase Z, Collier RW, Dickson, ML, Francois R, Hiscock MR, Honjo S, Marra J, Martin WR, Sambrotto RN, Syles FL, Sigmon DE (2002) Vertical budgets for organic carbon and biogenic silica in the Pacific sector of the Southern Ocean, 1996-1998. Deep-Sea Res Part II 49:1645–1674

Nowlin Jr WD, Klinck JM (1986) The physics of the Antarctic Circumpolar Current. Rev Geophys Space Phys 24:469–491

Olbers D, Gouretski V, Seiss G, Schofer J (1992) Hydrographic atlas of the Southern Ocean. Alfred Wegener Institute, Bremerhaven, Germany

Ono N (1968) Thermal properties of sea ice: Thermal constants of sea ice. Low Temperature Science, Series A 26:329–349

Orsi AH, Whitworth III T, Nowlin Jr WD (1995) On the meridional extent and fronts of the Antarctic Circumpolar Current. Deep-Sea Res I 42(5):641–673

Overland JE, Roach AT (1987) Northward flow in the Bering and Chukchi Sea. J Geophys Res 84:1155–1164

Padman L (1995) Small-scale physical processes in the Arctic Oceans. In: Walker Jr WO, Grebmeier JM, Arctic Oceanography: Marginal ice zones and continental shelves, American Geophysical Union, Washington DC, pp 97–130

Perovich DK, Tucker III WB, Ligett KA (2002) Aerial observations of the evolution of ice surface conditions during summer. J Geophys Res 107(C10):8048, doi:101029/2000JC000449

Perkins RG, Lewis EL (1984) Mixing in the West Spitsbergen Current. J Phys Oceanogr 14:1315–1325

Pritchard, R.S., 1988. Norton Sound and northeastern Bering Sea ice behavior: 1981–1982. In: Sodhi DS, Luk CH, Sinha NK (eds) Proceedings of the seventh international conference on offshore mechanics and Arctic engineering, Vol 4, Am. Soc. Mech. Eng., New York, p 60–74

Proshutinsky AY, Johnson MA (1997) Two circulation regimes of the wind-driven Arctic Ocean. J. Geophs. Res. 102(C6):12, 493–12, 514

Reid JL, Nowlin WD, Patzert WC (1977) On the characteristics and circulation of the southwestern Atlantic Ocean. J Phy Oceanogr 7:62–91

Rigor IG, Wallace JM, Colony RL (2002) Response of sea ice to the Arctic Oscillation. J Climate 15:2648–2663

Rudels B (1987) On the mass balance of the Polar Ocean, with special emphasis on the Fram Strait. Norsk Polarinstitut Skrifter 188:1–53

Rudels B, Joens EP, Anderson LG, Kattner G (1994) On the intermediate depth waters of the Arctic Ocean. In: Johannessen OM, Muench RD, Overland JE (eds) The role of the polar oceans in shaping the global climate, American Geophysical Union, Washington, DC

Rudels B, Anderson LG, Jones EP (1996) Formation and evolution of the surface mixed layer and the halocline of the Arctic Ocean. J Geophys Res 101:8807–8821
Sarmiento JL, Hughes TMC, Stouffer RJ, Manabe S (1998) Simulated response of the océan carbon cycle to anthropogenic climate warming. Nature 393:245–252
Schenke HW, Dijkstra S, Niederjasper F, Schone T, Hinze H, Hoppman B (1998) The new bathymetric charts of the Weddell Sea, AWI BCWS, In: Jacobs S, Weiss RF (eds) Ocean, Ice, and Atmosphere Interactions at the Antarctic Continental Margin, Antarctic Research Series, Vol 75, pp 371–380
Schmitz Jr WJ, McCartney MS (1993) On the North Atlantic Circulation. Rev Geophy 31:29–49
Schroeder M, Fahrbach E (1999) On the structure and transport of the eastern Weddell Gyre. Deep Sea Research, 46:501–527
Schwarzacher W (1959) Pack-ice studies in the Arctic Ocean. J Geophys Res 64(12):2357–2367
Smith Jr WO, Nelson DM (1990) Phytoplankton growth and new production in the Weddell Sea marginal ice zone in the austral spring and autumn. Limnology and Oceanography 35:809–821
Smith Jr W, Comiso JC (2008) The influence of sea ice primary production in the Southern Ocean: A satellite perspective. J Geophys Res 113, C05S93, doi: 10.1029/2007JC004251.
Smith W Jr, Comiso JC (2009) Southern ocean primary productivity: Variability and a view to the future. In: Krupnik I, Lang MA, Miller SE (eds) Smithsonian at the Poles: Contributions to the International Polar Year Science, Smithsonia Institution Scholarly Press, Washington, DC, pp 309–318
Smith WO Jr, Nelson DM (1985) Phytoplankton bloom produced by a receeding ice edge in the Ross Sea: Spatial coherence with the density field. Science 227:163–166
Smith WO Jr, Sakshaug E (1990) Polar phytoplankton. In: Smith WO Jr (ed) Polar Oceanography, Academic Press, San Diego
Stabeno PJ, Reed RK (1994) Circulation in the Bering Sea Basin observed by satellite-tracked drifters: 1986–1993. J Phys Ocean 24:848–854
Schmitz WJ Jr (1996) On the World Ocean Circulation: Vol. 1, Technical Report WHOI-96–30, p 141
Steele M, Morison JH, Curtin TB (1995) Halocline water formation in the Barents Sea. J Geophys Res 100(C6):881–894
Swift JH, Koltermann KP (1988) The origin of Norwegian Sea deep water. J Geophys Res 93:3563–3569
Sverdrup HU (1953) On the conditions for the vernal blooming of phytoplankton. J Cons Cons Int Explor Mer 18:287–295
Toudal L, Coon MD (2001) Interannual variability of the sea-ice induced salt flux in the Greenland Sea. Ann Glaciol 33:385–390
Thomas D (2004) The frozen oceans, Natural History Museum, London, UK
Thorndike AS, Colony R (1982) Sea ice motion in response to geostrophic winds. J Geophys Res 87(C8):5845–5852
Tucker WB, Govoni JW (1981) Morphological investigations of first year sea ice pressure ridge sails. Cold Regions Science and Technology 5(1):1–12
Tucker WB, Weeks WF, Kovacs A, Gow A (1980) Nearshore ice motion at Prudhoe Bay, Alaska. In: Pritchard RS (ed) Sea ice processes and models, University of Washington Press, Seattle, Washington
Tucker WB, Perovich DK, Gow AJ (1992) Physical properties of sea ice relevant to remote sensing. In: F. Carsey (ed) Microwave remote sensing of sea ice, Geophysical Monograph 68, American Geophysical Union, Washington DC
Tremblay JE, Smith Jr WO (2007) Phytoplankton processes in polynyas. In: Smith Jr WO, Barber DG (eds) Polynyas: Windows to the World's Oceans Elsevier, Amsterdam, pp 239–270
IOC, IHO and BODC, Centenary Edition of the GEBCO Digital Atlas, published on CD-ROM on behalf of the Intergovernmental Oceanographic Commission and the International Hydrographic

Organization as part of the General Bathymetric Chart of the Oceans, British Oceanographic Data Centre, Liverpool, UK, 2003
Untersteiner N (1961) On the mass and heat budget of Arctic sea ice. Arch Met Geophys Bioklim A(12):151–182
Untersteiner N (1964) Calculations of temperature regime and heat budget of sea ice in the Central Arctic. J Geophys Res 69(22):4755–4766
Vanney JR, Johnson GL (1985) GEBCO Bathymetric sheet 5.18 (Circum-Antarctic), In: Jacobs S (ed) Oceanology of the Antarctic Continental Shelf, Antarctic Research Series, Vol 43, pp 1–4
Vinje T, Finnekasa O (1986) The ice transport through the Fram Strait, Rep. 186, p 39, Norsk Polarinst., Oslo, Norway
Wadhams P (1980) A comparison of sonar and laser profiles along corresponding tracks in the Arctic Ocean. In: Pritchard (ed) Sea Ice Processes and Models, University of Washington Press, Seattle, WA
Wadhams P (2000) Ice in the Ocean. Gordon and Breach Science Publishers, London, UK
Wadhams P, Comiso JC (1992) The thickness distribution inferred using remote sensing techniques. In: Carsey F (ed) Microwave Remote Sensing of Sea Ice, American Geophysical Union, Geophysical Monograph 68, Washington, DC
Weeks WF, Ackley S (1986) The growth, structure and properties of sea ice. In: Untersteiner (ed) The Geophysics of Sea Ice, Plenum, New York
Weeks WF, Lee OS (1962) The salinity distribution in young sea ice, Arctic 15:92–108
Weeks W, Mellor M (1984) Mechanical properties of ice in the Arctic seas. In: Dyer I, Chryssostomidis C (eds) Arctic Technology and Policy, Hemisphere Publishing Corporation, New York, pp 235–259
Whitworth III T, Nowlin Jr WD (1987) Water masses and currents of the southern ocean at the Greenwich Meridian. J Geophys Res 92:6462–6476
Yen YC (1981) Review of thermal properties of snow, ice and sea ice. Cold Regions Research and Engineering Laboratory Report 81–10, Hanover, NH
Zubob NN (1945) L'dy Arktiki (Arctic Ice). Izdatel'stvo Glavsermorputi, Moscow (Translated to English in 1965, Translation AD426972, National Technical Information Service, Springfield, VA)

Chapter 3
Satellite Remote Sensing Techniques

Abstract There are currently many possibilities for remote sensing studies of various ocean and ice parameters. Satellite sensors are designed to take advantage of electromagnetic signals either originally from the surface or reflected from the surface. Multichannel and multipolarization systems in the visible, infrared and microwave frequencies have been developed to optimize the accuracy in measuring geophysical parameters of interest. The advantages and disadvantages of different types of systems and the preferred systems for measuring specific parameters are discussed. In many cases, there are ambiguities in the retrieval, and the concurrent use of data from more than one sensor would be highly desirable.

Keywords Electromagnetic spectrum • Passive sensors • Active sensors

3.1 Electromagnetic Spectrum and Channel Selections

The classical theory of electromagnetic radiation as developed by James Clerk Maxwell (1904) has been considered the single-most important discovery in physics during the nineteenth century. Maxwell's equations of electricity and magnetism are regarded as part of the fundamental laws of nature like Newton's Laws of Mechanics, but unlike the latter, they are valid even at relativistic speeds. Maxwell showed that the electromagnetic radiation consists of energy propagated as electromagnetic waves by oscillating electric and magnetic fields moving at right angle to each other. He also showed that the speed of propagation of electromagnetic radiation is identical to that of the speed of light (3×10^{10} cm/s). The puzzling duality in the behavior of light with some experimental results showing that light is a wave, while others indicating that light is a particle led to the development of the quantum theory, which provides a more general theory of electromagnetic radiation. According to the quantum theory, the radiation is both a wave and a particle with individual quantum of electromagnetic radiation being represented by a massless particle called a "photon." Electromagnetic radiation can thus be described as a stream of photons, each traveling in a wave-like pattern and moving at the speed of light.

J. Comiso, *Polar Oceans from Space*, Atmospheric and Oceanographic Sciences Library 41,
DOI 10.1007/978-0-387-68300-3_3,

At the same time, it can be referred to as a self-propagating wave with electric and magnetic components that oscillate in phase and at right angle to each other.

Matter with temperatures greater than absolute zero (i.e., > −273°C) is constantly emitting radiation, the intensity of which depends on temperature and the electrical properties of the material. Such radiation is the consequence of quantum electrodynamic processes that occur within atoms and molecules in the material. A familiar example is the emission of photons during the transition of an electronic orbit in an atom from one energy level to another. Such radiation is present almost everywhere and can come from a variety of sources including the surface of the earth and the sun. Also, the propagation of electromagnetic radiation does not require a material medium and can travel through a vacuum.

The distribution of radiative energy as a function of wavelength is called the electromagnetic spectrum that covers a very wide range of wavelength and frequency. The spectrum spans from the very short wavelength (10^{-14} m) to the very long (10^{8} m). A schematic diagram showing the entire spectrum as it is currently known, including emission processes, and measurement techniques, as will be discussed later, is presented in Fig. 3.1. The spectrum includes gamma and X-rays at one extreme through ultraviolet, visible and infrared radiation and up to microwave radiation. As indicated, visible light which is what we see with our own eyes is only a very small part of the spectrum. The scale is shown in units of frequency, ν, and wavelength, λ, which are related by $\nu = c/\lambda$, where c is the velocity of light.

The portion of the spectrum that is most pertinent to our study is from the ultraviolet to the microwave, which are commonly associated with emissions from

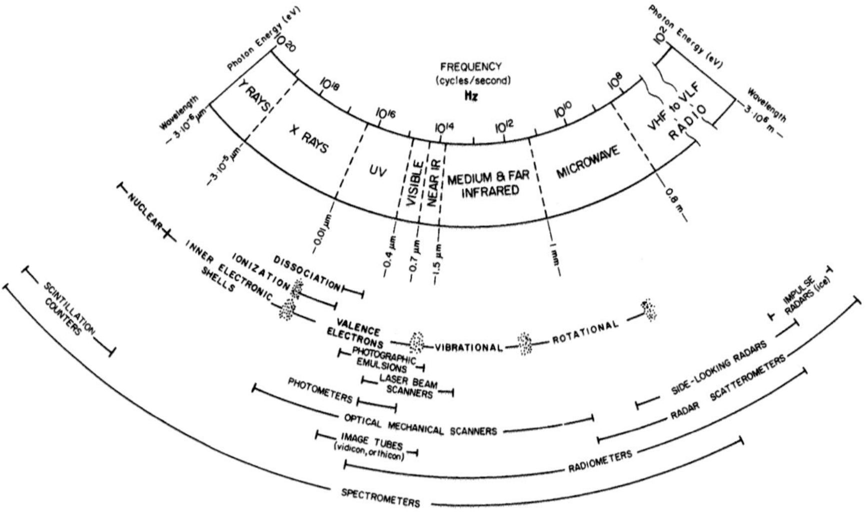

Fig. 3.1 The complete electromagnetic spectrum expressed in frequency and wavelength. Also shown are the natural emission processes and the instrumental techniques for measurement at various wavelengths. From Pogorzelski and Shapiro (1976) with permission of the Addison-Wesley Publishing Company

the Earth and the Sun. The blackbody radiation from the Sun, which has a temperature of about 6,000 K, reaches the Earth at a peak value of 0.6 μm. The main component of this radiation is what is referred to as shortwave radiation that ranges from about 0.2 to about 3.5 μm. On the other hand, radiation from the Earth's surface at temperatures of about 200 K to 300 K peaks between 9 and 15 μm. This corresponds to the thermal infrared region and has been called long wave radiation. At the higher end of the spectrum are microwaves and radio waves ranging in wavelengths from millimeter to kilometers.

Satellite sensors are designed to obtain accurate measurement of geophysical properties of interest, including those of the Earth's surface, subsurface and the atmosphere. The strategy is to measure the radiances or backscatters within the sensor field-of-view and use these measurements to infer the geophysical characteristics of the surface. Such measurements also enable the discrimination of different types of surfaces and to detect changes in the characteristics of these surfaces as a function of time. The sensors thus take advantage of unique emission and scattering characteristics that are known about the surfaces of interest. The sensors are usually multichannel systems and make measurements of surfaces at different wavelengths and polarizations to optimize information content as explained in Chapter 4.

3.2 Satellites and Sensors

Most of the satellites used for Earth science studies are usually launched on sun-synchronous orbit or polar orbit around the Earth at an altitude of about 500–1,000 km. By sun-synchronous orbit, we mean an orbit over polar regions following almost a longitudinal line and passing by a fixed location (e.g., the equator) about the same time of the day. The deviation from the longitudinal line is defined by the angle of inclination or the angle that the plane of the orbit makes with the equatorial plane. This angle is usually optimized to meet overall needs and applications for the satellite but would dictate whether all or only part of the polar regions is covered. The periods of the orbits are around 100 min, thereby allowing about 14 orbits a day and a repeat cycle of about 16 days. This means that 16 days are needed to have a complete coverage of the Earth by high resolution systems, like Landsat. Since the longitudes get closer to each other at high latitudes, the number of visits to a particular spot in the polar regions is as many as about six times a day for imaging systems. Imaging sensors are scanning sensors meant to provide good details of the spatial distribution of surfaces underneath the satellite sensor and can be cross-track scanning along nadir or conically scanning at a fixed angle with respect to nadir. The spatial coverage depends on the swath width of the scanner, which is sensor dependent as described below, while the resolution depends on the field-of-view of the satellite sensor. Non-imaging sensors take measurements at nadir or at a fixed angle.

The alternative orbital mode is the geostationary orbit in which the satellite angular velocity is the same as that of the Earth and the sensor would be designed to observe a specific surface of the Earth continuously. They are normally launched at an orbital height of 30,000–50,000 km and at an inclination of 1.9°–0.2°. At such an altitude, geometrical consideration would make the resolution relatively poor compared to those in sun-synchronous orbits for the same field of view. However, advances in technology have enabled geostationary sensors to have resolutions similar to those of polar orbiting satellites. Unfortunately, because of the inclination, the coverage is only up to a maximum latitude between 55° and 60° N (or S) and therefore, such systems are not useful for polar studies.

A comprehensive review of satellite systems and their history is provided by Kramer (2002). CEOS Earth Observation Handbook (2002) provides a very exhaustive list of available systems before it went into print but would now need to be updated. Updating is not trivial since launch dates are usually altered, and so many countries are now launching their own satellites or sensors, the final configurations of which are sometimes classified or may not be readily available. The key polar orbiting satellites that have been used for polar studies and associated orbital parameters are presented in Table 3.1. Among the parameters provided for each satellite are the altitude, inclination angle, time of equatorial crossing, launch date and the key sensors carried by the satellites.

3.2.1 Sensor Types

The primary instruments used for polar oceanography and polar process studies fall mainly in the visible, infrared and microwave frequencies with emphasis on those designed to detect the geophysical characteristics of the surface. They are also usually multichannel systems with the microwave instruments being single, dual-polarized or polarimetric with a view of enhancing information content of the measurements and minimizing ambiguity in the interpretation of the data, thereby optimizing the accuracy of the retrieved geophysical parameters. The ultraviolet region is primarily utilized in atmospheric studies and more specifically for quantitative assessments of the variability of the ozone hole in the stratosphere and aerosol, ozone and other pollutants in the troposphere.

An important component of sensor design is the choice of wavelength or frequency channel. This is, in part, dictated by the presence of the atmosphere, which severely limits the spectral bands that can be used for surface studies. Figure 3.2 provides a plot of the transmissivity of the atmosphere as a function of wavelength or frequency indicating large fluctuations of atmospheric opacity and the need to carefully identify the wavelength or frequency regions, called windows, in which a sensor would be most effective for certain measurements. Sensors for surface parameters are usually designed around wavelength or frequency channels in which the atmospheric transmissivity is high, while sensors for atmospheric parameters make use of channels in which the transmissivity is low. In practice, however, data from both types of channels are utilized in the retrieval of both surface and atmospheric data, since the retrieval of one parameter requires knowledge of the other parameter. Ideally, sensors should meet desired resolution and accuracy objectives. In practice, some compro-

Table 3.1 Polar orbiting satellites

Satellites	Altitude (km)	Inclination (degree)	Period (min)	Equatorial crossing	Launch date	Key sensors for polar studies
Nimbus-7	948	99.3	104.08	12:00	24 Oct 78	SMMR, CZCS, THIR
SeaSat	788	108.0	100.63	na	27 Jun 78	SMMR, SAR, Scat, Ralt
Landsat-7	705	98.2	99	10:00	15 Apr 99	ETM–+, MSS
Spot-4	832	98.7	101	10:30	24 May 98	DORIS, HRVIR
NOAA-16	870	98.8	102	13:54	21 Sep 00	AVHRR-3
DMSP-F8	830	98.7	101	05:55	01 Jun 87	SSM/I, OLS
OrbView-2	705	98.2	99	12:00	01 Aug 97	SeaWiFS
ERS-2	782	98.52	100.5	10:30	21 Apr 95	SAR, Scat, ATSR-2
RadarSat-1	798	98.59	100.7	18:00	04 Nov 95	SAR
Adeos1	803	98.6	101	10:30	17 Aug 96	OCTS, AVNIR, Polder
SeaStar	705	98.2	99	12:20	01 Aug 97	SeaWiFS
QuickScat	803	98.6	101	06:00	19 Jun 99	SeaWinds
Terra	705	98.2	98.88	10:30	19 Dec 99	MODIS, ASTER, MISR
Aqua	705	98.2	98.88	13:30	04 May 02	AMSR-E, MODIS, AIRS
Adeos2	803	99.0	101	10:30	14 Dec 02	AMSR, GLI
Envisat	782	98.5	100.5	10:30	01 Mar 02	ASAR, AATSR, MERIS
IceSat	600	94.0	97.0	na	12 Dec 02	GLAS
Aura	705	98.2	98.5	13:45	15 Jul 04	HIRDLS, MLS, TES
NOAA-18	845	98.74	102	13:43	20 May 05	AVHRR-3

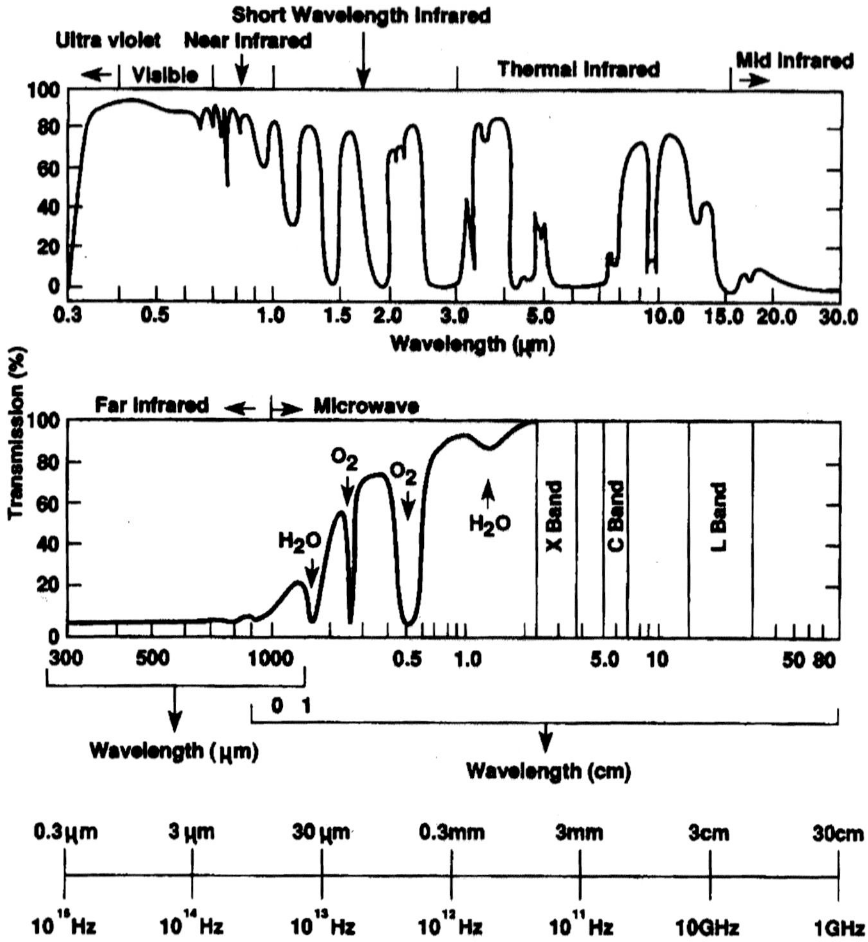

Fig. 3.2 Schematics of the Electromagnetic Spectrum from Ultraviolet to microwave and Atmospheric Transmission. Adapted from Elachi (1987), courtesy of NASA/JPL

mises have to be made because of cost considerations, different requirements in multidisciplinary projects and limitations in sensor technology.

Satellite remote sensing systems generally fall into two basic categories: passive and active. Passive systems are those in which the sensor measures radiation that is naturally emitted from a source through molecular, atomic, nuclear or other processes. Passive systems are in turn divided into visible, infrared and microwave systems. With visible systems, the radiation (called shortwave) usually originates directly from the sun and is reflected by the surface of interest. There are some that originate from the surface (e.g., by microorganisms through fluorescence), but while such contribution would be of interest the signal is sometimes too weak to detect. With infrared systems, the observed radiation (called longwave) is emitted directly from the surface of interest or from the atmosphere. To meet measurement objectives, the sensors are usually designed to have multiple channels, in part, to be able to assess and account for scattering and emission from an intervening surface or the atmosphere. For multiple applications and to take advantage of the synergy that the visible and infrared systems can offer, most modern systems are designed as combined visible and infrared systems. The passive microwave systems are also designed as multichannel systems to optimize utility and allow for a wider range of applications, but they are difficult to integrate with other systems because the wavelength of the radiation is long and large antennas are needed to meet resolution requirements. In particular, resolution is inversely proportional to wavelength and hence the requirement for a large configuration for a satellite microwave system. Some of the most popular passive systems from ultraviolet to microwave frequencies that have been used in the polar regions are summarized in Table 3.2 and will be discussed in the following sections. A more complete discussion of the sensors and their applications is provided by Ikeda and Dobson (1995), Kramer (2002) and Lubin and Massom (2006).

Active systems are different from passive systems, in that they generate their own signal, transmit this signal to the target, and receive and record the returned signal after the latter has interacted with the target. Such systems provide information associated with the physical and electrical properties of the surface, dynamics of the surface and/or topography of the surface. Earlier active systems have been dominated by microwave systems, which took advantage of mature designs of radar systems used by the military in World War II. Examples of this type of system include the Synthetic Aperture Radar (SAR), the Radar Altimeter (RA) and the Scatterometer (Scat or NSCAT) as listed in Tables 3.1 and 3.2. The advent of Lidars (light detection and ranging sensors) that followed the invention of the Laser (light amplification by stimulated emission of radiation) has also made it possible for visible channels to be used as active systems. There are many types of lidar instruments: (a) the backscatter lidar in which the backscattered or reflected laser beams give information of the scattering and extinction coefficients of the various atmospheric layers being probed; (b) the differential absorption lidar which determines densities of specific atmospheric constituents as well as water vapor and temperature profiles through analyzes of the returns from a tunable laser at different wavelengths; (c) Doppler lidar in which the Doppler shift of the light backscattered from aerosol particles transported by wind is measured, thereby allowing for the determination of wind velocity; and (d) the ranging and altimeter lidar which provides accurate measurements of surface topography.

Passive and active systems usually provide complementary information and are sometimes used in tandem to obtain a more complete characterization of the surface than that provided by a single system. For optimal utilization, information from different sensors needs to be nearly coincident and co-registered and also should be mapped in the same geographical format. This is, in most cases, a challenge because the different sensors are usually in different platforms and at times have vastly different resolutions. An important development is the advent of the A-train, as discussed later, which is a set of satellite systems that were launched in different periods but have the same orbital trajectories and are just a few minutes (or seconds) from each other. This allows for almost simultaneous coverage of the Earth's surface and atmosphere by as many sensors as in the system.

Table 3.2 Wavelengths and frequencies of recent and current satellite sensors. Number inside the parenthesis indicates the number of channels

	Ultraviolet	Visible	Near infrared	Thermal infrared	Microwave
Wavelength	0.1–0.4 μm	0.4–0.7 μm	0.8–1.4 μm	1.4–15.0 μm	1 mm to m
Frequency	$7.5–30 \times 10^{14}$	$4.3–7.5 \times 10^{14}$	$2.1–4.3 \times 10^{14}$	$0.2–2.1 \times 10^{14}$	1 MHz–200 GHz
Passive Systems					
	TOMS	Landsat (7)			ESMR (1)
		Spot (5)			SMMR (14)
		AVHRR (5)			SSM/I (7)
		MODIS[a] (36)			AMSR-E (14)
		GLI[a] (34)			TMI (2)
		MERIS[a]			
		SeaWIFS[a] (7)			WindSAT(?)
		Quickbird (2)			MSMR(8)
				THIR (2)	
				ASTER(8)	
				AATSR(4)	
Active Systems					
		GLAS (2)			SAR (1)
					RadarSAT (1)
					ASAR (2)
					ALOS/PALSAR (4)
					NSCAT (1)
					QuickScat (1)
					CryoSat

[a]With ocean color capabilities

3.2.2 *Passive Systems*

Passive instruments are basically photon counters, in that they provide measurements of the number of photons per unit time that goes through the field-of-view of the detector. There are three general types of detectors used by visible and infrared systems: photoemissive, photoconductive and photovoltaic. These detectors have discriminators that allow for the acquisition of data from different wavelengths into different channels or bands. The photo-emissive devices generally offer the greatest sensitivity in the visible through the near infrared. An example of such a device is the photo-multiplier tube in which the signal from a photon is amplified through the emission of electrons from a photocathode and with the electrons getting accelerated to a series of dynodes in which they rapidly multiply in numbers to produce an easily measurable current. Its high sensitivity and large linear range made this device the detector of choice for Landsat MMS (Multispectral Scanner), TOMS and other systems in the 1970s and 1980s. The photoconductive and photovoltaic devices are semiconductor systems that are fast, more compact and easier to use. The device consists of a p-n junction in which one section (i.e., n-type) has been doped with impurities to have excess electrons while the other section (p-type) has been adjusted to have a slight deficit of electrons, or an excess of positive "holes." When the two sections are brought in contact, electrons migrate from the n-type to the p-type side while the positive "holes" move in the opposite direction. Almost immediately, a potential gradient is formed at the junction and this impedes further movement of charge. There is a region at the junction, known as depletion zone, in which free charge carriers are swept away by the potential gradient. If the incident photon coming from the target of interest is energetic enough, it would excite an electron from the conduction band to the valance band and create an electron-hole pair. The electron-hole pair will be swept out of the depletion zone by the potential across the p-n junction and produce a measurable increase in the conductivity of the device. The photoconductive detectors are designed to take advantage of this phenomenon.

The photovoltaic devices make use of the same technique but with reverse bias for the potential across the p–n junction, yielding instead a detectable voltage. Photoconductive devices provide a fast response time of a few nanoseconds, the tradeoff being that it requires a relatively high bias voltage. On the other hand, the photovoltaic devices have a much slower response time (a few microseconds) but have the advantage of not having dark currents and requiring a more modest bias voltage. Of the two, photovoltaic devices are more commonly used but the choice needs to be based on the requirements for integration time and other engineering parameters. The most popular detectors are the silicon, for wavelengths less than 1 μm, indium gallium arsenide, for near-infrared wavelengths, between 2 and 5 μm, and mercury cadmium telluride, for the middle infrared to thermal infrared wavelengths, between 5 and 20 μm. Most of the new and advanced sensors like MODIS, ASTER and MERIS take advantage of these type of detectors.

At microwave frequencies, the wavelength of the radiation is too long and the energy of the photons too weak to be detected effectively by silicon or germanium detectors. The amount of radiation is also relatively weak since it is at the low end of the electromagnetic spectrum and it requires large fields-of-views (several kms) to be able to detect a good signal from the surface of interest. The passive microwave sensors make use of wave guides which are narrow rectangular cavities, the width of which are designed to allow only radiation that has the frequency band of interest. There are two general types of microwave radiometer systems that have been developed for satellite sensing: the Dicke type and the total-power radiometers. The Dicke type is a signal-modulated radiometer and had been commonly used because of its relative simplicity and insensitivity to receiver gain variations. However, it has a low intrinsic performance and its use has become less popular. The preferred system is now the total power radiometer because, although the system is more complex and difficult to design, they are more versatile and provide more accurate measurements. Also, improvements in technology led to substantial reduction in the manufacturing cost. The modern total-power radiometers provide a factor of two sensitivity improvement with respect to conventional Dicke-type switched system.

3.2.2.1 Visible Systems

Landsat, SPOT, and Quickbird

Among the most successful Earth remote sensing satellite system is the Earth Resources Technology Satellite (ERTS-1), which was developed and launched by NASA on 23 July 1972. The key satellite sensor is the Multispectral Scanner (MSS) with four channels covering bandwidths in the visible and reflective infrared regions, namely: 0.5–0.6, 0.6–0.7, 0.7–0.8, and 0.8–1.1 μm chosen primarily for land applications. The MSS sensor has a resolution of 80 m and a swath width of 185 km. On board the satellite is also a television-type system called Return-Beam Vidicon (RBV), which has three channels covering the same area at wavelengths of 0.475–0.475, 0.58–0.68, and 0.69–0.83 μm and a resolution of 79 m. The satellite had an inclination of 99.1°, orbital height of 903–921 km, equatorial crossing of 9:42 a.m. (local time) and a period of 103.27–103.28 min, providing a repeat coverage of an area near the equator of about 18 days.

The satellite was so successful for its intended application that it was renamed Landsat 1 before the launch of its successor, Landsat 2, on 22 January, 1975, with the latter equipped with basically identical sensors as those of the former. Landsat 2 was followed by Landsat 3 which was launched on 5 March 1978 and equipped with a slightly modified system. A thermal infrared channel (TIR) that has a resolution of 240 m was added to the MSS and a panchromatic band (wavelength of 0.505–0.75 μm) with a resolution of 40 m and a swath width of 98 km was added to the RBV. Unfortunately, the TIR in Landsat-3 failed soon after launch. The launch of Landsat-4 on 16 July 1982 and Landsat-5 on 1 March 1984 represented the advent of a more modern sensor called Thematic Mapper (TM) which has a resolution of 30 m and has six visible and near infrared channels (i.e., 0.45–0.52, 0.52–0.60, 0.63–0.69, 0.76–0.90, 1.55–1.75, 2.08–2.35 μm) and one thermal channel (i.e., 10.40–12.50 μm). The TM has more spectral, radiometric, and geometric sensitivity than the MSS, but the latter was also on board Landsat-4 and 5 to provide continuity of the data set that started in 1972. Landsat 4 and 5 were launched at an orbital height of 683–700 km and an inclination of 98.3°. Next in line was Landsat-6 which was launched in 1993 but failed to reach orbit. Fortunately, Landsat-7 was launched successfully on 15 April 1999 (Fig. 3.3) ensuring continuity with Landsat-5, which is still operating after collecting data for more than eight years beyond its lifetime. The advent of Landsat-7 also represented yet another improvement in technology with the introduction of the Enhanced Thematic Mapper (ETM), which has a panchromatic band at 15 m resolution. The ETM provided excellent data but in late May 2003 its scan line corrector (SLC) mechanism had a malfunction, and although coverage resumed in mid-July 2003, the quality of the data after this date was compromised. The ETM has a scan mirror that covers an image in a "zig-zag" manner and the SLC was meant to fill in the gaps in the form of alternating wedges in coverage. The loss of the SLC meant that about 22% of the data coverage is lost. A follow-on mission is scheduled for July 2011, and in the meantime, Landsat 7 images can be ordered with the option of having the gaps filled through a sophisticated gap filling software (e.g., Maxwell et al, 2007).

Following the successes of the Landsat Project, the first of the Systeme Pour l'Observation de la Terre (SPOT) was launched by the French Space Agency, CNES,

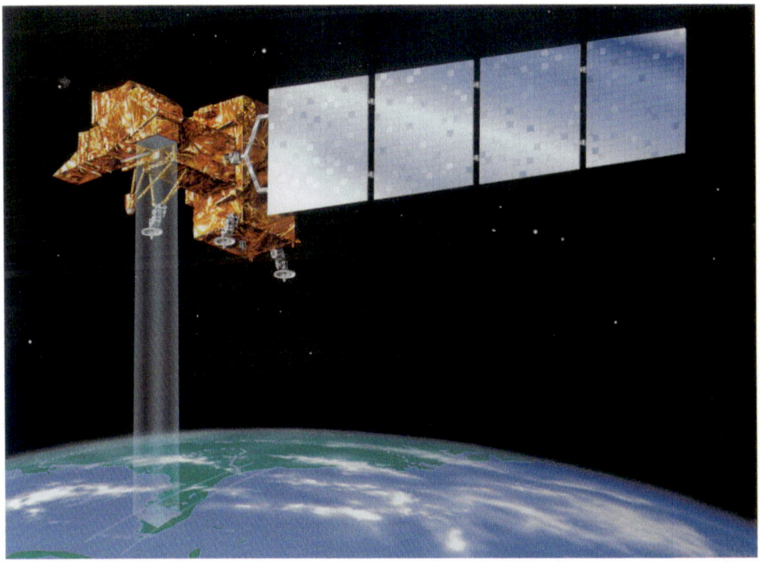

Fig. 3.3 An artist representation of the Landsat-7 satellite fully deployed and collecting data from space. It is the most recent version of the series which provide the longest record of the Earth's surface. New features of the sensor include a panchromatic band with 15 m spatial resolution, thermal infrared channel with high and low gain and a spatial resolution of 60 m and an on board radiometric calibration. Graphic courtesy of NASA

with participation of Sweden and Belgium, on 22 February 1986. The satellite has an orbital height of 824–828 km, an inclination of 98.7°, equatorial crossing of 10.30 a.m. (local time) and a period of 101.48 min. It usually takes 26 days to cover the same area on the surface near the equator but with its off-nadir viewing capabilities, the imaging interval can be shortened to an average of 2.5 days at 45° latitude. SPOT has many unique features including the ability to take stereoscopic images, an adjustable pointing capability up to 27° E or W off-nadir in 45 steps of 0.6° allowing for the observation of any area within a 950 km wide strip along its orbital path. The sensor has a panchromatic channel at 0.50–0.73 μm, with a resolution of 10 m, and the High Resolution Visible (HRV), which has three channels at 0.50–0.59 μm, 0.61–0.68 μm, and 0.78–0.89 μm at 20 m resolution. The second SPOT satellite (called SPOT-2), launched on 22 January 1990, and SPOT-3, launched on 26 September 1993, have payloads identical to those of the first SPOT. The three systems were launched as a constellation of satellites enabling revisit capabilities in which any point on 95% of the Earth's surface can be imaged any day by one of the three satellites. SPOT-4 was launched on 24 March 1998 with a monospectral channel at 0.61–0.68 μm replacing the panchromatic channel and the HRV modified with the addition of a short wave infrared (SWI) at 1.58–1.75 μm and a resolution of 20 m. SPOT-5 was launched in May 2002 with even higher resolution with the HRV further enhanced, such that all except the SWI channel has a resolution of 10 m and a panchromatic channel at 0.48–0.71 μm was introduced to replace the monospectral channel. SPOT-5 also has the High-Resolution Stereoscopic (HRS) imaging instrument which can take simultaneous stereopairs of a swath width of 120 km across and 600 km long. The stereopairs are acquired in panchromatic mode with a 10 m spatial resolution.

As high resolution visible systems became popular, it became apparent that even higher resolutions are needed for some scientific, commercial and military applications. To fill this need, some commercial ventures came about, one of which is the QuickBird by DigitalGlobe which was launched on 18 October 2001 on board a Delta II rocket. The satellite has an altitude of 450 km, an inclination of 98° and an orbital period of 93.4 min. The system has a panchromatic camera with a resolution of 60 cm and a multispectral system with four channels (i.e., 520–600 nm, 630–690 nm, and 760–900 nm) with 2.4 m resolution. The nominal swath width is 16.5 km at nadir which is distributed as 16.5 km by 16.5 km image or 16.5 by 165 km strip. The resolution is a big improvement to those of SPOT and Landsat and could be valuable in many mesoscale and validation studies in the polar regions.

AVHRR

One of the few satellite sensors that have provided comprehensive, continuous, and relatively long term data coverage suitable for interannual variability and decadal change studies is the Advanced Very High Resolution Radiometer (AVHRR). Despite the design lifetime of only a few years for each sensor, the accumulation of a relatively long record of the data was possible because the sensor is one of the key operational sensors needed for weather prediction and other applications.

Table 3.3 NOAA AVHRRs, launch date and dates of operation

Satellite name	Sensor (number of spectral channels)	Launch date	Dates of operation
TIROS-N	AVHRR-1 (4)	13 Oct 1978	19 Oct 1978 to 30 Jan 1980
NOAA-6	AVHRR-1 (4)	27 Jun 1979	27 Jun 1979 to 5 Mar 1983
			3 Jul 1984 to 16 Nov 1986
NOAA-7	AVHRR-2 (5)	23 Jun 1981	19 Aug 1981 to 7 Jun 1986
NOAA-8	AVHRR-1 (4)	28 Mar 1983	20 Jun 1983 to 12 Jun 1984
	AVHRR-1 (4)		1 Jul 1985 to 31 Oct 1985
NOAA-9	AVHRR-2 (5)	12 Dec 1984	25 Feb 1985 to 7 Nov 1988
			30 Oct 1994 to 15 Jan 1995
NOAA-10	AVHRR-1 (4)	17 Sep 1986	17 Nov 1986 to the present
NOAA-11	AVHRR-2 (5)	24 Sep 1988	8 Nov 1988 to 13 Sep 1994
NOAA-12	AVHRR-2 (5)	13 May 1991	14 May 1991 to 15 Dec 1994
NOAA-14	AVHRR-2 (5)	30 Dec 1994	30 Dec 1994 to the present
NOAA-15	AVHRR-3 (6)	13 May 1998	13 May 1998 to the present
NOAA-16	AVHRR-3 (6)	21 Sep 2000	21 Sep 2000 to the present
NOAA-17	AVHRR-3 (6)	24 Jun 2002	24 Jun 2002 to the present
NOAA-18	AVHRR-3 (6)	20 May 2005	20 May 2005 to the present
NOAA-19	AVHRR-3 (6)	6 Feb 2009	6 Feb 2009 to the present

Thus, before the end of the expected lifetime of a sensor, a new sensor is launched (see Table 3.3) for continuity in coverage. TIROS-N was the last of the series of TIROS research satellite systems developed by NASA and the AVHRR sensor on board this satellite was the first of the AVHRR operational sensors launched on 13 October 1978. A second sensor was launched a year after as NOAA took over the operation and was called NOAA-6. The series continued to the present time with the most current system in operation being NOAA-19 (Fig. 3.4). The history of the AVHRR launch dates and dates of operation are summarized Table 3.3. The altitude is 837.5 km with inclination angle of 102.3° and equator crossing time of 9:30 a.m. for even numbered satellites and approximately 2 p.m. local time for odd numbered satellites with some exceptions. With a swath width of 2580 km, much of the polar regions are covered within a single day, and temporal resolution is reasonably good.

The NOAA/Advanced Very High Resolution Radiometer (AVHRR) has been the workhorse in the monitoring of the large-scale characteristics of the atmosphere, ocean, sea ice and land features during the last 30 years. The sensor has 4–6 channels depending on version as indicated in Table 3.4 from the visible to infrared wavelength, two of which are window channels in the thermal infrared region. The basic resolution is 1 km at nadir, but because of very large storage requirements, considering the limitations of mass storage technology in the 1980s, only a 5 by 4 km subset, called GAC (Global Area Coverage), is automatically stored and archived for global and continuous coverage that started on 19 August, 1981. The 1 km data set, called LAC (Local Area Coverage), had been available only on a limited basis, until the 1990s when high capacity, compact, and affordable mass storage systems became available. The high resolution data are also downloaded in various AVHRR receiving stations around the world and are call High Resolution Picture

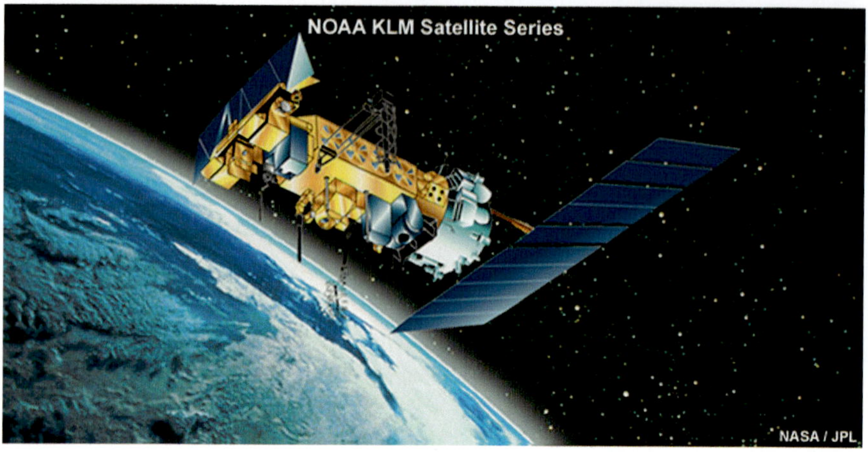

Fig. 3.4 Artist rendition of the NOAA-18 satellite which is a continuation of the KLM series represented by NOAA-15, 16 and 17. The prime sensor is AVHRR-3 which has 6 channels in the visible to thermal infrared regions. The AVHRR series provides continuous global coverage from August 1981 to the present. Graphic courtesy of NASA/JPL and NOAA

Table 3.4 Spectral ranges of AVHRR in the different satellites

Band number	IFOV (mrad)	NOAA AVHRR-1 6, 8, 10 (μm)	NOAA AVHRR-2 7, 9, 11, 12, 14 (μm)	NOAA AVHRR-3 15, 16, 17, 18, 19 (μm)
1	1.39	0.58–0.68	0.58–0.68	0.55–0.68
2	1.41	0.725–1.10	0.725–1.10	0.84–0.87
3a	1.30			1.58–1.64
3b	1.51	3.55–3.93	3.55–3.93	3.55–3.93
4	1.41	10.50–11.50	10.30–11.30	10.30–11.30
5	1.30	Repeat band 4	11.50–12.50	11.50–12.50

Transmission (HRPT) data to distinguish them from the similar LAC data. A comprehensive discussion of the details of the LAC, GAC, and HRPT data sets and applications is provided by Cracknell (1987). It is the GAC data that has been used mainly for seasonal and interannual variability studies because of longer record (e.g., Comiso 2003). LAC data were used mainly for operational use and for specific studies that require high spatial resolution.

MODIS, GLI and MERIS

The Moderate Resolution Imaging Spectroradiometer (MODIS) is a high-performing system designed to measure physical and biological characteristics of the Earth's surface and its atmosphere with complete global coverage every one or two days. Two MODIS systems are currently in operation, one launched on board Terra (Fig. 3.5) on 19 December, 1999, and the other on board Aqua (Fig 3.6) on 4 May,

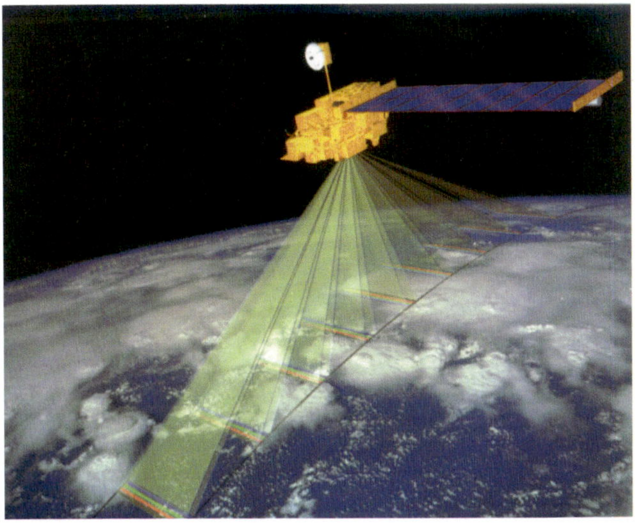

Fig. 3.5 Artist rendition of the EOS/Terra satellite which is the first of the three major satellite systems of NASA's Earth Observing System. Among the sensors on board are MODIS-T, MISR and ASTER. Graphic courtesy of NASA

Fig. 3.6 A fully deployed EOS/Aqua satellite with key EOS instruments that include MODIS-A, AMSR-E, and AIRS. Graphic courtesy of NASA

2002. Terra is a morning satellite with an equator crossing time of 10:30 a.m. while Aqua is an afternoon satellite with an equator crossing time of 1:30 p.m. With 36 channels covering the spectrum from 0.405 to 14.385 μm, and spatial resolutions ranging from 250 to 1,000 m, the instrument represents a more versatile and more advanced system than AVHRR. The swath width is comparable at 2,330 km across track for comprehensive coverage. The technical specification of each band and corresponding primary uses of each channel are given in Table 3.5. As indicated,

Table 3.5 Technical specifications of MODIS

Band	Bandwidth	Resolution (m)	Spectral radiance W/m^2-μm-st	Required SNR[a]	Primary use
1	620–670	250	21.8	128	Land/cloud (ice edge)
2	841–876	250	24.7	201	Boundaries (ice)
3	459–479	500	35.3	243	Land/cloud
4	545–565	500	29.0	228	Properties
5	1230–1250	500	5.4	74	Surface ice temperature
6	1628–1652	500	7.3	275	
7	2105–2155	500	1.0	110	
8	405–420	1,000	44.9	880	Ocean color/biogeochemistry
9	438–448	1,000	41.9	838	
10	483–493	1,000	32.1	802	
11	526–536	1,000	27.9	754	
12	546–556	1,000	21.0	750	
13	662–672	1,000	9.5	910	
14	673–683	1,000	8.7	1087	
15	743–753	1,000	10.2	586	
16	862–877	1,000	6.2	516	
17	890–920	1,000	10.0	167	Atmosphere water vapor
18	931–941	1,000	3.6	57	
19	915–965	1,000	15.0	250	
			Spectral radiance	Required NEΔT[b] (K)	
20	3.660–3.840	1,000	0.45	0.05	Surface/cloud
21	3.929–3.989	1,000	2.38	2.00	Temperature
22	3.929–3.989	1,000	0.67	0.07	
23	4.020–4.080	1,000	0.79	0.07	
24	4.433–4.498	1,000	0.17	0.25	Atmospheric temperature
25	4.492–4.549	1,000	0.59	0.25	
26	1.360–1.390	1,000	6.00	150	Cirus cloud
27	6.535–6.895	1,000	1.16	0.25	Water vapor
28	7.175–7.475	1,000	2.18	0.25	
29	8.400–8.700	1,000	9.58	0.05	
30	9.780–9.880	1,000	3.69	0.25	Ozone
31	10.780–11.280	1,000	9.55	0.05	Surface/cloud
32	11.770–12.270	1,000	8.94	0.05	Temperature
33	13.185–13.485	1,000	4.52	0.25	Cloud top altitude
34	13.485–13.785	1,000	3.76	0.25	
35	13.785–14.085	1,000	3.11	0.25	
36	14.085–14.385	1,000	2.08	0.35	

[a]*SNR* signal to noise ratio
[b]*NEΔT* Noise-equivalent temperature difference

the signal to noise ratio is greater than 500 at 1-km resolution (at solar zenith angle of 70°), and the absolute irradiance accuracy is ±5 percent from 0.4 to 3 μm (2% relative to the Sun) and 1% or better from 3 to 14.5 μm. The availability of 36 channels enables detailed characterization of clouds, including cloud types and cloud heights that, in turn provided the means to develop more accurate cloud masking algorithms for the polar regions where the signatures of snow and clouds have been difficult to discriminate. The channels were selected to enable measurements of many different parameters including several land and ocean parameters.

Despite the unusually large volume of data recorded each day globally, historical MODIS data are available for distribution through the NASA Earth Science Data and Information System Project (EOSDIS) archival centers called Distributed Active Archive Center (DAAC). The data can also be acquired real-time at receiving stations anywhere in the world with the appropriate antenna receiver. The availability of two bands at 250 m resolution and five bands at 500 m resolution is especially attractive and are suitable for many applications in the polar regions.

A Japanese sensor called Global Imager (GLI), which is very similar to MODIS in capabilities, was launched aboard ADEOS-II on 14 December 2002. The sensor has 34 channels most of which are approximately the same wavelengths as those of MODIS but with some channels slightly adjusted to optimize accuracy in obtaining some parameters. It has 22 channels in the visible and near-infrared region, five channels in the short wave-infrared region, and seven channels in the middle and thermal infrared region. The swath width is 1,600 km in the cross-track direction and is much less than that of MODIS. Unfortunately, the satellite had hardware problems and stopped providing data after it lost power on 24 October 2003 after 9 months of operation. However, with almost a year of data overlapping with those of MODIS, the GLI data set can be utilized for seasonal studies and for comparative analysis with MODIS data.

Another very capable system is the Medium Resolution Imaging Spectrometer (MERIS), launched on board the ESA/ENVISAT satellite in March 2002. The instrument provides high spectral resolution at wavelengths from 0.390 to 1.040 μm and has an unusual capability, in that after launch, a set of 15 spectral bands, different from the nominal set, can be selected by telecommand from the ground. The resolution can be selected as well, from 300 to 1,200 m (at nadir) depending on location and the imager covers a swath width of 1,150 km. The instrument makes use of the pushbroom observation technique in which a strip (i.e., a line of pixels) covering the Earth is imaged onto a slit whose length defines the across track dimension. Orbital two dimensional images are formed as data from series of strips are collected while the satellite is orbiting the earth. Another feature of the system is that it operates over a dynamic range that extends from the instrument noise level to a maximum signal such as those associated with high reflectance targets at the top of the atmosphere. The radiometric resolution requirements over various surfaces, such as the ocean, is at the same time maintained by making the gain of each selectable channel programmable and flexible to avoid reaching a saturation limit.

SeaWiFS

The Sea Viewing Wide Field-of-View Sensor (SeaWiFS) was launched on board a privately owned Seastar satellite on 1 August, 1997. The satellite was launched by a Pegasus XL rocket, which was carried by a plane to about 9.1 km, then dropped as the rocket engine fires and takes the satellite into orbit at an altitude of 705 km. The assembly of the instrument is shown in Fig. 3.7. The sensor has about noon time equatorial crossing, an orbital period of 99 min and a swath width of 2,801 km for LAC data that are collected up to a satellite scan angle of 58.3°. The GAC data utilize only the middle scans up to 45°, which correspond to a swath width of 1,502 km, and are stored at NASA Goddard Space Flight center where they are accessible to registered scientists.

Full data acquisition began on 18 September 1997. The sensor has eight bands intended primarily to monitor the color of the oceans: six in the visible (412, 443, 490, 555, and 670 μm) and two in the near infrared (765 and 865 μm). The visible channels were chosen to optimize the accuracy in the retrieval of phytoplankton concentration, while the near infrared channels were mainly intended for atmospheric correction. The data set that has been acquired is currently the longest continuous global data set on ocean color that is available. There has been some hardware issues in the last two years, and there are already several gaps in the daily data set for 2008 and 2009. The gaps are usually filled in by compatible MODIS data.

Fig. 3.7 The SeaWiFS sensor as being assembled before its launch on board the Seastar Satellite developed by Orbital Sciences Corporation. Courtesy of NASA

MISR

The Multi-angle Imaging Spectro-radiometer (MISR) was launched on board the Terra Satellite with the unique capability of providing measurements of the Earth's surface and its atmosphere at nine discrete view angles up to 70.5° relative to the local vertical. A full description of the instrument is provided by Diner et al. (1998), while the data products and analysis tools are discussed by Bothwell et al (2002). It is a pushbroom instrument (as in MERIS) with four spectral bands in the visible to near infrared (i.e., 0.446, 0.558, 0.672, and 0.866 μm). The different view angles taken simultaneously along the flight track using nine separate cameras enable observation of scattering effects of different parameters, including clouds, snow and ice. It has a swath width of 360 km and a resolution that can be changed (by command) from 275 by 275 m to 550 by 550 m or 1.10 by 1.10 m. The angles of observations are nadir and four angles (26.1, 45.6, 60.0 and 70.5°) in the fore and aft of nadir. With the different view angles available for the same surface, ability to characterize the surface is significantly improved allowing for more accurate surface classifications. An example of such classification is a MISR image of sea ice and land in the Canadian Archipelago as presented in Fig. 3.8.

Fig. 3.8 The sea ice cover and adjacent land areas in the visible channel and color composite as observed by MISR. The sensor sees the same surface at different look angles allowing for improved classification of surface and sea ice types. Courtesy of D.J. Diner, NASA/JPL

Two images of the same scene are presented with a very preliminary classification for sea ice (bottom left) shown in the color version of the original image. With the classification, thick ice floes and fast ice are easier to identify.

Infrared Sensors (THIR, AVHRR, ASTER, AASTR, MODIS, GLI, MERIS, AIRS)

Most of the current infrared instruments are basically integrated with visible instruments to form a complete set of co-registered channels from the visible to the infrared. This allows for synergistic utilization of complementary, co-registered and coincident data from the different channels. The wavelength of the radiation is longer than those of the visible but the basic principle for recording the photon signal is similar. In this section, we discuss some of the sensors that have been most utilized in the study of surface temperatures in the polar oceans. Since the start of the satellite era, the technology changed considerably not just in instrumentation and data acquisition but also in the ability to command the sensor and choose from selected options from the ground for optimal measurement capabilities. But the bottom line is the ability to intercalibrate the different sensors to ensure consistently derived geophysical parameters.

THIR

The first system that has been used for large-scale studies of surface temperature in the polar region was the Temperature Humidity Infrared Radiometer (THIR), which was on board the Nimbus-4, Nimbus-5, Nimbus-6, and Nimbus-7 research satellites. THIR is a scanning sensor with two channels: a 10.5 μm channel with a resolution of 20 km and meant to image the cloud cover and provide the temperature of ocean surfaces, land surfaces and cloud tops; and a 6.7 μm channel with a resolution of 6.7 km meant to provide information on the moisture and cirrus cloud content of the upper troposphere and stratosphere. Duplicate sensors were launched on board Nimbus 4, 5, 6, and 7, but it was only the version on board Nimbus 7 that had on-board digitization system that provided reliable and calibrated data. Nimbus 7 was launched at the Vandenberg Air Force Base in California on board a Delta rocket and operated at an altitude of 955 km in a sun-synchronous polar orbit with a noon (ascending) equatorial crossing. The satellite orbital period was 104.08 min with the orbits separated by 26.1° of longitude separation and at an angle of inclination of 99.3°. The Nimbus-7 system was different from those earlier in the series in that data were digitized on board the spacecraft. Thus, the Nimbus-7 THIR was the first system in the series that provided consistently good quality data that can be used for time series (1978–1986) analysis of polar parameters. The 6.7 μm channel was utilized primarily for quantitative assessment of the moisture content in the upper troposphere and in the stratosphere and also the location of jet streams and frontal systems. With no other channel to provide cloud cover information, cloud masking is not trivial, especially in the polar regions, and some special techniques have to be utilized to obtain reasonably accurate surface temperature data.

AVHRR, MODIS, GLI, MERIS

The most utilized infrared sensors for large scale spatial and temporal variability studies of surface temperature are the AVHRR, MODIS, GLI, and MERIS. The characteristics of these instruments have already been described previously. Each of them has at least a few channels in the shortwave infrared and thermal infrared bands that are useful for measuring surface temperatures. Although the least capable is likely the AVHRR system because of limited cloud masking capability due to limited number of channels, the record length of the data is now almost 30 years allows for a most meaningful long term variability studies. Ability to do comparative studies of AVHRR data with those of the other sensors also enables evaluation of the accuracy of the former and at the same time determine the presence of some biases, if any, that can be corrected.

ASTER

The Advanced Spaceborne Thermal Emission and Reflection Radiometer (ASTER) is a Japanese instrument launched on board the Terra satellite in December 1999 (Fig. 3.5). ASTER has a suite of three advanced optical radiometers operating in the visible to thermal infrared channels at resolutions from 15 to 90 m and a swath width of 60 km. It is the only high-resolution imager on board Terra and is meant to complement MODIS data in the study of processes that requires high resolution. The sensor has 14 channels, three in the visible to near infrared with a resolution of 15 m, six in the short wave infrared with a resolution of 30 m, and 5 in the thermal infrared with a resolution of 90 m. The five channels in thermal infrared centered at 8.30, 8.65, 9.10, 10.60, and 11.30 μm are meant to provide accurate surface temperatures, while the short wave infrared channels are especially useful for cloud cover studies and cloud masking. In addition, nadir- and aft-looking telescopes in the visible provide stereo images that can be combined with a digital elevation model to produce surface elevation topography in 3-dimension.

ATSR, AATSR

The first Along Track Scanning Radiometer (ATSR) sensor was launched on board the ESA/ERS-1 satellite in July 1991 and made use of infrared spectral channels 1.6, 3.7, 10.8, and 12 μm for a dedicated effort to measure the surface temperature. All channels are fully registered and have a resolution of 1 by 1 km at nadir and 1.5 by 2 km at a forward view (55° from Nadir), and a swath width of 500 km. A second ATSR was launched aboard ESA/ERS-1 in April 1995 with the same design but the addition of 0.55, 0.66 and 0.86 μm channels. The unique feature of the ATSR sensor is its ability to provide better atmospheric correction than other sensors. This is done through utilization of data at nadir and at a forward view that provide measurements of approximately the same surface but at two different atmospheric path lengths. Such configuration has been shown to retrieve SST at an accuracy of

0.3°C. Global coverage by this sensor, however, has been relatively sparse, especially in polar regions.

The spatial coverage, however, changed dramatically with the launch of AATSR, aboard the Envisat satellite launched in 2003, which currently collects data continuously around the globe. AATSR provides data similar to those collected by ATSR with the added advantage of an improved data rates and a high digital resolution (12 bits) for all channels. For a warming trend of 0.25 K per decade and over a period of 15 years, a stability of 0.1 K per decade is desired to be able to detect the change with good confidence. One of the key objectives of this sensor is to provide SST that meets this accuracy requirement.

AIRS

The Atmospheric Infrared Sounder (AIRS) is one of the key instruments of the Aqua satellite (Fig. 3.6), which was launched on 4 May, 2002. It measures upwelling radiances in 2,378 channels covering the infrared band from 3.74 to 15.4 μm. In addition, it has four channels in the visible to near infrared band (0.4–1.0 μm) to provide cloud cover and spatial variability characterization. The instrument is an imager with a swat width of 800 km and is primarily meant to provide clear-column air temperature profiles at high vertical resolution and accuracies comparable to those provided by radiosonde data. This is made possible through the use of the many channels each of which is sensitive to the temperature at a particular height of the atmosphere. The instrument is also useful for other applications including the measurements of humidity profiles, cloud characteristics and the amount of greenhouse gases. The data could thus be conveniently used for the retrieval of surface parameters using other sensors on board Aqua. AIRS also provides accurate measurements of land and sea surface temperatures and provides useful information about the snow cover.

3.2.2.2 Passive Microwave Sensors

ESMR

The Nimbus-5/ESMR was the first imaging microwave radiometer flown from space and was a relatively compact square-shaped Dicke-type system less than a meter in width. The sensor measures microwave radiation at 19 GHz and scans from −50° to + 50° with a ground resolution of about 30 km at nadir. It was flown in many aircraft missions before it was finally integrated and launched on board the Nimbus-5 research satellite in December 1972. It was the ESMR that provided the first spatially detailed distribution of sea ice in the polar regions (Zwally et al. 1983). The design lifetime of the sensor was for only about a year, but it provided good data up to early 1977. ESMR data were used in many applications but because only one channel was available, the geophysical parameters derived from the data were not very accurate.

The Nimbus-5/ESMR was followed by a two-channel Nimbus-6/ESMR sensor launched in July 1975. The latter was a dual polarized radiometer at 37 GHz and would have provided good complementary information to the 19 GHz ESMR-5 during the period of overlap data, but the Nimbus 6/ESMR had hardware, data loss, and calibration problems, and much of the data were either useless or difficult to interpret. The lack of good passive microwave data for more than a year following ESMR-5 was unfortunate because of a gap in the coverage of sea ice and other polar parameters.

SMMR

The launch of the Scanning Multichannel Microwave Radiometer (SMMR) on board Nimbus-7 on 24 October, 1978, marked the beginning of a series of high-performing multichannel passive microwave systems that have been providing reasonably good data up to the present. The Nimbus-7 had an orbital height of about 948 km, an orbital period of 104 min, an inclination of 99.3° and a repeat period for the orbits of about 6 days. Full descriptions of the satellite and its sensors are provided by the Nimbus-7 Users Guide and by Massom (1991). With ten channels of dual polarized data at five frequencies (6.63, 10.69, 18, 21, and 37 GHz) SMMR ushered the development of multi-channel algorithms for retrieving geophysical parameters. Unlike ESMR, it is a conically scanning system that receives data at a fix incidence angle of 50° and provides the same resolution for each data points within a scan line. SMMR was also launched on board the satellite SeaSat in July 1978, but SeaSat lasted only for 3 months for some unknown reasons. Nimbus-7 SMMR provided good data for almost 9 years, several years beyond its predicted lifetime of 5 years.

SMMR had a swath width of 783 km and ground resolutions that varied with frequency. The geometric footprint of the sensor was 136 by 89 km at 6.63 GHz, 87 by 58 km at 10.69 GHz, 54 by 35 km at 18 GHz, 44 by 29 km at 21 GHz and 28 by 18 km at 37 GHz. Because of limits in available power supplied by the Nimbus-7 solar panels, the sensor had to share power with the ocean color sensor called the Coastal Zone Color Scanner (CZCS) on board the same satellite. Thus, the SMMR and CZCS sensors took turns being turned on only every other day. These two sensors degraded in later years and the CZCS could not be turned on after continuous coverage by SMMR for two months in mid-1986. The daily coverage by SMMR continued until it was turned off in mid-August 1987.

SSM/I

The SMMR was succeeded by the Special Scanning Microwave Imager (SSM/I) launched on board a DMSP (Defense Meteorological Satellite Program) satellite on 19 June, 1987. The satellite had an orbital height of 830 km, inclination of 98.7° and a period of 101 min. The ascending node equatorial crossing time is 05:55 for the first sensor called "f8" with the follow-on sensors having an equatorial crossing time of 16:30. Like the SMMR, the SSM/I is a conically scanning radiometer with incidence angle of 55° and has dual polarized radiometers at 19.35, 37, and 85 GHz and a vertically polarized radiometer at 22.23 GHz. Full description of this sensor is also provided by Massom (1991). SSM/I has been the workhorse for polar research having operated continuously to the present and providing daily daytime and nighttime data sets. Being an operational satellite and with a design lifetime of about 5 years, the SSM/I data set is actually a compilation of data from similar sensors launched one after another.

The SMM/I has a swath width of 1,394 km with the swath organized such that it covers 64 pixels for the three lower frequencies and 128 pixels for the 85.5 GHz channels. The resolutions of the SSM/I radiometers are comparable to those of SMMR with the footprint being 70 by 45 km at 19.35 GHz, 60 by 40 km at 22.235 GHz , 38 by 30 km at 37 GHz and 16 by 14 km at 85.5 GHz. Most of the current data sets used for polar research are from the "f11" and "f13" satellites, but with the recent degradation of SSM/I in "f13," SSM/I data in "f17" data have been used instead starting in 2009. The data is archived and distributed by National Snow and Ice Data Center (NSIDC) of the University of Colorado.

AMSR-E

The launched of the Advanced Microwave Scanning Radiometer (AMSR-E) on 4 May 2002 on board the EOS-Aqua satellite (Fig. 3.6) ushered a new era in satellite passive microwave remote sensing of sea ice. As indicated earlier, Aqua is an afternoon satellite with equatorial crossing of 1:30 p.m. local time and an orbital height of 705 km. The sensor has a swath width of 1,445 km, which is slightly larger than that of SSM/I. The primary improvements over the SSM/I are higher resolutions at all frequencies and a wider spectral range. In particular, AMSR-E has resolutions of 73.0 by 43.1 km at 6.9 GHz, 49.8 by 29.6 km at 10.65 GHz, 26.2 by 16.5 km at 18.7 GHz, 30.8 by 19.0 km at 23.8 GHz, 13.7 by 10.3km at 36.5 GHz, and 6.0 by 4.9 km at 89 GHz. The lower frequency channels provide the ability to retrieve Sea Surface Temperature (SST) and Surface Ice Temperature (SIT) that are useful not only as climate data sets but also in removing ambiguities in the retrievals due to atmospheric and surface temperature effects. Furthermore, the higher resolution minimizes the uncertainties associated with the use of mixing algorithms to retrieve some geophysical parameters.

For almost a year, there is also complementary information from Adeos2 AMSR sensor which was launched in December 2002, but unfortunately ceased operation in October 2003 because of a hardware problem with the satellite. A description of AMSR and AMSR-E and data products from these sensors are provided by Shibata et al. (2003). Since the Adeos2 is a morning satellite while the Aqua is an afternoon satellite data from the two provide ability to evaluate diurnal changes of surface parameters including the emissivity and other radiative characteristics of the surface. On board the Aqua satellite are two other sensors, namely, MODIS and AIRS as described earlier, which provide excellent ancillary data that can be used to further improve the retrievals from AMSR-E and characterization of polar parameters. Data from MODIS with wavelengths ranging from the visible to the infrared is especially useful because some channels have resolutions of 250 m, which can be used for validation or enhancing the interpretation of AMSR-E data. Data from AIRS are also useful for assessing how variations in atmospheric conditions affect the observed brightness temperatures of sea ice at different frequencies.

3.2.3 Active Systems

3.2.3.1 High Resolution Imaging Systems

The Synthetic Aperture Radar (SAR) is the only high resolution imaging system that operates in the microwave region. Unlike the visible and infrared high-resolution imaging systems, it provides useful surface information under day/night almost all weather conditions as with passive microwave radiometers. Also, with its long wavelength, it can penetrate some surfaces like vegetation, ice sheet, and snow, thereby providing information not available from the visible and infrared sensors. SAR is basically an enhanced version of the real aperture radar imager, i.e., the Side-Looking Radar (SLR), in which the two-image dimensions are the range and the azimuth with the range going perpendicular while the azimuth going along the radar flight path. The spatial resolution in range depends on the bandwidth of the transmitted radar pulse, with high resolution being achieved through pulse compression techniques. The spatial resolution in the azimuth depends on the beamwidth of the antenna and is typically about a km, depending on the distance between the sensor and the surface. SAR gained its name and popularity from the special technique in which it is able to obtain high resolution along the azimuth. Detailed discussions behind the operation of the SAR are provided by Elachi (1987) and Martin (2004). In simple terms, the instrument synthesizes the along track signal through the use of additional information associated with the movement of the satellite and the Doppler shift of the backscattered signal as it reaches the satellite detector. An example of a Doppler shift is the change in frequency when a train is approaching as compared to a train that is receding. The Doppler shift is different for different surfaces and allows the moving satellite sensor to obtain along track resolutions for the data as good as 15 m. In principle, data from an area are collected and placed into computer memory, line by line, in time sequence. These data are then processed to form the image. The radar pulses from SAR are polarized and can be horizontally polarized (H) or vertically polarized (V). A SAR system in which the antenna broadcast in H and received in V polarization is usually referred to as operating in the HV mode. There are thus a possibility of four modes, namely: HH, VV, HV, and VH. Older SAR systems have operated only in one mode which is determined by application. But more recent systems provide the capability of changing from one mode to another as directed by ground operators. Some operates simultaneously in two modes while others have used the ultimate which is the use of quad polarization or fully polarimetric mode. In this case, the radar measures the amplitude and phase of the backscattered wave for the four transmit and receive combinations. Such capability optimizes the utility of the system in most geophysical applications.

Seasat/SAR

The Seasat/SAR was the first spaceborne SAR system and provided the first demonstration of the kind of detail and potential of such an imaging system. The satellite had an orbital height of 788 km, inclination angle of 108 degrees, and an orbital period of 101 minutes. Although it lasted for only 90 days, the collected SAR data served as the baseline performance standard leading to modern systems. The radar was an L-band (23.5 cm, 1.275 GHz) with a 10.7 by 2.36 m antenna and an HH polarization. It has an incidence angle of 23° with 4 looks, range and azimuth resolutions of 25 cm and a swath width of 100 km. It operated at an altitude of 800 km with inclination angle of 108° but it was not sun-synchronous. Coverage is limited to only 10 min per orbit, and during its lifetime, some images were taken in the Arctic region but none in the Antarctic.

ERS1/SAR and ERS2/SAR

The first and second European Remote Sensing (ERS1 and ERS2) SAR systems launched on 16 July 1991 and 21, April 1995, respectively are both C-band (5.7 cm, 5.3 GHz) systems with antenna size of 10 by 1 m and VV polarization. It operates at an incidence angle of 23.0°, six looks, and has a range resolution of 26 m, an azimuth resolution of 28 m and a swath width of 100 km. ERS1/SAR provided the first spaceborne SAR image after the demise of Seasat/SAR, which means that there was a period of about 13 years of no spaceborne SAR data. A contract between NASA and ESA also led to creation of the Alaska SAR Facility, which housed and processed thousands of ERS1 and ERS2 SAR images in the Arctic.

NASDA/JERS1

The Japanese Earth Resources Satellite, (JERS-1)/SAR, launched in 1992 was an L-band SAR (23.5 cm, 1.275 GHz) with antenna size of 11.9 by 2.4 m and HH polarization. The capabilities are therefore similar to those of Seasat/SAR but it had an incidence angle of 39°, range and azimuth resolutions of 18 m, 3 looks, and a swath width of 75 km. The deployment of the SAR antenna was delayed for 2 months because of a stubborn mechanism, but it was finally successfully deployed. However, the transmitter was not able to achieve full radiated power. As a result, the operational σ° was 6 db larger than design value. Also, the HH reflectivity was relatively weak because of a rather large incidence angle of 39°. These two factors combined together to contribute to the generally weak performance of the JERS1/SAR system.

Radarsat/SAR

The Radarsat/SAR is Canada's first earth resources remote sensing satellite and was launched on 1 November, 2003. This sensor operates at C band (5.7 cm, 5.3 GHz) HH polarization with an antenna size of 15 by 1.5 m. Several imaging modes are available and range in resolution of 25 by 28 m for range and azimuth with a swath width of 100 m, to 100 by 100 m with a swath width of 500 m. The launching vehicle was provided by NASA, and in return, Canada agreed to provide SAR data primarily in the polar regions and to be processed at the Alaska SAR facility. The baseline configuration is for the sensor to be pointing to the north side of the orbital plane. This allowed the sensor to cover much of the Arctic including the North Pole but kept a large part of the Antarctic from being covered. For two periods during the first two years of the mission, the satellite was rotated about it's yaw-axis to set the antenna beam pointing to the south side of the orbital plane and therefore to obtain a complete SAR map of Antarctica.

ENVISAT/ASAR

The advent of multipolarization systems with the launch of ENVISAT/ASAR (Fig. 3.9) on 1 March, 2002 is an exciting and most welcomed development. More research needs to be done but the new capability may provide the means to improve our ability to classify different types of surfaces including sea ice types and snow cover. ASAR operates at C band (5.7 cm, 5.3 GHz) and as with Radarsat, the system is programmable and has many modes available. One is the Alternating Polarization

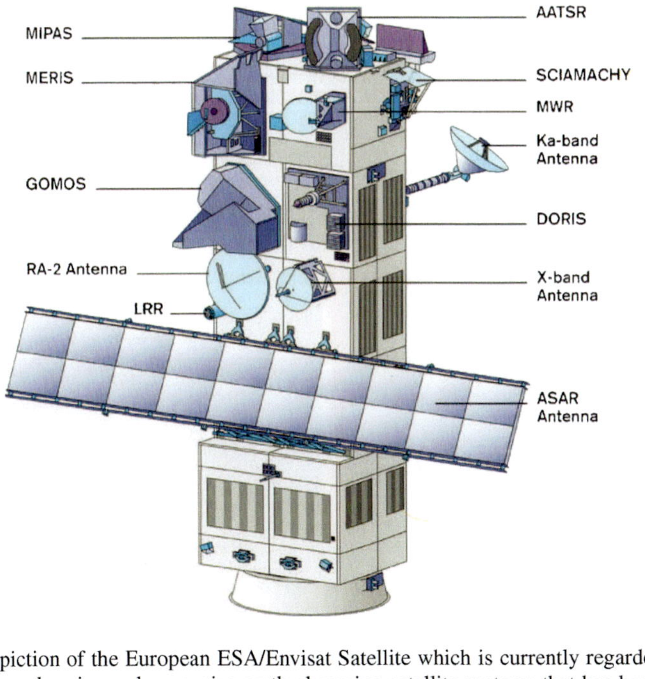

Fig. 3.9 Depiction of the European ESA/Envisat Satellite which is currently regarded as one of the most comprehensive and expensive earth observing satellite systems that has been launched. Among the key sensors are ASAR, AATSR, Radar Altimeter (2) and MERIS as shown in the diagram. Graphics courtesy of ESA

(A) mode in which the ASAR has a resolution of 30 m and a swath width of 100 km and is configured in one of three ways: (a) VV and HH; (b) HH and HV; and VV and VH. Another mode is the wide-swath mode in which continuous image at a resolution of 150 m and a swathwidth of 405 km is obtained. The global monitoring mode provides comprehensive along-track coverage at a reduced resolution of 1000 m and a swath of 405 km. On the other hand, if extremely high resolution is desired, the sensor can be in the wave mode, which provides imaging of small areas, about 5 km by 5 km in sizes, at a resolution of less than 10 m.

ALOS PALSAR

The Phased Array L-band SAR (PALSAR) which was launched on board the Japanese Advanced Land Observing Satellite (ALOS) in 2005 represent yet another big innovation in active imaging systems (Fig. 3.10). It is the first fully polarimetric SAR system, and as the name implies, it operates at L band (23.6 cm, 1.270 GHz). In full polarimetric mode, PALSAR can collect, simultaneously, data at HH, VV, HV, and VH polarizations, and also can measure the relative phase between channels. Coverage in this mode is limited by narrow swath width and the satellite duty cycle. In the ScanSAR mode, single polarization data at HH or VV can be collected at a resolution of 100 m and a swath width of 250–350 km. The nominal high resolution mode provides data at 10 m (two looks) resolution at a swath width of 70 km in either HH, VV, HH+HV or VV+VH. The highest latitude at which data can be obtained, however, is 81°. This means no coverage in a large part of the Arctic Ocean.

TerraSAR

Among the latest of the active imaging systems is the TerraSAR-X system launched in 2007, which provides the highest resolution in an active space system. Two versions will be flying in tandem but at different frequencies: TerraSAR-X which is a

Fig. 3.10 An artist's depiction of the Japanese/JAXA's ALOS Satellite as deployed in space. The three sensors on board are the PALSAR, AVNIR-2 and PRISM (Courtesy of JAXA)

German System and TerraSAR-L which is a planned European system. The X-band TerraSAR-X operates at 3 cm wavelength or 9.65 GHz frequency. The system can operate at the spotlight mode to acquire 1–2 m resolution data across a 5–10 km swath width or in "stripmap" mode which provides 3 m resolution and 30 km swath, or "ScanSar" mode for 16 m resolution and 100 km swath width. TerraSAR-X also provides single, dual, or full polarimetric modes. The satellite has an orbital inclination of 97.44° with a repeat period of 11 days.

3.2.3.2 Scatterometers and Altimeters

Scatterometers

A scatterometer is an active microwave radar that operates the same way as a regular radar but scans the surface along a 360° elliptical track. It has been meant primarily to measure wind vectors on the surface of the ocean. By looking at the backscatter from different directions over liquid ocean surfaces, the data collected can be analyzed to obtain wind strength and direction along the surface. Statistically, the backscatter is strongest at two opposite directions and this information is used to retrieve the wind velocity. The strength of the backscatter provides a measure of the magnitude of wind speed. Recently, it has been used as an imager for large scale studies and has found many applications for sea ice covered surfaces as well.

The first satellite based scatterometer was the system on board the Seasat satellite launched on 27 June 1978. The scatterometer operated at 14.595 GHz with a band widgth of 500 kHz and incidence angles of 25 to 65 degrees. It was designed to detect wind speeds ranging from 7 to 50 m/sec. The successor to this system are the two basically identical altimeters on board the European ERS1 and ERS2 satellites which operated at 5.3 GHz and has an incidence angle of 25 to 59 degrees. A NASA scatterometer, called NSCAT, was launched on board ADEOS-1 and ADEOS-2 satellites which both failed as indicated earlier after about 9 months in orbit. NSCAT was a variable doppler system operating at 14 GHz with incidence angle of 17 to 60 degrees. Because of the importance of wind ocean data in weather prediction and other applications a quick replacement was put together and launched on 19 June 1999 after the failure of ADEOS-1. The satellite, called QuikSCAT, has an orbital altitude of 800 km and an orbital period of 101 min. The sensor, called SeaWinds, produces a pencil-beam radar pulse at 13.6 GHz at incident angles of 48 and 54 degrees and a resolution of 25km. As with the other altimeters, the radar pulse is fed to a set of antenna oriented at different directions. During operation, the frequency of the backscatter is changed by Doppler shift, and the data at different angles are processed to obtain the strength and direction of the wind. The sensors are capable of detecting wind speeds ranging from 3 to 20 m/sec with an accuracy of about 2 m/sec.

Radar Altimeters

The radar altimeter is designed to measure precisely the wave heights. This is done through the use of a very precise time-of-flight technique. An original signal is sent out at nadir by the satellite sensor and the backscattered signal is detected. Because of finite beam width, when the signal hits the surface, the radiation spreads out to a finite area that is then backscattered to the sensor. For a relatively horizontal surface, the nadir point data would have reach the detector first and after that signals from concentric circles at increasing radius from nadir would be detected within the sensor field of view. Thus, instead of detecting a single signal coming back, it is a family of signals which is called waveforms. For ocean surfaces, the signal is relatively uniform altered only by different states of the ocean as influenced by wind, ocean current and tides. The system has thus been used to derive the significant wave heights (SWH) for ocean waves and reconstruct the dynamic topography of the surface. It has also been used successfuly for monitoring the topography of the ice sheet and the changes in the thickness of the sea ice cover. Detailed discussion about the system and its applications can be found in Elachi (1987) and Martin (2004).

The first space based radar altimeter was launch on 10 April 1974 on board the Geodetic Earth Obserbation Satellite-3 (GEOS-3). The satellite had an orbital height of about 846 km, a period of 101.8 minutes and an inclination of 115 degrees. The sensor operated at 13.9 GHz and has a footprint diameter of about 3.8 km in high intensity mode and 14.3 km in global mode. This system was followed by a similar altimeter on board the Seasat satellite which operated at 13.5 GHz with a beamwidth of 1.6 degrees and a footprint diameter that ranges from 2.4 km for calm seas to 12 km for rough seas. The satellite lasted only for about three months from 27 June 1978 to 10 October 1978 and there was no successor until the launch of ERS1 and ERS2 in 1991 and 1995, respectively. The ERS-1 and ERS2 altimeters were basically identical and operated at 13.8 GHz with a footprint diameter of 16 to 20 km depending on the type of surface. A doppler system called Cryosat that is meant primarily for cryospheric sciences studies was launched in 2005 but not successfully. A similar system called Cryosat2 is scheduled for launch in February 2010.

Laser Altimeters (ICESAT)

The Geoscience Laser Altimeter System (GLAS), which was launched on board ICESat in 2003, is the very first satellite lidar system that has been developed and launched successfully (Fig. 3.10). It operates at two wavelengths: one at 1064 nm and the other at 532 nm. The first is meant as a ranging instrument or an altimeter to study topography of the surfaces, including land, ice sheet and sea ice. The other is primarily for cloud and atmospheric aerosol studies. The footprint of the laser has a diameter of approximately 70 m and the separation of the measurements is about 170 m. In the ranging mode, the receiver uses a 1 m diameter telescope to collect the reflected 1064 nm laser light and a detector that provides the precise timing of the outgoing and reflected pulses. The transmitted and reflected pulses are recorded by a digitizer at a resolution of one nanosecond (ns) providing an estimated ranging accuracy of 5 cm. To measure the vertical distribution of clouds and aerosols, data from both the 1,064 nm and 532 nm lasers are utilized. Data from the 1,064 detector are used to measure the height and echo pulse shape from thicker clouds, while data from the 532 nm detector are used to measure the vertical extent of thinner clouds and the height of the atmospheric boundary layer.

A precision GPS receiver is also on board the ICESat to establish the exact location of the satellite while in orbit and to obtain an accurate estimate of the laser's pointing angle relative to inertial space. The latter is used to minimize uncertainty in the measurements over sloping surfaces such as those in the ice sheets. The system also uses a stellar reference system (SRS) to measure the pointing angle of each laser pulse relative to selected bright stars that are basically static reference points in space. Each laser pulse is measured relative to a high precision star camera with a laser reference system (LRS). The reflected 1,064 nm and 532 nm pulse data, along with the GPS and LRS data, are used in the final determination of the distance to the reflecting surface, the degree of pulse spreading due to atmospheric conditions and vertical distribution of any surface vegetation.

3.2.3.3 Interferometric and Polarimetric Systems

The SAR sensors can provide not just backscatter information but also time of flight information. As with radar altimeters, the time of flight information converts into topography information. If you put together two co-registered data of this type for the same location, the two images should look identical. However, if parts of the surface have been distorted because of earthquake, volcanic eruptions, mudslides, or avalanches, the two images would not match. By taking the difference of these two images, one can then infer the change in surface elevation for each pixel and the magnitude of the change. Such changes are usually exhibited in terms of fringes. When the fringes are close to each other, it means there is a radical change in the region. When they are far apart, the changes are less dramatic. This technique has been used in many applications in ice sheet and coastal studies but its utility in ocean and sea ice studies has so far been very limited.

As indicated earlier, a system that measures all the polarization components (i.e., VV, VH, HH, and HV) concurrently is called a polarimetric system. We have previously mentioned the use of a polarimetric SAR like ALOS but there are also some so called polarimetric passive microwave. Having all components provides additional information that enables parameters to be derived more accurately if not more completely. An example of the latter is the WindSat which is a passive microwave polarimetric system that is designed to measure not just scaler winds but also wind velocity. WindSat was launched on 6 January 2003 onboard the US Naval Research Laboratory's Coriolis satellite and is the first spaceborne polarimetric microwave radiometer. The sensor operates in discrete frequencies at 6.8, 10.7, 18.7, 23.8 and 37 GHz but two of these (i.e., 6.8 and 23.8 GHz) are only dual-polarized while the others are fully-polarimetic. The resolution ranges from 40 by 60 km at 6.8 GHz to 8 by 13 km at 37 GHz. The sensor was designed to obtain wind speed over the range of 3 to 25 m/sec with an accuracy of 2 m/sec but validation studies are still in progress. Undoubtedly, the data will find some applications in sea ice and ocean studies and in particular in sea ice surface classification. The use of polarimetric SAR data such as those provided by ALOS is equally attractive and has the potential of enhancing mesoscal studies in the polar regions.

3.2.4 Combined Systems

As will be discussed further in the next chapter, there are ambiguities in the retrieval of geophysical parameters from satellite data and sometimes the errors are greater than the maximum allowed to meet science requirements. One way to improve the accuracy is to use data from one sensor in combination with those from other sensors. Since different sensors provide different information, some of the ambiguities can be resolved through joint analysis of data from different sensors. The process of combining data, however, can be quite complex because different types of sensors have different resolutions, penetration depths, and sensitivities to atmospheric and surface effects. For example, SSTs from infrared sensors may not match SSTs from the microwave sensors because of different penetration depths and the measurements may actually come from two different layers of the ocean. It is thus important that parameters derived from the different sensors are well-defined and completely understood. Also, different sensors even of the same type may operate during different times of the day and therefore under different atmospheric conditions. Thus, when combining or comparing sea ice cover images, for example, the time difference should be identified, and the movement of sea ice during interval between the acquisition periods should be taken into account. Furthermore, averages from different types of sensors for the same parameter may not necessarily match because of different sampling periods. Microwave systems acquire surface data continuously, but infrared and visible data could acquire surface data only during cloud-free conditions. Therefore, weekly averages of a parameter from one system may not match those of other systems.

The differences in resolution is usually overcome by degrading the data with good resolution to match that of the poorest resolution. While this can be done without any problem in some applications, there are cases where such procedure would not work. For example, the desired parameter may require the highest resolution and therefore degrading the data would cause the loss of vital information. In this case, data from the other sensors should only be used as ancillary data to help in the interpretation. Other consideration includes an understanding of the field-of-view of the various sensors. With some types of sensors (e.g., passive microwave), primary signal is accompanied by other signals, such as the "side lobes," which are data taken at different angles and contaminates the desired signal. This effect changes with frequency and sensor design but cause some smearing in the data especially when edges of a feature like an island or an ice edge are involved. It should also be noted that for some sensors the resolution is defined as the area where 60% of the signal originates. This should be taken into consideration when comparative analysis is done.

3.2.4.1 Multisensor Systems

Multisensor platforms were initially meant to enable several sensors, each intended for different purposes, to be launched together on board the same launch vehicle with cost effectiveness as the primary reason. For example, on board Nimbus-7 are SMMR, CZCS, THIR, TOMS, ERB, SAMS, SBUV, and LIMS, all for different measurement objectives, some to measure parameters on the surface of the Earth and some to measure the chemical composition of the atmosphere. Science teams were organized around the different sensor projects and these teams seldom talk to each other. There is, however, a general recognition that data from one sensor are seldom easy to interpret, and ancillary data are required for proper interpretation. SMMR data were used for sea ice studies but the resolution is poor and there is no data in the same platform that could be used to validate retrievals of ice concentration. The option was to get near simultaneous data from AVHRR or Landsat to ensure that the retrieval is accurate. But even a few hours of difference in acquisition time can mean much since the ice is so dynamic and some ice features can come and go within this time period. The importance of getting data that are related from the same platform was quickly recognized. With the launch of more modern systems like Aqua and ENVISAT, the idea of synergy between sensors of the same platform becomes more apparent. With Aqua for example, sensors like AMSR-E, MODIS, and AIRS provide information that related and should be examined concurrently for potential relationships of the variables derived from these different systems. For example, sea ice cover and sea surface temperature can be acquired from both Aqua and MODIS data. The use of these two data set concurrently can improve retrievals of sea ice parameters derived from AMSR-E since MODIS has the resolution to provide accurate open water information during clear skies condition. An improved AMSR-E data could then be used to fill in the gaps in the MODIS data that are attributed to cloudy conditions. Furthermore, there are uncertainties in the AMSR-E and MODIS data that are associated with changing atmospheric conditions. Through concurrent use of AMSR-E, MODIS, and AIRS data, some of the ambiguities can be resolved since AIRS can provide the much needed atmospheric information.

3.2.4.2 Multiplatform and the A-Train

The A-Train was conceived with objectives similar to those of a multisensor systems and at the same time minimize the problems associated with typical multiplatform systems. The multi-sensor system has limits in size and weight and hence the number of instruments it can hold. Also, as they get to be very complex and expensive, the time it takes to get them launched after approval of the project can be very lengthy and there is always the risk of an unsuccessful launch that would put to waste a huge investment in resources and many years of preparation. The alternative is to expand the capability by getting only a modest number of sensors in a platform and get them to be banded together like a train, following the same orbital track but with one behind the other by

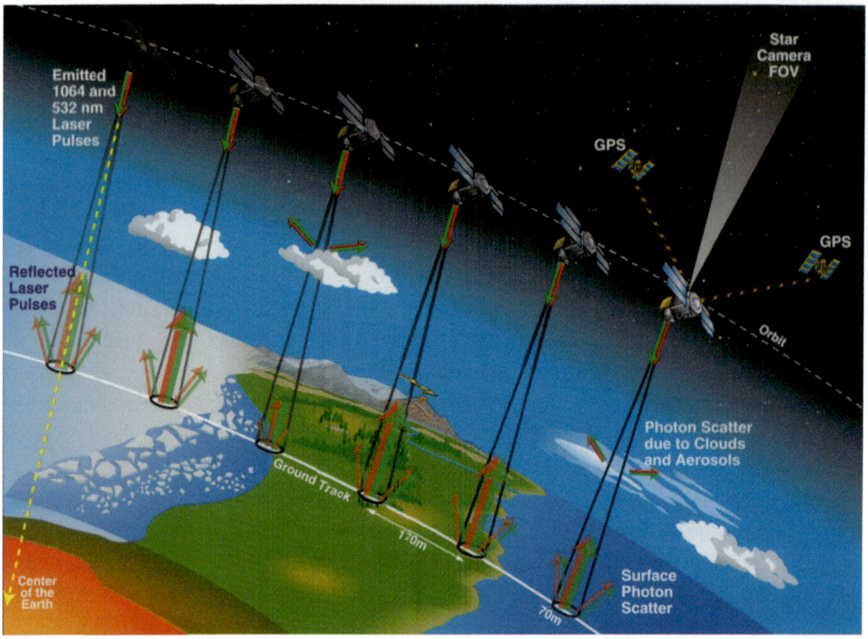

Fig. 3.11 Schematics (as provided by Deborah McLean) illustrating how the GLAS instrument on board ICESat makes measurements of the surface of the Earth. The instrument makes use of a laser system at 1064 and 532 nm to obtain ice sheet elevations, height profiles of clouds and aerosols, land elevations and approximate sea ice thickness. Graphics courtesy of NASA

only a few seconds or minutes. The first and only system of this kind so far is the "A-train" with the sensors being in the "afternoon" orbit and with the main components being "Aqua" and "Aura" satellites and include CloudSat, Cloud-Aerosol Lidar and Infrared Pathfinder Satellite Observations (CALIPSO), Polarization and Anisotropy of Reflectances for Atmospheric Sciences coupled with Observations from a Lidar (PARASOL) and Orbiting Carbon Observatory (OCO).

A schematic of the constellation system is shown in Fig. 3.12. The implementation of the A-train is not trivial since it requires very careful maneuvering, and there is always the chance for collisions since the vehicles are so close to each other. But by 2008, all the satellites except OCO have been successfully launched and positioned as planned. OCO was launched in 2009, but unfortunately, it did not reach orbit and ended up in the Southern Ocean. Nevertheless, the key sensors have been deployed and the potential for multiplatform studies can be explored. There are obvious applications for polar studies including a more accurate assessment of the cloud cover as can be provided by CloudSat and this would lead to better retrievals of surface parameters as derived from MODIS, AMSR-E and other sensors from Aqua and other satellites.

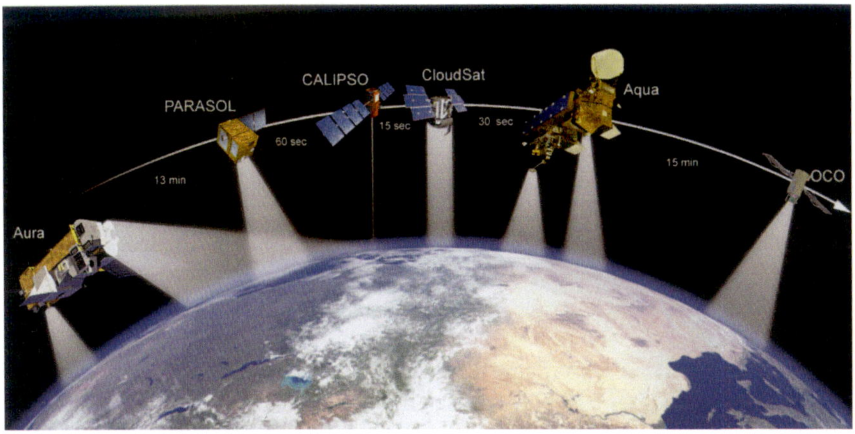

Fig. 3.12 Schematics of the A-Train concept with satellites flying in formation to make near-simultaneous observation of the Earth. All satellites in the figure are currently in operation except OCO which was not launched successfully in 2009. Graphics courtesy of NASA

3.2.4.3 Upcoming Systems

One of the most anticipated of the new systems is the National Polar-orbiting Operational Environmental Satellite System (NPOESS) series which are supposed to replace the Terra, Aqua and Aura as key Earth Science monitoring systems of the United States. The NPOESS systems will have sensors that are designed to exceed the performance of current sensors. NPP is expected to provide a continuation of the key measurements before NPOESS is ready for launch. The launch of these satellites has been delayed and it is fortuitous that the sensors they are going to replace are still active and are providing good data. Although the design lifetime of these systems is about 3–5 years, it is usually the case (as with SMMR and SeaWiFS) that they can last a lot longer (by about 2–3 times).

There are also some upcoming systems that are supposed to provide new capabilities. One is the Aquarius mission, which is part of the Earth System Science Pathfinder program and is scheduled for launch in 2010 on board a Delta II rocket. It is a joint program of NASA and CONAE of Argentina. This mission is specially designed to provide global measurements of sea surface salinity at the upper thin layer of the ocean surface. Among the regions where its capabilities can be effectively used is in the polar regions where the detection of areal extent of melt water would allow the study of the impact of the latter to plankton concentrations. Combined with sea surface temperature, the data can also be used to estimate sea surface density as well. In addition, because its primary channels have relatively long wavelength, it may provide means to measure the thickness of some types of sea ice, and in combination with other systems (i.e., AMSR-E), it may provide improved ability to discriminate first year ice from the older ice types. The polar regions would also be among the best regions to validate the surface salinity derived from the data because it is the region where the largest variations in salinity can be found.

Another upcoming mission is the ESA/CryoSat-2, which is a dual frequency Doppler radar altimeter system meant primarily for sea ice thickness distribution studies and to detect small changes on the surface of the ice sheets. CryoSat-2 will have a resolution of 250 m in the along-track direction using Synthetic Aperture technique and an estimated accuracy of less than a cm for the detection of the free-board elevation. Other applications for the sensor will likely emerge such as the ability to quantify ridging and floe sizes more accurately.

The successful launch of ESA's gravity mission GOCE (Gravity Field and Steady-State Ocean Circulation Explorer) on 17 March, 2009, the current estimates of ocean dynamic topography are likely to improve since GOCE will address a completely new range of spatial scales. GOCE data are expected to provide the means to generate a model of the Earth's gravity field with unprecedented accuracy and resolution.

It is impossible to mention all missions that are planned for the future. India, China, Korea and Taiwan have joined the ranks of the US, ESA, Canada and Japan in launching satellites that are meant to observe and monitor the Earth's system and sometimes satellite projects from these countries gets too quickly implemented that the satellites get launched even before they are advertised to the world.

3.3 Data Products and Sources

Satellite studies of the Earth's Climate System require the storage of extremely large volume of data. This was initially thought to be a very expensive and complex activity knowing that such studies require the storage of many surface and atmospheric parameters over very demanding spatial and temporal scales. One of the best known system put together to address this problem is NASA's Data and Information System (EOSDIS). Fortuitously, advanced technology now includes high capacity storage systems, and together with high performing computers it has become possible for global satellite data to be stored in data centers and for such data to be readily available to users. The data management and user services of EOSDIS are the Distributed Active Archive Centers (DAACs), which does the processing, archiving, documentation, and distribution of data from NASA's past and current Earth-observing satellites and field measurement programs. There are nine centers as follows, each dedicated to an Earth system science area of expertise:

Alaska Satellite Facility (ASF) DAAC (Univ. of Alaska, Fairbanks) – in charge with Synthetic Aperture Radar (SAR), Sea Ice, Polar Processes and Geophysics;

EROS Data Center DAAC (Sioux Falls, South Dakota) – Land Processes

GSFC Earth Sciences (GES) DAAC (NASA Goddard Space Flight Center, Greenbelt, MD) – in charge with atmospheric composition, atmospheric dynamics, global precipitation, ocean biology, ocean dynamics and solar irradiance;

Global Hydrology Resource Center (DHRC) DACC (Marshall Space Flight Center, Huntsville, AL) – Hydrologic Cycle, Lightning
Jet Propulsion Laboratory DAAC (Pasadena, CA) – Physical Oceanogrphy
Langley Research Center DAAC (Langley, VA) – Radiation budget, aerosols, tropospheric chemistry, clouds
National Snow and Ice Data Center DAAC (University of Colorado, Boulder, CO) - Snow and Ice, Cryosphere and Climate
Oak Ridge National Laboratory DAAC (Oak Ridge, TN) – Biogeochemical dynamics
Socioeconomic Data and Application Center DAAC (CIESIN at Columbia University, Palisades, NY) – human interactions in the environment

The surface area of the Earth is quite large and estimated to be $5.10 \times 10^{14}\,m^2$. The storage of satellite data that covers completely the whole Earth at 1m resolution (e.g., Quickbird data) would require 1,020 terra-bytes storage capacity. A 35-channel satellite sensor with 1-m resolution would require 13×10^6 terra-bytes of disk storage to archive one year of gridded daily average global data set with 1-m spacing. The data requirement can go up very fast because there are many parameters of interest and different levels of data are required. In the 1960s when mass storage systems were very expensive and very limited in capacity, storing large volumes of data needed for global studies was almost impossible. Current technology now allows for the storage of terra-bytes of data in relatively compact disk drives. But even with such capabilities, it is difficult to keep up and maintain the data sets. It is thus necessary to be prudent and make sure that resolution needs are carefully assessed. Most global satellite data sets do not have such high resolutions. With AMSR-E passive microwave data, the requirement for global coverage is much less since the storage requirement would only be about 6.6 mega-bytes for a daily average data at 12.5 km resolution. The polar data sets used in this book make use of a polar stereographic- grid with the projection plane crossing the Earth at 70° Latitude. The gridding technique is discussed in Gloersen et al. (1992).

3.4 Data Fusion and Data Assimilation

Satellite data have revealed that many of the phenomena and processes that we intend to study are more complex than previously expected. Many studies can thus benefit from multiple synergistic observations that make simultaneous use of data from all sources including satellite, aircraft and in situ measurements. Ability to link these measurements with numerical models would also have great advantages.

By data fusion for satellite data, we mean the process of combining data from different sensors onto a common grid or format for concurrent analysis and optimal utilization of satellite data. For polar studies, the polar stereographic grid adapted for SSM/I sea ice data has had a long history and is the primary grid that has been utilized in the analyzes of satellite polar data. The grid resolution, however, has been made flexible to take advantage of their finer resolution of other satellite data. For example, although the standard grid for SSM/I sea ice data is at 25 by 25 km resolution, the grid for AMSR data has been set to 12.5 by 12.5 km resolution, while those of AVHRR, SeaWiFS, and MODIS data is 6.25 by 6.25 km to take advantage of the finer resolution of the latter. Time averaging is made as is practicable as possible with basic data being daily data but with night time (descending) data usually separated from daytime (ascending) data to enable investigation of diurnal effects. The common grid or format also facilitates comparative analyzes, and allows pixel-by-pixel correlation studies for the various parameters.

Data fusion can be done in many ways one of which is the fusion of data from several types of sensors (e.g., visible, infrared, and microwave) for the same geophysical parameter. This type of fusion allows for a geophysical parameter to be examined in such a way that some of the ambiguities in interpretation is minimized. For example, sea ice concentration is usually derived from passive microwave data, but the resolution of this data is coarse and some mixing algorithms has to be utilized to interpret the temporal and spatial changes in the brightness temperature of the surface. With the use of co-registered data from the visible and infrared, one can evaluate to what extent the mixing algorithm is providing the correct information.

Another type of data fusion involves the synergistic analyzes of the relationships of different parameters within a system. This is especially important in the evaluation of environmental effects that leads to observed variability of a certain parameter. For example, when analyzing the variability of phytoplankton distributions in a certain region, it is good to have concurrent study of the variability of other parameters like surface temperatures, sea ice, cloud cover, and wind. Moreover, observed variability of sea ice could also be conveniently studied with the same data set to better understand its temporal and spatial variability.

3.5 Limitations and Future Needs

The use of satellite data is fairly new, and as we get more and more involved with their utilization, there is a realization that currently available data do not exactly fit actual needs for process and modeling studies. We try to get the most out of the data as discussed in Chapter 4, and oftentimes, there is a need to compromise with the required accuracy.

The aforementioned improvements in capability provide not only the means to improve the accuracy in the retrieval of polar parameters but also the potential of retrieving complementary information. Needs are usually dictated by the requirements of phenomenological and modeling studies. Ability to meet these requirements is in turn dependent on the limitations associated with the natural characteristics of electromagnetic radiation and by what current technology can provide at reasonable cost. Undoubtedly, current instruments will evolved into much more capable systems that provide more accurate data. We also hope that technology will evolve such that it will become possible to measure some important parameters that have not been possible to measure.

References

Bothwell GW, Hanser EG, Vargo RE, Miller KC (2002) The multi-angle imaging spectroradiometer science system, its products, tools and performance. IEEE Trans Geosci Remote Sens, 40(7): 1467–1476

CEOS (Committee on Earth Observation Satellites) (2002) Earth observation handbook ESA, Paris France

Comiso JC (2003) Warming trends in the Arctic. J Climate 16(21):3498–3510

Comiso JC, Zwally HJ (1989) Polar Microwave Brightness Temperatures from Nimbus-7 SMMR, Time Series of Daily and Monthly Maps from 1978 to 1987, NASA RP1223

Comiso JC, Yang J, Honjo S, Krishfield RA (2003) The detection of chang in the Arctic using satellite and buoy data. J Geophys Res 108(C12), 3384, doi:1029-2002JC001247

Cracknell AP (1987) The advanced very high resolution radiometer, Taylor and Francis, London

Diner DJ, Beckert J, Reilly T, Bruegge C, Conel J, Kahn R, Martonchik J, Ackerman T, Davies R, Gersti S, Gordon H, Muller JP, Myeni R, Sellers P, Pinty B, Verstracte M (1998) Multiangle imaging spectroradiometer (MISR) instrument description and experiment overview. IEEE Trans Geosci Remote Sens, 36: 1072–1087

Elachi C (1987) Introduction to the Physics of Remote Sensing. John Wiley and Sons, New York

Gloersen P, Campbell W, Cavalieri DJ, Comiso JC, Parkinson CL, HZwally HJ (1992) Arctic and Antarctic Sea Ice, 1978-1987: Satellite Passive Microwave Observations and Analysis. NASA Spec. Publ. 511

Ikeda M, Dobson FW (eds) (1995) Oceanographic applications of remote sensing, CRC Press, Boca Raton

Kramer HJ (2002) Observation of the earth and its environment: survey of missions and sensors, Springer, Berlin

Lubin D, Massom R (2006) Polar remote sensing, Springer-Praxis, Berlin

Martin S (2004) An introduction to Ocean Remote Sensing. Cambridge University Press, Cambridge, UK

Massom R (1991) Satellite remote sensing of polar regions volume 1: applications, limitations and data availability, Belhaven Press, London

Maxwell JC (1904) A treatise on Electricity and Magnetism. 3rd Ed., Oxford Univ. Press, Oxford, UK

Maxwell SK, Schmidt GL, Storey JC (2007) A multi-scale segmentation approach to filling gaps in Landsat ETM+ SLC-off images. Int J Rem Sen 28(23):5339–5356

Pogorszelski RJ, Shapiro KA (1976) Introduction to the physics of remote sensing. In: Lintz Jr J, Simonett DS (eds) Remote sensing of environment, Addison Wesley Publishing Company, Reading, Massachusetts

Shibata A, Imaoka K, Koike T (2003) AMSR/AMSR-E level 2 and 3 algorithm developments and data validation plans of NASDA. IEEE Trans Geosci Remote Sens, 41: 195–203

Zwally HJ, Comiso JC, Parkinson CL, Campbell WJ, Carsey FD and Gloersen P (1983) Antarctic Sea Ice 1973–1976 from Satellite Passive Microwave Observations, NASA Spec.Publ., 459

Chapter 4
Geophysical Parameters and Algorithms for High Latitude Applications

Abstract The parameters of interest for polar studies that can be derived from satellite data are surface temperature, sea ice concentration, albedo, cloud cover, winds, sea ice drift, and sea ice thickness. The basic radiative and physical characteristics that enable the detection and characterization of various surfaces and parameters are discussed. The different types of sensors are evaluated with a view of assessing which one is best suited for the retrieval of each parameter. The algorithms that convert digital data to geophysical parameters are presented and assessments of errors in the retrieval are provided. There are limitations in the use of these data, but some of these can be overcome through the concurrent use of ancillary data that include in situ measurements and modeling results.

Keywords Optical properties • Dielectric characteristics • Emissivity

4.1 Introduction

An appropriate algorithm is required to be able to convert satellite data into useful geophysical data or information. The basic assumption of these algorithms is that the radiative energy that emanates or is scattered from the material of interest provides enough information to enable the retrieval of the desired geophysical parameter. There are three main classes of satellite algorithms, namely: radiative transfer, statistical, and neural network. The first involves the use of a radiative transfer model that reconstructs numerically, the emission, absorption, and scattering of the flux of radiation as it propagates through an intermediate medium (e.g., snow and atmosphere) before it reaches the detector. This enables the assessment of the deviation of the measured signal from the original signal on account of the presence of the intervening media and allows an accurate estimate of the parameter of interest. The statistical technique makes use of satellite data in a phenomenological analysis in conjunction with in situ data and known information about the surface to arrive at a set of statistical parameters that converts satellite radiances directly to geophysical parameters. This technique is not so elegant, but it

J. Comiso, *Polar Oceans from Space*, Atmospheric and Oceanographic Sciences Library 41,
DOI 10.1007/978-0-387-68300-3_4,

is quite popular and produces realistic results, even when absolute calibration of the satellite data is not available. Sometimes, a combination of the first and second technique is used to optimize accuracy. The neural network technique also takes advantage of in situ data, which are utilized as a "training data set" for a conceptual network model that utilizes available multichannel data as input to extract as much information as possible from the satellite data. The interconnectivity of information from the input data is exploited through the use of nodes in a multilayer system to generate a set of weighting functions that is in turn used to derive the surface parameter of interest. Each of the different techniques has advantages as well as disadvantages and it is usually the accuracy requirement, quality of the calibrated data, and ease of use that determine which one to utilize in the end.

Although the propagation of radiation through matter is basically a well-known phenomenon (Chandrasekhar 1960), assessing and accounting for the effects of an unpredictable and highly variable atmosphere are still a big challenge in the retrieval of surface parameters. Through emission, absorption, and scattering, the atmosphere alters the surface signal before it reaches the satellite sensor. The impact of emission and absorption depends on the electrical properties of the intervening material, while the impact of scattering depends on the wavelength of the radiation and the size of the scattering particles. The effectiveness in the ability to assess the contribution of the atmosphere depends on how well the characteristics of the atmosphere are known when the satellite measurement was made. These measurements are made continuously, but over different surfaces at different locations and therefore different atmosphere.

In the case of passive microwave data, for example, the brightness temperature, (T_B), recorded by the satellite sensor at a given wavelength can be estimated using the basic radiative transfer equation given by

$$T_B = \varepsilon T_S e^{-\tau} + \int_0^{\tau} T(z)\zeta(z)e^{-\tau+\tau'(z)}d\tau'(z) + (1-\varepsilon)\kappa\, e^{-\tau}\int_0^{\tau} T(z)\zeta(z)e^{-\tau'(z)}d\tau'(z) \quad (4.1)$$

where ε is the emissivity of the surface, T_s is the physical temperature of the surface, $\tau'(z)$ and τ are the atmospheric opacities from the surface to a height z and from the surface to the satellite height, respectively, κ is an estimate of the diffusiveness of the surface reflection, and $T(z)$ and $\zeta(z)$ are the temperature and the emittance at z respectively. In (4.1), the first term represents radiation directly from the earth's surface, which is the dominant contribution for measurements at microwave frequencies. The second term represents satellite observed radiation that emanates directly from the atmosphere, while the third term represents downwelling radiation from the atmosphere that has been reflected by the surface before it reaches the satellite detector. A fourth term that takes into account the reflected contribution of radiation from free space (i.e., the cosmological 2.7 K contribution from the Big Bang), which is an additive contribution, is usually negligible and not included in (4.1). The radiative transfer equations for visible and infrared passive sensors are similar, but more complex while those for active systems are a little different because the original signal comes from the sensor itself.

Radiative transfer models have been developed for the retrieval of many geophysical parameters (e.g., Swift et al. 1985; Svenson et al. 1987; Winebrenner et al. 1992; Kumerow 1993). Implementation of a radiative transfer model to retrieve geophysical parameters from satellite data is effective only if the atmospheric parameters that are used as input provide good approximations of the true state of the atmosphere when the measurement was made. In particular, the models require knowledge of the surface material of interest and the physical characteristics of the intervening atmosphere, which changes in space and time. For example, in (4.1), one has to know the dependence on height of many atmospheric parameters including temperature, humidity, pressure, particle size, and particle type. The problem may not be so serious in polar regions where the atmosphere is regarded as arid and having relatively low humidity. But generally, good approximations of the input parameters over an annual cycle in the two polar regions are hard to come by. Some of the information is provided by ground-based measurements, such as radiosonde launches, which are done only in some places at certain times. Unfortunately, radiosonde and similar data are not available at good enough spatial resolution to take into account the effect of regional and temporal variability of the atmosphere. Moreover, the radiative characteristics of surfaces (e.g., their emissivities) are usually not well defined and/or not known within the field-of-view of the sensor as they vary from region to region. The radiative transfer technique is generally suited as a regional technique and more specifically, in areas where the input parameters are well known.

Satellite data provide global coverage and ideally, global algorithms should be designed to efficiently generate geophysical data in any location on the surface of the Earth and for any time. For polar studies, the use of a separate algorithm for the Northern and Southern Hemispheres would be acceptable, especially if it means more accurate retrievals, since the delineation of boundaries is not an issue. Most algorithms take advantage of the availability of multichannel data that can be used in part to account for the spatial and temporal variability of atmospheric and surface effects. This is done by taking advantage of different sensitivities of radiation at different frequencies and polarizations to different atmospheric and surface conditions. Such effects usually follow certain patterns that can be incorporated in algorithms through statistical and other techniques, making sure at the same time that results are consistent with in situ data or well-known information about the surface. While not necessarily the most suitable for regional studies, global algorithms allow the processing of global data sets that can be used for global and modeling studies. The alternative is to put together a global data set using regional data provided by regional algorithms, but such endeavor leads to many complex issues such as getting boundaries between adjacent regions consistent. Available in situ data are also limited in spatial and temporal coverage and regional algorithms in many parts of the globe may be difficult to develop. It should be emphasized, however, that for regional studies regional algorithms should be used, especially if optimal accuracy is required.

The use of neural networks (or artificial intelligence) has also gained some popularity both for global and regional studies because of their versatility and also because they can be further enhanced through the incorporation of a radiative transfer model. In the polar regions, however, there is a serious lack of adequate training data sets that can effectively account for large spatial and temporal variations in surface and atmospheric conditions in the region. This lack of data will likely change in the near future because of growing interest in the large changes observed in the region that may be associated with anthropogenic climate change. The use of the technique can also be done concurrently with the fusion of data from many different sensors. Such endeavor would maximize information content that can be extracted from satellite data. While the neural network algorithm would likely evolve to become the algorithm of the future, we used for this book techniques that have been validated and are currently being utilized for deriving the geophysical parameters.

4.2 Radiative Characteristics of Surface Parameters

Satellite sensors provide the synoptic and large-scale coverage needed to study the polar oceans. Many useful parameters can be derived from data provided by these sensors, but there are limitations in the ability to completely characterize the oceans because of the natural characteristics of the electromagnetic radiation in the region. In particular, visible sensors can detect signals from ice free oceans down to only a few meters beneath the surface, while infrared and microwave sensors can detect only the top micron to cm layer (skin depth). In sea ice covered regions, visible and infrared sensors can observe only the top layer as well while passive microwave could detect radiation emanating several centimeters to a few meters from the surface depending on ice type and snow cover, which influence the dielectric properties of the material. Thus, the set of polar parameters that can be derived directly from satellite data are those that are related to the limited radiative information that can be detected by satellite sensors.

The emission and scattering characteristics of sea ice and liquid water are also quite difficult to model because of the paucity of accurate input data that are consistent with the specific characteristics of the sensors. In particular, the emission and backscatter characteristics of open ocean surfaces as they are affected by waves, sun glint, foam and weather, and of sea ice as affected by snow surface condition, age, thickness, salinity, and liquid content must be known at the specific frequencies and polarizations utilized by the satellite sensors. Critical input parameters are sometimes derived using simple linear interpolation from existing data and may compromise the utility of the modeling efforts.

The key to a successful algorithm is how well it is able to reproduce nature. This is the reason why statistical techniques are usually popular. Validation efforts have also been done to establish the magnitude of the errors and the utility of the derived parameters. Overall, the retrieved satellite products have been relatively accurate and generally useful for process and global change studies. They have also provided new insights into our understanding of the polar oceans and their sea ice cover. In the following sections, we provide an overview of the physical and radiative characteristics of the ocean and ice surfaces that are relevant to the interpretation of remote sensing data, the techniques used to retrieve the data, and a discussion of some of the problems and errors associated with the retrieved data.

4.2.1 Optical Properties of Liquid Water and Sea Ice

Our current knowledge of the optical properties of the ocean is the result of studies that started over a century ago. Detailed description of the optics of both liquid and solid portions of the ocean has been presented in many review articles and textbooks (Williams 1972; Morel 1974; Warren 1984; Perovich 1998; Curry and Webster 1999; Martin et al. 2004; Lubin and Massom 2006). While the two are basically interchangeable, we follow the convention of using the index of refraction and related variables to discuss the optical characteristics of water and ice while we use the dielectric properties to discuss the microwave radiative characteristics of the same material.

Maxwell's equations form the basis of all electromagnetic phenomena and have been used to study how radiation propagates through various medium, including sea water, sea ice, and the atmosphere. The equations provide the means to evaluate how the signal emitted by (or scattered from) a surface material is altered by an intervening media before it gets detected by the satellite sensor. They also enable estimates to be made of relevant information such as optical depth and opacity of a material, and how they vary with frequency and polarization of the radiation.

The propagation of electromagnetic radiation, E, through a dielectric medium as a function of displacement z, assuming no time dependence, can be expressed in a simple form as follows:

$$E(z) = E_0 e^{-\gamma z} \tag{4.2}$$

where E_0 is the electric field intensity in a vacuum while γ is the propagation constant. In a vacuum, γ is zero and the field intensity would not change, but in a medium with thickness z, the field gets attenuated depending on the propagation constant as indicated in (4.2). The propagation constant is a complex variable defined by:

$$\gamma = \alpha + i\beta \tag{4.3}$$

where α and β are the absorption and phase constants of the medium, respectively. These parameters are related to the refraction index of the material as follows:

$$\alpha = k_0 n'' \tag{4.4}$$

$$\beta = k_0 n' \tag{4.5}$$

where n' and n'' are the real and imaginary parts of the complex index of refraction and $k_0 = 2\pi/\lambda_0$ is the wave number in free space, and λ_0 is the free-space wavelength in meters.

Radiation propagating from one dielectric medium to another is affected by three key processes, namely, reflection, absorption, and scattering. Briefly, reflection refers to the fraction of the radiation that is reflected by the surface while absorption refers to the fraction that interacts directly with the molecules in the medium and scattering is the fraction diverted from the original direction by the medium. The amount of radiation reflected at the surface depends on surface roughness, incidence angle, and the intensity of available radiation. This amount is further enhanced by radiation scattered below the ocean surface, which goes in the same direction as reflected light. The surface signal is usually difficult to differentiate from the subsurface signal. When the surface is disturbed, the surface area of the roughened surface increases, and the angle of incidence varies with time leading to occasional appearance of glare or glitter.

The overall effect of roughness on the reflectance as observed by satellite sensors is on the average relatively small when compared with that caused by changes in the solar zenith angle. The solar zenith angle and other angles of interest are schematically depicted in Fig. 4.1 and defined as follows. The angle made by the satellite scan point with nadir from the satellite sensor is called the zenith angle θ_Z while the angle made by the sun and a vertical line at the satellite scan point is the solar zenith angle θ_S. The angle θ is usually referred to as look angle or angle of

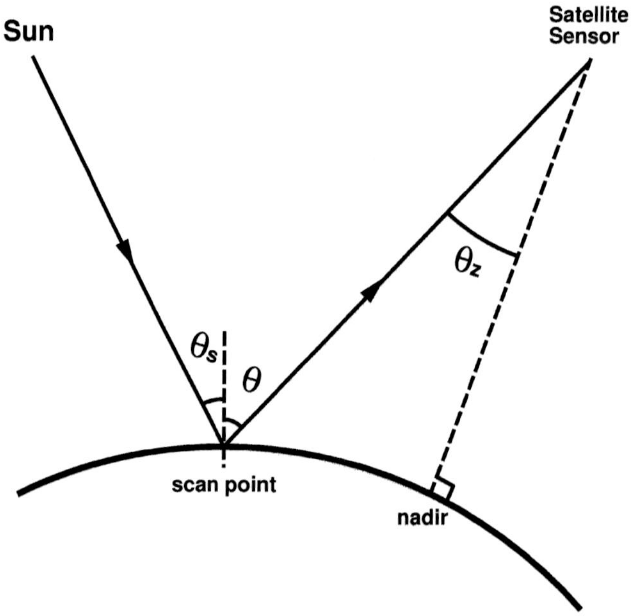

Fig. 4.1 Schematics showing the solar zenith angle, θ_S, the incidence or look angle, θ, and the satellite zenith angle, θ_Z

incidence and defined as the angle made by a vertical line at the satellite scan point and a line from this point to the satellite sensor.

The reflectance, r, from a smooth surface can be estimated from the Fresnel's equation:

$$r = \frac{1}{2}\left(\frac{\tan^2(\theta - \varphi)}{\tan^2(\theta + \varphi)} + \frac{\sin^2(\theta - \varphi)}{\sin^2(\theta + \varphi)} \right) \tag{4.6}$$

where r is the reflectance, θ is the angle of incidence, and φ is the angle of refraction or the angle to the normal made by the refracted radiation. At nadir, this equation reduces to

$$r = \left(\frac{n-1}{n+1} \right)^2 \tag{4.7}$$

where n is the relative refractive index, related to θ and φ through Snell's equation given by $n = \sin\,\theta / \sin\,\varphi$. Using this equation and $n = 1.34$ for water, the reflectance when the sun is at nadir is about 2%, increasing slowly to 3.5% at solar zenith angle of 50° and to 100% when the solar zenith angle is 90°. It should be noted, however, that a large portion of the radiation reaching the ocean surface is in the form of scattered light usually referred to as diffuse radiation. This varies with cloud cover condition, but even in the absence of clouds, a significant portion of incoming solar energy is diffused with the percentage varying from 27% to 67% during summer and 15% to 59% during winter for solar zenith angles from 50° to 80°. During a completely overcast situation, the reflectance has been estimated to be about 7% (Newmann and Pierson 1966).

Among the most relevant parameter associated with the reflectance or reflectivity of the surface is the albedo, which is usually defined as the ratio of diffusely reflected radiation to the incident radiation. Thus, a perfectly white diffuse reflector would have an albedo of 1, while a blackbody would have an albedo of 0. The term reflectance means basically the same as albedo but only for incoming radiation at a single incidence angle. Surfaces in which the emitted radiance is independent of direction have been called Lambertian surfaces. Examples of such surfaces are reflectors, such as foam and clouds, which provide basically the same intensity of radiation in the visible and infrared frequencies regardless of direction.

The albedo of liquid water and sea ice has been studied by many investigators (e.g., Willams 1973; Warren 1982; Allison et al. 1993). The albedo of ice free sea surface is among the lowest on Earth with values ranging from 0.03 to 0.30 while that of ice and snow is the highest with values from 0.40 to 0.95. The range for ice and snow is quite large because of so many surface types. In particular, the albedo of old snow and sea ice is between 0.5 and 0.6, that of seasonal snow is about 0.7–0.8 while that of newly fallen snow and clouds is about 0.9 or higher. For comparison, the albedo of grassland, green crops and trees is between 0.15 and 0.25 while that of sand is between 0.18 and 0.28. The albedo of a given surface may also change with time as the physical and radiative characteristics of the surface changes. The albedo of the open ocean is occasionally altered by the changing strength and direction of winds that dictate the state of the surface and the occurrence of big waves and foam (Williams 1973). Also, the albedo of sea ice is dependent on age, thickness, roughness, liquid content, and state of its snow cover (Warren 1982; Allison et al. 1993). In general, the planetary albedo, which is the percent of incoming radiation from the sun that is reflected back to space from all sources on the Earth's surface, is about 0.30.

The reflectance is influenced primarily by the index of refraction, which is frequency or wavelength dependent as indicated in Fig. 4.2. The real part is shown to decrease with increase in wavelength from 0.4 to 2.5 μm for both liquid water surface and ice covered surface. However, the liquid surface has significantly higher values than the ice surface. The imaginary part, on the other hand, increases with increase in wavelength (Fig. 4.2b) and it appears that both water and ice have very similar values except between 1.0 and 1.8 μm.

The optical properties of sea water are governed by the same physical principles and electromagnetic equations as other materials. Sea water has been characterized as having marginal ability to support the transport of electromagnetic energy. Pure sea water has stable optical characteristics and changes only slightly with salinity, temperature, and depth. However, the optics of sea water have been the center of special attention in recent years because it is the habitat of living organisms. Life in the sea is directly or indirectly dependent on tiny aquatic plants, which in general could not exist without the availability of light. The region of primary interest is what is called the "euphotic zone," which is the layer from the surface of the ocean down to where light energy is reduced to 1% of its surface value. In general, the thickness of this layer is about 100 m, but in clear waters, it can be as thick as 150 m. It turned out that the optical characteristics of the sea is largely affected by the tiny aquatic plants and also by suspended sediments, yellow substance, detritus,

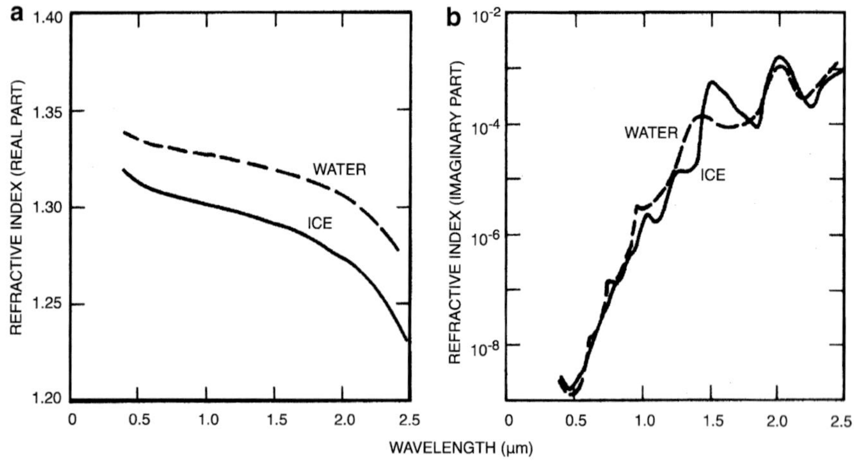

Fig. 4.2 Complex refractive index of pure ice and water as a function of wavelength in the visible to mid-infrared channels for (**a**) real part of n and (**b**) imaginary part of n. From NASA (1987). Courtesy of NASA

and biological processes that occur in the euphotic zone. The presence of these tiny plants and contaminants has also been associated with the color of the ocean.

The theory behind the color of the ocean has been attributed mainly to the work of Kalle (1949) who correctly explained that the blue color of clear open ocean water is caused by molecular scattering and that the observed color of coastal waters is due to the addition of a yellow substance (called gelbstoff) that absorbs the blue light yielding a brownish yellow color to the water. He also suggested that the larger particles in the ocean, such as phytoplankton, could impart color to the water. Many studies have indeed confirmed that the color of sea water is mainly altered by dissolved organic matter, suspended particulates, and phytoplankton (e.g., Hoepffner and Sthyendranath 1993). The yellow substance is what is referred to as colored dissolved organic material (CDOM), which can have terrestrial or oceanic origins. The terrestrial components are from river or land-based run-offs consisting mainly of dissolved humic and fulvic acid from decayed or decaying forms of vegetation. The suspended particulates can be organic or inorganic. The organic particulates consist mainly of phytoplankton and zooplankton cell fragments and zooplankton fecal pellets and are generally called detritus, while the inorganic ones consist of sand and dust coming from the erosion of land-based rocks and soils. Phytoplanktons are single-celled organisms that include diatoms and other algae and are among the most important components of the marine ecosystem. They are at the bottom of the food chain and their concentration is reduced as they are consumed by other organisms (referred to as grazing) or natural decomposition through a process called photolysis.

The standard parameter retrieved from the ocean color sensors is chlorophyll-*a* concentration (C), which is a measure of the abundance of phytoplankton. This parameter has been the focus of interest because it plays a central role in the estimates for primary production and light attenuation coefficients. It is also the parameter that is normally measured in situ at sea and is therefore convenient to validate. As indicated earlier, chlorophyll-*a* is just one of the many constituents that make up the inherent optical properties (IOP) of the ocean. The IOP has been the subject of intensive research (e.g., Shifrin 1988; Kirk 1984; Mobley 1994) since it is the primary determinant of the color of the ocean as detected by sensors. SeaWiFS has been the key instrument for estimating chlorophyll-*a* concentration, but this parameter can be derived as well from newer sensors such as MODIS and MERIS, that also have additional channels that enable the detection of other characteristics of the water (e.g., fluorescensing properties) that has not been possible to measure before.

The beam attenuation coefficient (b), which determines the inherent optical property of the liquid is generally defined as the sum of the absorption (a) and scattering (s) coefficients which are the key parameters relevant to the reflectance:

$$b(\lambda) = a(\lambda) + s(\lambda) \tag{4.8}$$

The absorption coefficient can be expressed in terms of the various components as

$$a(\lambda) = a_w(\lambda) + a_{ph}(\lambda) + a_{pd}(\lambda) + a_g(\lambda) \tag{4.9}$$

where a_w, a_{ph}, a_{pd}, and a_g represent water, algal, nonalgal, and dissolved (gelbstoff) particulate matter components, respectively. Of the various components, only a_{ph} is strongly related with chlorophyll-*a* while the other components are only weakly related. However, it has been shown (Mobley et al. 2004; Lee and Hu 2006) that this is not necessarily the case. The complication is that there are generally two types of water, as recognized by Morel and Prieur (1977), each providing different optical information: Case 1 water in which the reflectance is determined by photosynthetic pigment and their covarying detrital pigment absorptions and Case 2 water in which the reflectance is determined by suspended particulates and dissolved organic matter (DOM, yellow substance or gelbstoff). In Case 1 water, it is usually assumed and generally accepted that all nonwater components vary closely with chlorophyll-*a*. In coastal areas where Case 2 water is dominant, the same assumptions would lead to incorrect results.

The theoretical aspect of the scattering of light has been discussed extensively in the literature (e.g., Van de Hulst 1981; Schuerman 1980). The scattering coefficient can be decomposed into its forward, s_f and backward s_b components as follows:

$$s(\lambda) = s_f(\lambda) + s_b(\lambda) \tag{4.10}$$

where

$$s_f = 2\pi \int_0^{\pi/2} \beta(\theta)\sin(\theta)\mathrm{d}\theta \tag{4.11}$$

and

$$s_b = 2\pi \int_{\pi/2}^{\pi} \beta(\theta)\sin(\theta)\mathrm{d}\theta \tag{4.12}$$

where β is the volume scattering function in units of $m^{-1}\,sr^{-1}$. The forward and backward components can be further separated as

$$s_f(\lambda) = s_{fw}(\lambda) + s_{fp}(\lambda) \tag{4.13}$$

$$s_b(\lambda) = s_{bw}(\lambda) + s_{bp}(\lambda) \tag{4.14}$$

where the subscripts *w* and *p* correspond to scattering by water molecules and particulate matter, respectively. The particulate components include the algal, nonalgal, and dissolved particulate matter. The beam attenuation coefficient, which determines the optical property of the water is thus dependent on many variables with the algal particulate contribution being only one of them.

For ocean color studies the variables which are most relevant are the algal term for absorption, a_{ph}, and the backward (and algal) component of the scattering of the particulate matter s_{bp}. Radiative transfer modeling activities would require sufficient knowledge of these parameters.

4.2.2 Dielectric Properties and of Sea Water and Ice

The radiative characteristics of sea water and sea ice at infrared and microwave frequencies have been studied extensively by many investigators and results have been discussed in many review articles (e.g., Irvine and Pollack 1968; Swift 1980; Ulaby et al. 1986; Carsey 1992). At infrared and microwave frequencies, it is more of a convention to use the dielectric constant, ε, which is the parameter associated with the basic electrical property of the material. It is the latter that provides the connection between the intensity of radiation (called radiance) measured by the satellite sensor at different wavelengths and polarizations and the physical characteristics of the source of radiation which is the material or parameter of interest. The dielectric constant is also a complex variable given by

$$\varepsilon = \varepsilon' - i\varepsilon'' \tag{4.15}$$

with the real component, ε', the permittivity of the material, and the imaginary component, ε'', the loss factor. The dielectric constant is related to the index of refraction n through

$$\varepsilon = n^2 = (n' + in'')^2 \tag{4.16}$$

Thus, the two variables are basically interchangeable with

$$\varepsilon' = (n')^2 - (n'')^2 \tag{4.17}$$

and

$$\varepsilon'' = 2n'n'' \tag{4.18}$$

Knowledge of the dielectric and refractive properties of a medium is important since it is these properties that control the emission and scattering characteristics that in turn determine the amount of radiation that goes through the medium.

The dielectric constants of pure and fresh water at microwave frequencies are well known and could be approximated with good accuracy from the Debye equation (Debye 1929) as follows:

$$\varepsilon = \frac{\varepsilon_\infty + (\varepsilon_S - \varepsilon_\infty)}{(1 + i\omega\tau)} - \frac{i\sigma}{\omega\varepsilon_0} \tag{4.19}$$

where ω is the angular frequency ($=2\pi f$, f is frequency in Hz), ε_∞ is the permittivity at high frequencies, ε_S is the static permittivity, τ is the relaxation time in seconds, σ is the ionic conductivity in mhos m^{-1} and ε_0 is the permittivity in outer space and is approximately 8.854×10^{-12} F m^{-1}. The magnitude of ε_∞ has been estimated by Lane and Saxton (1952) to be about 4.9. The static permittivity, ε_S, is temperature dependent and can be inferred using the following regression equation derived by Klein and Swift (1977):

$$\varepsilon_S = 88.045 - 0.4147\,T + 6.295 \times 10^{-4}\,T^2 + 1.075 \times 10^{-5}T^3 \tag{4.20}$$

where T is the temperature of the material. The equation was inferred from measurements at 1.43 to 2.65 GHz but has been found to be appropriate for other frequencies as well. If T is known, the real and imaginary parts of the dielectric constant can be estimated using (4.19).

The dielectric constant of sea water can also be estimated from the Debye equation as long as the dependence of the various parameters (e.g., ε_S, σ, and τ) on salinity is taken into account (Swift 1980; Wentz 1983). The dependencies of the real and imaginary parts of the dielectric constant for fresh and sea water and for 0°C and 20°C surface temperatures as a function of frequency are illustrated in Fig. 4.3. The plot in Fig. 4.3 shows little sensitivity of the real part to changes in

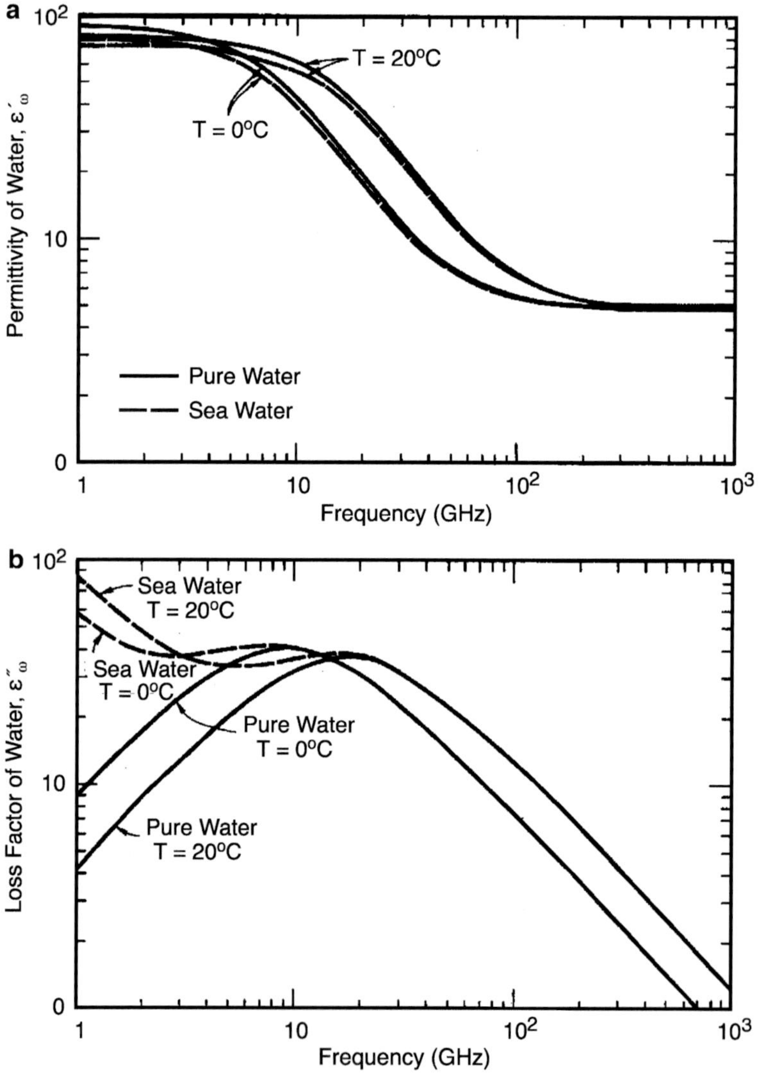

Fig 4.3 Dielectric constant for fresh and sea water as a function of frequency at 0oC and 20oC surface temperature. From Ulaby et al. (1986) with permission of the Artech House, Inc fig E.1, p. 2023)

salinity, but considerable sensitivity to changes in temperature between 8 and 300 GHz. It is apparent that the imaginary part shows strong sensitivity to changes in salinity at less than 20 GHz and strong sensitivity to changes in temperature at all frequencies. The accuracy of the calculated dielectric constants for sea water using the Debye equation has been considered to be very good with errors estimated to be around 0.2% (Klein and Swift 1977; Swift 1980).

It should also be noted that in open water regions away from the ice pack, the dielectric constant could vary spatially primarily because of wind effects that cause the formation of large waves, foam and upwelling. The presence of foam provides an additional dielectric layer that alters the signature of the surface while upwelling could cause a change in surface temperature. Within the ice pack where the surface temperature and salinity are basically constant and big waves are not an issue, the dielectric properties of ice free areas are expected to be more uniform.

Estimates of the dielectric properties of fresh ice benefit from the simplification that can be made at microwave frequencies since the relaxation of pure ice, takes place in the kilohertz region. The Debye equation thus simplifies to

$$\varepsilon' = \varepsilon_\infty \tag{4.21}$$

$$\varepsilon'' = \frac{(\varepsilon_s - \varepsilon_\infty)}{1 + i\omega\tau} \tag{4.22}$$

Experimental measurements indicate little or no dependence of ε' to both temperature (below 0°C) and frequency in the microwave region and that the value is relatively constant at about 3.15 (Cummings 1952). On the other hand, ε'' is quite sensitive to both parameters and estimates using the Debye equation at 10 GHz of 2×10^{-5} is two orders of magnitude smaller than actual measurements of 1.2×10^{-3} by Lamb (1946) and 2.7×10^{-3} by Cumming (1952). The Debye equation also incorrectly predicts the frequency dependence of ε''. For example, in the equation ε'' would decrease with increase in frequency while measurements show a change in slope at greater than 10 GHz. This observed departure from the Debye equation has been attributed to contributions by the infrared absorption spectrum of ice (Evans 1965).

For sea ice, the aforementioned simplification cannot be done because of complications, both structurally and electromagnetically. Sea ice is an inhomogeneous material with brine and air pocket components as discussed in Chapter 2, Section 2.3.1. The presence of the brine inclusions strongly alters the dielectric properties of the material since the dielectric constant of brine is higher when compared with that of ice. This is especially relevant to seasonal ice which has relatively high dielectric constants despite a salinity of only around 5–12 psu. The brine content declines with age of the ice and is significantly different for ice that has gone through a summer season and turn into old ice. Measurements of the dielectric properties of seasonal ice and old ice indicate dramatic differences with the values for the latter being similar to that of fresh (or zero salinity) ice (Vant et al. 1974). Sea ice is most saline when it is newly formed, the salinity being about 12 psu. Some measurements of the imaginary part, ε'', have yielded values of 0.03 for first year ice and 0.003 for multiyear ice at 19 GHz. However, ε'' is sensitive to many

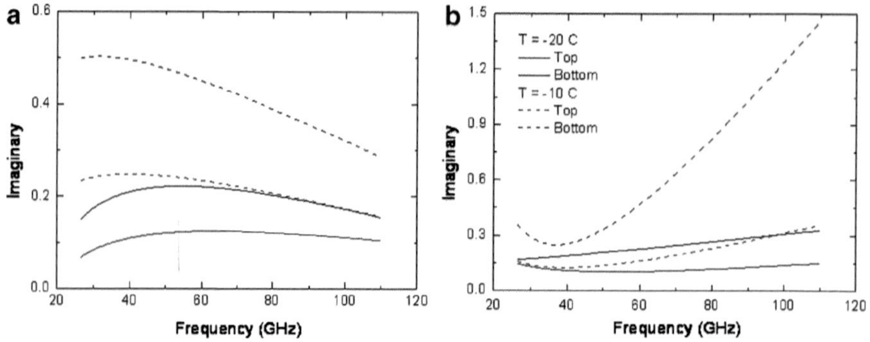

Fig. 4.4 Imaginary part of the dielectric constant of Antarctic sea ice as a function of frequency measured at temperatures of -10°C and -20°C. Results are plotted separately for (**a**) typical granular ice and (**b**) low density granular ice. Courtesy of G. Koh and D. Perovich, 2009

factors including temperature, brine volume and frequency and measurements have been confined to only few observations (Vant et al. 1978; Arcone et al. 1986; Koh 1992; Hallikainen and Winebrenner 1992). Recent measurements in the Antarctic by Koh and Perovich (private communication in 2009) are presented in Fig. 4.4 for typical granular and low density granular ice at −10 and −20°C and the values are about 0.2 for the frequency range from 26 to 112 GHz for the top of the ice, which is most relevant to remote sensing. The measurements also show that the ice has significantly higher values when it is colder by about 10 degrees.

The dielectric constant is usually expressed as

$$\varepsilon = \varepsilon' \,(1 - i\tan\delta) \tag{4.23}$$

where $\tan\delta$ is called the loss tangent and is the ratio $\varepsilon''/\varepsilon'$. Materials with relatively high values of ε'' have high values of the loss tangent and are usually called lossy materials. Thus, sea water and seasonal ice are lossy, while multiyear and fresh ice are not and are instead transparent.

Among the quantities most relevant to remote sensing, especially for sea ice, is the penetration depth, which is defined as the depth at which the incident energy is reduced by $1/e$ (=0.369). It represents approximately the maximum depth of a material that contributes to its brightness temperature or its backscatter coefficient. For a scatter-free and homogeneous medium, it can be estimated from

$$\delta_p = 1/\kappa_a \tag{4.24}$$

where κ_a is the power absorption coefficient equal to 2α, where α is as defined in (4.4). If $\varepsilon''/\varepsilon' << 1$,

$$\delta_p = \sqrt{\frac{\varepsilon'}{k_o \varepsilon''}} \tag{4.25}$$

The equation shows that the penetration depth for snow and ice depends largely on the value of ε'' since ε' is relatively constant with values close to 3.15. Thus, when

ε'' is high, as with first year ice, the penetration depth is small and the material is usually referred to as opaque. Alternatively, when ε'' is low as with fresh and multiyear ice, the penetration depth is large and the material is considered transparent. Such difference in the penetration depth allows for discrimination of ice types as will be explained in the next section.

4.2.3 *Emissivity and Backscatter*

The emissivity of a material at a given wavelength is defined as the ratio of radiation emitted by this material to that emitted by a blackbody at the same temperature. A blackbody is a hypothetical material that absorbs 100% of the incident radiation and emits the maximum amount of radiation possible over all wavelengths. For passive infrared and microwave sensors, it is the emissivity, ϵ, of the material that provides the critical information about surface characteristics that enables the determination of surface parameters from satellite passive measurements in the infrared and microwave regions. The total radiation is determined by Planck's law that states that the monochromatic irradiance (at a given wavelength λ) of energy E_λ emitted by a blackbody is uniquely determined by its absolute temperature (T) from the equation

$$E_\lambda = \frac{c_1 \lambda^{-5}}{e^{\frac{c_2}{\lambda T}} - 1} \tag{4.26}$$

where c_1 and c_2 are the first and second radiation constants with values 3.74×10^{16} W m^{-2} and 1.44×10^{-2} m K, respectively.

Most materials are nonblackbody and the amount of radiation, R, from matter can be derived from (4.26) by integrating over all wavelengths. The result is the Stefan-Boltzmann's Law given by

$$R = \epsilon \sigma T_S^4 \tag{4.27}$$

where ϵ is the surface emissivity, σ is the Stefan–Boltzmann constant and T_S is the surface temperature. As indicated later, the emissivity of sea water, snow and sea ice is relatively stable and close to that of a blackbody. Such stability is fortuitous since it makes it possible for infrared observations to be used for measuring surface temperature from these surfaces.

At the microwave frequencies, the emitted radiance varies linearly with surface temperatures and can be estimated using the Rayleigh–Jeans approximation which is given by

$$T_B = \epsilon T_S \tag{4.28}$$

where T_B is the radiance expressed in units of temperature called "brightness temperature," T_S is the temperature of the emitting material and ε is the emissivity of the material. Equation (4.28) is actually the long wavelength limit of the Planck's

Law. Radiances from satellite passive microwave data are usually expressed in units of brightness temperatures. The emissivity of any surface is frequency and polarization dependent and varies as a function of its absorption and scattering coefficient, wavelength, and temperature. Different surfaces may thus be uniquely identified through their emissivity and with the use of a combination of sensor data from different frequencies and polarizations.

At microwave frequencies, the emissivities of sea water and sea ice have been studied by several investigators (e.g., Klein and Swift 1977; Eppler et al. 1992; Grenfell et al. 1994). Sea water is opaque at microwave frequencies and the penetration depth is less than the wavelength of the radiation. The emissivity is relatively uniform and predictable at low frequency channels and the effects of winds can be taken into account making it possible to estimate obtain sea surface temperatures from microwave data (Wentz et al. 2000; Shibata 2004).

The microwave emissivity of sea ice is considerably more variable and less predictable than that of sea water. This is not unexpected since the dielectric properties of sea ice vary so much during different stages of growth and decay and are affected by thickness, snow cover, roughness, and surface wetness. In particular, the emissivity of the surface varies from close to that of liquid water when it is covered with grease (frazil) ice to a maximum value of about 0.95 when the surface is covered with young ice about 20 cm thick without snow cover. The emissivity tends to be the highest for young ice because of its high salinity and goes down as the ice acquires a snow cover and starts to be desalinated. The snow serves to lower the effective emissivity because the snow particles scatters some of the radiation that emanates primarily from an ice layer at the snow-ice interface. The closer the wavelength of the radiation is to the size of the snow particles, the more likely the radiation gets scattered by snow particles. With further desalination, especially as the ice goes through the summer period and becomes second or older ice types, the material becomes more transparent to the radiation and the emissivity declines further because of additional scattering within the ice. In this case, the scattering is usually caused by air pockets and other inhomogeneities within the ice.

To gain an understanding of this phenomenon, radiative transfer modeling that makes use of known absorption and scattering properties of the dielectric is needed. Matter is both the source and the scatterer of radiation. The net radiation, $\boldsymbol{n}$, is wavelength dependent and is the difference of the amount of radiation emitted and the fraction of this emitted radiation that has been scattered. The magnitude of emitted and scattered radiation depends on the extinction and scattering coefficients, respectively, of the material. More specifically,

$$\boldsymbol{n}(\lambda) = \boldsymbol{e}(\lambda) + \boldsymbol{s}(\lambda) \tag{4.29}$$

where $\boldsymbol{e}$ is the extinction coefficient and $\boldsymbol{s}$ is the scattering coefficient. The standard technique of radiative transfer is to simulate in a numerical model the propagation of the radiation from the deepest source through the material. The material is usually divided into thin scattering layers in which the parameters in (4.29) are assumed known. Since sea ice is a mixture of ice, brine, salt and air, it would be necessary to know the dielectric constant, volume fraction and sizes of each of the component.

As the radiation propagates through each layer, the net gain and loss of radiation is estimated and the net radiation emitted from the surface of the material is derived. Multiple scattering is usually accounted for through an iterative process. The modeling is quite complex and results have not been so reliable because of the general lack of adequate information about each layer and the absence of a good scattering model. However, such models have provided useful insights into the physics of the propagation of radiation through ice and snow. Advances in the technique include the introduction of the matrix doubling method (Tjuatja et al. 1993) and the use of the strong fluctuation theory, as described by Stogryn (1988).

The difference in the physical and radiative characteristics of the two key types of thick ice cover, namely, first year (FY) and multiyear (MY) ice, is schematically illustrated in Fig. 4.5. Since first year ice is saline and lossy, the brightness temperature measured by the satellite sensor is emitted basically from a thin layer at the snow ice interface and scattering within the ice is not a big factor. The multiyear ice type, on the other hand, is much less saline and therefore more transparent with a large fraction of the measured radiation emanating from below the ice surface and therefore subject to scattering within the ice. Such scattering reduces the net radiation observed by the sensor and therefore the effective emissivity of the material. The probability for scattering depends on the wavelength of the radiation. Radiation with wavelengths smaller or similar to the size of the scattering particles have, higher probability to be scattered than those of longer wavelengths.

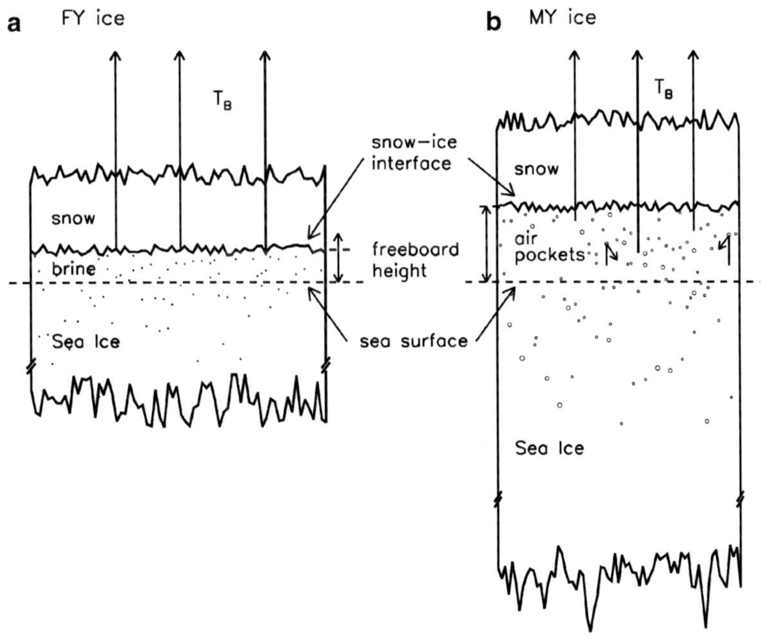

Fig. 4.5 Schematics of the emission characteristics of (**a**) first year and (**b**) multiyear ice. The key difference is the high salinity of first year ice which makes it lossy while the much fresher multiyear ice is more transparent to radiation

The variability of the emissivity of these two dominant ice types and also of sea water at various frequencies and polarizations for both winter and summer as observed in field measurements (Svendsen et al. 1983; Matzler et al. 1984) is illustrated in Fig. 4.6. The distribution covers the frequencies used by SMMR, SSM/I and AMSR-E sensors. The frequency dependences of these two ice types are apparently very different during the dry winter period (Fig. 4.6a). The emissivity of first year ice is shown to vary only slightly with wavelength (or frequency) with the emissivities at horizontal polarization slightly lower than those of the vertical polarization. The lack of frequency dependence of the emissivity for first year ice is consistent with the opaque and lossy nature of this saline ice type. Since much of the observed radiation comes from the lossy layer at the snow ice interface, scattering within the ice is negligible. Scattering can occur at the snow layer and the atmosphere, but the effect is significant only when the wavelength of the radiation is comparable to the size of the particles or scatterers in the snow and the atmosphere. For multiyear ice, the frequency or wavelength dependence is quite large, as shown in Fig. 4.6a, and the emissivity is shown to decrease with increase in frequency (or decrease in wavelength). This is consistent with the multiyear ice being fresh and transparent and with the observed radiation affected by scattering within the ice. At the higher frequencies, the wavelength of the radiation becomes more comparable to the size of air pockets and other scatterers in the multiyear ice and the radiation is more likely to get scattered and hence the lower emissivity. As the snow starts to melt in early spring even just a small fraction (e.g., 3%) of liquid water in the snow would make the surface opaque

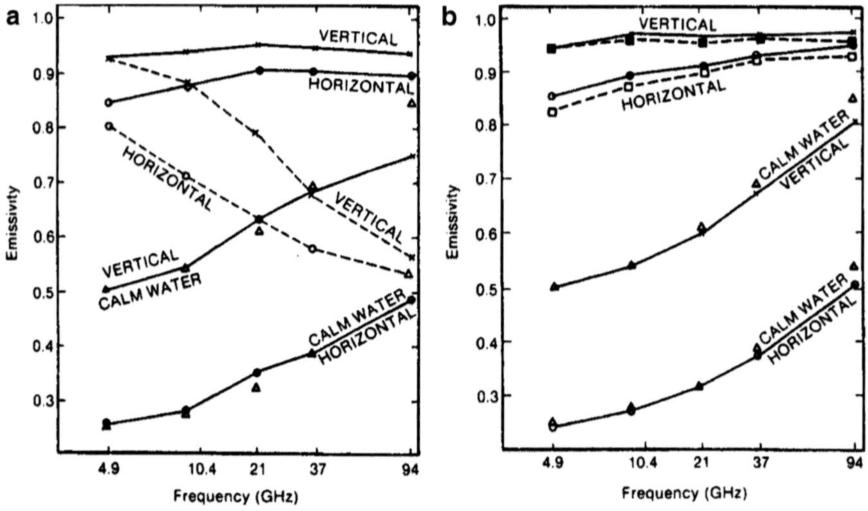

Fig. 4.6 Microwave emissivity as a function of frequency for first-year ice (solid line), multiyear ice (dash line) and calm water (solid line) for both vertical and horizontal polarizations. Triangles indicate theoretical emissivity of calm water. Left plot is for fall-winter conditions (adapted from Svendsen, 1983) while right plot is for summer conditions. From Matzler et al. (1984) with permission of the IEEE

and the emissivity of the surface independent of wavelength or frequency and close to that of a blackbody (Fig. 4.6b). Since the measured radiation emanates from a thin snow layer at the surface of either first year or multiyear ice, the signature of these two ice types becomes almost identical and impossible to discriminate.

To be able to develop a technique to discriminate sea ice covered from ice free surfaces, it is also important to know the emissivity of ice free water and the way it changes with frequency. The observed emissivity of open water is shown in Fig. 4.6 and it is apparent that during both winter and summer it increases with frequency. This is somewhat unexpected since the imaginary part of the dielectric constant decreases with frequency for sea water, as shown in Fig. 4.3, but other factors like penetration depth are likely more important to cause the observed increase in emissivity with frequency for sea water. Such increase in emissivity with frequency for sea water is unfortunate since the contrast of the brightness temperature of sea ice and that of open water is reduced at the frequencies where the sensor resolution is the highest. Such reduction in contrast causes higher uncertainties in the retrieval of ice concentration at these frequencies. Note that the emissivities for open water do not change much from winter to summer.

4.3 Geophysical Parameters and Retrieval Techniques

4.3.1 Cloud Cover and Binary Classification

Clouds are ubiquitous on the surface of the planet and in the polar regions, they are more prevalent and persistent than in most other areas. Estimates have indicated that the cloud cover in the polar regions ranges from 60 to 80% (Spinhirne et al. 2004; Intieri et al. 2002). Although it is not a surface parameter, we include clouds as one of our parameters in our study because of their important role in the climate system and their impacts on processes in the polar regions. Cloud is made up mainly of H_2O in various forms (gas, tiny liquid droplets or ice crystals) and is an important part of the hydrological cycle since it is the focal point in the conversion of water vapor to rain and snow. The importance of clouds in climate and hydrological studies have been discussed in many publications (e.g., Arking 1991; Curry and Webster 1999). The realization of the lack of good quality data has also led to the design and launch of new instruments (e.g., CloudSat) dedicated to their study. Full characterization of the cloud cover is a complex endeavor and is beyond the scope of this book. What we will present is mainly cloud statistics associated with the requirement to detect the cloud free (clear skies) areas that are needed for estimating surface parameters, such as surface albedo and surface temperature. The study is thus plainly an analysis of the presence or absence of clouds (i.e., binary distributions) as detected by satellite sensors. We will also confine our discussion on cloud detection with AVHRR data as the main source with the purpose of providing an idea about interannual variability and trend.

What was expected to be the simplest characterization of the cloud cover, which is the detection of clouds (or no clouds), turned out to be already a relatively complicated task especially over sea ice and snow covered surfaces. To optimize accuracy, cloud analysis was done separately over ice or snow covered region and ice free ocean region. To implement this, sea ice covered areas in the oceans were identified using ice concentration data from passive microwave data. This strategy allows for the optimization of a technique that would be suitable primarily for the discrimination of clouds from sea ice or snow covered areas.

In open ocean regions and during daylight conditions, the visible channels are useful for cloud detection since the albedo of clouds are distinctly different from those of ocean surface. At night when albedo information is not available, discrimination is also possible since cloud free ocean surfaces are usually warmer than cloud covered areas and the thermal infrared channels can be used for cloud/surface classification. In addition, the use of spectral and bispectral techniques has been shown to work very well as, described by Minnis and Harrison (1984). There are however some problems associated with this technique because one of the key channels (i.e., 3.7 μm channel) is noisy especially at low temperatures. Efforts to minimize this problem has been done by several investigators with limited success (e.g., Simpson and Yhann 1994). An alternative is the use of a modified thresholding technique sometimes referred to as standard deviation technique. The scatter diagram shown in Fig. 4.7 of standard deviation of a group of pixels (5 by 5) vs. the mean temperature of these pixels shows a distribution of data points in which cloud covered areas can be discriminated. In this technique, used previously by Coakley and Bretherton (1982), the sea surface and an opaque (100%) cloud cover are assumed to have a much smaller standard deviation than cloud covered surfaces. Thus in the diagram, surfaces at the warmer end (between 272 and 278 K) with

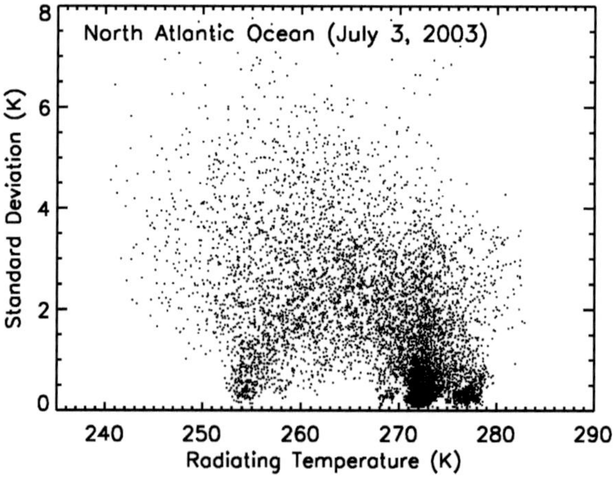

Fig. 4.7 PPlot of standard deviation of radiating temperatures at 11 μm μm versus the radiating temperature in the North Atlantic Ocean between 55°N and 60°N following Coakley and Bretherton (1982)). Cloud free areas are the dark clusters near zero standard deviations at around 272 and 277 K

near zero standard deviation correspond to cloud free surfaces, while those with near zero standard deviation at the colder end (253 K) correspond to 100% cloud cover. Those with higher standard deviations are considered mixtures of clouds and no clouds or different types of clouds. This technique appears powerful, but is not always effective in an automated procedure because the higher end could vary spatially, as indicated, and sometimes the desired threshold is not predictable.

Among the bi-spectral techniques is the use of the difference of the brightness temperatures at 3.7 μm (*T*(3.7 μm)) and at 11 μm (*T*(11 μm)) vs. that of *T*(11 μm). During daytime (sunlight) conditions, the use of brightness temperature difference channels appears to be specially effective. In particular, a scatter plot of the difference in the brightness temperatures of *T*(3.7 μm) – *T*(11 μm) vs. that of *T*(11 μm) for the same data set as in Fig. 4.7 is shown in Fig. 4.8a. For comparison, a scatter plot of differences of brightness temperatures at 11 and 12 μm vs. the brightness temperatures at 11 μm is shown in Fig. 4.8b. The concept behind this technique is that optically thick clouds are expected to have signatures similar to a black body and the difference would be close to zero and lower than that for cloud free surfaces. However, the range of values for the difference for cloud covered regions is large and the overlap with cloud free region is considerable. Also, Fig. 4.8b shows that the

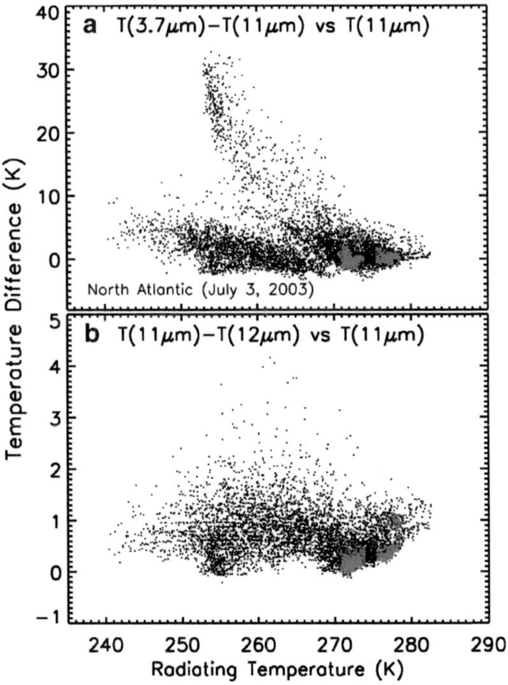

Fig. 4.8 Cloud masking using two sets of channel difference threshold technique: (**a**) *T*(3.7 μm) – *T*(11 μm) vs. *T*(11 μm) and (**b**) *T*(11 μm) – *T*(12 μm) versus *T*(11 tm) in North Atlantic using the same study area as in Fig. 4.7. The gray dots correspond to cloud free areas identified in Fig. 4.7

assumption is not necessarily true. Assuming that the standard deviation technique provides good cloud free data, the pixels with T(11 μm) values between 271 and 278 K and less than 0.5 K standard deviation in Fig. 4.7 have been identified as cloud free pixels and are shown in gray in Figs. 4.8a and b. It appears in Fig. 4.8a that a cloud mask similar to that in Fig. 4.7 can be used by using a simple upper limit threshold of 2 K and a lower limit of −2 K on the difference (ordinate) and a lower limit threshold of 271 K on T(11 μm) (abscissa). It is apparent that some cloud covered areas will be classified as cloud free using this procedure. In Fig. 4.8b, a lower threshold of −0.1 K and an upper threshold of 0.5 K would separate most of the cloud free areas identified in Fig. 4.7. Again, the technique would include some cloud covered areas and also exclude some cloud free areas. The required threshold could vary greatly depending on snow surface, solar zenith angle and viewing angle.

At high latitudes, the problems of cloud detection is quite serious even for ice free oceans and for both day and night conditions because the signature of clouds is even more variable and the thresholds are more unpredictable. There are also other issues like fog and cloud shadows that are not easy to identify. Furthermore, the 3.7 μm channel (channel 3) does not work well at the low temperature end, and therefore discrimination using this channel cannot be used. The accuracy requirements for SST is high and extreme steps have been utilized such as the "erosion technique" (Casey and Cornillion 1999) to minimize errors associated with the contamination of the cloud free data by cloud covered data elements that were not masked by standard procedures. The technique assumes that the extent of the clouds as detected is less than actual cover and therefore adds additional pixels around the edges of the retrieved cloud covered areas and consider them as cloud covered areas. This technique indeed minimizes the errors but may cause unnecessary deletion of useful surface data and in addition would cause a bias in the cloud statistics. Instead of the erosion technique for our final cloud filter for ice free oceans, we adapted the technique used by the AVHRR pathfinder project that requires that the derived sea surface temperatures agree with the Reynold's SST within ±2 K. The data points that do not meet this requirement are considered cloud covered data. A most useful procedure especially in ice covered areas is the use the differencing technique as discussed in the following paragraphs.

Over ice/snow covered areas, the techniques used for the open ocean usually do not work well because oftentimes, as indicated earlier, there is not enough contrast of the signatures of clouds and sea ice/snow covered areas. Also, clouds usually have a darker background than snow in the visible but not over bare ice, thin ice, pancakes, and meltponded ice (Stowe et al. 1989). Thresholding techniques using the 3.74 μm data have been used (Kidder and Wu 1984; Yamanouchi et al. 1987) while others use the difference (or ratio) of data from T(3.7 μm) and T(11 μm) (Maxson 1992; Massom and Comiso 1994) with limited success. Moreover, the T(3.7 μm) data has been noisy, especially at the low temperature end (Simpson and Yhann 1994). A filtering technique was proposed by Simpson and Yhann (1994) for the T3.7um data, but this has been tested mainly for SSTs retrieval and not for snow and ice areas. Other techniques for the detection of clouds includes the use of fuzzy logic segmentation (e.g., Simpson and Keller 1995; Key et al. 1989a, b) and neural network (Yhann and Simpson 1995; Lee et al. 1990). Texture techniques in combination with spectral methods have also been utilized (Welch et al. 1990; Key et al. 1989a, b). These techniques have been implemented but only with mixed success in part because of the lack of adequate training data sets and the limitations in the number of available channels for cloud masking in the AVHRR instrument.

In our study, we use all the thresholding techniques where they make a difference in cloud detection. However, we find that among the most effective is the cloud differencing technique as described in Comiso (2000, 2003) that takes advantage of the availability of a time series of data. We know that cloud cover is seldom stationary and because they move, the time difference of two observations provides information about their presence. The T(11 μm) data from an orbit for a particular day, n, are processed concurrently with data from an almost overlapping orbit the day before (i.e., $n-1$) and the day after (i.e., $n+1$). Two difference maps are then created, one with the day before and the other with the day after. Analysis of both maps provides the means to assess the cloud movement pattern. The difference of surfaces that are cloud covered both days are likely smaller than the difference of surfaces that are cloud covered one day but not the other day. This information is to be utilized as the basis of the detection (and masking) of clouds for the particular orbital data on day n. Subsequent orbits are processed in a similar fashion and the masked data are mapped to a common polar stereographic grid used for all the data sets discussed in this book. The technique is effective in identifying about 70% of the cloud covered areas. One problem is that although clouds are generally colder than the surface, this is not necessarily the case all the time. In particular, there could be atmospheric inversion which has been found to be a frequent occurrence in winter at high elevations in the Antarctic ice sheets (Phillpot and Zillman 1970). To minimize use two other techniques to supplement the differencing technique. One is to use a running standard deviation analysis on the daily maps and set a limit in the standard deviation of the temperatures of each pixel and neighboring grid. This assumes that partially cloud covered areas have higher spatial variability and hence higher standard deviation than cloud free areas as illustrated in Fig. 4.6. The standard deviation is calculated from a 3 by 3 grid with the pixel in consideration in the middle. If the standard deviation is greater than 1 K for ocean and 3 K for ice or snow, we consider the pixel as a cloud covered area.

Despite efforts to get rid of cloud covered areas in the data, some residuals remain. For optimum accuracy over sea ice and continental regions, a modified version of the technique imposed for SST by the AVHRR pathfinder algorithm was adapted. The intent is to remove values that are highly abnormal. In the absence of data similar to that of the Reynolds data set, we created a daily climatology map in which each pixel is the average of historical data from 1978 to 2007. Each pixel in the grid for each day of data is compared with the corresponding value in the daily climatology and if the data being processed have values that deviate by 10 K from climatology, they are considered cloud covered areas. This technique was validated by comparative analysis with visible channel data and by visual inspection and it appears that it is able to get rid of most of the extreme values in the data that are likely associated with clouds or radiometer noise.

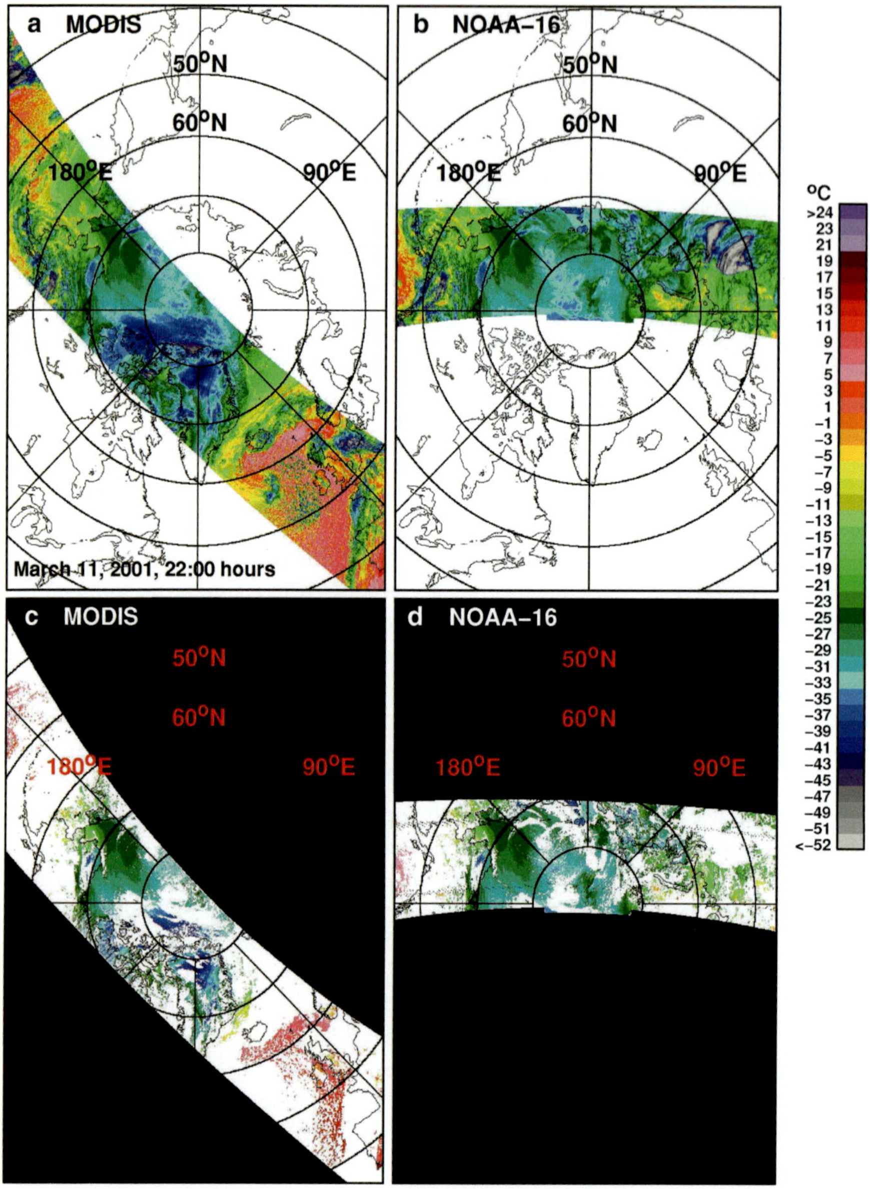

Fig. 4.9 Comparison of two orbits using Aqua/MODIS and NOAA-16/AVHRR data before (**a** and **b**) and after (**c** and **d**) cloud masking. Areas identified as clouds are shown in white. Common areas show similarity of cloud masked areas

To illustrate how effectively clouds are detected and masked, Fig. 4.9 shows orbital data for Aqua/MODIS and NOAA-16/AVHRR in the Arctic region both before and after cloud masking. The separate orbits which occurred at about the same time on 11

March 2001 show some common areas in Alaska and neighboring areas and it appears that the radiances are almost identical where there is coverage in both images. Similar cloud masks were applied on both sets of image and are presented in Figs. 4.9c and d for MODIS and AVHRR, respectively. It is also apparent that the same cloud cover is masked in the two different data sets. In Fig. 4.10, we show visible and infrared versions of the NOAA-11/AVHRR orbital images in the Antarctic (outline of the continent is shown for reference). The visible channel data in Figs. 4.10a and c provide the means to visually assess where the cloud cover is located. The corresponding masked 11 μm data presented in Figs. 4.10b and d, show that the areas identified as clouds (shown in white color) using the technique discussed above are consistent with what are recognizable as clouds in the visible images. Although the cloud masking used for

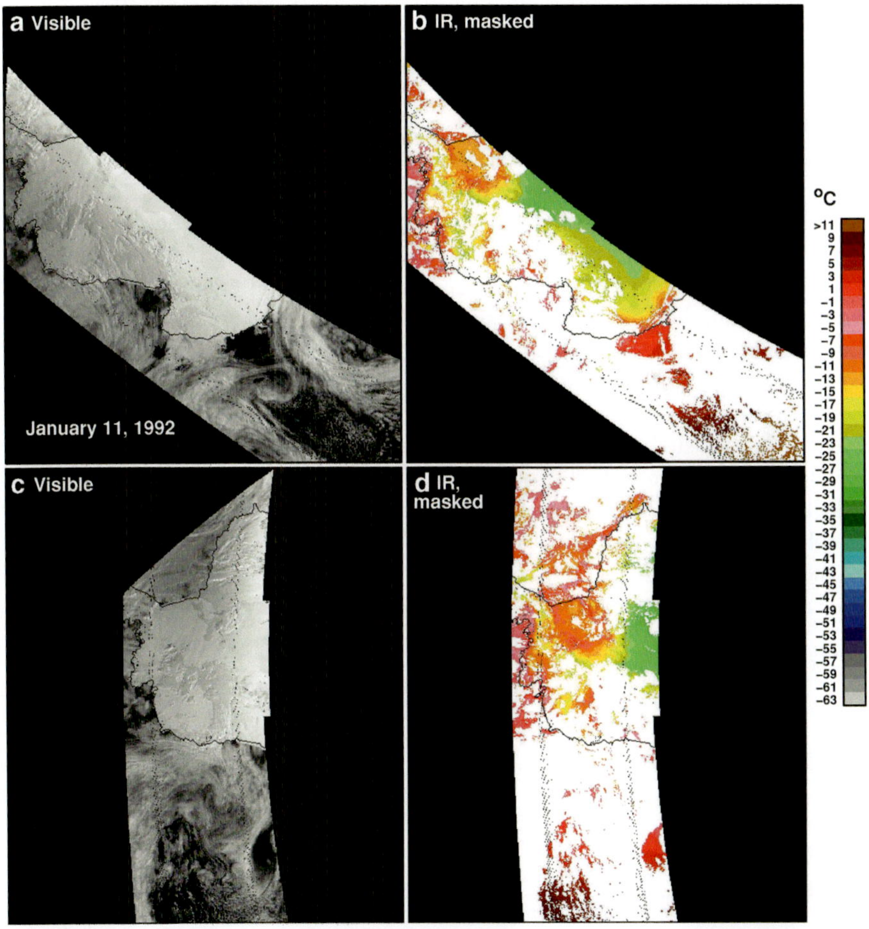

Fig. 4.10 (**a**) Visible and (**b**) cloud masked infrared (11.5 μm) data for an AVHRR orbits on January 20, 1992. A similar set but for another orbit on the same day are shown in (**c**) and (**d**)

AVHRR is not perfect and many issues remain, the images shown in Figs. 4.9 and 4.10 indicate that the technique generally works.

4.3.2 Albedo

The fraction of incident short wave radiation at a certain frequency that is reflected from a surface is usually referred to as the narrow-band albedo of the surface. A more general term is what is called total or broadband albedo, α, which is defined as the fraction of the downward solar irradiance incident on the surface that is reflected. In particular, it is the integral of the contribution from the entire visible spectrum which for some is from 400 to 1,500 nm (Perovich et al. 1998) while for others, it is from 300 to 3,000 nm (Hanesiak et al. 2001). There are currently many sensors that have been used to measure the albedo of the Earth's surface. They include Landsat, Spot, Terra/MODIS, Aqua/MODIS, MERIS, and AVHRR. We are mainly interested in the estimate of albedo from AVHRR data since the latter provide the longest record and the means to examine long term variability. Similar techniques can be used for data from the other sensors.

As the satellite orbits around the Earth, the AVHRR provides synoptic coverage of the surface (see Figs. 4.9 and 4.10) continuously at different latitudes and times of the year. The sensor also scans cross-track from nadir to a maximum angle of 55°. The recorded radiances represent measurements at different view angles, different solar illumination angles and different environmental conditions. Thus, even for the same surface, the values observed by the sensor could vary significantly even if the surface characteristics remain basically the same. Some corrections are therefore necessary to convert the data to a format that is useful for scientific research. In our study, only the middle scans up to a maximum angle of 47° are used to minimize the use of larger and distorted footprint and markedly different reflectivities at high scan angles, but the scans are extended as necessary to have complete coverage at the poles.

The basic technique that has been used in previous studies to convert the reflectance as measured by AVHRR to narrow-band albedo and then to broadband albedo

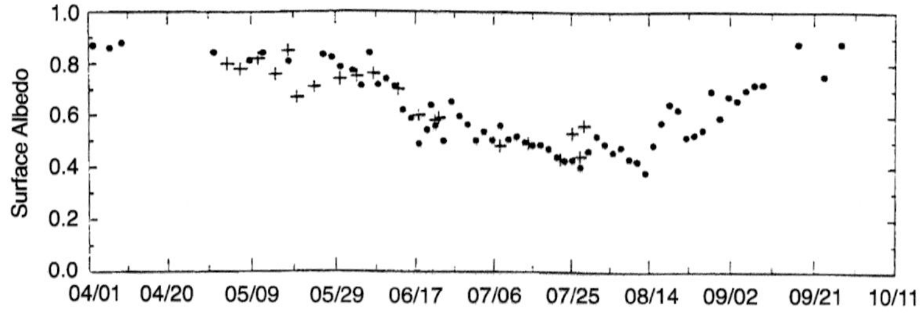

Fig. 4.11 Albedo measurements at the SHEBA ice stations (*circles*) and from a C130 aircraft in the Arctic from Curry et al. (2001) with permission of the Academic Press

(e.g., Lindsay and Rothrock 1994; Curry et al. 1995; Stroeve et al. 2001; Xiong et al. 2002) can be summarized as follows: (a) convert digital data to calibrated radiance data; (b) correct for solar zenith angle and calculate top of the atmosphere reflectance; (c) account for nonisotropic reflectance of ice and atmosphere; (d) correct for atmospheric interference; and (e) convert the narrow band albedos to broad band albedo and (f) normalize for temporal consistency of the data.

The calibration is done through the use of calibration parameters provided by Rao and Chen (1996) or by the NOAA/NASA AVHRR Pathfinder Calibration Working Group. Each data element is geolocated using ephemeris data and corrected for drifts in clock time. The calibrated radiance data is then normalized to obtain a corrected narrow band reflectance using the relationship:

$$R_i(\theta_s,\theta,\varphi) = \pi W_i L_i (S_i \cos\pi\, \theta_s) \tag{4.30}$$

where S_i is the solar spectral irradiance (in W m^{-2}), W_i is the equivalent width in μm as provided by Kidwell (1991) and θ_s is the solar zenith angle.

In the polar regions, there exists a strong bidirectional reflectance over ice and snow covered areas and an anisotropic correction is required to obtain the top of the atmosphere (TOA) narrowband albedo. This is done through the use of the bidirectional reflectance function (BDRF) of the surface, $F_i\,(\theta_s,\ \theta,\ \varphi)$, which is defined as the ratio of the reflected radiance (W m^{-2}sr^{-1}) in a particular direction (θ_r, φ_r) to the incident irradiance (W m^{-2}) from direction (θ_i, φ_i). The BDRF is related to the narrow band albedo by the equation

$$\alpha_i(0,\ \mathrm{TOF}) = F_i(\theta_s,\ \theta,\ \varphi)\ /f_i(\theta_s,\ \theta,\ \varphi) \tag{4.31}$$

where f_i is the narrowband anisotrophic reflectance factor. Direct information about f_i is not available, but broadband values of f have been compiled by Taylor and Stowe (1984) and further improved by Suttles et al. (1988) both using Nimbus-7 Earth Radiation Budget (ERB) data. Using the latter, the narrow band albedo at the top of the atmosphere can be derived using

$$\alpha_i(0,\ \mathrm{TOF}) = R_i(\theta_s,\ \theta,\ \varphi)\ /f_i(\theta_s,\ \theta,\ \varphi) \tag{4.32}$$

where i refers to the channel number.

Atmospheric effects can be taken into account using a linear relationship, inferred empirically by Koepke (1989), between the top of the atmosphere albedo, α_i, and surface albedo α_{si}. The relationship is of form

$$\alpha_i = a + b\alpha_{si} \tag{4.33}$$

where a and b are the regression parameters provided by Koepke (1989). Further normalization of the data is needed for temporal consistency of the satellite records since after the regular calibration, there is an apparent mismatch in the values during the switch from one AVHRR sensor to another as explained below.

The broadband albedo is derived from the AVHRR narrow band visible and near infrared albedos as follows

$$\alpha = C_1\alpha_{vis} + C_2\alpha_{nir} + C_3 \tag{4.34}$$

where C_1, C_2 and C_3 are coefficients that can be calculated. Estimates of these coefficients have been made (e.g., Xiong et al. 2002; Stroeve et al. 2001; Lindsay and Rothrock 1994) and used to obtain broadband albedo.

With only two channels available in the AVHRR data, the concern is how well the derived data represent actual broadband albedo. The main problem is that good validation data are rarely available to enable the validation of current estimates and make assessments about the accuracy of the retrievals. Airborne observations of albedo have been attempted, but are beset with challenges due to aircraft orientation. Also, previous airborne observations (e.g., Tschudi et al. 2001) have been limited to a few campaigns in different locations and show high variability in summer sea ice albedo. Recent developments in stabilized platforms may help address this problem, but are not yet available for UAV deployment. A number of investigators have performed surface-based measurements of wavelength-integrated and spectral albedos for a wide variety of ice types and conditions (e.g., Curry et al. 2001; Perovich et al. 2002a, b; Hanesiak et al. 2001). These measurements, an example of which is shown in Fig. 4.11, can be very useful for evaluating and interpreting satellite data that in turn can be used for pan Arctic or global studies.

The AVHRR sensors provide continuous record of 0.6 and 0.9 μm data from 1981 to the present that can be used concurrently with the thermal infrared data. As mentioned above, there are serious issues associated with the accuracy of the calibration and the stability in the performance of each sensor Apparent drift in the measurements associated with sensor degradation and changes in orbital parameters has also been observed. A key problem is the lack of good surface measurements that can be used to refine the calibration in order to make them temporally consistent during the lifetime of each sensor and to correct for mismatches in the calibration of the different sensors as was done with surface temperature. Such a correction was applied by Lindsay and Rothrock (1994) for the polar regions but only for a season and 1 year of data and was done after the calibration process. A continuous calibration target is needed for the period from 1981 to 2008 and what we did for our study is to use a fix target in Greenland as the high end calibration point and an adjacent ice free ocean area in the North Atlantic at about the same latitude as the low end calibration point. The target in Greenland is a 600 km by 1,300 km area in the central part of the ice sheet. We then created a monthly climatology based on a similar product from the better calibrated and more temporally consistent Aqua/MODIS narrow-band albedo data from 2002 to 2008. We assume that the monthly averages of albedo in the Greenland ice sheet and adjacent open water area are relatively stable and that the interannual variability in these regions can be neglected. Each monthly AVHRR data is then regressed vs. MODIS narrow-band albedo data using this Greenland study area and an open water area in the North Atlantic and used the parameters to normalize the AVHRR data for the pan-Arctic region. A relatively large study area was chosen so that normalization data for a whole year can be obtained. However, this provided albedo values which were relatively variable even in the ice sheet. We therefore used only the average of the highest 30% albedo values for each month from the Greenland study area in the regression analysis.

The normalization assumes that these high albedo values do not change much for the same month from 1981 to 2008 as was observed for the MODIS data from 2002 to 2008. To gain insight into the effectiveness of the procedure, plots of monthly averages of narrow-band albedo in the study area in Greenland are presented in Fig. 4.12 both before and after the normalization. It is apparent that the monthly distribution of albedo was greatly improved after the normalization. The monthly data in the plots do not show identical values at the top because in this case we used averages over the entire study area and not just those for the highest 30% albedos.

There are concerns that even with a vastly improved data set after the normalization, an absolute calibration may still be lacking. The available visible and near infrared AVHRR data do not exactly match the corresponding MODIS data in wavelength, bandwidth and other specifications. The albedo of Greenland as measured by AVHRR is also more variable (not shown) than that measured by MODIS. This may in part explain the interannual variability of the albedo in Greenland as depicted in Fig. 4.12b. Thus, even with the normalization of the AVHRR data, we don't have enough confidence that the results from estimates of broadband albedo will be very useful. In this book, we will present results from analysis of the normalized 0.6 μm channel data and will most of the time simply call it narrow-band albedo. Although the entire procedure outlined above was not followed, we assume that the normalization to MODIS narrow-band albedo will care of most of the linear corrections, including eq. 4.33.

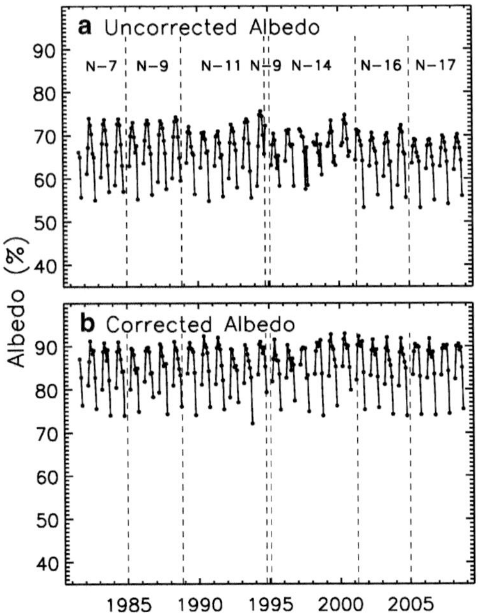

Fig. 4.12 Monthly averages of the 0.6 μm narrow-band albedo in Greenland from 1981 to 2008 (**a**) before and (**b**) after the normalization of the albedo of the Greenland ice sheet using climatological AVHRR/MODIS data in the region

4.3.3 *Wind Speed, Wave Heights and Dynamic Topography*

NCEP and ECMWF (European Centre for Medium-Range Weather Forcasting) reanalysis winds are currently the only data set available to study the direction and strength of winds in both land and ocean regions. Wind vectors over open ocean have been derived at good accuracy using scatterometers, but currently, there are no satellite instruments that can be used to derive winds over sea ice or land. Winds over land and consolidated ice are needed to study processes in the polar regions including coastal polynyas, divergence, convergence and the transport of dust or iron to the Arctic and Antarctic regions.

To illustrate how well NCEP reanalysis wind data reproduce satellite observations of winds, such as those provided by Quikscat data, Fig. 4.13 compares winds from these two sources in the Southern Ocean. It is apparent that wind directions are very similar and that the magnitudes are generally consistent except for a few exceptions. Both data sets capture the circumpolar winds around the continent in an almost identical manner. The maps are for the summer when sea ice has retreated to almost their minimum values. At the boundaries between sea ice and open ocean, it is apparent that NCEP actually provides more accurate values than QuikSCAT in part because of errors introduced because of the presence of ice or land in the field-of-view of the sensor. A good illustration for this is the wind distribution in the Weddell Sea near 45°W and 68°S. It is also apparent that the NCEP winds in the open ocean blends very well with winds in the Antarctic continent as would be expected. At the coastal regions, the NCEP winds are more likely to be more accurate and more useful for polynya and heat flux studies than QuikSCAT data. Color-coded sea level pressure data are superimposed in both images and it appears that the wind directions are consistent with those expected from the pressure data. Overall, the NCEP reanalysis data provide a good representation of the distribution of winds in the polar regions.

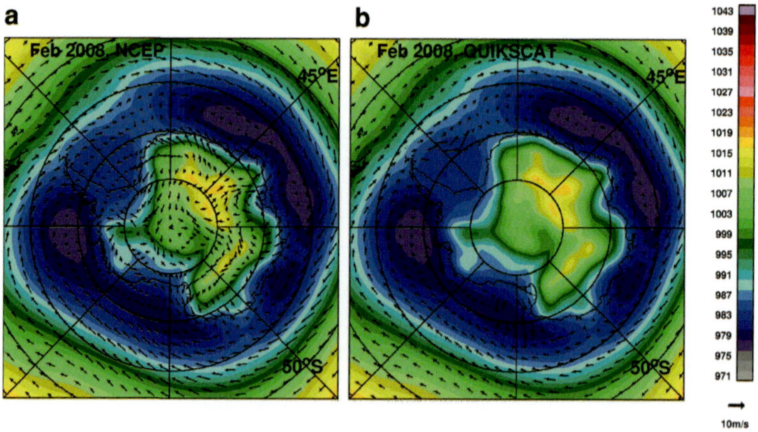

Fig. 4.13 Monthly average sea level pressure distribution in the Southern Hemisphere in February 2008 and a comparison of (**a**) NCEP and (**b**) QuikSCAT winds. Sea level pressures from NCEP are used in both images. Wind data is available from QuikSCAT only in ice free oceans

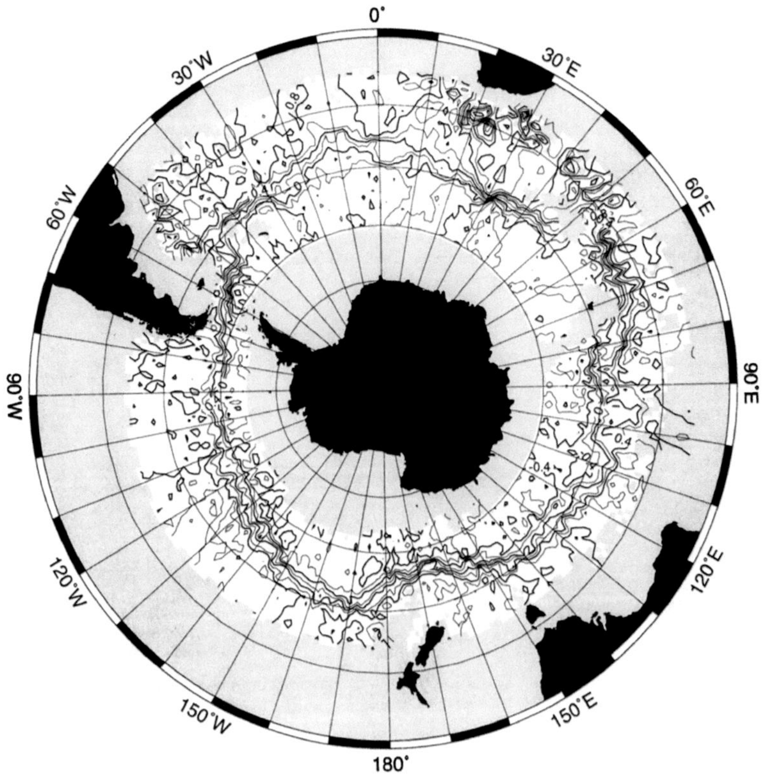

Fig. 4.14 Synoptic map of sea surface height in the Southern Ocean for December 28, 1986 as reconstructed from Geosat altimeter data. The contour interval is 0.1 m and the regions with no data or with height errors greater than 0.07 are in gray. Over most regions, the errors are between 0.05 and 0.07 m. From Gille and Kelly (1996) with permission of the American Geophysical Union

Comparative studies of QuikSCAT wind data, with NCEP, NCEP/NAM, and ECMWF reanalysis data have been done at extra polar latitudes and the agreement has been generally good as well (e.g., Chelton et al. 2006).

A question that has been of interest is how the Antarctic Circumpolar current respond on a global scale to large scale changes in wind forcing. Getting an answer using in situ measurements would be extremely difficult if not impossible. Instantaneous sea surface heights in the Southern Ocean as derived from the radar altimeter data using the technique described by Gille (1994) is shown in Fig. 4.14. Such data are useful in assessing topographical features that may be caused by eddies and tides. They have been utilized by Gille and Kelly (1996) to establish the scales of spatial and temporal variability of the surface that may be correlated with surface winds (Fig. 4.14). Empirical Orthogonal Function (EOF) analysis of the data suggested that the scales of motion are relatively short when compared with those of larger scale processes that may be associated with geostrophic wind. The observed mesoscale features may thus be determined primarily by local instability mechanisms.

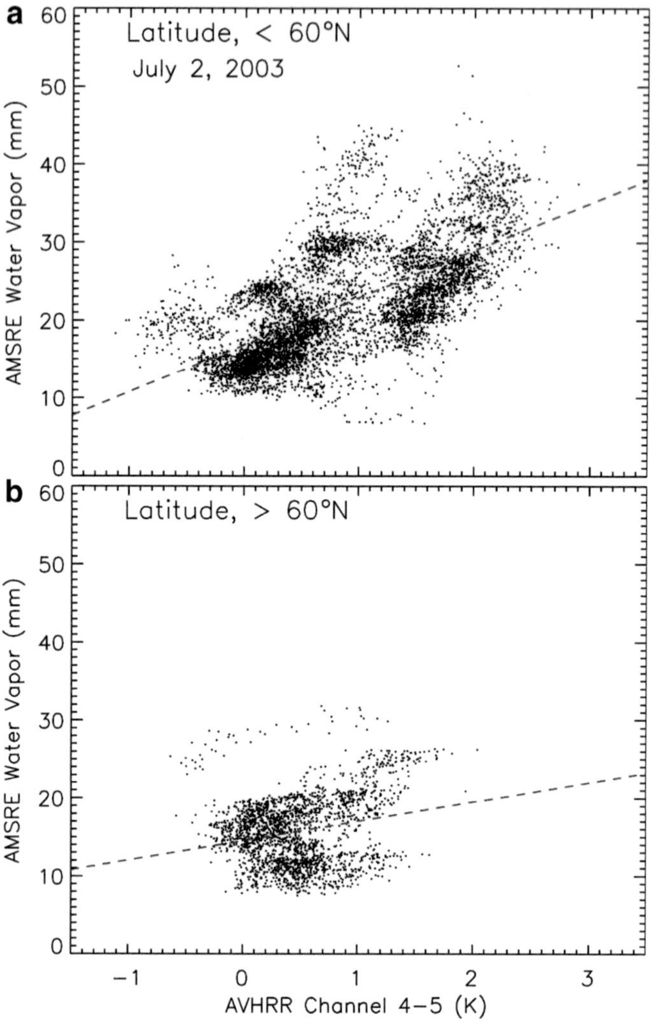

Fig. 4.15 Scatter plot of total water vapor content from SSM/I data vs. the difference in brightness temperature of Channel 4 and Channel 5 AVHRR data for (**a**) intermediate latitudes, < 60°N; and (**b**) high latitudes, >60°N. The plots indicate no relationship of water vapor with the difference of the AVHRR channels 4 and 5 and show consistency with Kumar et al (2003)

4.3.4 SST and SIT

Surface temperature is routinely observed in thousands of meteorological and other stations around the world on a daily basis. These data have been supplemented by ship and buoy data over the oceans. However, there are many areas worldwide in which in situ data are quite sparse, among which are the polar regions. Because of limited accessibility and harsh environmental conditions, not much can be done to improve in situ coverage and the only practical way to study the large scale varability of surface temperature in polar regions is through the use of satellite remote sensing. Although satellite sensors do not provide as good accuracy and temporal resolution, they provide synoptic information and detailed spatial distribution of surface temperature. Among the proven sensors for this purpose are the thermal infrared and microwave sensors as described in Chapter 3. Thermal infrared systems are quite useful for both open ocean and ice (or snow) covered surfaces because the emissivity of ice, snow and open ocean at this wavelength is relatively uniform (Comiso 2000). Passive microwave systems can be used for snow, ice and land studies, but only in the limited regions where the emissivity of the surface is well known (Shuman et al. 1995; Schneider and Steig 2002; Shuman and Comiso 2002). Because of the large spatial and temporal variability of the microwave emissivity of snow, ice and land surfaces, they have not been used to generate spatially detailed maps of surface temperature. However, in regions where the emissivity is known, the satellite microwave data can be used for temporally continuous coverage and would greatly supplement the infrared data. In the open ocean where the microwave emissivity is much more stable and can be estimated with good enough accuracy, passive microwave systems have been shown to have enormous potential for sea surface temperature studies, especially with the advent of AMSR-E data (Wentz et al. 2000; Shibata 2004). The infrared system was previously assessed to provide more accurate sea surface temperatures (SST) than passive microwave data (Bernstein 1982). However, with the day/night almost all weather coverage, the passive microwave data have some distinct advantages such as providing surface temperatures during cloudy conditions and therefore true daily and monthly averages over oceans and specific areas over land. Because of differences in penetration depths (microns for infrared and mms for passive microwave), and resolution (i.e., 1 km for infrared and 50 km for passive microwave) the synergistic use of the two is desirable for optimum temperature information.

The thermal infrared sensors are good for estimates of surface temperatures in the polar regions because the emissivities of ice, snow and water are relatively constant and quite high, approaching that of a blackbody. The infrared emissivity for ice and snow is on the average about 0.995 while that of liquid water is 0.988 (Rees 1993). Surface temperature can be inferred from the measured infrared radiance, R, by using eq. 4.27. Solving for T_S in this equation, the emissivity ends up being raised to the 1/4th power which makes the retrieval of surface temperature even less dependent on emissivity. In particular, the emissivity raised to the 1/4th is 0.999 for snow and ice and 0.997 for open water. However, the accuracy requirements for surface temperature is high (especially for SST) and even a slight fluctuation in emissivity could mean a significant error since the radiance in this equation is expressed in units of Kelvin. For example, a slight change in open water emissivity from 0.988 to 0.978 would cause a change in retrieved temperature from 280 to 279.3 K or a variation of 0.7 K which is significant since the accuracy requirements for SST which is sometimes set at less than 0.7 K.

The radiation emitted from the surface is modified by the intervening atmosphere, mainly because of water vapor that attenuates the radiation before it reaches the satellite sensor. Other atmospheric contributions, such as those due to aerosols, ozone and other gases, are usually not accounted for because of the lack of adequate data. The appropriate technique for making atmospheric correction would be to make use of a radiative transfer model to simulate atmospheric effects which is primarily that of water vapor. This has been implemented but with mixed success because of the spatial and temporal variability of the required input parameters and the need for absolute calibration of the data. A more efficient and more common procedure used for processing AVHRR data is to take advantage of the two thermal channels and use the split window technique in a regression analysis. This is based on the concept that the observed brightness temperature at 12 μm (Channel 5) includes more absorption due to atmospheric water vapor than the observed brightness temperature at 11 μm (Channel 4). The initial attempt to make this correction using the split window technique and was done with a simple linear equation given by the equation:

$$T_S = a + bT_4 + c(T_4 - T_5) \tag{4.35}$$

The constant parameters *a*, *b*, and *c* were derived through the use of radiosonde data to model and determine the effects of atmospheric water vapor (Barton 1995; Llewellyn-Jones et al. 1984). The other techniques include the regression of concurrent satellite data and in situ data (McMillan and Crosbey 1984) while a third one is to make use of the regression results to refine the model for the atmospheric water vapor (McClain et al. 1985; Kilpatrick et al. 2001). A second version of the algorithm takes into account the different path length of the radiation for different scan angle and includes another term as follows:

$$T_s = a + bT_4 + c(T_4 - T_5) + d(T_4 - T_5)(\sec\theta - 1) \tag{4.36}$$

where θ is the solar zenith angle. This algorithm provided accurate SST in the tropics, but not so accurate in the polar regions with a bias of about 2.4 K and a standard deviation of about 0.95 K (Minnett et al. 2001). A more robust SST algorithm was developed at the University of Miami under the Pathfinder program (Kilpatrick et al. 2001; Evans and Podesta 1996) making use of the nonlinear SST (NLSST) algorithm developed by Walton et al. (1998) which is of the form:

$$T_S = C_1 + C_2T_4 + C_3T_{sfc}(T - T_5) + C4(T_4 - T_5)(\sec\theta - 1) \tag{4.37}$$

where T_{sfc} is an estimate of SST. The difference between the pathfinder method and the NLSST is that the former accounts for the difference between dry and moist atmosphere as detected by the difference between the two channels. Thus, if $T_4 - T_5 < 0.7$ K which is assumed for dry surface, a different set of coefficient is used than when $T_4 - T_5 > 0.7$ K, which is regarded as a condition for a moist atmosphere. These adjustments, however, improved the accuracy of SST retrievals up to 60°N. It turned out that while $T_4 - T_5$ show linear dependence and good correlation with water vapor in the tropics and mid-latitudes, the correlation is poor at high latitude regions (Kumar et al. 2003). The phenomenon is illustrated in Fig. 4.15 which shows that at latitudes <60°N, there are some relationships with water vapor, where in this case water vapor is derived from Aqua/AMSR-E, while at >60°N, the relationship is basically very weak. Kumar et al. (2003) studied the effect for different seasons and for autumn, winter and spring, the scatter plots show no relationships between the two variables at high latitudes. This phenomenon was also studied by Vincent et al. (2008) at the North Open Water (NOW) near Greenland and in this study they also observed high values of $T_4 - T_5$ which were attributed to the presence of ice fog near the surface. Noting that the window-difference technique is unreliable in the polar regions, Vincent et al. (2008) adapted a regression technique with in situ data to obtain surface parameters and the results were more effective than those derived using a radiative transfer model (e.g., Key et al. 1997). In this study, we adapted the Pathfinder Program algorithm (i.e., 4.37) for SST retrievals with the Reynold's data at latitudes between 50° and 60° used as in situ data and in part to account for mismatches in the calibration of the different AVHRR sensors.

Results from previous analysis of sea surface temperature (SST) and Surface Ice Temperature (SIT) as derived from the AVHRR thermal infrared data in the polar regions have been reported (Comiso 2000; Comiso 2003) using a technique similar to that used by Vincent et al. (2008). It should be pointed out that an independent technique is needed to discriminate ice free water from ice covered water since data from the AVHRR sensor are sometimes not able to provide the location of the ice edge. In our case, ice concentrations from passive microwave data (see Chap. 7) are used to masked out areas where there is sea ice cover. Comparative analysis of surface temperatures derived from AVHRR data with coregistered and coincident in situ measurements confirms that the thermal infrared data capture the same variability in surface temperature as the ground values (Comiso 2000; Comiso 2003). The results of some of the comparisons of in situ and AVHRR surface temperatures in both Northern and Southern Hemispheres are presented in Figs. 4.16. Errors associated with the retrieval of surface temperatures have been evaluated and estimated to range from 2 to 3 K (Steffen et al. 1993; Key and Haefliger 1993; Comiso 2003; Shuman and Comiso 2002). The temperature data set that we use is an enhanced version of that used in the Comiso (2000) and Comiso (2003) studies with cloud masking that has been improved and the time series and updated to 2008. In the Northern Hemisphere, if the station data are compared separately with AVHRR data, the results shown in Fig. 4.16a show good consistency with standard deviation of about 2.6 K. Not all the stations provide accurate results especially since some of the thermistors can be buried by snow at some periods when unattended. To minimize this effect, we use the average of temperatures at all stations for each month and compare them with the corresponding average using AVHRR data, the results are actually much better as shown in Fig. 4.16b with the standard deviation being about 0.7 K. The trend line for the monthly averages is also almost along the diagonal, showing very slight bias. Similar comparison regarding using data in the Southern Hemisphere is presented in Fig. 4.16c. When monthly values from individual stations are compared, the standard deviation is about 2.5 while with the monthly averages compared, this improves to 0.6 K.

It is important to note that the thermal infrared sensors measure the temperature of the top layer (usually referred to as "skin depth") of the water, snow or ice surface. The actual temperature of the skin surface has been reported by Minnett et al. (2001) to be significantly different from the bucket temperatures used in ship measurements. This was done through direct comparison of temperatures from a thermal infrared radiometer from on board a ship and the bucket temperature of the sea

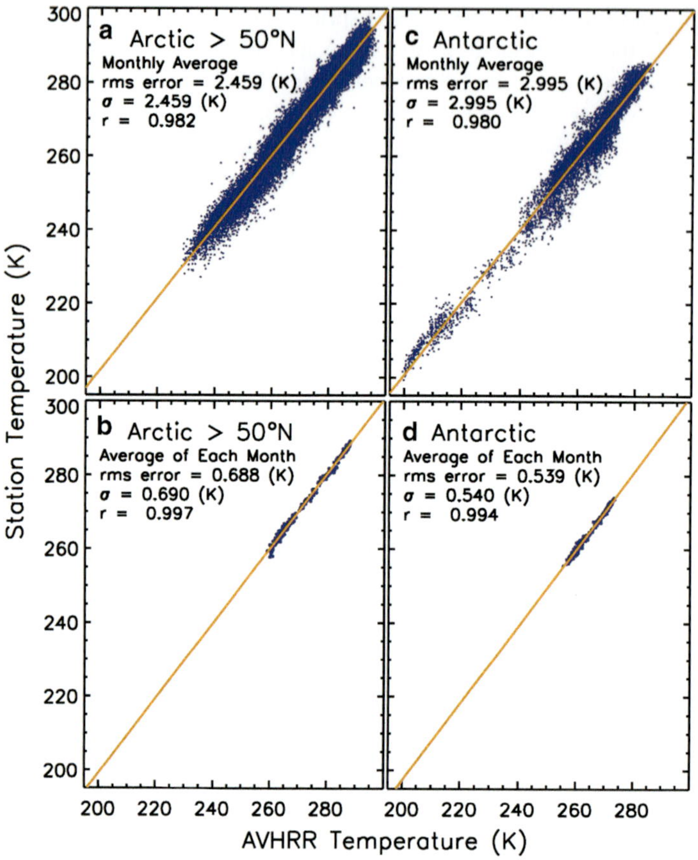

Fig. 4.16 Comparison of in situ surface ice temperatures as provided by Jones et al. (1999) with those derived from AVHRR data for data located >50oN in the Arctic using (**a**) monthly data plotted separately for each station; (**b**) monthly data plotted as averages from all stations; and in the Antarctic continent using (**c**) monthly data plotted separately for each station; and (**d**) monthly data plotted as averages from all stations

surface taken at the same time. More in depth validation studies with a view of improving the retrieval of SST was reported by Donlon et al. (2002). The sea surface temperatures that we derived had been normalized to be generally consistent with the in situ data set put together by Reynolds et al. (2002) and which we will simply call Reynolds data. This was done following the studies of Vincent et al. (2008) and Kumar et al. (2003) as indicated earlier. A comparison of sea surface temperature as derived using AVHRR, MODIS, AMSR-E and Reynold's data are presented in Fig. 4.17. The figures show general consistency of the different data sets. The AVHRR and MODIS data have similar resolution and capture the same mesoscale distributions although there are slight differences likely on account of slightly different cloud masks. Although the resolution is not as good as those of the AVHRR and MODIS, the spatial distribution from the AMSR-E data shows

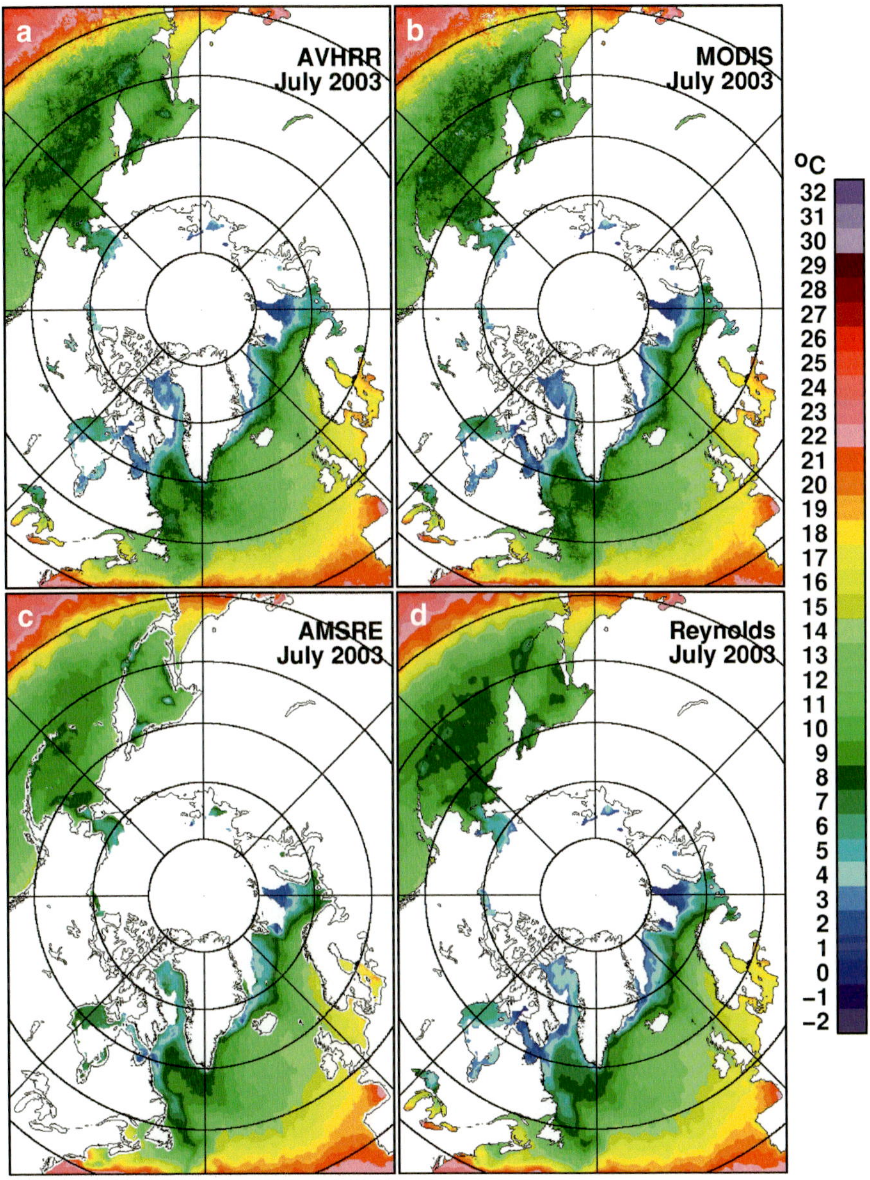

Fig. 4.17 Comparison of surface temperature maps as derived from (**a**) AVHRR; (**b**) MODIS; (**c**) AMSR-E, and (**d**) Reynolds SST data as described by Reynolds et al. (2002)

good consistency with the other two. This is encouraging since there are no gaps in the AMSR-E data associated with clouds and the data are expected to provide the true monthly average. The AMSR-E surface temperatures are shown to be slightly higher than those of the other satellite images. A possible reason for this is that the cloud cover is generally known to cause a warming effect and that AMSR-E measures the temperature from a thicker (and usually warmer) layer of water than the infrared sensors. However, it is not known at this time if these factors can completely account for the difference. The Reynold's data which are based mainly on buoy and ship data (Fig. 4.17d) show better agreement in magnitude with the AVHRR and MODIS data but do not show the same spatial detail.

4.3.5 Plankton Concentration, PAR and Primary Productivity

The retrieval of chlorophyll *a* (CHl *a*) concentration, which provide information about the distribution of phytoplanktons, provides the spatial distribution of the is usually done using either the empirical or the analytical technique. The empirical technique has been the one adapted for global retrievals primarily because of overall simplicity and ease of implementation. The use of a global algorithm that accounts for regional changes in the optical properties of water and atmospheric effects would be quite complex especially because only about ten percent of the radiance that is observed by the satellite is associated with the Chl *a* concentration. A schematic of the various sources of radiances received by the satellite sensor is shown in Fig. 4.18. The radiance of interest is the water leaving radiance (solid line) associated with scattering from the planktons. The terms that contaminate the signals are those scattered from air molecules (dotted line) and aerosol molecules (dash-dot line) in the atmosphere, waves, foam and sun glint (long dash line) at the surface, and water molecules (short dash line) underneath the surface. As discussed in Chap. 2, there are other complications such as the existence of two optically different types of water, namely, Case 1 and Case 2 water.

Case 2 water is found mainly in coastal regions and shallow shelf areas while Case 1 water is overwhelmingly dominant and found in open seas and the vast oceans. Nevertheless, Case 2 water is important to study because of their proximity to human habitat and their potential impact on the ecology of the region. The boundaries of Case 1 and Case 2 waters are not well defined and mixtures of these two types of water occur in large areas between the coast line and the deep water. One of the current challenges in ocean color remote sensing is to be able to correctly identify where both are prevalent and be able to quantify the relative concentration of each type in these areas.

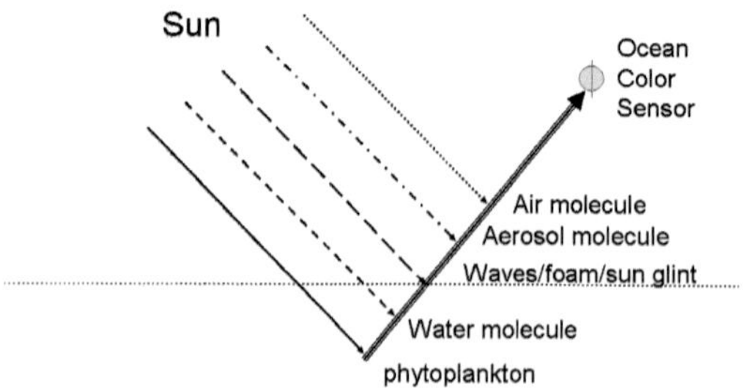

Fig. 4.18 Schematics of the various components of scattered radiation from the sun measured by an ocean color sensorr

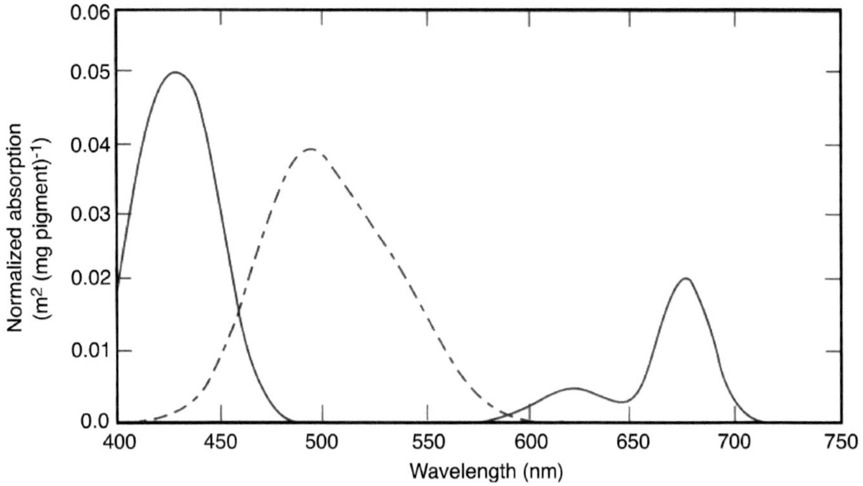

Fig. 4.19 Normalized absorption for Chlorophyll *a* (*solid line*) and carotenoids (*dashed line*). (Adapted from Hoepffner and Sathyendranath 1993 by Martin 2004 with permission from Cambridge University Press)

The algorithms that have been used for SeaWiFS, MODIS and MERIS data are primarily meant for Case 1 water only and are usually applied globally including coastal regions. Algorithms for Case 2 water are currently under consideration but have not matured to the point where they could be used in the processing of standard satellite data products.

To illustrate how the signature of Chl *a* is represented in the ocean color data, a plot of normalized absorption coefficient for Chl *a* vs. wavelength is presented in Fig. 4.19. The Chl *a* curve (in solid line) shows two major absorption peaks, a blue maximum near 440 nm, and a red maximum centered at about 683 nm. At wavelengths between 550 and 650 nm, the absorption approaches zero, giving chlorophyll-rich water its characteristics green color. The peak to the farthest right is also the fluorescence peak at 683 nm, which is a region not covered by SeaWiFS but is covered by both MODIS and MERIS sensors. The dashed line is the corresponding absorption coefficient plot for carotenoids which are organic pigments occurring in the chloroplast and chromoplast. The plot for carotenoids includes contributions from both the photosynthetic and photoprotective carotenoids and shows a peak near 500 nm and a bandwidth that covers 450–550 nm.

The radiances observed for different concentrations of Chl *a* are shown in Fig. 4.20. The small horizontal boxes (with numbers) at the top represent the location of SeaWiFS channels while those at the bottom represent the location of the MODIS channels. It is apparent that the radiances show large variability at 443 and at 650 nm, while the variability is small at 555 nm. The empirical SeaWiFS algorithm is based on regression of the ratio of radiances with in situ measurements. Example of plots of such ratios for wavelengths 443/555, 490/555 and 510/555 vs. in situ Chl a data are presented in Figs. 4.21a, b, and c, respectively, as adapted from O'Reilly et al. (1998).

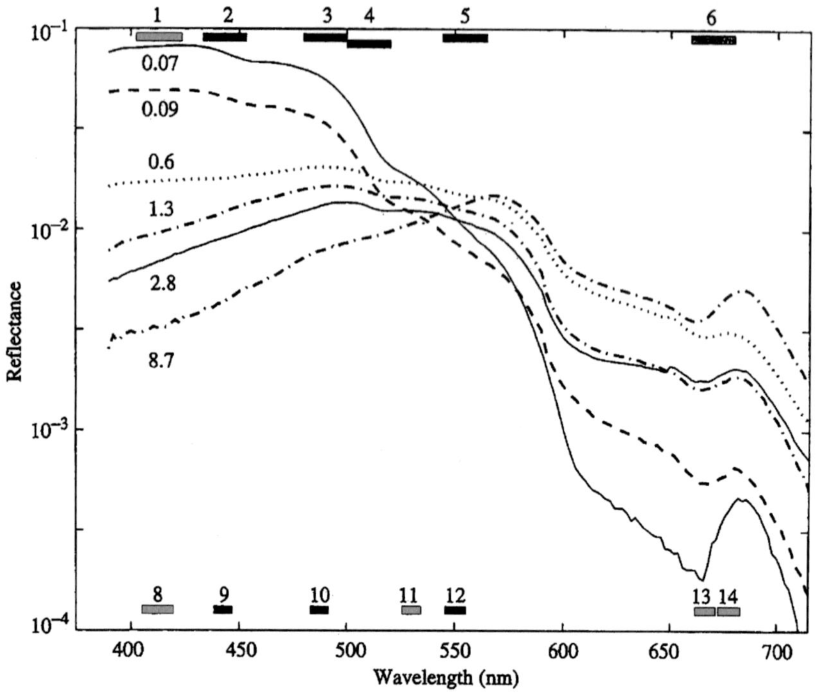

Fig. 4.20 Dependence of subsurface reflectance with wavelength for various values of chlorophyll concentrations in units of mg m^{-3}. The horizontal bars at the top represent the bands used by MODIS with those used in the algorithm shown in black. The horizontal bars at the bottom are those used by SeaWiFS with those used for the algorithm in black. (From Martin 2004 with permission from Cambridge University Press)

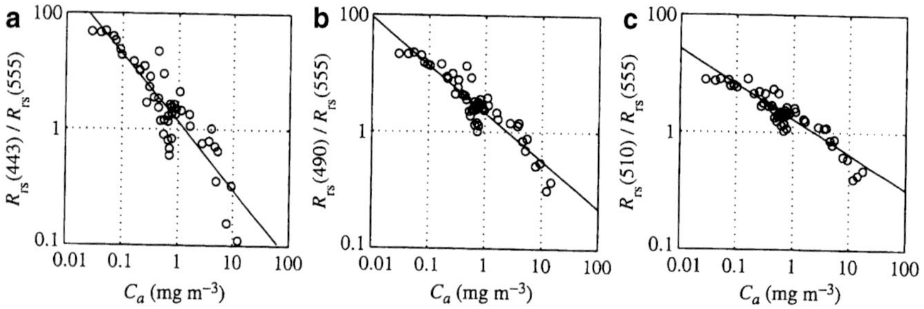

Fig. 4.21 Regression plots using in situ data from Moby data. SeaWiFS algorithm uses 3 ratios while MODIS algorithm uses 2 ratios. (adapted from Aiken et al. 1995 by Martin 2004 with permission from Cambridge University Press)

The SeaWiFS algorithm makes use of all three ratios, while the MODIS algorithm makes use of only the first two and the CZCS algorithm used the first and the third. The straight lines are linear least squares fit to the data points and the parameters of the fit are used to convert the ratios as measured by the satellite sensor to chlorophyll values. The use of 3 ratios is meant to optimize the accuracy in the retrieval of pigment concentration over the entire range. Using different ratios for different intervals of chl *a* values as was done previously caused discontinuities (or abrupt shifts) from one interval to another. A new scheme adapted for the OC4 version of the global algorithm makes use of the so called maximum band ratio (MBR). In this scheme, the ratio that provides the highest value in each data element is the one selected for the retrieval. In this scheme, O'Reilly et al. (1998) found out that the 443/555 was maximal predominantly from lowest chlorophyll concentration, C, to 0.3 $\mu g\ l^{-1}$; 490/44 was maximal predominantly between 0.3 and 1.5 $\mu g\ l^{-1}$; and 510/555 was maximal predominantly when C exceeded 1.5 $\mu g\ l^{-1}$. In this scheme, the problem of discontinuity is minimized because there is an overlap of 10–30% at the boundary between ratios for the selected ratios. This allows for much smoother transition from one interval to another.

The main source of ocean color data that was used in this study is the recently reprocessed version 5 of SeaWiFS and called OC4v5 (Platt et al. 2003). The SeaWiFS data have been more carefully calibrated and widely validated than any other ocean color satellite (Hooker and McClain 2000). The data have been used in many studies, and have been shown to be spatially and temporally coherent (McClain et al. 1993, Gregg and Conkright 2002). Comparative studies of chlorophyll concentrations derived from SeaWiFS with those measured in situ in the Arctic have indicated that the in situ data diverge significantly from global retrievals (Cota et al. 2004; Wang et al. 2005). Retrievals from SeaWiFS global OC4v5 algorithm (O'Reilly et al. 1998) overestimate low chlorophyll concentrations (~0.7 mg m^{-3}), but underestimate higher concentrations almost twofold (Cota et al. 2004) using data in the Bering, Chukchi, Beaufort and the Canadian Archipelago. These results are consistent with the previous reports indicating that the optical properties of water in the polar regions are different from those at lower latitudes (Mitchell 1991; Sullivan et al. 1993). It is however not clear that the algorithm used in Cota et al. (2004) is a pan-Arctic algorithm. The SeaWiFS global algorithm has been shown to overestimate the plankton concentration in the Kara and Barents Seas when compared with in situ data in the region (L. Ametistova, private communication, 2003). This becomes a predicament since the use of the Cota et al. (2004) algorithm for the entire Arctic region, shows apparent improvements in the Bering, Chukchi and Beaufort Seas, but makes it worse at the Kara and Barents Seas. Given that the Cota et al. (2004) algorithm is highly nonlinear and many not be suitable for the pan Arctic region and the Southern Ocean, we find it more appropriate to use data provided by the global algorithm for the polar oceans in both hemispheres. The use of regional data was avoided because the assembly of such data to create a global product has many problems including getting the values consistent at the boundaries of the different regions.

Since ocean color data are available only during clear sky conditions, there are likely some biases associated with the data and especially in the monthly averages. In the Arctic, the cloud cover can be very persistent, and can be difficult to discriminate from the sea ice cover. Another problem associated with the process of cloud masking is the other extreme in which more clouds were masked than is necessary. Some areas in the Arctic were identified during the processing of the SeaWiFS data as either sea ice or cloud covered area. Our study shows evidence that cloud and sea ice masking in ice free areas in the Arctic during the summer is sometimes overdone, and may cause the loss of valuable data and possible bias in the statistics. Unfortunately for this case, the only way to correct this problem is to reprocess the entire SeaWiFS data set with an improved cloud and sea ice mask. Such a reprocessing is a big endeavor, but will be considered during future reprocessing of the data. Also, along the same lines, averaging for ocean color data is done only for pixels with good data during the averaging period. Thus, the average in some pixels may be from as little as one or two measurements, while the average from other pixels may be from as many as a hundred measurements. Such an averaging procedure is employed for lack of a better technique and would certainly cause some biases with respect to the true average unless the value did not change much during the period. Criticisms applicable to cloud masking and averaging are endemic to most ocean color (and also albedo and infrared temperature) investigations. Knowing that there are all these sources of errors to consider, it should be noted that the data are shown to be spatially and temporally coherent and are likely the best set of global data currently available for pigment concentration and productivity studies.

With the oceans covering more than 70% of the surface of the Earth, the role of the planktons in the global carbon cycle can be significant. During the photosynthesis, CO_2 is removed from surface waters and gets transformed into organic carbon and initially bound within the phytoplankton. A significant fraction of this organic carbon eventually sinks through the thermocline to the deep ocean. Such process has been called "biological pump" and has been regarded as the reason why the concentration of total CO_2 is generally higher in deep waters than in surface water. The rate of assimilation of carbon by plankton has been called primary productivity, which can be estimated using chlorophyll concentrations in conjunction with temperature, Photosynthetically Active Radiation (PAR) and a vertically generalized primary production model. Several promising techniques for estimating primary production have been developed (e.g., El-Sayed et al. 1983; Platt and Sathyendranath, 1988; Carder et al. 1991; Morel et al. 1996). We adapt the model of Behrenfeld and Falkowski (1997), in which primary productivity (PP_{eu}, in units of mg C m^{-2} d^{-1}) is calculated using the following equation:

$$PP_{eu} = 0.66125 \times P^{B}_{opt} \frac{E_o}{E_o + 4.1} C_{Sat} \times Z_{eu} \times D_{Irr} \tag{4.38}$$

where P^{B}_{opt} is the optimal rate of photosynthesis within the water column (mg C (mg chl)$^{-1}$ h^{-1}) and is regulated by temperature, E_o is the surface daily PAR (mol photons m^{-2} d^{-1}), C_{Sat} is the surface chlorophyll concentration (mg chl m^{-3}) determined by satellite, Z_{eu} is the depth of the euphotic zone in meters, and D_{Irr} is the photoperiod (h). P^{B}_{opt} was estimated from sea surface temperatures by the polynomial equation of Behrenfeld and Falkowski (1997), and all values at temperatures less than −1.0°C were set to 1.13. Productivity was calculated on a daily basis, and binned in a manner similar to that of chlorophyll. The gridding technique (the so called "drop in a bucket" procedure) and the presence of clouds caused a large fraction of data elements (pixels) in the daily maps to have missing data. In the case where a single empty or voided pixel is surrounded by pixels with data, a simple interpolation technique is utilized to fill the empty pixel. For slightly larger data gaps of a few pixels, a combination of spatial and temporal interpolation was utilized. Such interpolation filled only a very small fraction of missing data in the daily map, and for time-series studies weekly averages were produced as the basic product.

Irradiance has two components: photoperiod and absolute irradiance impinging on the sea surface. Photoperiod can be relatively easily modeled, as it is solely a function of latitude. In the polar regions, photoperiod varies from 0 to 24 h within 1 year, and from 16.3 to 24 h at the seasonal maximum. Surface PAR can be either measured via satellite as discussed by Arrigo et al. (1998) or it can be computed using a clear-sky model (e.g., Gregg and Carter 1990). Modeled PAR can be generated at any resolution and any location, while measured PAR is biased by cloud cover and is limited to the resolution of the sensor. Both modeled and measured estimates of PAR were tested in the productivity model, and surprisingly it was found that there was little difference between the two. As a result, we used the modeled PAR for this study especially since the surface PAR data derived from the SeaWiFS have limited coverage in the polar regions because the sea ice mask that was used extended much farther into the open ocean than the actual ice cover. Modeled PAR ranged from 0 to 70 mol photons $m^{-2} d^{-1}$ within the year, and maxima ranged from 63 to 68 mol photons $m^{-2} d^{-1}$ among the selected study regions. It was also approximately the same as the scalar irradiance calculated by others (e.g., Mitchell and Holm-Hansen 1991).

4.3.6 Sea Ice Concentration

Sea ice concentration is usually defined as the fraction of ice covered area within an observational field, which in our case is the footprint of the satellite sensor. The need to quantify ice concentration came about after the discovery that in ice covered areas beyond the ice edge, there are open water areas of varying coverage. Most of early field and aircraft projects in sea ice covered regions were done mainly in spring and summer when these regions were easily accessible for observational studies. The term ice concentration has been associated with observations during these seasons when a clearcut/binary distinction of solid and liquid surfaces can be easily made. It is now generally the term used as the means to quantify the large scale characteristics of the sea ice cover.

The sea ice concentration algorithm that we will discuss is that used that for passive microwave data which have so far provided the most comprehensive information about the large scale distribution, variability and trends in the sea ice cover.

Similar techniques can be used for other types of data especially those where the resolution if not good enough to detect the individual features in the ice cover. With the binary characterization of the surface in mind, ice concentration, C_I, can be quantified using the brightness temperature, T_B, of a surface as observed by satellite data, by using a mixing algorithm given by

$$T_B = T_O C_O + T_I C_I \tag{4.39}$$

where T_O and T_I are the brightness temperatures of 100% ice-free and 100% sea ice covered oceans, respectively, and C_O is the concentration of open water (i.e., $C_O = 1 - C_I$). The equation would apply for any type of observation regardless of frequency as long as the observation allows for a clearcut discrimination of sea ice and open water areas.

In sea ice covered areas, the radiation observed within the field-of-view of the passive microwave sensor comes from ice, open water, or a combination of both. We can solve for C_I in (4.2) and get the expression

$$C_I = \frac{T_B - T_O}{T_I - T_O} \tag{4.40}$$

which is the fundamental equation for ice concentration algorithms, assuming a binary observational surface of either sea ice or water. The challenge has been how to obtain accurate estimates of T_O and T_I which are both functions of emissivity (ε), surface temperature (T_S), and atmospheric opacities (τ, and τ'), as indicated in (4.1).

An ideal algorithm for retrieving the ice concentration would be one that calculates accurately the second and third terms in the right side of (4.1) and the opacity τ of the atmosphere for each satellite measurement of the brightness temperature, T_B, of the surface. The actual brightness temperature of the surface can be inferred from the first term in the right side of (4.1) which is then used for estimating sea ice concentrations, using (4.40). To be able to make this estimate effectively, one would need atmospheric and surface emissivity data, as indicated in (4.1), including the atmospheric opacity parameter τ, for every satellite measurement. Satellite Aqua/AIRS data, as described in Chap. 3, could be very useful in this regard even if the fraction of useful data from this sensor is limited by the presence of persistent clouds in the polar regions. Earlier algorithms used radiosonde data at some Arctic regions and assumed constant atmospheric and surface conditions and the results provided good results regionally and for the period atmospheric data were available (Svendsen et al. 1983; Swift et al. 1985) but generally, there were problems when they were used as global algorithms.

The algorithms that are currently being used are those that take advantage of the multichannel capability for obtaining the required input parameters in (4.40). Two of these algorithms are the Bootstrap Algorithm and the Nimbus-7 Team Algorithm (now called the NASA team algorithm), both developed at the NASA Goddard Space Flight Center (Cavalieri et al. 1984; Comiso 1986). The Bootstrap Algorithm took advantage of the unique distribution of brightness temperature (or emissivity) data points in two or three dimensional space to assess and evaluate the general distribution of consolidated ice and use the pattern to obtain the unknown values of

T_I and T_O in (4.40). The original Team Algorithm used three tie points, namely, two for two ice types and one for open water, and employs polarization and gradient ratios to minimize the effect of varying surface temperatures. The different techniques for accounting for spatial changes in ice temperature and emissivity and the use of different sets of channels yielded different results from the two algorithms in large areas of the polar regions (Comiso et al. 1997; Comiso and Steffen 2001). The different sensitivity of the different frequency and polarization channels to surface effects, as discussed in Matzler et al. (1984), has been regarded as the primary reason for the discrepancies in the retrievals. The original NASA Team Algorithm (now called NT1) has been substantially revised to make use of the 89 GHz in combination with other channels and is currently called NT2 (Markus and Cavalieri 2000). With NT2, a radiative transfer model has to be used to correct for the high sensitivity of the 89 GHz data to atmospheric effects. It also requires using the proper surface emissivity of sea ice which is spatially very variable at this frequency. The ice concentration data utilized to obtain results presented in this book were generated using the Bootstrap Algorithm. To acquaint the reader on how the geophysical ice parameter is retrieved from satellite data, we will present the basic principles and attributes of the algorithm. A complete description of the basic algorithm is provided in Comiso (1995) while some enhancements for improved performance are described in Comiso (2009).

The unique clustering of data from consolidated sea ice regions in a multichannel brightness temperature space was first recognized when SMMR data became available (Comiso et al. 1984; Comiso and Sullivan 1986; Comiso 1986) and was used as the basis for the Bootstrap Algorithm. When the brightness temperatures (or emissivities) at one passive microwave channel are plotted vs. those of another channel in ice covered regions, most of the data points in the consolidated ice region where

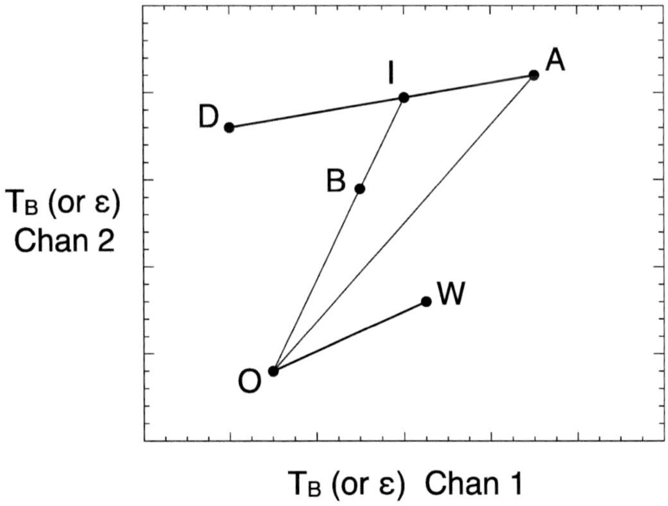

Fig. 4.22 Schematic Diagram of the technique used by the Bootstrap Algorithm the line AD represents consolidated ice while OW represents open water (or 0% ice concentration. The ice concentration corresponding to data at point at B is the ratio of the segments OB and OI

the concentration is about 95% or more are clustered along a line, AD, as schematically illustrated in Fig. 4.22. The line AD is thus assumed to represent T_{I} in (4.21). Also, most of the data points in the open water and also in ice free regions within the pack are clustered along OW. The data points that represent ice free surfaces within the ice pack, or T_{O}, are the ones corresponding to the lowest brightness temperatures (or emissivities) along OW and located near the label O since these are the ones that correspond to relatively calm surfaces. Waves and atmospheric disturbances over open ocean regions tend to change the values closer to W along the line OW. The spatial variation in ice emissivity is reflected by the range of the AD cluster along each coordinate which usually gets wider as the frequency gets higher. This is an indication that the key reason for the variability is the difference in wavelength (or frequency) since radiation at shorter wavelengths (or higher frequency) tends to scatter more than those at longer wavelengths (Eppler et al. 1992). The scatter of the data points along the AD cluster has been attributed primarily to the spatial variations in surface ice temperature and in part to atmospheric effects and to the presence of different ice types. The width of this scatter provides a measure of the accuracy in the determination of T_{I} and hence in the retrieval. The accuracy can thus be optimized through the choice of the appropriate sets of channels as discussed below.

Data points located between the line AD, which represents 100% ice and the tie point at O, which represents 100% open water, have ice concentrations between 0 and 100%. For any type of ice surface represented by a data point I, along the line AD, different concentrations of this ice type would be represented by data points along the line OI. Thus, for any data element with a set of values located in point B in the plot, the ice concentration can be derived by extending the line along OB, and find the intersection point with the line AD. The location of I is determined by finding the common points of two equations, one for the line along AD and the other along OB, as explained in Comiso (1995). The intersection point, I, corresponds to the value T_{I}, in (4.40) and together with T_{O}, the ice concentration can be calculated. Any set of channels can be used once the tie points are determined but when T_{I} and T_{O} are close or equal to each other, as can happen when using 36 GHz and higher frequencies, both numerator and denominator in (4.40) get close or equal to zero and the ice concentration cannot be determined. To avoid this possibility, the ratio of OB to OI is usually used, which provides the same value for C_{I} in (4.40). Ice concentration is not calculated below a cut-off of about 8% because these values are not possible to discriminate from those of open water in the passive microwave data. The cut-off is made using the pattern associated with open water and is usually a line approximately parallel to OW. Further refinements to this technique have been made to improve accuracy as will be explained later. The line AD is determined by applying regression analysis of data points along the line and a positive bias is applied to account for the known presence of open water (of about 2–3%) in consolidated ice regions.

Among the criteria for choosing the appropriate channels to be utilized in an algorithm are to optimize the resolution and to minimize the use of ancillary data in deriving the data product. Thus, the set of 36 GHz channels at horizontal and vertical polarizations (called HV36) is used because of reasonably good resolution and because the corresponding data in consolidated ice usually form a linear cluster with a well defined slope for AD, the value of which is approximately equal to one

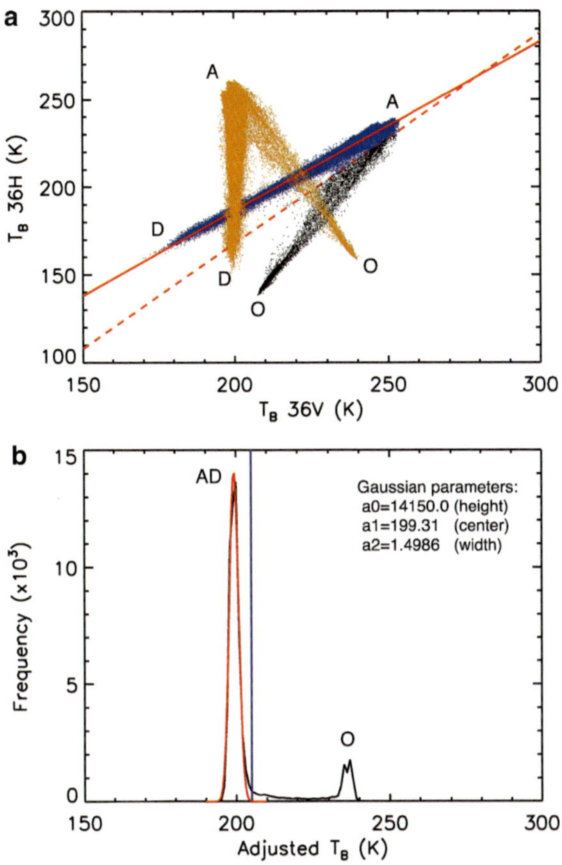

Fig. 4.23 (**a**) Scatter plot of vertically polarized vs. horizontally polarized data at 36 GHz and (**b**) plot of the frequency distribution of a rotated data points (in orange in (**a**)) A Gaussian fit to the plot in (**b**) is also shown with a standard deviation of 1.5 K. The data points along the cluster AD which correspond to 100% ice concentration is shown to provide varying estimates of ice concentration because of spatially varying emissivity of the surface but the associated error is only about 2 to 2.5% in ice concentration. From Comiso and Nishio (2008) with permission of the American Geophysical Union

as illustrated in Fig. 4.23a. The cluster of data points along AD is compact and linear because the changes in brightness temperature in the two channels due to changes in temperature, emissivity, snow cover, atmospheric water vapor, are basically the same. The slope is about one for the same reason. It also means that the retrieval of ice concentration is not sensitive to spatial changes in ice temperature, T_I, since the net effect of such a change is just to cause the data points to slide along the line AD and therefore, no impact on the accuracy in the retrieval of ice concentration. The width of the cluster of points along the line AD is estimated by rotating the line such that it is along the vertical as indicated in Fig. 4.23a and binning the values along the horizontal to produce the frequency plot shown in Fig. 4.23b. The distribution in the plot shows two peaks, one corresponding to consolidated ice at around 200 K and the other corresponding to open water at around 236 K. The standard deviation (width) of a fitted Gaussian curve on the first peak is 1.5 K which provides the uncertainty in the estimate of T_I. Such an uncertainty in brightness temperature for the ice tie-point corresponds approximately to an uncertainty of about 2–2.5% in the estimate of ice concentration. This is an important result because it represents the highest accuracy in ice concentration that can be obtained from passive microwave data.

The sole use of 36 GHz channels provides some ambiguities in the determination of ice concentration since for this set of channels, the open ocean data cluster (i.e., the line OW in Fig. 4.22) is either along the line OA in Fig. 4.23 or to the left of this line. Furthermore, the horizontal channel is more sensitive to layering and other surface effects than the vertical channel (Matzler et al. 1984) and in some ice covered areas, the data points fall below or to the right of the line AD. Both problems are resolved through the additional use of a set of channels that utilizes the 18 (or 19) GHz in combination with the 36 (or 37) GHz channel at vertical polarization (thereafter called the V1836 set) which provides an even easier discrimination because of the higher contrast in the emissivity of ice and water at 18 (or 19) GHz. Although the V1836 set provides a good mask for open ocean areas, the additional use of the 22 GHz channel significantly improves the effectiveness, as described in Comiso (1995). The use of the V1836 set shows some sensitivity to variations in surface temperature, but the associated error in ice concentration introduced is usually about 3% because the observed standard deviation of ice temperatures is only about 2.5% for snow covered surfaces. As indicated earlier, the error is minimized through the use of the HV36 set of channels which is the set that is first utilized. The scheme is that all data points with values that are along AD – 3 K and above are processed using the HV36 set. The rest are processed using the V1836 set.

The Arctic and the Antarctic regions are quite different in that sea ice is surrounded by land in the Arctic, while sea ice surrounds land in the Antarctic. In the winter, the Arctic is basically covered by consolidated ice that are more confined, thicker and colder than those in the Antarctic. In the Arctic, the ice floes can be as old as 7 years, while in the Antarctic, it is rarely the case that an ice floe is older than 2 years, the reason being that the remnants of summer ice gets flushed out of the original locations by strong ocean currents (e.g., Weddell gyre) during autumn and winter. Also, there is more divergence in the Antarctic and no limit in the advance of sea ice to the north because of the lack of a northern boundary. It is thus not surprising that the seasonality and microwave signatures of the sea ice cover in the two hemispheres are different. We use the same algorithm for the two hemispheres but with different tie points to account for the differences in physical and radiative characteristics of ice in the two regions.

4.3.6.1 Northern Hemisphere

For climate change studies, long historical records are always desired and this is possible for satellite measurements only by combining data from different satellites.

It is thus important to find out how the ice concentrations derived using the AMSR-E Bootstrap Algorithm (ABA) compare with those derived using the SSM/I Bootstrap Algorithm (SBA). AMSR-E data are now considered the baseline for ice concentration data because of the much higher resolution of the sensor that those of other sensors. On the other hand SSM/I has been providing continuous sea ice data since July 1987. ABA and SBA are actually identical algorithms and used on brightness temperatures that have been intercalibrated. Color-coded ice concentration maps using ABA and SBA for a summer day in the Northern Hemisphere are presented in Figs. 4.24a and b to illustrate how AMSR-E and SSM/I data compares. As indicated, the patterns of high and low ice concentrations values in the two maps are very similar and the two maps are quite consistent. There are slight differences, as indicated in the difference map of the two data sets shown in Fig. 4.24c, but the differences are small and within the errors of the retrieval. The differences are in part associated with the difference in the equatorial crossing of the satellite orbits which is about 4 h. Within such time, significant changes in the ice cover can occur.

The utility of the AMSR-E data, however, goes beyond extending the data set being produced by SSM/I and other similar sensors. The comparison in Fig. 4.24 is done for the 25 km grid that has been used as standard for SSM/I data. The higher resolution of the AMSR-E sensor enabled the use of a 12.5 km grid for the standard

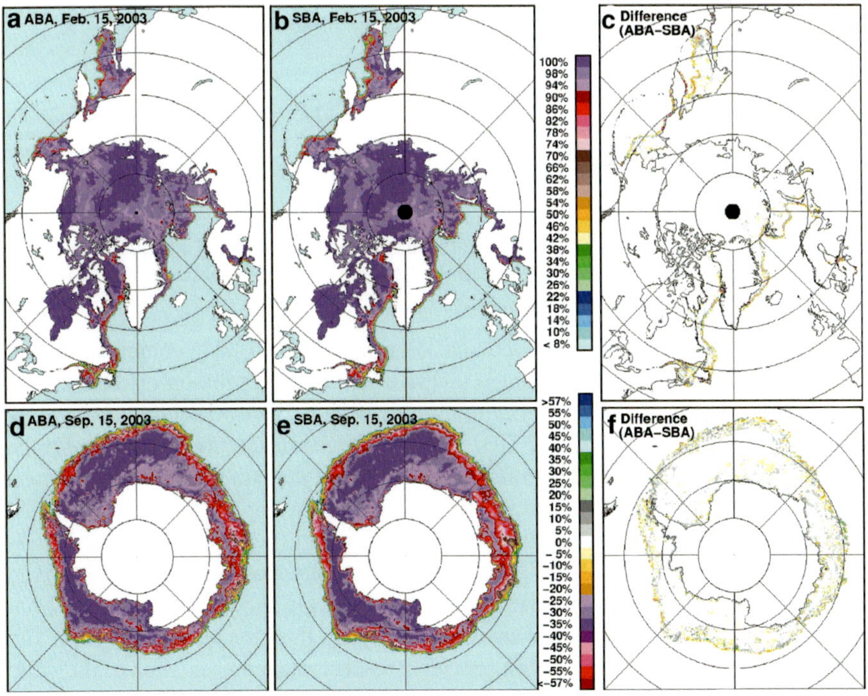

Fig. 4.24 Sea ice concentration maps in the (**a**) Arctic using AMSR-E data; (**b**) Arctic using SSM/I data; (**d**) Antarctic using AMSR-E data and (**e**) Antarctic using SSM/I data. Difference maps between AMSR-E and SSM/I data are presented for (**c**) the Arctic and (**f**) the Antarctic From Comiso and Nishio (2008) with permission of the American Geophysical Union

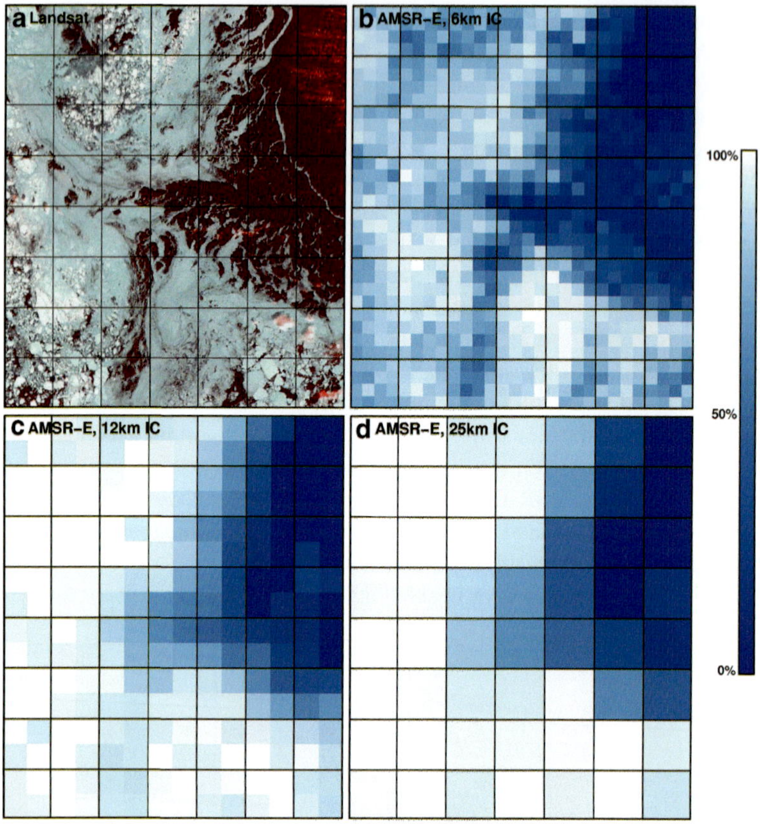

Fig. 4.25 (**a**) Landsat data with sea ice concentrations using (**b**) 89 GHz, 6 km resolution data; (**c**) 12 km resolution data using 18 and 37 GHz channels; and (**d**) 25 km data using 18 and 37 GHz channels. From Comiso (2009) with permission of the Remote Sensing Society of Japan. Copyright (2009)

product which can be easily degraded to a 25 km grid when used in combination with SSM/I data. Also, we took advantage of the even higher resolution of the 89 GHz channel to produce special ice concentration maps at 6.25 km resolution using only the two-channel dual-polarized 89 GHz data. This higher resolution data allow for the detection of many mesoscale features of the ice cover that are not possible to detect with the coarser data set. To illustrate this new capability, Fig. 4.25 shows a comparison of Landsat data with retrievals from AMSR-E data using the 6.25 and 12.5 grids and SSM/I data using the 25 km grid. It is apparent that the 6.25-km AMSR-E data capture the unique features associated with the ice margin and the interior of the pack as depicted by the Landsat data significantly better than the other data sets. The AMSR-E data at 12.5-km resolution provide a respectable representation of the same features but not as well, while the 25-km grid smooth out much of these features. For many mesoscale process studies in the polar regions, such as those involving the quantification of heat and salinity fluxes in polnyas, leads and marginal ice zones, the advantage of having data at the higher resolution grids is obvious. Since fluxes from open water and thin ice can be an order of magnitude higher than those of thick ice, ability to quantify such features accurately is highly desired. For large scale studies, however, the advantages may be more limited. More quantitative comparative analysis between coregistered Landsat and AMSR-E data yielded correlation coefficients of 0.929, 0.928, and 0.979 for the 6.25, 12.5, and 25-km data, respectively, which are all very high, but indicate that the 25-km data provide good representation of the large scale features of the ice cover. The lower correlation at the higher resolution grids is partly due to imperfect registration between the two systems. It is also important to point out that the 6.25 km data are generated using only the 89 GHz data (vertical and horizontal polarizations) which are sensitive to atmospheric and surface effects. The 89 GHz data are therefore most useful during clear skies and good weather conditions. In general, the 89 GHz data overestimates the ice concentration during adverse weather conditions and it is always desirable to compare the retrieved 89 GHz data with those of the coarser resolution data that made use of lower frequency channels for consistency.

In the middle of winter, the ice concentration in the inner zone is basically 100% or close to 100%. This is especially the case in the Central Arctic region where the sea ice cover is surrounded mainly by land. Occasionally strong winds and ocean upwelling would cause the formation of leads and polynyas but these

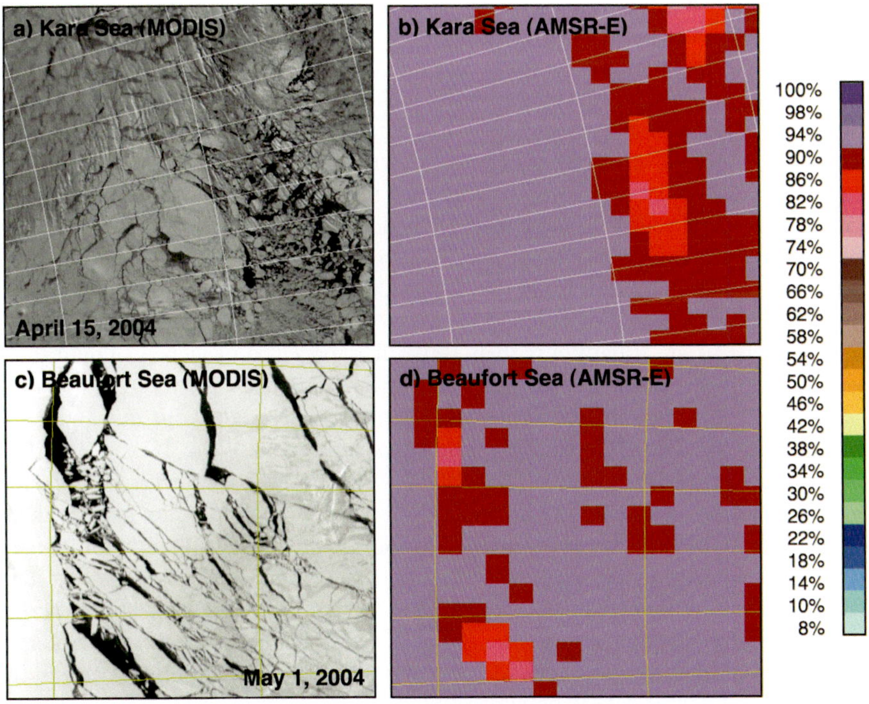

Fig. 4.26 Comparison of Aqua/MODIS and AMSR-E data in the Kara Sea (**a** and **b**) and in the Beaufort Sea (**c** and **d**)From Comiso (2009) with permission of the Remote Sensing Society of Japan. Copyright (2009)

features acquire an ice cover in a matter of hours. In Fig. 4.26, we compare derived sea ice concentrations from AMSR-E with near coincident MODIS data in the Kara Sea and Beaufort Sea. The MODIS image at the Kara Sea shows an ice cover area that is mainly consolidated, but part of which (i.e., right half of the image) have broken up ice floes with either open water or thin ice. The corresponding AMSR-E image shows 100% concentration in most areas except in the region that has some open water areas as can be identified in the MODIS data. The visible and microwave images are obviously sensitive to the same features within the ice pack. In the Beaufort Sea, the ice cover is slightly different and dominated by the presence of distinct linear features referred to as leads. Some of the leads are small and gets frozen within hours after they are formed, but are still visible because the ice cover is thin and not covered by snow. Such leads are not detectable in the AMSR-E data. In areas where the leads become bigger, a corresponding feature is observed in the AMSR-E data in the form of a reduced ice concentration. It is apparent that the AMSR-E image captures most of the regions with significant open water areas.

4.3.6.2 Southern Hemisphere

The sea ice cover in the Southern Hemisphere can be characterized by a divergent and generally loose ice cover in the marginal ice zone, some deep ocean polynyas and numerous leads within the pack, and coastal polynyas near the continent. The ice pack is also more vulnerable to winds, tides and ocean currents than that of the Northern Hemisphere because of the lack of a northern boundary. The ice cover thus consists of a more diversed types of seasonal ice cover. Because of the frequency of the calving of big icebergs in recent years, the spatial patterns in the distribution of ice have also been altered in some places. The icebergs are much more massive than the sea ice cover and their drift patterns, if not grounded, are very different, causing open water and thin ice to form in its vicinity. Also, since the emissivity of icebergs is significantly lower that of saline sea ice, ice concentration estimates of areas covered by icebergs are usually less than 100%. The presence of icebergs could thus cause a bias in the estimates of fluxes in ice covered regions.

The ice concentration Bootstrap Algorithm used previously for the Southern Hemisphere was a two channel algorithm that utilized only the 18 and 36 GHz channels at vertical polarization. The 18 GHz channel was used with SSM/I data because it provided a better contrast in emissivity between the ocean and ice than the higher frequency channels. With AMSR-E data, the availability of 6 and 10 GHz channels provides data with even better contrast between ocean and ice. The use of one of these channels in combination with the 36 GHz channel at vertical polarization yielded almost the same ice concentration as the V1836 set. Although the algorithm provides ice concentrations that are comparable to those derived from Landsat data (Comiso and Steffen 2001), there are issues, especially near the coastal regions. In the autumn and winter, strong southerly winds at the coastal boundaries cause the advection of sea ice to the north leaving open water behind. The open water in turn gets frozen almost right away and as the new ice gets thicker, they get advected to the north as well and the cycle repeats itself. At the coastal boundaries, there are thus a progression of ice of different thickness with the thinnest being the closest to the continent. Depending on thickness, thin ice has lower emissivity than that of thick ice. The temperature of the ice can also be significantly lower than average ice temperature when there is no snow cover. Both factors cause underestimates in ice concentration in these regions. To overcome this problem, the additional use of the set of 36 GHz channels at both polarizations (called VH36) is introduced since this set is less susceptible to surface temperature and ice thickness variations. It also makes the Antarctic algorithm practically identical to the Arctic ice concentration algorithm.

A comparison of a daily ice concentration map derived from AMSR-E and that from SSM/I in the Southern Ocean is presented in Figs. 4.24d and e. As with the retrievals in the Arctic region, the ice concentration map derived from AMSR-E is very similar to that derived from SSM/I. The difference map shown in Fig. 4.24f indicates that much of the discrepancies occur at the sea ice marginal ice region. As discussed in Comiso and Steffen (2001), there are merits for the two different techniques in that the two channel algorithm provides more information about thin ice distributions while the three channel version (VH36 and V1836) provides values closer to the true concentration as depicted by high resolution visible channel data.

The general characteristics of the ice cover during the same period as recorded in one infrared channel of MODIS is shown in Fig. 4.27. In the image, different thicknesses of new and young ice are usually represented by different gray levels (high temperature) with white (low temperature) representing the thicker ice types. The presence of clouds and cloud shadows makes this type of data useless for sea ice studies in some areas but where the areas are cloud free, the data provide the means to identify divergence areas that are also identified by the 2 channel algorithms. The open water or new ice near the coastlines and icebergs are examples of features that can be identified by the 2 channel algorithm, but not with the 3 channel algorithm. The coincident EOS-Aqua MODIS data may thus provide ancillary data to enhance physical characterization of the sea ice cover.

The higher resolution AMSR data provide better geophysical parameters in many aspects of ice studies. Figure 4.27 shows values of brightness temperature across the marginal ice zone at both vertical and horizontal polarizations for 23 July 2002. It is apparent that despite differences in resolution for the different frequencies, the profiles along the ice edge are very similar and the brightness temperatures go up at about the same time for all channels. This provides confidence in the quantitative determination of the location of the ice margin. A similar plot (not shown) for SSM/I shows slight discrepancies in the profiles mainly associated with the larger footprints and a different side lobes that can slightly alter the values at the ice edge. The derived ice concentrations for both AMSR-E and SSM/I are shown in Fig. 4.27c and indeed, the location of the 15% ice concentration point can be different by about 12–25 km. In other locations, the disagreements were not as bad, but these results indicate that the AMSR-E data provide a new standard for ice data and for assessing the quality of ice products derived from other radiometers. The brightness temperatures at 89 GHz channels are also shown to have very little contrast between ice and water and can be very variable within the ice pack.

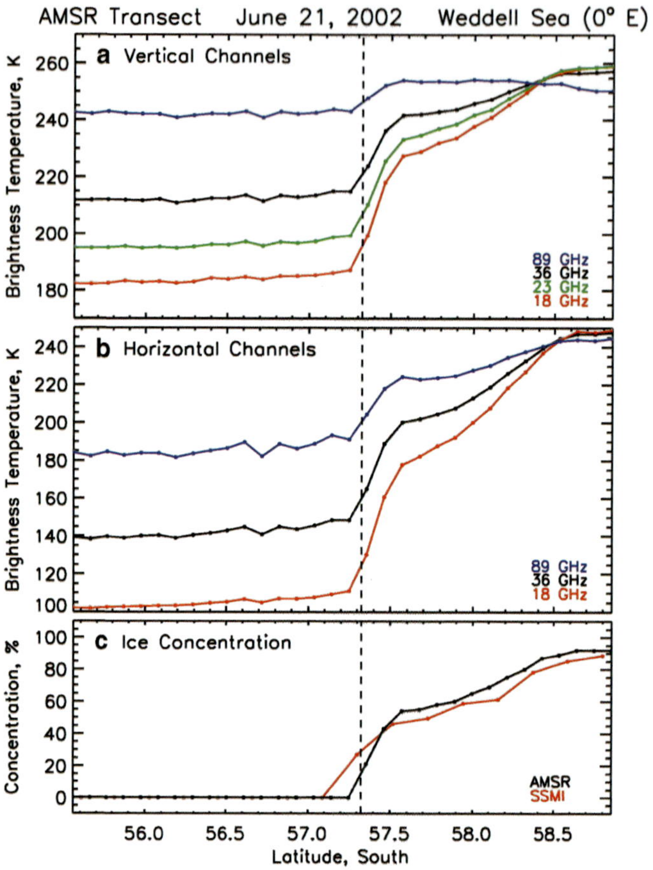

Fig. 4.27 Transects from open water into the ice pack using (**a**) all vertically polarized channels of AMSR-E, (**b**) all horizontally polarized channels of AMSR-E, and (**c**) derived ice concentration from AMSR-E and SSM/I

4.3.7 Sea Ice Thickness and Topography

Our current knowledge about the thickness distribution of sea ice have come from a few decades of upward looking sonar data. The data have been acquired routinely by nuclear submarines and were generally classified with a few exceptions until in recent years. The sonar data provide a measure of the distance between the submarine and the bottom of the ice floes (sometimes referred to as called keels). This measurements can be converted to actual ice thickness using estimates of the average density for the sea ice and the thickness of its snow cover. The data from submarines were extensive when compared with a few measurements done manually with hand and hot water drills during field programs, but they are nevertheless lacking in spatial coverage and much more in temporal coverage. Other techniques for measuring ice thickness includes the use of electromagnetic induction system that is normally installed aboard helicopters. Again, with a very limited range of helicopters, such technique does not provide good spatial coverage.

The only practical way to monitor the thickness of the global sea ice is actually through the use of satellite techniques. Radar altimeters have been used for monitoring the wave heights and surface topography of the liquid ocean and also the topography of ice sheets in Greenland and the Antarctic (Zwally et al. 2005; Thomas et al. 2001) since the early 1990s. Its use for monitoring the sea ice thickness has been introduced only recently (Laxon et al. 2003), but coverage is still not so good, especially in the Arctic where there is no data north of 82°N. The launched of GLAS aboard ICESAT in 2003 provided a significant improvement in spatial coverage but because of hardware problems, temporal coverage has been lacking. A new system dedicated to the measurement of the thickness of sea ice call CryoSAT was unsuccessfully launched in 2005. A follow-on system is however scheduled to be launched in a few years.

That the thickness of sea ice can be derived from freeboard data was demonstrated during a P3 overflight project over a submarine in the Arctic in 1987. Figure 4.28 shows a plot of the probability density of ice draft data as observed from a submarine when compared with a normalized plot of freeboard data observed by a laser ranging system on boar a P3. The normalization factor is 7.91 which is approximately what would be expected to account for the difference in density of sea ice and sea water. The draft data and the freeboard elevation data

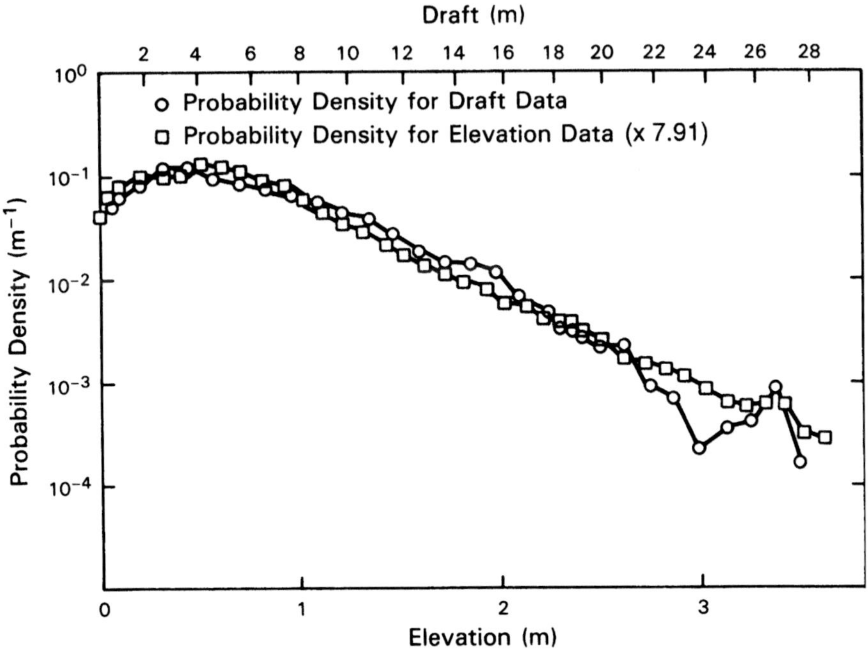

Fig. 4.28 Comparison of ice drafts as measured by an upward looking sonar on board a nuclear submarine and normalized ice freeboard elevation inferred from Lidar altimeter measurements on board a NASA P3 aircraft. The freeboard elevation distribution for 60 km of track of data is normalized by a factor of 7.91 (mean draft/elevation ratio) and plotted against corresponding draft distribution. From Comiso et al. (1991) with permission of the American Geophysical Union

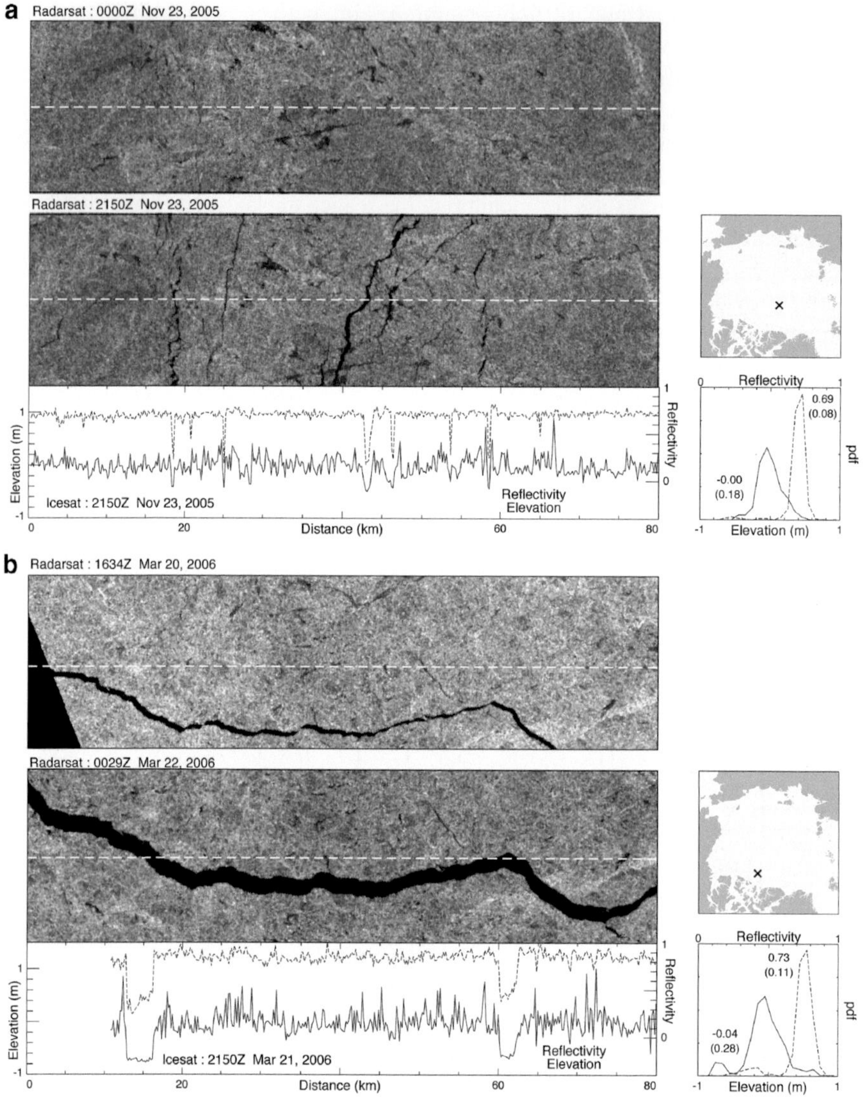

Fig. 4.29 Examples of new leads in the Arctic as identified by near-coincident RADARSAT and ICESat observations on (**a**) 23 November 2005 and (**b**) 22 March 2006. The leads correspond to the dark (low backscatter) areas in the RADARSAT image while reductions in elevation and reflectivity in the ICESat data. A cross mark identifies the location of the RADARSAT image and the ICESAT 80 km profile. From Kwok et al. (2007) with permission of the American Geophysical Union

clearly agrees indicating the potential of getting ice thickness from freeboard measurements. Studies discussing the retrieval of ice freeboard and thickness using ICESat freeboard data have been reported by Kwok et al. (2007) and Zwally et al. (2008). An example of ICESat data of freeboard elevation in the Arctic over an 80 km track is shown together with near coincident SAR data in Fig. 4.29. The elevation data is shown to be highly variable consistent with the track being in a heavily ridged region of the Arctic ice pack. Such ridging is also evident in the SAR data. Some work still needs to be done to accurately establish the actual thickness as derived from ICESat data. Among the current problems are the lack of knowledge about the snow cover and the density of the ice. Unfortunately, coincident submarine draft data are not available to get a good estimate of these parameters. Some estimates have been made using historical in situ measurements and the data has been used to study changes in thickness from 1 year to another.

4.3.8 Sea Ice Drift from Passive Microwave, Visible and SAR

The drift of sea ice is an important variable to monitor because it represents the energy and momentum associated with the sea ice cover. The dynamics of sea ice is influenced primarily by wind but also by ocean current, tides, and the Corriolis force. Ice motion has been inferred from drifting buoys, but spatial coverage is lacking with the number of buoys at any one time in the Arctic being less than 10 and is significantly less in the Antarctic.

Because of its high resolution, SAR data have been used for monitoring sea ice drift (Kwok et al. 1990). AVHRR data have also been used but at a more limited extent because of cloud cover problems. It was a pleasant surprise when it was discovered that passive microwave data provide drift information as well. It was initially tried on the highest resolution data available, and that means the SSM/I 89 GHz brightness temperature data. But it was soon determined that the other channels (e.g., 37 GHz channels) provide drift information as well.

Two techniques were derived for extracting sea ice drift velocity information from passive microwave data. One is the cross correlation technique that has been used for SAR and AVHRR data (Emery et al. 1991) and the other is the Wavelet technique (Liu and Cavalieri 1998). In cross correlation technique, which was developed initially using AVHRR data, the end points of the vectors are determined by locating the maximum cross correlations between sequencial subsets (windows) of image pairs. A series of statistical and spatial coherency filters were also developed to automatically remove vectors that appeared erroneously derived due to cloud cover and other causes not related to ice drift. A pleasant surprise was that the same technique was successfully applied to the coarser resolution passive microwave data in which cloud cover was not a problem. A validation of this automated technique using both AMSR-E and SSM/I data was done by Yaguchi and Cho (2009) by comparing the results with vectors manually derived from the higher resolution MODIS data. An example of such comparison is shown in Fig. 4.30 and it is apparent that the patterns are generally consistent. Some mismatches are apparent, but these are usually deleted by the filtering technique.

In the wavelet technique, an isolated ice feature is determined by a close contour derived by setting a 5% threshold above the minimum of a wavelet transform of passive microwave data. The closed contour is framed in a rectangular window, called a

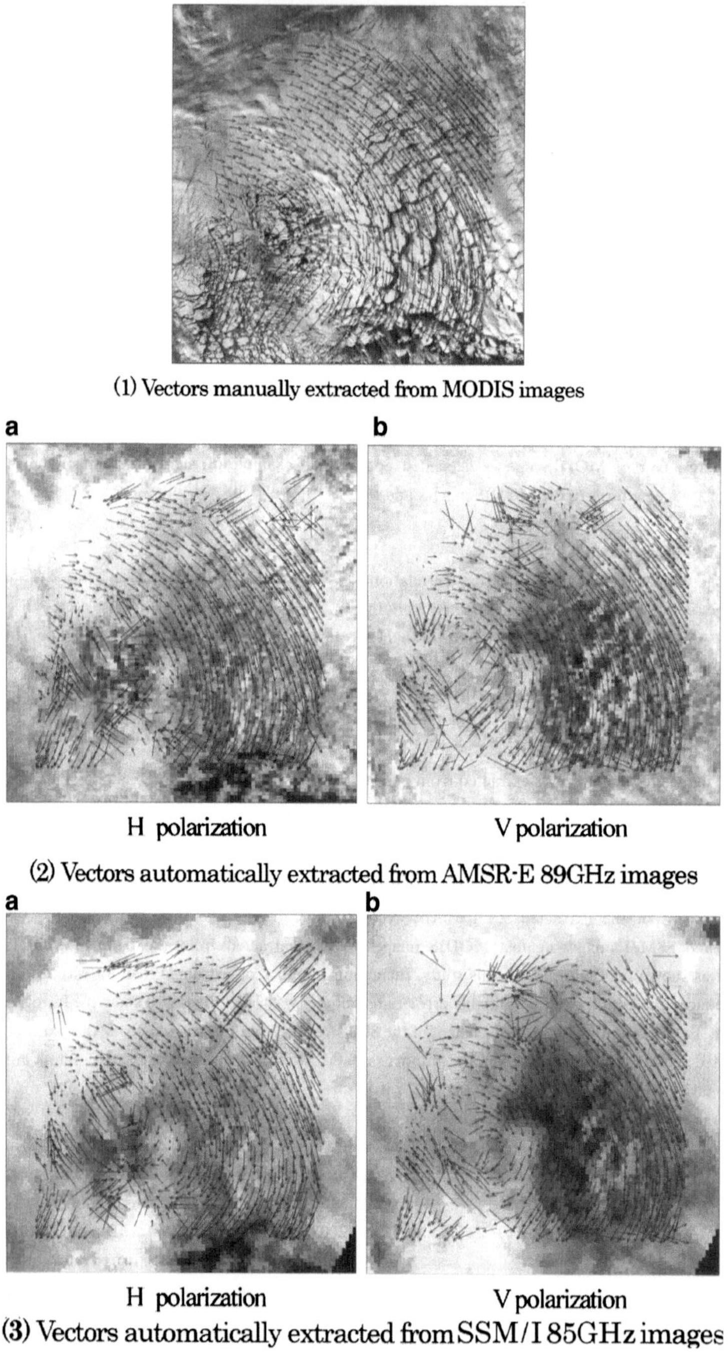

Fig. 4.30 Validation of sea ice drift vectors derived from AMSR-E and SSM/I data in the Sea of Okhotsk on 11 February 2007 and 13 February 2007 using cross-correlation technique with manually extracted vectors from a sequence of MODIS data. From Yaguchi and Cho (2009) with permission of the Remote Sensing Society of Japan. Copyright (2009)

template, with its four sides tangent to the extreme locations of the contour. The vectors are then derived by matching such template with a corresponding template using data from approximately the same location but two days later. Comparative analyses of result with buoy data have shown good agreement (Liu et al. 1998).

4.4 Error Analysis and Accuracy Requirements

The biggest source of error in the retrieval of surface parameters including surface temperature, albedo, and phytoplankton concentration is likely the cloud masking. Clouds occur in too many types, heights and distributions that it is nearly impossible to catch all of them even with a sensor that has a good cloud detection system. The detection of cloud in the polar regions is especially difficult because of similar signatures of clouds and snow or ice. The limitations of the AVHRR data in terms of the lack of adequate channels for cloud masking only exacerbate the problem. Despite these problems, the retrieved parameters appear to be relatively accurate. This is not to say that there is no room for improvement. For example, the effect of occurrence of inversions which actually gets more frequent at high altitudes (Phillpot and Zillman 1970) should be examined in greater details. The reason is that one of the cloud masking techniques used takes advantage of the fact that clouds are generally colder than the surface. It would be useful to be able to detect occurrences of inversions and be able to evaluate when a cloud masking technique cannot be used effectively.

One of the critical issues associated with the use of AVHRR data for time series studies is the inconsistency in the calibration of the different AVHRR sensors. The relatively short lifespan (typically 5 years) of each AVHRR sensor made it necessary to launch many different AVHRR sensor to have reliable data real time for operational analysis. The archived composite record is thus a collection of data from several AVHRR sensors with no overlap coverage between sensors. The lack of overlap has made it difficult to compare the performance and compatibility of the different AVHRRs. For lack of a better solution, in situ data have been used to improve calibration and ensure consistencies in the derived temperatures from the different sensors. We try to overcome this shortcomings by using in situ data for calibrating the thermal infrared channels. Unfortunately, there is a general paucity in the number of stations because of general inaccessibility and the high cost of deployment and maintenance. The installation of more automated weather stations in recent years has improved coverage, but sometimes the quality is sacrificed because unattended sensors can provide inaccurate readings due to instrumental (e.g., calibration change) and environmental problems (e.g., thermistors buried by snow). Even in manned stations, the maintenance of accurate record continuously is not trivial because of the harsh conditions that often do not allow periodic checks of the conditions of the sensors. Time series data, however, usually provides ability to assess the quality and eliminate erroneous data sets. For surface temperature retrievals in the Arctic, the RMS error is estimated to be 2.46 K when individual stations are compared, but this improves to 0.69 K when monthly averages of temperatures from these stations are compared with AVHRR data. This indicates that there are outliers in the individual station data set that causes the larger RMS errors when individual stations are compared. In the Antarctic, the RMS error is 3.00 K when monthly stations are used individually and this goes down to 0.54 when monthly averages are used. The outliers are likely in more stations in the Antarctic, in part because of harsher weather conditions and less monitoring of the stations.

Another key issue is that infrared data are useful for surface temperature measurements only during clear sky conditions. The cloud masking techniques used for AVHRR data was discussed previously, but some refinements are desired to improve accuracy in the retrieved surface parameters. Although clouds are generally persistent in the polar regions and only a small fraction of the data is considered cloud free, the frequency of observation by polar orbiting satellite sensors at high latitudes is relatively high because of wide swath and a relatively dense orbital tracks. Thus, the chance of getting cloud free surface measurements in a particular area in a few days is very good. Weekly and monthly averages of the data have been generated with practically no gaps over ice covered areas. The good agreement of the monthly averages of in situ and AVHRR surface temperatures indicates that the "clear-sky" sampling bias is likely to be small. However, the resulting maps should still be regarded as surface temperature maps for "clear-sky" conditions only.

Quantifying atmospheric effects in the retrieved surface temperature data, including those of water vapor and aerosol, are also the challenges since the effect could vary from one pixel to another. Radiosonde measurements over the polar regions are very rare and to be able to make this correction properly, we need to have the parameters on a pixel by pixel basis. The use of sounding data such as those from Aqua/AIRS which are available during cloud free conditions would help improve the accuracy of the retrievals.

The fusion of infrared and passive microwave data would also help improve accuracy and spatial as well as temporal coverage. The microwave data provide accurate and good temporal resolution when done over areas where the emissivity of the surface is well known. Through the construction of good emissivity maps, it may be possible to get spatially detailed measurements from passive microwave data that could be used concurrently with infrared data to obtain surface temperature maps that have high temporal resolution and spatially consistent values.

The errors in ice concentration has been estimated to vary from 5 to 10% during dry and cold conditions to 10–20% during the melt seasons. These estimates are based on validation studies using multisensor techniques in conjunction with available in situ data (Cavalieri, 1992). In principle, the accuracy can be as high as 3% during the dry and cold conditions. Also, errors during the melt seasons can be minimized through adjustment of the tie points or use of other frequency channels (Markus and Dokken 2002; Comiso 2009). But there are other sources of errors that are not related to the algorithm utilized. One of this is the land mask which is intended to eliminate the problem of discriminating land ice from sea ice. The accuracy of the land mask depends on the accuracy of available data and relatively frequent calving of large icebergs makes it necessary to have such a mask updated. A second one is the ambiguity in the detection of the exact location of the ice edge.

This is especially true during stormy weather where the signature of open water is similar to that of ice covered areas and therefore the retrieval of an ice cover in areas where there is no ice. When the feature is far from the ice edge, they are automatically mask using climatological surface temperatures. However, when they are close to the ice edge, it becomes difficult to detect where the ice edge ends and where the weather problem begins. A third problem is the land/ocean spill-over problem associated with the coarse resolution and antenna side-lobe. Some techniques have been developed to overcome this (e.g., Cho et al. 1996), but they do not always work.

A fundamental problem associated with sea ice algorithm is in the definition of ice concentration itself. Sea ice changes its emissivity as it goes through the different stages of developments. The emissivity of grease ice and some other new ice types, for example, is much closer to that of open water than that of thick ice. Most new ice types have emissivities between that of open water and thick ice. If ice concentration is defined such that a region that is covered by any type of ice is 100%, the resulting ice map will be mainly 100% ice in the autumn/winter period and the data won't be able to show areas of divergence and polynya regions. Fortuitously, current techniques provides data that show these important features.

Errors in albedo estimates from AVHRR data are difficult to establish because of inconsistencies in the calibration of the different AVHRR sensors, the lack of overlapping data, sensor degradation and changes in orbital parameters. Because of the lack of a better technique we resort to the use of MODIS data which has been shown to have good calibration as the means to improve the quality of the AVHRR data. The assumption is that the high albedo values of the Greenland ice sheet do not change much from 1 year to another. Intuitively, this is a valid assumption but nevertheless it needs to be validated. The precision of the enhanced data should be good and could provide a unique time series that enables quantitative assessment of interannual variablity and trends of the albedo of the polar regions.

Errors in the estimates of pigment concentration has been established through a dedicated period of intense validation activities for SeaWiFS data. At the center of this activity is the Sensor Intercalibration and Merger for Biological and Interdisciplinary Studies (SIMBIOS) with the primary goal of providing the science community with well calibrated and high quality ocean color time series data set (McClain et al. 2004). The algorithm used for the polar regions, is the same as that used globally and tested in the tropics. There is reason to suspect that the optical properties of polar oceans are different from those of extra polar regions (Sullivan et al. 1993). Regional algorithms, however, are difficult to develop especially if in situ data are sparse if not lacking. Combining results from regional algorithms would also cause problems of mismatches at the boundary of the different regions. For lack of better information, we assume that the errors are not any worse that estimate from the validation activities.

Despite aforementioned problems satellite data have been extremely useful. They provide information about large scale variability of many parameters and trends that would have been impossible to obtain otherwise. We have observed many unexpected changes in surface temperature, sea ice cover, and plankton distribution. We have also managed to successfully use the data to extend the length of a time series using satellite data in combination with in situ and a statistical model as was done in Steig et al. (2009). We also continue to improve the design and capabilities of new sensors making the measurements to be almost as good eventually as those taken from the ground.

References

Aiken J, Moore GF, Trees CC, Hooker SB, Clark, DK (1995) The SeaWiFS CZCS-Type Pigment Algorithm. SeaWiFS Technical Report Series, Vol. 29, Hooker SB, Firestone ER (eds) NASA Tech. Memo 104566, Greebelt MD

Allison I, Brandt RE, Warren SG (1993) East Antarctic sea ice: albedo, thickness distribution, and snow cover. J Geophys Res 98(C7):12,417–12429

Arcone SA, Gow AJ, McGrew S (1986) Structure and dielectric properties at 4.8 and 9.5 GHz of saline ice. J Geophys Res 91:14,281–14,303

Arking A (1991) The radiative effects of clouds and their impact on climate. Bull Amer Meteorol Soc 71:795–813

Arrigo KR, Worthen DL, Schnell A, Lizzote MP (1998) Primary production in Southern Ocean waters. J Geophys Res 103:15,587–15,600

Barton IJ (1995) Satellite-derived sea surface temperatures: current status. J Geophys Res 100:8777–8790

Behrenfeld MJ Falkowski PG (1997) A consumer's guide to phytoplankton primary productivity models. Limnol Oceanogr, 42:1479–1491

Bernstein RL (1982) Sea surface temperature estimation using the NOAA 6 satellite advanced very high resolution radiometer. J Geophys Res 87(C12):9455–9465

Brown JW, Brown OB, Evans RH (1993) Calibration of Advanced Very High Resolution Radiometer infrared channels: a new approach to nonlinear correction. J Geophys Res 98(18):18257–18268

Carsey F (1992) Microwave Remote Sensing of Sea Ice, AGU Monograph 68, Washington, DC

Casey KS, Cornillon P (1999) A comparison of satellite and in situ-based sea surface temperature climatologies. J Climate 12:1848–1863

Cavalieri, DJ (1992) The validation of geophysical products using multisensory data. In: Carsey F (ed) Microwave remote sensing of sea ice, Geophysical Monograph 68, American Geophysical Union, Washington DC

Cavalieri DJ, Gloersen P, Campbell WJ (1984) Determination of sea ice parameters with the Nimbus 7 SMMR. J Geophys Res 89:5355–5369

Chandrasekhar S (1960) Radiative transfer. Dover Publications, Inc., New York

Chelton DB, Freilich MH, Sienkiewicz JM, Von Ahn JM (2006) On the use of QuikScat Scatterometer measurements of surface winds for marine weather prediction. Monthly Weather Rep 134:2055–2071

Cho K, Sasaki N, Shimoda H, Sakata T, Nishio F (1996) Evaluation and improvement of SSM/I sea ice concentration algorithms for the Sea of Okhotsk. J Remote Sensing of Japan 16(2):47–58

Coakley JA Jr, Bretherton FP (1982) Cloud cover from high resolution scanner data: Detecting and allowing for partially filled fields of view. J Geophys Res 87(C7):4917–4932

Comiso JC (1986) Characteristics of winter sea ice from satellite multispectral microwave observations. J Geophys Rev 91(C1):975–994

Comiso JC (1995) SSM/I sea ice concentrations using the Bootstrap Algorithm, NASA RP1380

Comiso JC (2000) Variability and trends in Antarctic surface temperatures from in situ and satellite infrared measurements. J Clim 13(10):1674–1696

Comiso JC (2003) Warming Trends in the Arctic. J Clim 16(21):3498–3510

Comiso JC (2009) Enhanced sea ice concentrations and ice extents from AMSR-E data, J. Remote Sens Soc Japan 29:199–215

Comiso JC, Nishio F (2008) Trends in the sea ice cover using enhanced and compatible AMSR-E, SSM/I, and SMMR data. J Geophys Res 113:C02S07, doi: 10.1029/2007JC004257

Comiso JC, Steffen K (2001) Studies of Antarctic sea ice concentrations from satellite data and their applications. J Geophys Res 106(C12):31361–31385

Comiso JC, Sullivan CW (1986) Satellite microwave and in-situ observations of the Weddell Sea ice cover and its marginal ice zone. J Geophys Res 91(C8):9663–9681

Comiso JC, Ackley SF, Gordon AL (1984) Antarctic Sea Ice Microwave Signature and their correlation with In-Situ Ice Observations. J Geophys Res 89(C1):662–672

Comiso JC, Cavalieri DJ, Markus T (2003) Sea ice concentration, ice temperature, and snow depth, using AMSR-E data. IEEE TGRS, 41(2), 243–252

Comiso JC, Cavalieri D, Parkinson C, Gloersen P (1997) Passive microwave algorithms for sea ice concentrations. Remote Sensing Environ 60(3):357–384

Comiso JC, Wadhams P, Krabill W, Swift R, Crawford J, Tucker W (1991) Top/Bottom multisensor remote sensing of Arctic sea ice. J Geophy Res 96(C2):2693–2711

Cota G, Wang G, Comiso JC (2004) Transformation of global satellite chlorophyll retrievals with a regionally tuned algorithm. Remote Sensing Environ 90:373–377

Cumming WA (1952) The dielectric properties of ice and snow at 3.2 centimeters. J App Phuys 23(7):768–774

Curry JA, Webster PJ (1999) Thermodynamics of Atmospheres and Oceans, International Geophysics Series, vol 65. Academic Press, London s4.2.1

Curry JA, Schramm JL, Perovich DK, Pinto JO (2001) Applications of SHEBA/FIRE data to evaluation of snow/ice albedo parametrizations. J Geophys Res 106(D14):15,245–15,355

Cummings W (1952) The dielectric properties of ice and snow at 3.2 cm. J Appl Phys 23:768–773

Debye P (1929) Polar molecules, Dover, New York

Donlon CJ, Minnett, PJ, Gentelmann C, Nightingale TJ, Barton IJ, Ward B (2002) Toward improved validation of satellite sea surface skin temperature measurements for climate research. J Clim 15:353–369

El-Sayed SZ, Biggs DC Holm-Hansen (1983) Phytoplankton standing crop, primary productivity and near-surface nitrogenenous nutirient fields in the Ross Sea, Antarctica. Deep-Sea Res, Part A 30:871–886

Emery WJ, Fowler CW, Hawkins J, Preller RH, (1991) Fram Strait image-derived ice motions. J Geophys Res 96:4751–4768

Eppler D, Farmer LD, Lohanick AW, Anderson MR, Cavalieri DJ, Comiso JC, Gloersen P, Garrity C, Grenfell T, Hallikainen M, Maslanik JA, Maetzler C, Melloh RA, Rubinstein I, Swift CT (1992) Passive microwave signatures of sea ice. In: Carsey F (ed) Chapter 4, Microwave Remote Sensing of Sea Ice, American Geophysical Union, Washington, DC, pp 47–71

Evans S (1965) Dielectric properties of snow and ice – A review. J Glaciol 5(42):773–792

Evans R, and Podesta G (1996) AVHRR Pathfinder SST approach and results. Eos Trans Amer Geophys Union 77(Suppl):F354

Gille ST (1994) Mean surface height of the Antarctic circumpolar current from Geosat data: Method and application. J Geophys Res 99:18,255–18,273

Gille ST, Kelly KA (1996) Scales of spatial and temporal variability in the Southern Ocean. J Geophys. Res 101(C4): 8759–8773

Giles KA, Laxon SW, Worby AP (2008) Antarctic sea ice elevation from satellite radar altimetry. Geophys Res Letts 35, L03503, doi:10.1029/2007GL031572

Gregg WW, Carter KL (1990) A simple spectral solar irradiance model for cloudless marine atmosphere. Limnol Oceanogr 35(8):1657–1675

Gregg WW, Conkright ME (2002) Decadal changes in global ocean chlorophyll. Geophys Res Lett 29(15), doi:10.1029/2002GL014689

Grenfell TC, Comiso JC, Lange MA, Eicken H, Wenshahan MR (1994) Passive microwave observations of the Weddell Sea during austral winter and early spring. J Geophys Res 99(C5):9995–10,010

Hallikainen M, Winebrenner DP (1992) The physical basis for sea ice remote sensing. In: Carsey FD (ed) Microwave Remote Sensing of Sea Ice, Geophysical Monograph 68, AGU, Washington, DC, pp 9–28

Hanesiak JM, Barber DG, De Abreu RA, Yackel JJ (2001) Local and regional albedo observations of first-year sea ice during meltponding. J Geophys Res 106(C1):1005–1016
Hoepffner N, Sathyendranath S (1993) Determination of the major groups of phytoplankton pigments from the absorption spectra of total particulate matter. J Geophys Res 98:22,789–803
Hooker SB, McClain CR (2000) The calibration and validation of SeaWiFS data. Prog Oceanogr 45:427–465
Intieri JM, Shupe MD, Uttal T, McCarty BJ (2002) An annual cycle of Arctic cloud characteristics observed by radar and lidar at Sheba. J Geophys Res 107(C), 8090, doi:10.1029/2000JC000423
Irvine WM, Pollack JC (1968) Infrared optical properties of water and ice spheres. Icarus 8:324
Jones PD, New M, Parker DE, Martin S, Rigor IG (1999) Surface air temperature and its changes over the past 150 years. Rev Geophys 37:173–199
Kalle K (1949) Fluoreszenz und Gelbstoff im bottnischen und finnischen Meerbusen. Deut Hydrograph 2.2:117–124
Key J, Haefliger M (1992) Arctic ice surface temperature retrieval from AVHRR thermal channels. 97(D5):5855–5893
Key JR et al (1989a) Cloud classification from satellite data using a fuzzy sets algorithm. Int J Remote Sens 10(12):1823–1842
Key J, Maslanik JA, Schwiger AJ (1989b) Classification of merged AVHRR and SMMR arctic data with neural networks. Photogramm Eng Remote Sens 55:1331–1338
Key JR, Schweiger AJ, Stone RS (1997) Expected uncertainty in satellite-derived estimates of the surface radiation budget at high latitudes. J Geophys Res 102(C7):15,837–15,847
Kidder SQ, Wu HT (1984) Dramatic contrast between low clouds and snow cover if daytime 3.7 imagery. Monthly Weather Review 112:2345–2346
Kidwell KB (1991) NOAA polar orbiter data users guide. NOAA/NESDIS, Washington DC
Kilpatrick KA, Podesta GP, Evans R (2001) Overview of the NOAA/NASA advanced very high resolution radiometer Pathfinder algorithm for sea surface temperature and associated matchup database. J Geophys Res 106(C5):9179–9197
Kirk JTO (1984) Dependence of relationship between inherent and apparent optical properties of water on solar altitude. Limnol Oceanogr 29:350–356
Kirk JTO (1994) Light and Photosynthesis in aquatic ecosystems. University Press, Cambridge
Klein LA, Swift CT (1977) An improved model for the dielectric constant of sea water at microwave frequencies. IEEE Trans Antennas Propagation AP-25(1):104–111
Koepke P (1989) Removal of atmospheric effects from AVHRR albedos. J Appl Meteorol 28:1341–1348
Koh G (1992) Dielectric constant of ice at 26.5–40 GHz. J Appl Phys 71:5119–5122
Kumar A, Minnett PJ, Podesta G, Evans RH (2003) Error Characteristics of the atmosphere correction algorithms used in retrieval of sea surface temperatures from infrared satellite measurements: Global and regional aspects. J Atmos Sci 60:575–585
Kumerow C (1993) On the accuracy of the Eddington approximation for radiative transfer in the microwave frequencies. J Geophys Res 98:2757–2765
Kwok R, Curlander JC, McConnell R, Pang SS (1990) An ice motion tracking system at the Alaska SAR Facility. IEEE J Ocean Eng 15(1):44–54
Kwok R, Cunningham GF, Zwally HJ, Yi D (2007) Ice, cloud and land elevation satellite (ICESat) over Arctic sea ice: retrieval of freeboard. J Geophys Res 112:C121013, doi: 10.1029/2006JC003978
Lamb J (1946) Dielectric Constants. Trans Fraday Soc. 52A:238-244
Lane JA, Saxton JA (1952) Dielectric dispersion in pur polar liquids at very high radiofrequencies: 1. Measurements on water, metyl and ethyl alcohols. Proc Roy Soc 213:400–408
Laxon S, Peacock N, Smith D (2003) High interannual variability of sea ice thickness in the Arctic region. Nature 425:947–950
Lee ZP, Hu C (2006) Global distribution of Case-1 waters: An analysis from SeaWiFS measurements. Remote Sens Environ 101:270–276
Lee J et al (1990) A neural network approach to cloud classification. IEEE Trans Geosci Remote Sens 28:846–855

Lewis M, Cullen J, Platt T (1983) Phytoplankton and thermal structure in the upper ocean: consequences of nonuniformity of chlorophyll profile. J Geophys Res 88:2565–2570
Lindsay RW, Rothrock DA (1994) Arctic sea ice albedo from AVHRR. J Clim 7(11):1737–1749
Liu AK, Cavalieri DJ (1998) On sea ice drift from the wavelet analysis of the Defense Meteorological Satellite Program (DMSP) Special Sensor Microwave Imager (SSM/I) data. Int J Remote Sens 19(7):1415–1423
Liu AK, Zhao Y, Liu WT (1998) Sea-ice motion derived from satellite agrees with buoy observations. EOS Transactions 79(30):353–359
Lubin D, Massom R (2006) Polar Remote Sensing Vol. I, Springer-Praxis, Berlin s 4.2.1
Markus T, Cavalieri DJ (2000) An enhancement of the NASA team sea ice algorithm. IEEE Trans Geosci Remote Sensing 38:1387–1398
Markus T, Dokken ST (2002) Evaluation of late summer passive microwave Arctic sea ice retrievals. IEEE Trans Geosci Remote Sensing 40:348–356
Martin S (2004) An introduction to ocean remote sensing. Cambridge University Press, Cambridge
Martin S, Drucker R, Kwok R, Holt B (2004) Estimation of the thin ice thickness and heat flux for the Chukchi Sea Alaskan coast polynya from SSM/I data, 1990–2001. J Geophys Res 109:C10012, doi: 10.1029/2004/JC002428
Massom R, Comiso JC (1994) Sea ice classification and surface temperature determination using Advanced Very High Resolution Radiometer satellite data. J Geophys Res 99(C3):5201–5218
Matzler C, Ramseier RO, Svendsen E (1984) Polarization effects in sea ice signatures. IEEE J Oceanic Eng OE-9:333–338
Maxson RW (1992) Comparison of areal extent of snow as determined by AVHRR and SSM/I satellite image. MS Thesis, Naval Postgraduate School, Monterey, California
McClain EP, Pichel WG, Walton CC (1985) Comparative performance of AVHRR-based multichannel sea surface temperature 90(C6):11,587–11,601
McClain CR, Feldman G, Esaias W (1993) A review of the Nimbus-7 coastal zone color scanner data set and remote sensing of biological oceanic productivity. In: Parkinson C, Foster J, Gurney R (eds) Global Change Atlas, Cambridge University Press, New York
McClain CR, Feldman GC, Hooker SB (2004) An overview of the SeaWiFS project and strategies for producing a climate research quality global ocean bio-optical time series. Deep-Sea Research Part II, 51:5–42
Minnett PJ, Knuteson RO, Best RA, Osborne BJ, Hanafin JA, Brown OB (2001) The marine-atmospheric emitted radiance interferometer: A high accuracy seagoing infrared spectroradiometer. J Atm Oceanic Tech 18:994–1013
Minnis P, Harrison EF (1984) Diurnal variability of regional cloud and clear-sky radiative parameters derived from GOES data. Part III: November 1978 Radiative Parameters. J Climate Appl Meteor 23:1032–1050
Mitchell BG, Holm-Hansen O (1991) Bio-optical properties of Antarctic Peninsula waters: Differentiation from temperate ocean models. Deep-Sea Research, 38:1009–1028
Mitchell BG, Brody E, Yeh EN, McClain C, Comiso JC, Maynard NC (1991) Meridional zonation of the Barents Sea ecosystem inferred from satellite remote sensing and in situ Bio-optical observations. Pro Mare Symp Polar Res 10(1):147–162
Mobley CD (1994) Light and water: radiative transfer in natural waters. Academic Press, New York
Mobley CD, Stramski D, Bissett WP, Boss E (2004) Optical modeling of ocean waters: Is the Case1-Case2 classification still useful? Oceanography 17(2):60–67
Morel A (1974) Optical properties of pur water and pur sea water. In: Jerlov NG, Nielsen ES (eds) Optical aspects of oceanography, Academic New York, pp 1–24
Morel A, Prieur L (1977) Analysis of variations in ocean color. Limnol Oceanogr 22(4):709–722
Morel A, Antoine D, Babin M, Dandonneau Y (1996) Measured and modeled primary production in the northeast Atlantic (EUMELI JGOFS program): the impact of natural variations in photosynthetic parameters on model predictive skill. Deep-Sea Res I 43:1273–1304

Neumann Pierson (1966) Principles of physical oceanography. Prentice-Hall, Englewood Cliffs, NJ
O'Reilly JE, Maritorena S, Mitchell BG, Siegel DA, Carder KL, Garver SA, Kahru M, McClain C (1998) Ocean color chlorophyll algorithms for SeaWiFS. J Geophys Res 103(C11):24,937–24,953
Perovich DK, Roesler CS, Pegau WS (1998) Variability in Arctic sea ice optical properties. J Geophysical Res 103:1193–1208
Perovich DK, Grenfell TC, Light B, Hobbs PV (2002a) Seasonal evolution of the albedo of multiyear Arctic sea ice. J Geophys Res 107(C10):8044, doi:10.1029/2000JC000438
Perovich DK, Tucker WB III, Ligett KA (2002b) Aerial observations of the evolution of ice surface conditions. J Geophy Res 197(C10):doi: 10.1029/2000JC000449. 4.2.1
Phillpot HR, Zillman JW (1970) The surface temperature inversion over the Antarctic Continent. J Geophys Res 75:4161–4169
Platt T, Sathyendranath S (1988) Oceanic primary production: estimation by remote sensing at local and regional scales. Science 241:1613–1620
Platt T, Fuentes-Yaco C, Frank KT (2003) Spring algal bloom and larval fish survival. Nature 423:398–399 4.3.5
Rao CRN, Chen J (1996) Revised post-launch calibration of the visible and near-infrared channels of the Advanced Very High Resolution Radiometer on the NOAA-14 Spacecraft. Int J Remote Sensing 20(18):3485–3491
Rees WG (1993) Infrared emissivities of Arctic land cover types. Int J Remote Sens 14(5):1013–1017
Reynolds RW, Rayner NA, Smith TM, Stokes DC, Wang W (2002) An improved in situ and satellite SST analysis for climate. J Clim 15:1609–1625
Sathyendranath S, Cota G, Stuart V, Maass M, Platt T (2001) Remote sensing of phytoplankton pigments: a comparison of empirical and theoretical approaches. Appl Opt 37:2216–2227
Shibata A (2004) AMSR/AMSR-E SST algorithm developments: Removal of ocean wind effect. Ital J Remote Sens 30/31, 131–142
Shifrin KS (1988) Physical optics of ocean water. American Institute of Physics, New York
Shuman C, Comiso JC (2002) In situ and satellite surface temperature records in Antarctica. Ann Glaciol 34:113–120
Shuman CA, Alley RB, Anandakrishnana S (1995) An empirical technique for estimating near-surface temperature trends in central Greenland from SSM/I brightness temperatures. Remote Sens Environ 51:245–252
Simpson JJ, Keller RH (1995) An improved fuzzy logic segmentation of sea ice, clouds, and ocean in remotely sensed Arctic Imagery. Remote Sens Environ 54:290–312 4.3.1
Simpson JJ, Yhann SR (1994) Reduction of noise in AVHRR Channel 3 data with minimum distortion. IEEE Trans Geosci Remote Sens 32(2): 315–328 4.3.1
Schneider DP and Steig EJ (2002) Spatial and temporal variability of Antarctic ice sheet microwave brightness temperatures. Geophys Res Lett 29, 1984, doi:10.129/2002GL15490
Schneider DP, Steig EJ, Comiso JC (2004) Recent climate variability in Antarctica from satellite derived temperature data. J Clim 17:1569–1583
Schuerman DW (1980) Light scattering by irregularly shaped particles. Plenum Press, New York
Shifrin KS and Oliver D (1988) Physical optics of ocean water. American Institute of Physics, New York
Spinhirne JD, Palm SP, Hart WD, Hlavka DL (2005) Cloud and aerosol measurements from GLAS: Overview and initial results. Geophy Res Lett 32, L22S03, doi:10.1029/2005GL023507
Steig EJ, Schneider DP, Rutherford SD, Mann ME, Comiso JC, Shindell DT (2009) Warming of the Antarctic ice sheet surface since the 1957 International Geophysical Year. Nature 457:459–463
Steffen K, Bindchadler R, Casassa C, Comiso JC et al (1993) Snow and ice applications of AVHRR in polar regions. Ann Glaciol 17:1–16
Stogryn A (1988) Investigation of extensions to the distorted Born approximation in strong fluctuation theory, RN 9316, Aerojet Electrosystem, Azusa, CA
Stowe LL, Yeh HYM, Eck TF, Wellemeyer CG, Kyle HL (1989) Nimbus-7 global cloud climatology, Part I: First-year results. J Climate 2:671–681

Stroeve JC, Box JE, Fowler C, Haran T, Key J (2001) Intercomparison between in situ and AVHRR polar pathfinder-derived surface albedo over Greenland. Rem Sens Environ 25:360–374

Sullivan CW, Arrigo KR, McClain R, Comiso JC, Firestone J (1993) Distributions of phytoplankton blooms in the Southern Ocean. Science 262:1832-1837

Suttles JT, Green RN, Minnis P et al (1988) Angular radiation models for Earth-atmosphere system. Volume I-Shortwave radiation. NASA Reference Publication 1184, NASA, Washington DC

Svendsen E, Matzler C, Grenfell TC (1987) A model for retrieving total sea ice concentration from a spaceborne dual-plarized passive microwave instrument operating near 90 GHz. Int J Remote Sens 8:1479–14877

Svendsen EK, Kloster K, Farrelly B, Johannessen OM, Johhannessen JA, Campbell WJ, Gloersen P, Cavalieri DJ, Matzler C (1983) Norwegian Remote Sensing Experiment: Evaluation of the Nimbus 7 Scanning Multichannel Microwave Radiometer. J Geophys Res 88(C5): 2781–2792

Swift CT (1980) Passive microwave remote sensing of the ocean – a review. Boundary-Layer Meteorol 18:25–54

Swift CT, Fedor LS, Ramseier RO (1985) An algorithm to measure sea ice concentration with microwave radiometers. J Geophys Res 90(C1):1087–1099

Taylor VR, Stowe LL (1984) Atlas of reflectance patterns for uniform earth and cloud surfaces. NOAA/NESDIS Technical Report, Washington DC

Tschudi M, Curry JA, Maslanik JA (2001) Airborne observations of summertime surface features and their effect on surface albedo during SHEBA. J Geophys Res 106:15335–15344

Tjuatja S, Fung AK, Dawson MS (1993) An analysis of scattering and emission from sea ice. Remote Sens Rev 7:83–106

Ulaby FT, Moore RK, Fung AK (1986) Microwave remote sending, Active and Passive: Theory and Applications. Artech House, Inc., Dedham, MA USA

Van de Hulst (1981) Light scattering by small particles. Dover Publications Inc., New York

Vant MR, Gray RB, Ramseier RO, Makios V (1974) Dielectric properties of fresh and sea ice at 10 and 35 GHz. J Appl Phys 45(11):4712–4717

Vant MR, Ramseier RO, Makios V (1978) The complex-dielectric constant of sea ice at frequencies in the range 0.1-40 GHz. J Appl Phys 49(3):1234–1280 et al 1978

Vincent RF, Marsden RF, Minnett PJ, Buckley (2008) Arctic waters and marginal ice zones: 2. An investigation of Arctic atmospheric infrared absorption for AVHRR sea surface temperture estimates. J Geophys Res 113, C08044, doi:10.1029/2007JC004354

Walton CC, Pichel WG, Sapper JF (1998) The development and operational application of nonlinear algorithms for the measurement of sea surface temperatures with the NOAA polar-orbiting environmental satellites. J Geophys Res 103:27,999–28,012

Wang J, Cota GF, Comiso JC (2005) Phytoplankton in the Beaufort and Chukchi Seas; Distributions, dynamics, and environmental forcing. Deep Sea Res II 52:3355–3368

Warren SG (1982) Optical properties of snow. Rev Geophys Space Phys 20:67–89

Warren SG (1984) Optical constants of ice from ultraviolet to the microwave. Applied Optics 23:1206–1225

Welch RM, Kuo KS, Sengupta SK (1990) Cloud and surface textural features in polar regions. IEEE Trans Geosci Remote Sens 28:520–528 4.3.1

Welch RM, Sengupta SK, Goroch AK, Rabindra P, Ragaraj N, Navar MS (1992) AVHRR imagery: an intercomparison of methods. J Appl Meteorl 31:405–420

Wentz JF (1983) A model function for ocean microwave brightness temperature. J Geophys Res 898(C3):1892–1908

Wentz JF (1992) Measurement of oceanic wind vector using satellite microwave radiometer. IEEE Trans Geosci Remote Sens 30(5):960–972

Wentz FJ, Gentemann C, Smith D, Chelton D (2000) Satellite measurements of sea surface temperature through clouds. Science 288:847–850

Williams J (1973) Optical properties of the ocean. Rep Prog Phys 36:1567-1608

Winebrenner DP, Bredow J, Fiung AK, Drinkwater MR, Nghiem S, Gow AJ, Perovich DK, Grenfell TC, Han HC, Kong JA, Lee JK, Mudaliar S, Onstott RV, Tsang L, West RD (1992)

Microwave sea ice signature modeling. In: Carsey F (Ed) Microwave Remote Sensing of Sea Ice, Geophysical Monography 68, American Geophysical Union, Washington DC
Xiong X, Stamnes K, Lubin D (2002) Surface albedo over the Arctic Ocean derived from AVHRR and its validation with SHEBA data. J Appl Meteorol 41:413–425
Yaguchi R, Cho K (2009) Validation of sea ice drift vector extraction from AMSR-E and SSM/I data using MODIS data. J Remote Sens Jpn 29(1):242–252
Yamonouchi T, Suzuki K, Kauragouchi S (1987) Detection of clouds in Antarctica from infrared multispectral data of AVHRR. J Meteorol Soc Jpn 65:949–961
Yhann SR, Simpson JJ (1995) Application of neural networks to AVHRR cloud segmentation. IEEE Trans Geosci Remote Sens 33(3):590–604
Zwally JZ, Yi D, Kwok R, Zhao Y (2008) ICESat measurements of sea ice freeboard and estimates of sea ice thickness in the Weddell Sea. J Geophys Res 113: doi: 10.1029/2007/JC004284

Chapter 5
Variability of Surface Pressure, Winds, and Clouds

Abstract The variability of sea level pressure and winds is studied using National Centers for Environmental Prediction (NCEP) reanalysis monthly-average data from 1979 to 2008. Results of analyses of monthly distributions of pressures and winds indicate that on average the seasonal patterns are consistent and coherent, but large interannual variabilities in the locations of lows and highs are observed. For a given month the interannual changes in wind circulation patterns can be very large and thereby could affect the spatial distribution of surface temperature and sea ice cover. Cloud frequency distributions as derived from AVHRR data also show large interannual variability and a slight decadal decline in both hemispheres

Keywords Sea level pressure • Hemisphere

5.1 Introduction

Among the fundamental parameters that govern the climate system are sea level pressure, winds, clouds, surface temperature, and albedo. These five variables have special and important roles in many of the physical processes occurring in the polar regions. Pressure and winds are the primary drivers of near-surface atmospheric circulation and the distribution and transport of heat and humidity around the Earth. These two variables are strongly coupled, and it is usually advantageous to discuss them in tandem. Differences in atmospheric pressure cause the movement of air masses referred to as winds, the strength and direction of which depend on the location and the magnitude of the differences in pressure as explained below. It is wind that drives the surface circulation of the oceans and the ice cover and that causes big waves, the advection of sea ice, and the ridging and rafting of ice within the pack. For example, it has been observed that in the Central Arctic, geostrophic wind accounts for 60–80% of the variance of the drift of sea ice during the year (Thorndike and Colony 1982; Serreze and Barry 1988). Also, it is wind that brings in warm air from lower latitudes to the poles, causing enhanced temperature in the region. Winds also cause the transport of cold air from high to low latitudes.

J. Comiso, *Polar Oceans from Space*, Atmospheric and Oceanographic Sciences Library 41,
DOI 10.1007/978-0-387-68300-3_5,

Identification of the location and patterns of high- and low-pressure systems and how they vary with time are needed to improve our understanding of the atmospheric circulation in the polar regions and the distribution of surface temperatures and other parameters (Chapman and Walsh 1993).

Likewise, cloud cover plays a very important role in the Earth's radiation balance and is among the most important component of the climate system. Because of their high albedo, clouds limit the amount of shortwave solar radiation that reaches the Earth's surface and at the same time, they trap longwave radiation coming from the surface and keep it from exiting to outer space. The high albedo is due to the high efficiency of the cloud particles (especially ice crystals and small drops) in scattering visible radiation. In general, thick clouds reflect much of the solar radiation and cool the Earth while thin and high clouds (or no clouds) allow the radiation to reach the surface and warm the Earth. Clouds are also good absorbers of longwave radiation emitted from the Earth's surface, and such radiation can be emitted in the downward direction and warms the Earth as well. The overall effect depends on cloud type, height, and characteristics of cloud particles, and although there is a balance between the cooling and warming effects, the cooling effect usually predominates. The impact of clouds on the radiation and heat balance in the polar regions is expected to be strong since for most of the year, the cloud cover in the regions is known to be persistent, with the cloud frequency estimates being in the range of 60–85% (Rossow 1989; Schweiger 2004; Spinhirne et al. 2005). Cloud cover is also a dominant component of the hydrological cycle since it causes the transport of water vertically and horizontally and the removal of water vapor from the atmosphere through precipitation. It also plays an active role in many chemical reactions in the atmosphere and the vertical transport of chemicals.

In this chapter, we will discuss the general characteristics and variability of sea level pressure, winds, and clouds in the polar regions. Surface temperature and albedo will be discussed separately and in detail in Chap. 6. Sea level pressure is a parameter that is currently not possible to measure directly from satellite sensors. We usually rely on surface measurements and radiosondes launched from meteorological stations, ships, and other places around the world in conjunction and depend on models to obtain spatial distributions of surface pressure. Vector winds over the open ocean are now routinely measured by scatterometers (Freilich and Dumbar 1999) an example of which is the QuikScat/SeaWinds launched on June 19, 1999. Also, scalar winds are being derived from current and historical passive microwave data (Wentz 1992). With the launch of WindSat in January 2003, measurement of vector winds over ocean regions from passive microwave systems has also become possible. However, we still do not have any satellite system that can be used to reliably measure winds over sea ice and land. We make use of National Centers for Environmental Prediction (NCEP) reanalysis pressure and wind data (Kalnay et al. 1996) provided by the National Center for Atmospheric Research (NCAR) because it is the type of data set that provides continuous, consistent and long-term global coverage. Since 2002, the NCEP numerical model has been using satellite scatterometer wind data (e.g., QuikSCAT data) through assimilation techniques, and the resulting reanalysis data on winds have been shown to be generally accurate and are consistent with satellite data (Atlas et al. 2001; Chelton and Freilich 2006). The liquid and solid portions of the polar oceans are the main focus of this work, but the land portions are included in some of our analyses and discussion because of their relevance to ocean and sea ice process studies.

The cloud parameter that we analyze is cloud frequency, which we define as the fraction of the time that cloud is detected in each data element. The AVHRR data provide continuous information about the Earth's surface and its cloud cover, but the observation of surface parameters is possible only during clear sky conditions. We use the "clouds" or "no clouds" binary classification on each element of AVHRR data as discussed in Chap. 4 as the basic data used to determine cloud frequency. The cloud frequency is determined by keeping track of the binary classification and taking the ratio of the number of times cloud cover is detected in a data element and the total number of measurements in the data element. There are other sources of cloud data that probably provide more accurate results, including MODIS, MERIS, ICESAT, and Cloudsat. We use AVHRR cloud frequency data because of the relatively long historical record that enables variability and trend studies. This also enables comparison of cloud statistics with results from analysis of surface temperature, albedo, and other variables coming from the same sensor. We reiterate, however, that the detection of clouds in the polar regions with AVHRR data is difficult, as discussed by Simpson and Keller (1995), and the errors in the retrieval of clouds over snow and ice covered areas can be relatively large.

5.2 Sea Level Pressures and Winds

Pressure is usually defined as force per unit area and has been expressed in terms of a standard unit called bar, which is equivalent to 10^6 dynes per cm^2. Atmospheric pressure measurements are usually expressed in millibar (mb), or one thousandth of a bar. At a given altitude and location, the pressure in the atmosphere is determined by the weight of the air within an imaginary column along the vertical divided by the cross-sectional area of the column. Atmospheric sea level pressure has been observed to range from 850 to 1,084 mb with an average value of 1,013 mb, which is usually referred to as the standard atmosphere. The pressure depends on the temperature and density of the atmospheric gas above each surface area. Low- and high-pressure areas have been the focus of analysis in pressure maps because they relate to synoptic, horizontal comparison of atmospheric pressure. The centers of these areas are usually identified by connecting points at equal pressure called isobars. The relationship between pressures and wind is well known and governed by a concept called geostrophy. In this concept, the Earth's rotation and the effects of the Coriolis force and other factors are considered, and in the Northern Hemisphere, air flowing along a direction approximately parallel to isobars moves clockwise along high-pressure areas and counterclockwise along low-pressure areas. The direction of the air flow would be the reverse in the Southern Hemisphere. The direction can be altered by many factors including frictional drag over rough surfaces and topographical features. Near low-pressure areas, water vapor in air tends to form clouds as rising air expands and cools to saturation. Conversely, near high-pressure areas, the sinking motion compresses and warms the air causing the evaporation of liquid water droplet and a clear sky condition. Using this information, some high- and low-pressure systems can be identified indirectly using satellite visible and infrared data. When examining sea level pressure maps, the focus of interest is usually the horizontal pressure gradient that produces the net force that causes air masses to move from high-pressure areas to low-pressure areas. Strong winds are associated with high-pressure gradients or with isobars that are close to each other while weak winds occur where the isobars are farther apart.

There are areas on the surface of the earth that are historically regarded as areas where lows or highs are persistent. Among these are the Islandic Low in the Northern Atlantic and the Aleutian Low in the Northern Pacific. These areas have been known as stormy areas with high winds and extensive cloud cover. In the Southern Hemisphere a low-pressure system along a belt following approximately along the 60°S latitude is also quite known and has been the site of cyclones or intense weather conditions. The system is in part responsible for the strength and intensity of the Antarctic Circumpolar Current (ACC) as mentioned in Chap. 2.

5.2.1 Northern Hemisphere

The atmospheric circulation is defined mainly by pressure and wind patterns that change constantly from one season to another and from one year to the next. Typical sea level pressure and surface wind patterns for each month of the year from January to December are depicted in the images shown in Fig. 5.1. Each image is referred to as the climatology for the month and represents the average of NCEP reanalysis wind and pressure data for this month from 1979 to 2008. The data have been gridded in the standard polar stereographic format as described in Chap. 4. Two distinct low pressure regions are apparent in the January image, namely, one located at the Bering Sea (top left) and the other around the Greenland Sea (bottom middle-right) surrounding a high-pressure region over the Greenland ice sheet. These two low-pressure regions, which have been referred to as the Aleutian and Islandic lows earlier, basically control the wind circulation at high latitudes producing on the average two semicyclonic circulation patterns. The low-pressure patterns are shown to be most intense in the winter and to weaken considerably in spring and summer. The low-pressure systems start to build up at the end of summer in September and then grow in size and magnitude in October and November and reach their lowest values at around December or January. The low-pressure systems in these regions gradually weaken (i.e., become lighter blue) from February through March to April, and by May, the low-pressure areas shift by about 40°–90° to the east in a clockwise direction with respect to the North Pole. In July, the low-pressure system extends from Russia through the North Pole to Canada.

The locations of the high-pressure patterns in winter are consistently over the same general areas (i.e., mainly in Asia extending to the Arctic and North America, and the Atlantic Ocean and Southern Europe) but not as well defined as the

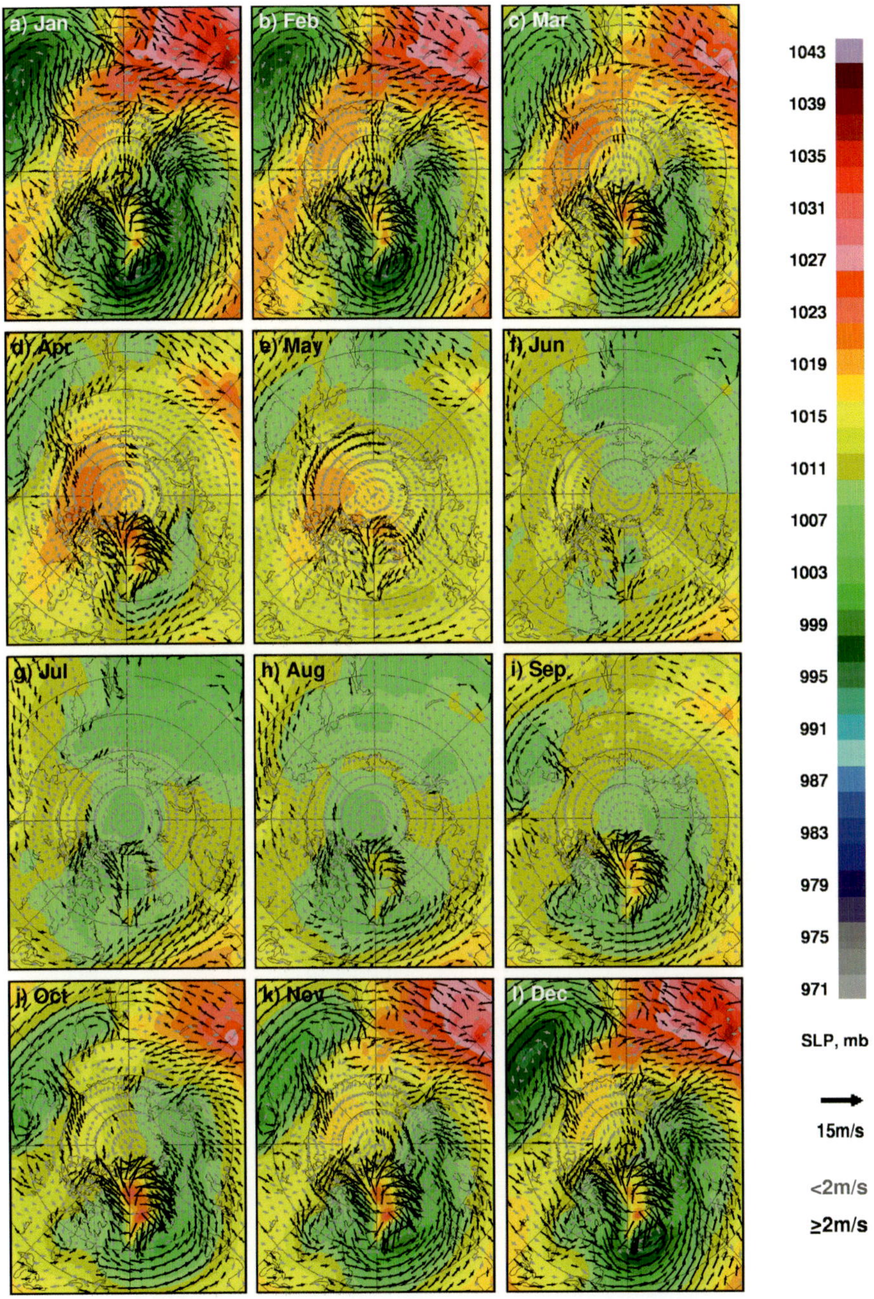

Fig. 5.1 Monthly Climatology of sea level pressure and wind as derived from NCEP reanalysis data as described by Kalnay et al. (1996) in the Northern Hemisphere (1979 to 2008)

low-pressure systems. During the summer, high-pressure areas occur only at relatively low latitudes in the Atlantic Ocean. Overall, the patterns are generally coherent from one month to the next, and it appears that pressure and winds in the Arctic would be predictable. However, as discussed below, there are large interannual variations, and for a given month, pressure patterns in one year can be very different from those of another year.

The monthly wind vectors as depicted in Fig. 5.1 follow the patterns that would be expected from geostrophy. Cyclonic wind patterns are apparent around the center of the lows in the Greenland Sea and Bering Sea, gaining strength as the lows deepen in October and November and reaching maximum values in December and January. The pattern is maintained through February and March and starts to weaken after that. Anticyclonic winds are also apparent around the high pressure system in Greenland. Winds are not so well defined in the high-pressure system in Asia (top right). In the summer, easterly winds are apparent at around 50°N to 60°N. It is interesting to note that in July and August, southerly winds are apparent at the Bering Strait, which could cause the flow of surface water from the Pacific Ocean to the Arctic while starting in September, the direction is reversed. The change in direction is associated with the deepening of the lows in the region during the period and the formation of cyclonic wind patterns. Strong winds along the Lomonosov Ridge (see Fig. 2.1) from Russia to Greenland are also apparent in December, January, and February. This could facilitate the transport of Arctic sea ice to the Fram Strait and the Greenland Sea.

Monthly averages of the pressures and winds in a winter month (i.e., February) are presented in Figs. 5.2a, 5.3, and 5.4. The series of figures is aimed at providing all the monthly data from 1979 to 2008 but at different levels of magnification to optimize the ability to interpret the data. Figure 5.2 shows a blown-up version of the images for both February 2008 and September 2008 to acquaint the reader with the gridding format and the geographical orientation of the maps and to show the

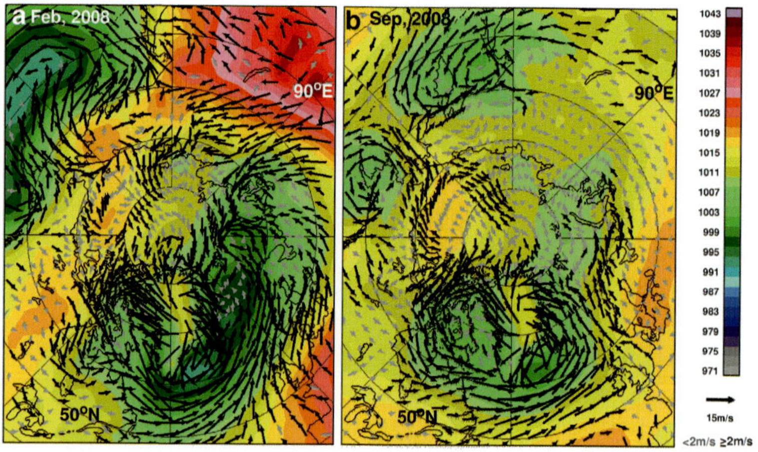

Fig. 5.2 Monthly averages of sea level pressure and wind in the Northern Hemisphere as derived from NCEP reanalysis data as described by Kalnay et al. (1996) in (**a**) February 2008 and (**b**) September 2008

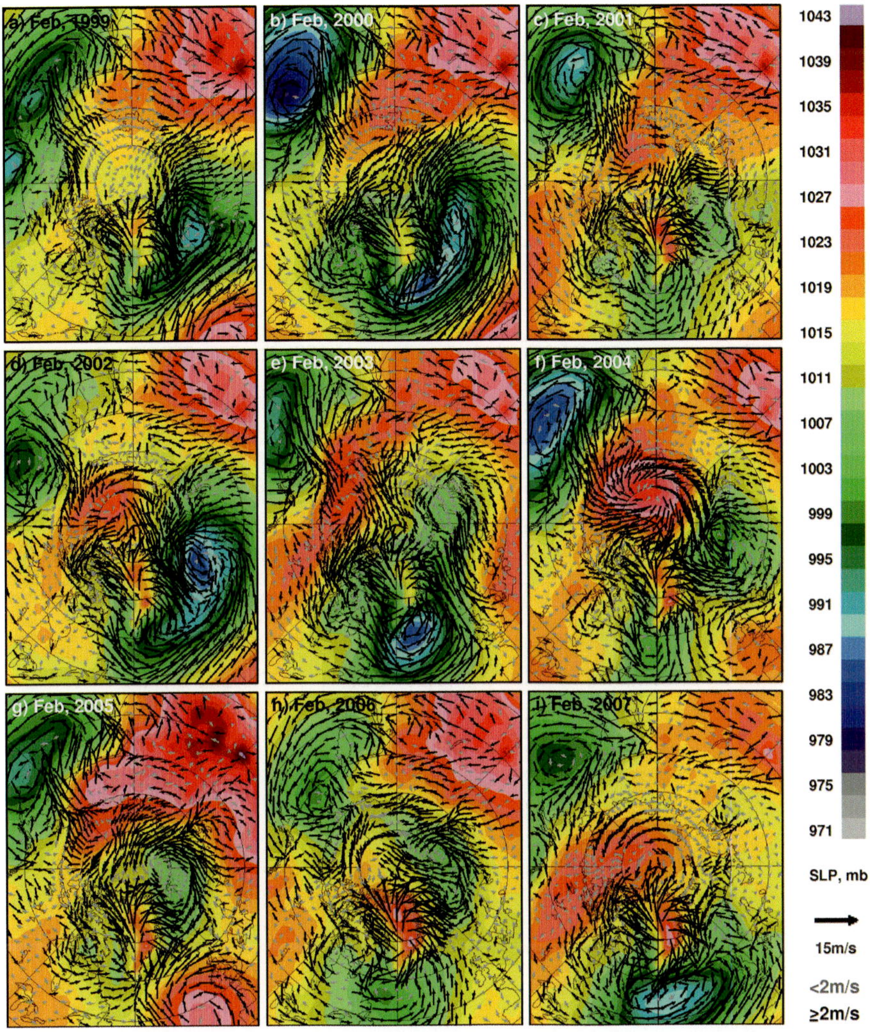

Fig. 5.3 Monthly averages of sea level pressure and wind in the Northern Hemisphere as derived from NCEP reanalysis data as described by Kalnay et al. (1996) in winter (February) for the years 1999 to 2007

spatial details of major features of interest. Labels for some latitudinal and longitudinal lines and outlines of the continents are also provided, but a useful area of reference is the relatively high pressure area in Greenland for most of the months. Generally, the pressure and wind patterns in February 2008 are very similar to those of the February climatology, as presented in Fig. 5.1, with the locations and extents of the lows around Greenland and the Bering Sea area almost identical. The set of images presented in Fig. 5.3 show the pressure and wind patterns for February in recent years (i.e., 1999–2007) and are presented in a format large enough for proper interpretation of most of the details in the data. It is apparent from this set that there is

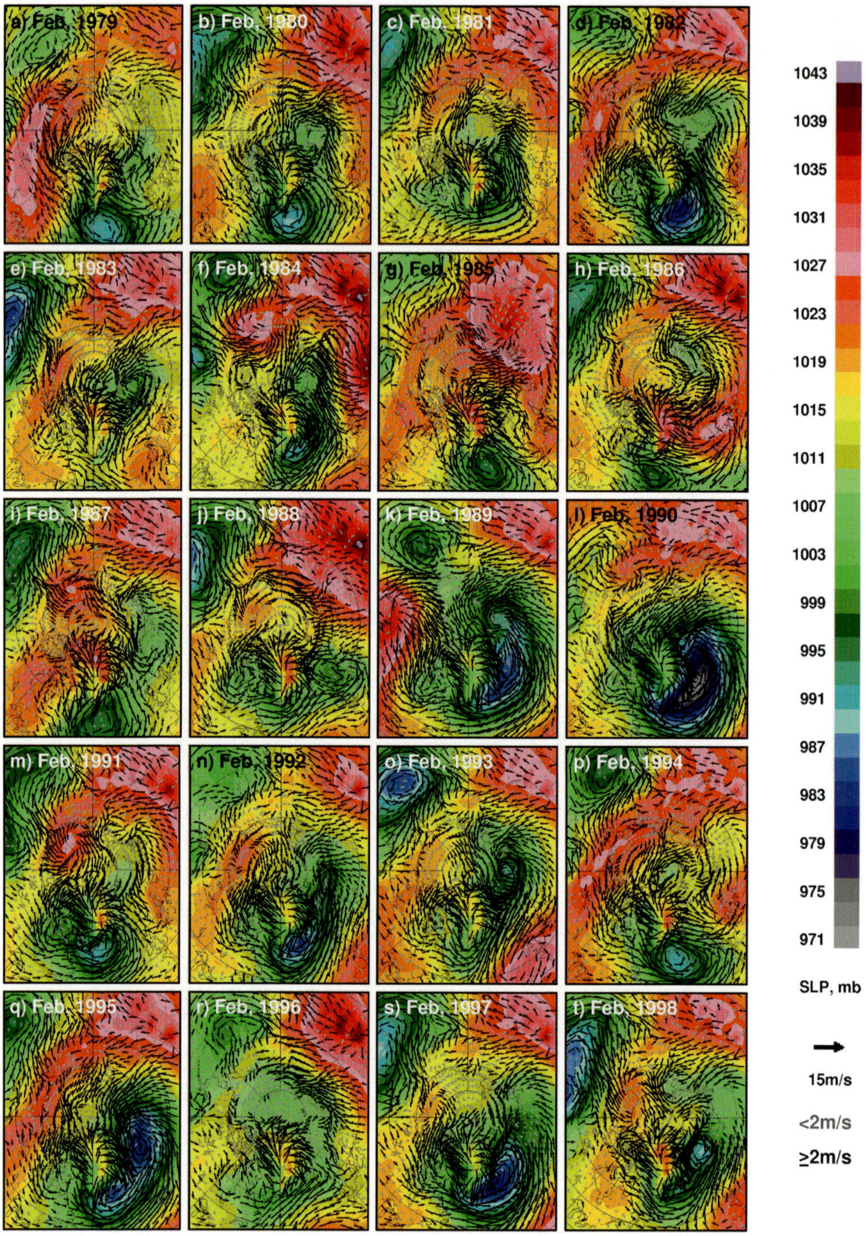

Fig. 5.4 Monthly averages of sea level pressure and wind in the Northern Hemisphere as derived from NCEP reanalysis data as described by Kalnay et al. (1996) in winter (February) for the years 1979 to 1998

large interannual variabilities in pressure patterns and winds during this period. The low-pressure system in the Bering Sea was unusually low in 2000 and 2004, and as expected the cyclonic winds in the region were relatively strong during these years. In the Greenland Sea, extremely low pressure systems occurred in 2000, 2002, and 2003, and winds in the region were also very strong. The pressure patterns in the Arctic Basin also change from one year to another with a strong high-pressure system apparent in 2002, 2004, and 2007. The wind strength and direction are also shown to vary considerably as illustrated by the change from a relatively weak to a very strong pattern from 1999 to 2000.

The set of images shown in Fig. 5.4 show data for the rest of the time series (i.e., 1979–1998), and it is clear that the pressure and wind patterns in earlier years were not any more stable than in the more recent years. The locations of the lows are generally in the same regions and consistent with the climatology, but it is obvious that there are significant interannual changes in shape, size, and magnitude as well as location of the pressure systems. The low-pressure systems are shown to be located farther to the north in the Central Arctic Basin in 1982, 1983, 1984, 1989, 1990, 1992, 1993, 1995, 1996, and 1998 than in other years. The wind patterns during these years were also generally cyclonic in the Arctic Basin. Note that even relatively minor changes in the location and distribution of the pressure patterns from February 1999 to February 2000 led to dramatic changes in wind circulations. The changes in pattern and magnitude of low-pressure areas from February 2000 to February 2001 also led to a completely different wind circulation in the Fram Strait/Greenland Sea region. In the Arctic Basin, the occurrence of a relatively confined high-pressure system around the North Pole in February 2002 and February 2004 led to a well-defined anticyclonic circulation in the region. The other years when the highs also intruded into the Arctic Basin were 1985, 1987, 1995, 2001, 2003, 2005, and 2007, and in these cases, the pressure and wind patterns were all different and also differed from the February climatology. The large interannual variability makes it difficult to predict the atmospheric circulation patterns despite the strong temporal coherence depicted in the climatology shown in Figure 5.1.

The corresponding sets of images but for each month at the end of the summer (i.e., September) are presented in Figs. 5.2b and 5.5–5.6. Because of relatively weaker wind strength, the scale has been changed for wind vectors to better illustrate the actual direction of the wind circulation. It is apparent that the contrast between the highs and the lows is not as strong in the summer as it is in the winter. Although not as deep as in February, the low-pressure systems were much more prevalent in the Arctic Basin in September especially in 1980, 1982, 1988, 1989, and 1997–2006. Quite often, when the low-pressure systems were centered in some locations in the Arctic Basin (e.g., 1988, 1989, 1997, 1999, 2000, 2002, 2003, and 2006) intense cyclonic circulation occured in the region. There are a few times (e.g., 1979, 1996, and 2007) when a high-pressure system intruded in the region, causing a reversal in wind circulation. In the North Pacific Ocean, a low-pressure system is usually found near the Bering Sea, but occasionally high-pressure systems would move in as in 1983, 1992, and 1996. In the North Atlantic Ocean and near the Greenland Sea, the low-pressure system gets shifted around, and there are times when it gets significantly weaker than normal (e.g., 1992, 1996, and 2001).

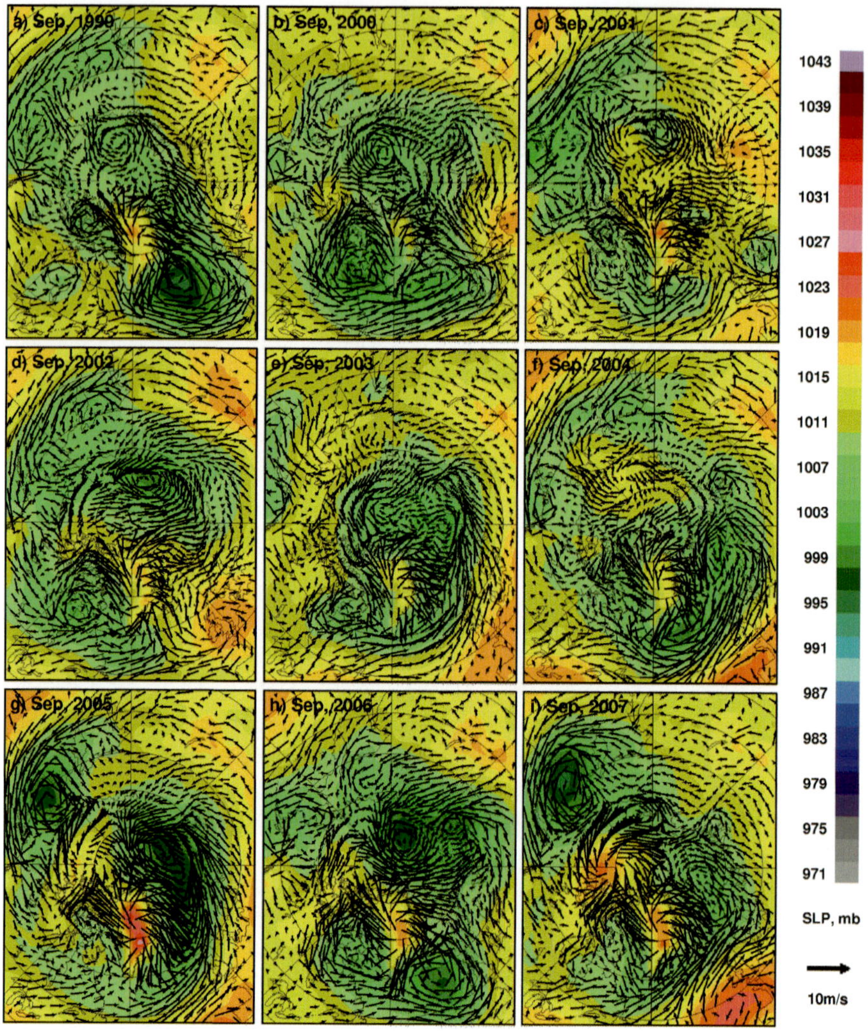

Fig. 5.5 Monthly averages of sea level pressure and wind in the Northern Hemisphere as derived from NCEP reanalysis data as described by Kalnay et al. (1996) at the end of summer (September) for the years 1999 to 2007

The wind circulation patterns are shown to change substantially from one year to another and from the Arctic Basin to the peripheral seas depending on the location and strength of the pressure systems. The large spatial variability makes it difficult to evaluate interannual changes and trends in the winds and pressure system in the Arctic region. But it would be good to know if the persistence of the lows in the late 1990s and in the 2000s is connected to the changes in the sea ice cover in the Arctic region. Note that the pressure patterns for September 2007 when the record low perennial ice cover occurred and for September 2005 when the previous record low

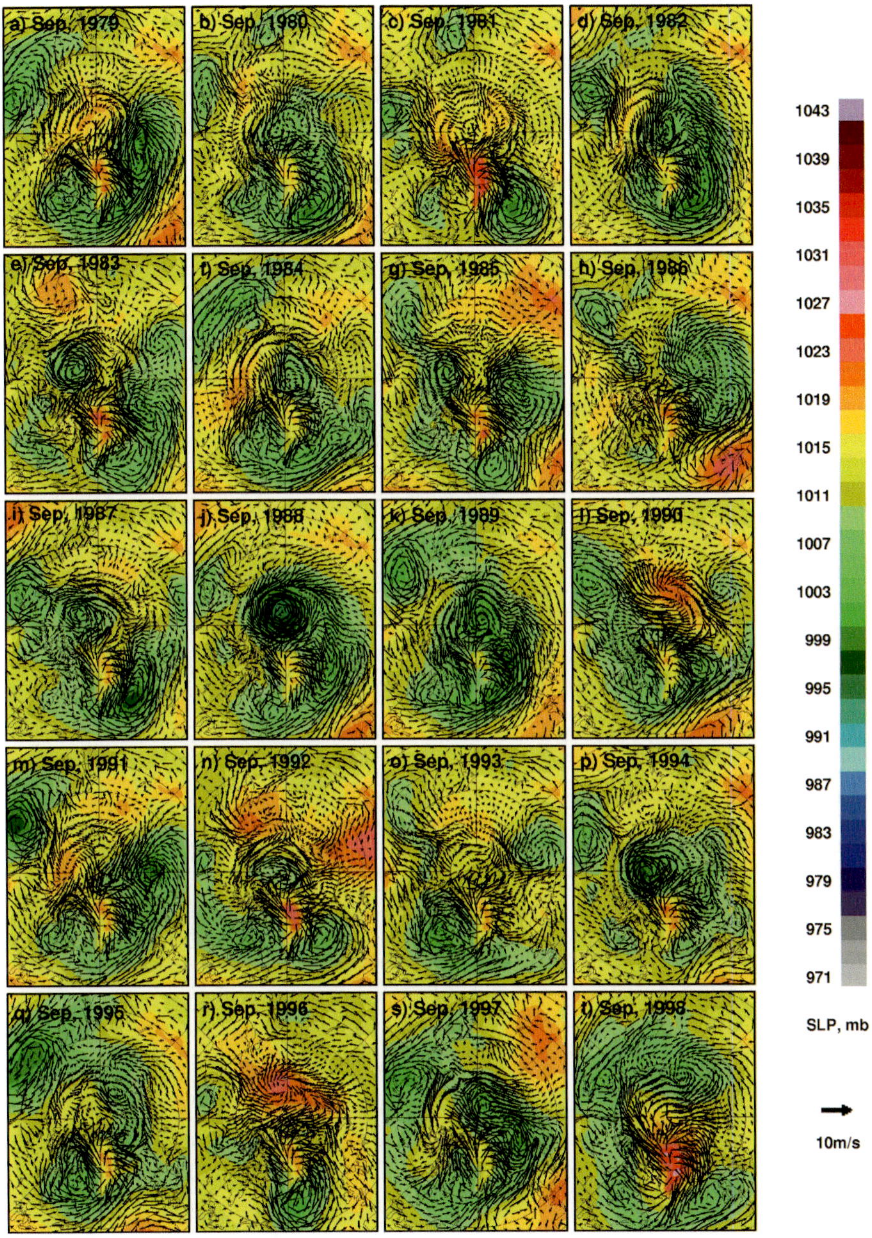

Fig. 5.6 Monthly averages of sea level pressure and wind in the Northern Hemisphere as derived from NCEP reanalysis data as described by Kalnay et al. (1996) at the end of summer (September) for the years 1979 to 1998

ice cover occurred are very similar. Some pressure and wind circulation patterns would be more favorable than other patterns to the flushing of Arctic sea ice cover at the Fram Strait or the migration of warm water to the north.

The large interannual and seasonal variabilities of the pressure and wind circulation system in the Northern Hemisphere, as revealed by the aforementioned images, are manifestations of a complex Arctic climate system. Efforts to understand this system and its variability have been going on for decades. Insights into the impact of the Icelandic low-pressure system have been provided by Hurrell (1995, 1996) who took the difference between normalized pressures at Lisbon (Portugal), representing the Azores Highs, and Stykkisholmur/Reykjavik (Iceland), representing the Icelandic lows. The monthly differences, referred to as "indices" as provided by the NOAA Climate Prediction Center, for the period 1979–2007 are presented in Fig. 5.7a (in gray with running averages in black). The indices have been called the North Atlantic Oscillation (NAO) indices, after an empirical orthogonal function analysis of the data from 1864 to 1983 revealed an oscillatory pattern. The data for the winter period as presented by Hurrell (1995) show predominantly negative

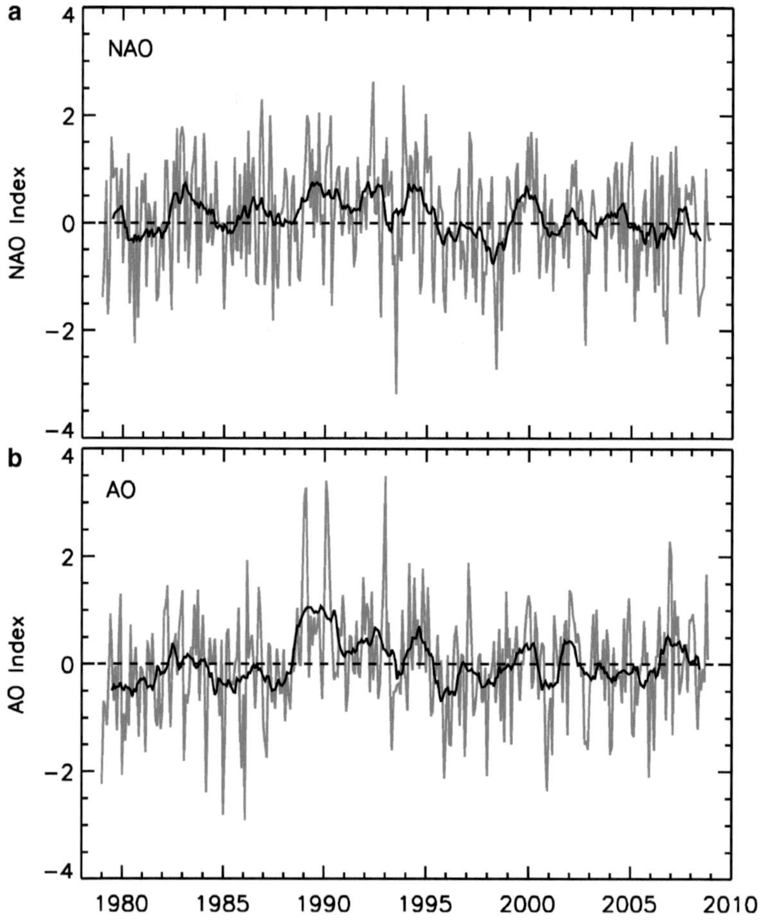

Fig. 5.7 (**a**) Monthly NAO indices from 1979 to 2007 using data described by Hurrell et al (1995); (**b**) Monthly AO indices from 1979 to 2007 using data described by Thompson and Wallace (1998)

indices in the late 1800s, positive indices in the early 1900s, negative indices in the mid-1900s, and positive indices in the late 1900s. The NAO has since been regarded as the dominant mode of atmospheric behavior in the North Atlantic region throughout the year. The period 1979–2007 shows mainly the positive part of the NAO index time series. However, it is clear from the continuous monthly data in Fig. 5.7a that there are positive as well as negative phases during this period with the running average indices (bold line) very close to zero from 2002 to the present. An updated version of the Hurrell data is presented in Fig. 5.8a for the winter periods from 1951 to 2008, and it appears that the periods from 1950 to 1970 were generally negative (with some positives from 1972 to 1975) and were positive from 1980 to the present. Winter in this case includes the months of December, January, February, and March. The peak values for the positive indices occurred between 1989 and 1994 and since then, the values were less positive or near neutral.

The NAO has been cited by Thompson and Wallace (1998, 2000) as a regional manifestation of a pan-Arctic system, which characterizes the atmospheric system for the entire Northern Hemisphere at high latitudes. This system has been named the Arctic Oscillation (AO) or more recently by some, the Northern Hemisphere Annular

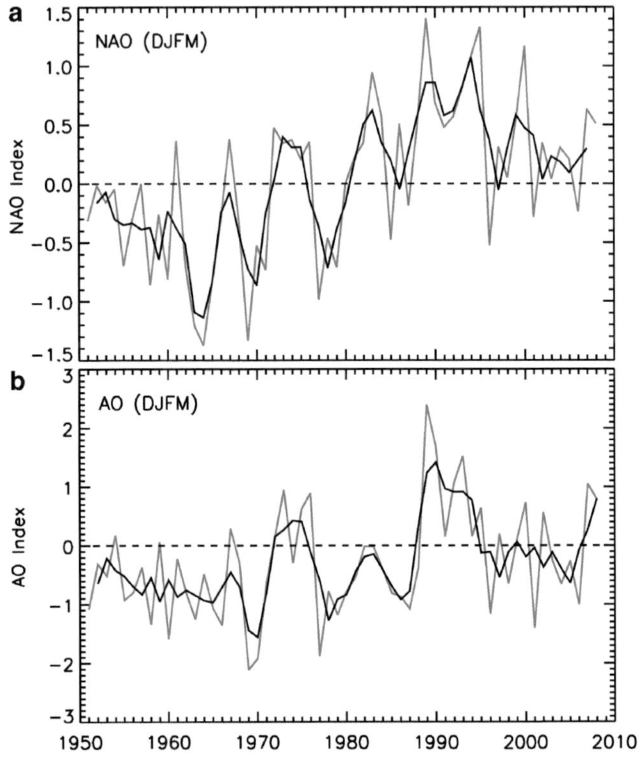

Fig. 5.8 (**a**) Yearly winter (DJFM) NAO indices from 1951 to 2008 using data described by Hurrell et al (1995); (**b**) Yearly winter AO indices from 1951 to 2008 using data described by Thompson and Wallace (1998)

Mode (NAM), to better reflect the annular character of the data utilized. The system is represented by AO indices, which are based on the difference in pressure between those inside the Arctic basin and the average over an annular area around the basin. The AO is expected to define the variability of the Arctic atmospheric circulation system including the observed seesaw of atmospheric air masses between the polar ice cap and the middle latitudes that includes both the Atlantic and Pacific Ocean basins. The monthly variability of the AO indices for the period 1979–2008 is depicted in Fig. 5.7b. It turned out that for the long-term period, the NAO and the AO are very similar and highly correlated (Dickson et al. 2000) with the correlation coefficient of monthly anomalies over the Northern Hemisphere during the cold season being about 0.95 (Deser et al. 2000). An updated version of the AO for the period 1951 to 2008 is presented in Fig. 5.8b, and indeed, there are similarities between the AO and NAO variabilities. However, it is apparent that there are more negatives in the AO than in the NAO, and for the period 1995–2008, the AO was mainly negative until it became positive in 2007 and 2008. The two plots are, however, coherent and show good correlation.

The variability of the AO and NAO has been assumed to be closely linked to the variability of the Arctic climate system and is expected to be closely connected to the variability of surface temperature, clouds, precipitation, and the sea ice cover. The signatures of the AO and NAO on local temperatures and precipitations have been found to be very similar, and the effect of one also reflects the effect of the other (Thompson and Wallace 2000; Hurrell 1995). Analysis of Arctic pressure data by Walsh et al. (1996) has indicated, as illustrated in Fig. 5.9, that the sea-level pressure (SLP) in the Arctic declined significantly during the period 1950–1994. This suggests that the AO and NAO must have been changing as well over this time period. Rigor et al. (2002) analyzed Arctic buoy ice drift data from the 1970s to 1998, and his results suggested that during the period of extremely high AO indices (i.e., 1989–1990) a large fraction of the thick multiyear ice floes might have been advected out of the Arctic basin. This was because the Arctic circulation favored the advection of multiyear ice into the Greenland Sea through Fram Strait during this time period. The pressure and wind data for February 1989 and 1990 (Fig. 5.4) support this phenomenon. It is interesting to note that the perennial ice cover, which represents mainly multiyear ice floes, was basically stable up to 1997 after which it started to decline more rapidly (Comiso 2002; Comiso et al. 2008) while in the meantime, the AO has become neutral and has changed in phase (Overland and Wang 2005). Such changes in the AO have actually been attributed to significant shifts in the centers of the low-pressure systems (Zhang et al. 2008).

5.2.2 Southern Hemisphere

The climatological monthly averages from January to December of sea-level pressure and wind patterns in the Southern Hemisphere are depicted in the set of images shown in Fig. 5.10. As with the maps in the Northern Hemisphere, the averaging was done using data from 1979 to 2008. The pressure patterns are shown to be very

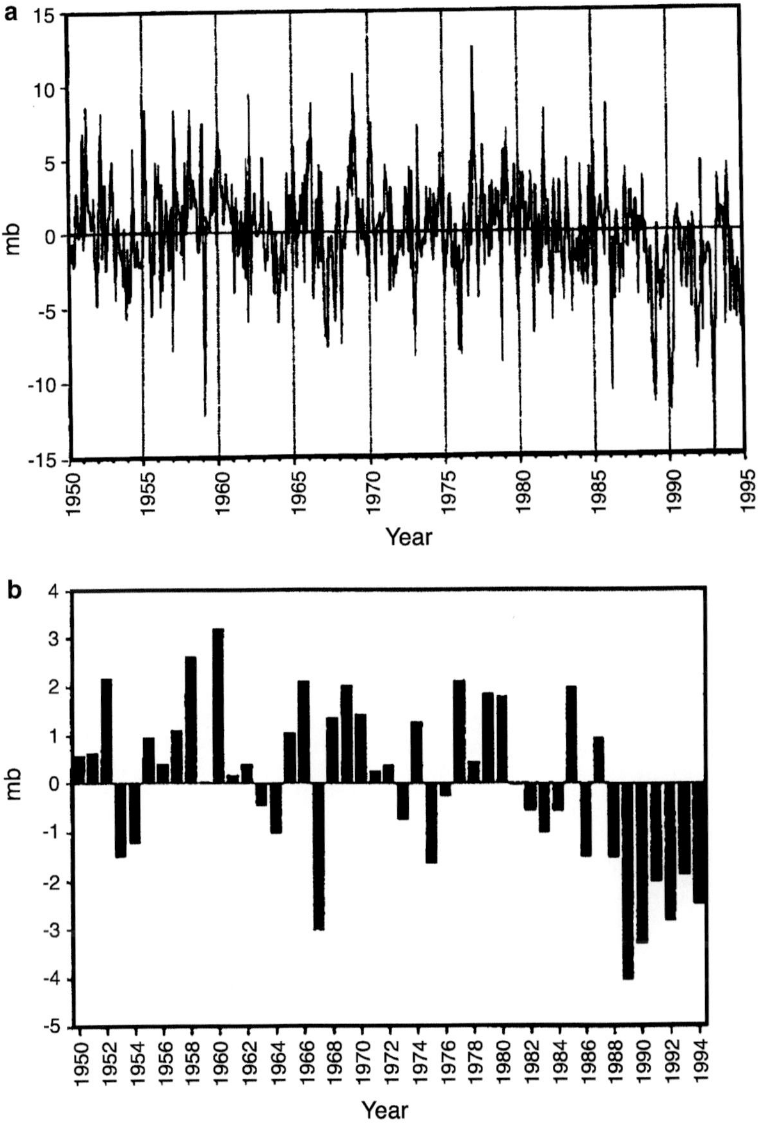

Fig. 5.9 (**a**) Monthly Sea level pressure (a) monthly averages from 1950 to 1994 and (**b**) yearly anomalies relative to the 1950 to 1994 mean from 70o to 90oN. From Walsh et al. (1996) with permission from the American Meteorological Society

seasonal with the contrast between the highs and the lows gradually increasing from summer to winter. The winds are therefore very seasonal especially over the Antarctic continent with the strongest winds occurring in mid-winter months. The circumpolar nature of the low-pressure system especially at around 50°S to 60°S is apparent. Along this latitude band, relatively low pressure alternates with relatively

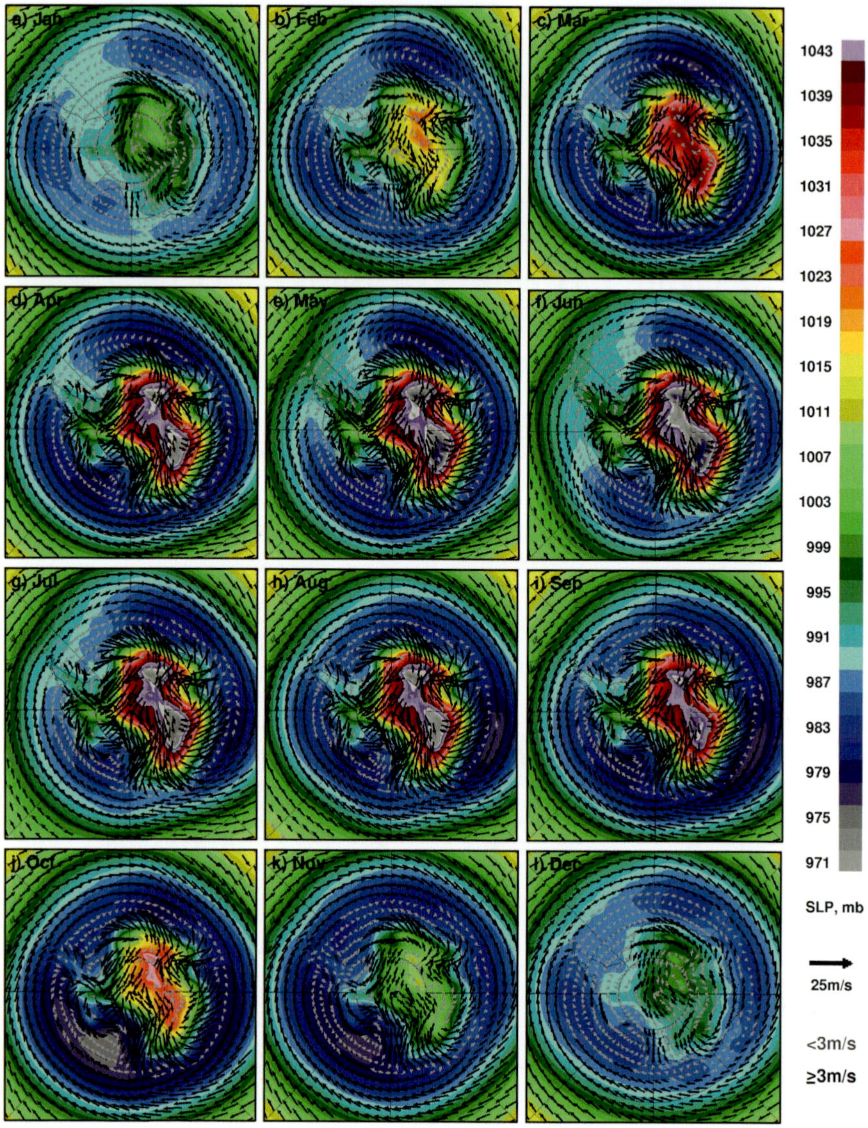

Fig. 5.10 Monthly climatology of sea level pressure and winds as derived from NCEP reanalysis data as described by Kalnay et al. (1996) in the Southern Hemisphere (1979 to 2008)

high pressure, but the pattern is not so predictable. A low-pressure system is apparent in the Western Pacific at approximately 130°E in August and September. The system deepened and moved further to the east at around 200°E in October and November. The lows become more moderate in December and January. The highs are shown to be consistently located in the continent and centered at the East

Antarctic plateau. The pattern of the pressure is similar to that of the topography of the continent. The sea level pressure in the continent is shown to be relatively low in January going up to more moderate values in February and March and becomes relatively high in April. It stays high from April to September, peaking at around the month of June.

The monthly climatology shows that Southern Ocean is dominated by westerly winds that generally follow the contour of the low-pressure systems around the continent. These relatively strong winds are the primary driver of the ACC. Closer to the continent, the winds gets diverted by the low-pressure systems and form well-defined gyres, such as the Weddell gyre, as discussed in Chapter 2), that cause winds to go counterclockwise along the continental boundary. Geostrophic and katabatic winds from the continent also serve to strengthen the southerly winds across the boundary.

The year-to-year variability of pressure and winds during a summer month is illustrated in the February monthly averages for 2008, 1999–2007, and 1979–1998 in Figs. 5.11a, 5.12, and 5.13, respectively. The monthly averages for each year show much more defined pressure contours and a higher contrast of high and low values than the February climatology shown in Fig. 5.10. The contours are also shown to vary considerably in magnitude and location from one year to another. The most active region is the latitude band between 50°S and 60°S. In this region, extremely low values occurred during some years as in 1989, 1990, 1994, 2002, 2005, and 2008 while much more moderate values occurred in 1980, 1983, 1984, 1985, 1986, and 2004. The pressure patterns in the continent are usually very similar, but interannual variations in magnitude can be considerable. The sea-level pressure was relatively high in East Antarctica in 1986, 1991, 1992, 1996, 1997, 1999, and 2001 while it was much more moderate in the region in 1985, 1988, 1989, 2005, and 2008.

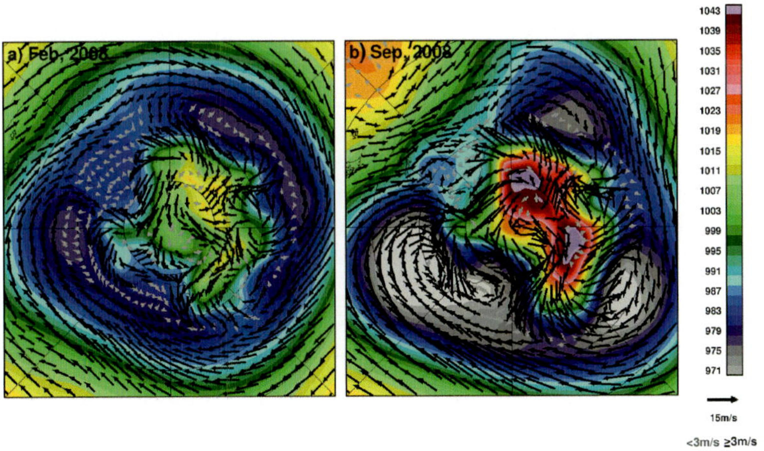

Fig. 5.11 Monthly averages of sea level pressure and winds in the Southern Hemisphere as derived from NCEP reanalysis data as described by Kalnay et al. (1996) in (**a**) February 2008 and (**b**) September 2008

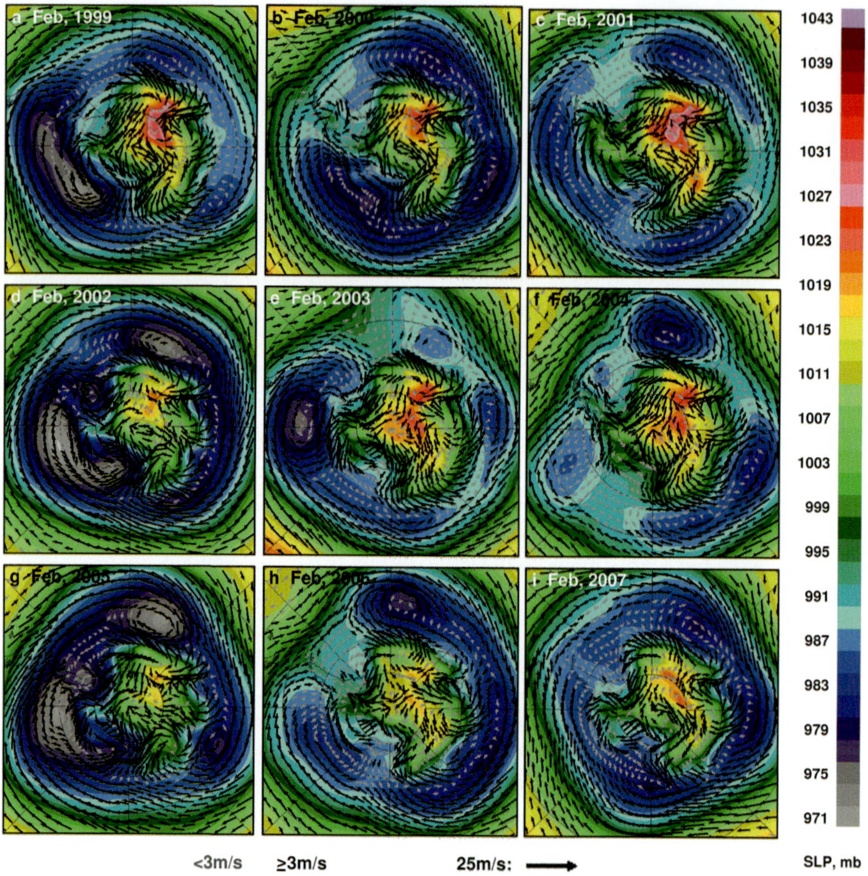

Fig. 5.12 Monthly averages of sea level pressure and winds in the Southern Hemisphere as derived from NCEP reanalysis data as described by Kalnay et al. (1996) in summer (February) for the years 1999 to 2007

An alternating pattern of low and intermediate pressures around the continent is observed, but generally they are not uniformly spaced and sometimes the lows have extended spatial coverage as in 1998, 2002, and 2008 making them almost circumpolar. The monthly wind circulation pattern in the region is also shown to have high interannual variability, as expected, since the wind pattern is directly linked to the pressure pattern. The pressure pattern has been analyzed statistically and has been identified as a propagating mode-2 pattern by White and Peterson (1996). A mode-2 system would have two lows on opposite sides of the continent and two relatively higher pressures in between. Examination of the monthly pressure maps indicated that the occurrence of mode-2 systems is not really the norm and that the patterns may be more complex as suggested by Comiso (2000) and Yuan and Li (2008). The pattern in February 1994 shows the classic mode-3 system where there are

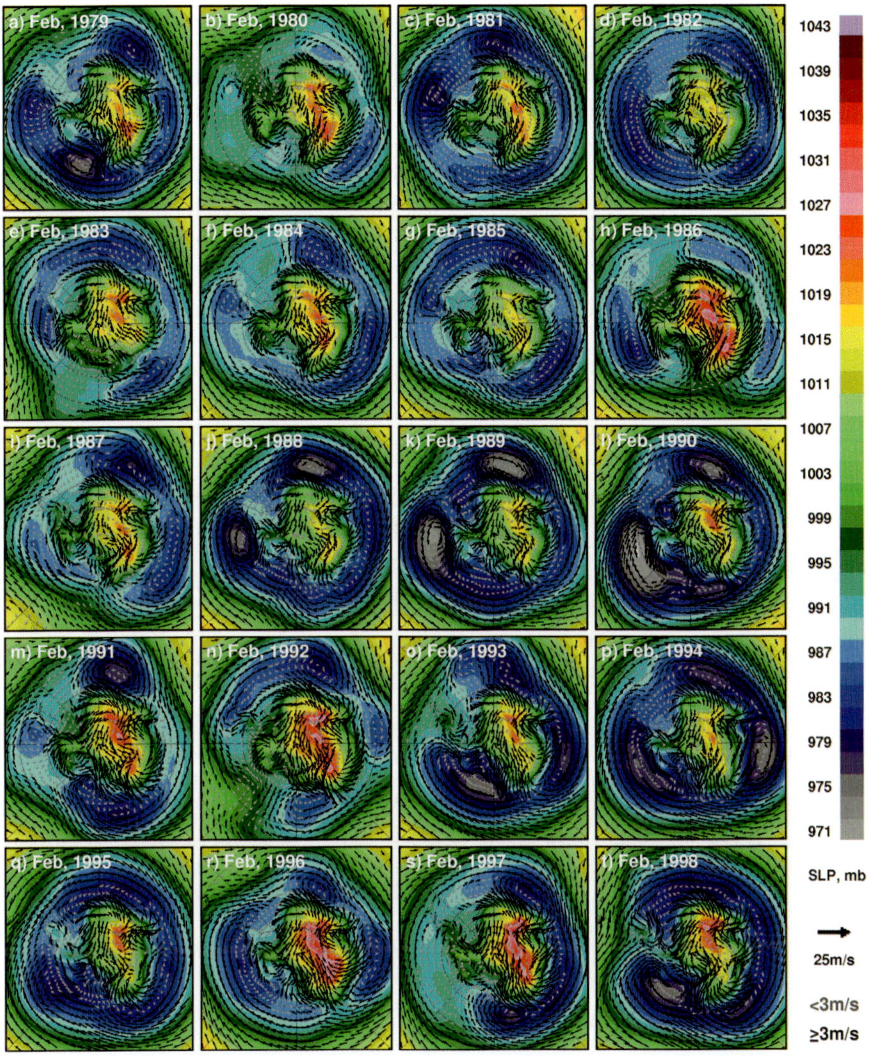

Fig. 5.13 Monthly averages of sea level pressure and winds in the Southern Hemisphere as derived from NCEP reanalysis data as described by Kalnay et al. (1996) in summer (February) for the years 1979 to 1998

three lows almost evenly spaced and three intermediate values in between. This is not the kind of mode that is associated with the Antarctic Circumpolar Wave (ACW) as described by White and Peterson (1996). The pressure maps in 1991 and 2008 are somewhat good examples of a mode-2 pattern. Many of the other years exhibited a semi-mode-3 pattern such as in 1988, 1993, 1996, 2000, 2001, 2002, 2004, and 2005, but the modal distributions are not as clear-cut as in 1994. During some others years, the mode is not obvious as in February 1998 when the system was in either mode 3 or mode 4. The ACW is also manifested in a similar manner in the surface temperature and sea ice data as discussed in Chapters 6, 7 and 8. A constant shifting of the location of the center of the lows is obvious and could cause interannual shifts in the location of the centroid and strength of the gyres around the Antarctic. However, there are areas such as the Amundsen Sea where the lows appear to be relatively persistent. Note that the high-pressure patterns in the continent are approximately in the same general location but the magnitude changes from one year to another.

The corresponding maps for a winter month (i.e., September) from 1979 to 2008 are presented in Figs. 5.11b, 5.14, and 5.15. The contrast between the highs and the lows is much higher during this period, and the winds are thus generally stronger, especially at high latitudes. The patterns of low and intermediate pressures around the continent are also apparent but almost as unpredictable as during the summer months.

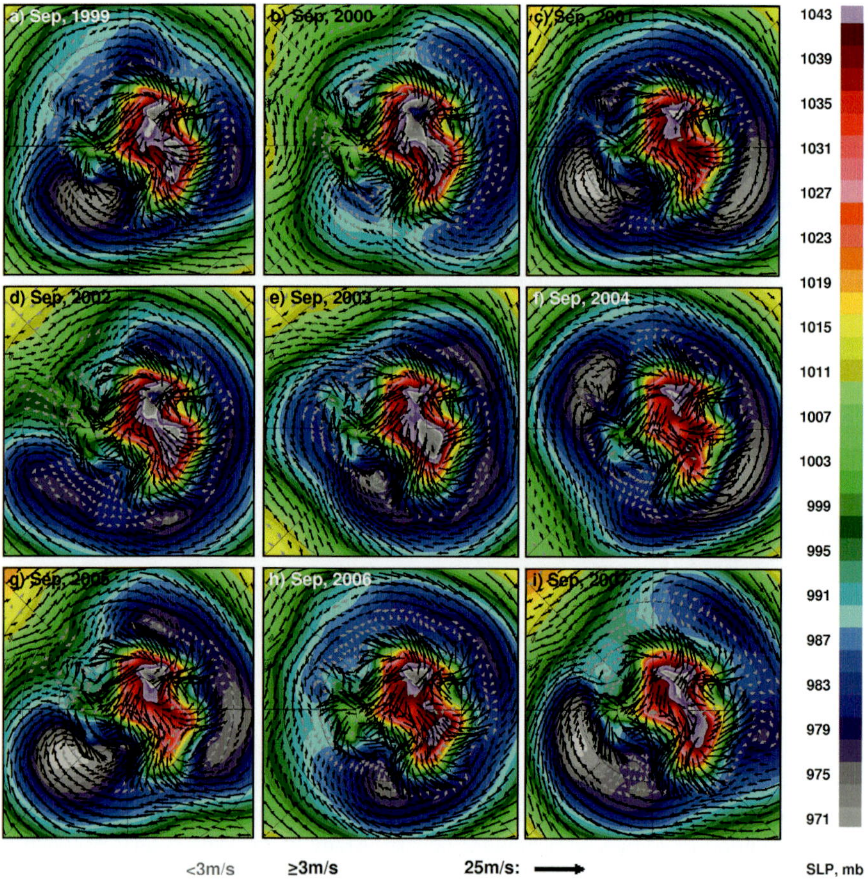

Fig. 5.14 Monthly averages of sea level pressure and winds in the Southern Hemisphere as derived from NCEP reanalysis data as described by Kalnay et al. (1996) in winter (September) for the years 1999 to 2007

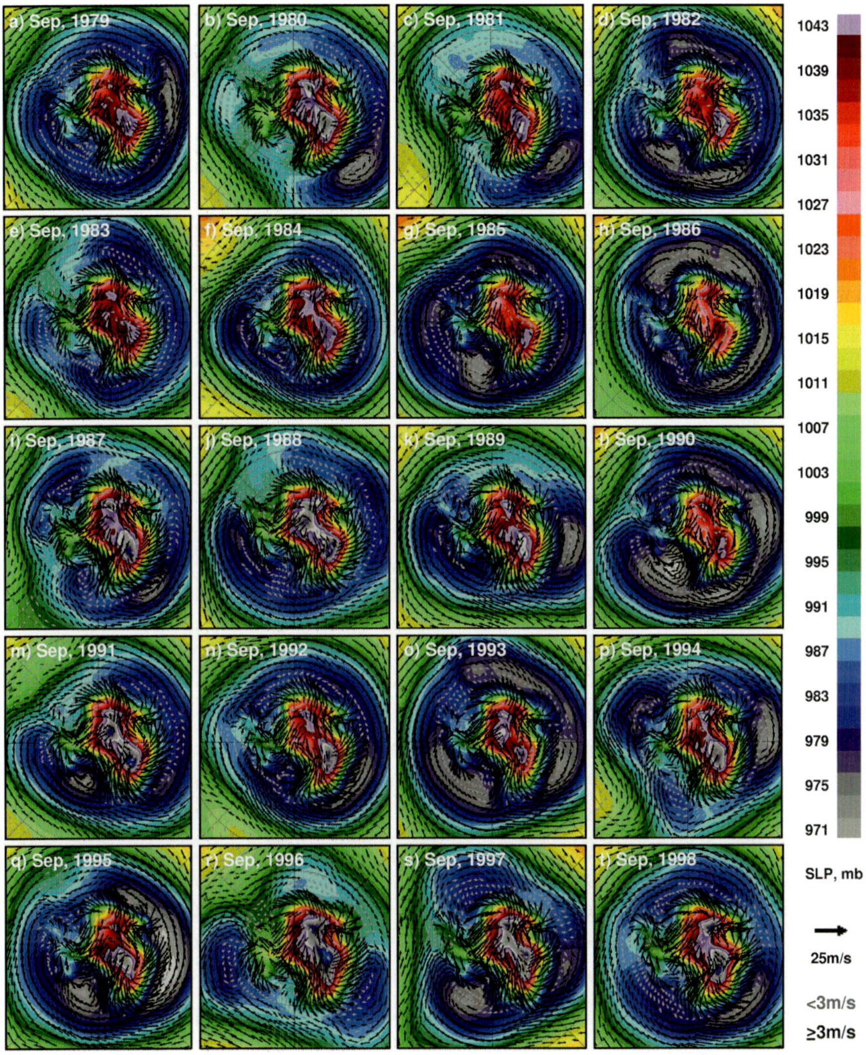

Fig. 5.15 Monthly averages of sea level pressure and winds in the Southern Hemisphere as derived from NCEP reanalysis data as described by Kalnay et al. (1996) in winter (September) for the years 1979 to 1998

Some mode-3 events can be identified (e.g., for September 1994, 1997, 1999, and 2005 events), but for some years, it is not straightforward to make the classification because the low-pressure systems can cover extended areas as in 1986 and 1990 or just a very limited area as in 1980 and 1981. The patterns in September 1993, 1995, and 2004 appear to be good candidates for mode-2 events. The patterns at intermediate latitudes are generally spatially symmetric but there are a few exceptions including September 1980, 1981, 1994, 1996, 1997, 1999, 2001, 2002, 2005, 2007, and 2008. Note that the Ross Sea is the site of a strong low-pressure system in 1985, 1990, 1993,

1995, 1997, 1999, 2001, 2003, 2005, 2006, and 2007. The higher frequency of extremely low pressure systems is consistent with stronger cyclonic flow in the region that favors more ice production in the Ross Sea (Turner et al. 2009). Strong lows were also apparent in the Western Pacific Ocean region at around 110°E in 1986, 1989, 1990, 1993, 1994, 1995, 1997, 1998, 2001, 2002, 2004, 2005, and 2008. The region is also a site of a few coastal polynyas that can serve as ice factories as discussed in Chapter 8, Section 8.2.2.

The atmospheric circulation at high latitudes in the Southern Hemisphere is governed mainly by the Antarctic Oscillation (AAO), sometimes called Southern Annular Mode (SAM) which has some resemblance to the Arctic Oscillation (or NAM). As described by Gong and Wang (1999) and Thompson and Wallace (2000), SAM represents the atmospheric interaction between air masses in mid-latitudes and those at high latitudes in the Southern Hemisphere. The interannual characteristics and variability is usually quantified through the use of SAM indices which are the difference of zonal mean sea level pressure between 40oS and 65oS. Much of the variability in the ocean circulation and sea ice distribution have been attributed to SAM (Visbeck and Hall 2004; Krahmann and Visbeck 2003). Relationships of sea ice, surface temperature and ACW patterns with SAM and ENSO have been studied and quantified by various investigators (e.g, Ledley and Huang, 1997; Kwok and Comiso, 2002; Peterson and White, 1998; Stamerjohn et al., 2008). The relationship varies with region and can be unpredictable as discussed by Kwok and Comiso (2002) and illustrated in Fig. 5.16. The correlation study of ice edge versus SAM shows relatively low correlation that varies in sign with longitude and is primarily negative at the Ross Sea and primarily positive at the Bellingshausen/Amundsen Seas and the Weddell Seas. For comparison, the correlations of the ice edge with the Southern Oscillation show generally similar fluctuations but the correlations are more significantly negative at the Ross Sea and more significantly positive at the Bellingshausen, Amundsen and Weddell Seas. The negative correlations are associated with colder temperatures while the positive correlations are associated with warmer temperatures. These are consistent with the trends in surface temperature and ice concentration in these areas as will be shown in Chapters 6 and 7.

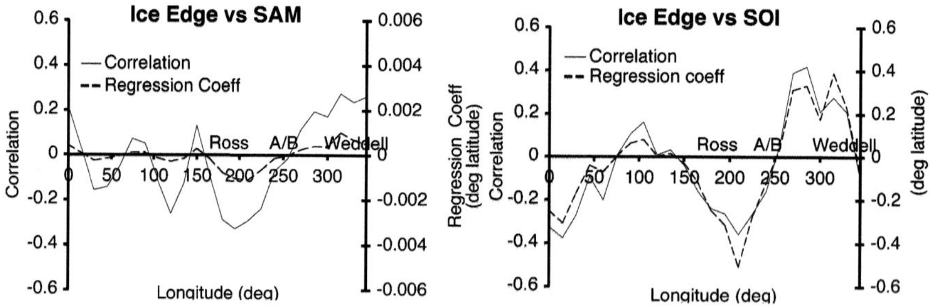

Fig. 5.16 Results of regression analysis of sea ice edge anomalies (in degrees of latitude) with SAM and SOI indices for the period 1982 to 1998. The correlation coefficients area also shown. From Kwok and Comiso (2002) with permission from the American Geophysical Union

5.3 Clouds

Clouds are formed from the condensation of water vapor and consist of tiny liquid droplets or solid ice crystals, which are about 0.01 mm in diameter. They are the most visible part of the atmosphere and the most distinctive feature of the Earth when viewed from space. It is interesting to note, however, that by volume, water vapor actually constitutes only less than 4% of atmospheric gases with nitrogen (78%) and oxygen (21%) providing the primary bulk. Dense thick clouds usually have a high reflectance of about 70–90% and generally look white while the thinner ones usually have much lower reflectance and have almost the same color as the surface when viewed from space. There are three main types of clouds, namely, cirrus, cumulus, and stratus, which are identified according to shape, height, and whether they are precipitating or not. The principal cloud shapes are the curly or fibrous clouds (e.g., cirrus clouds), the layered or stratified clouds (e.g., stratus clouds), and the lumpy or heaped clouds (e.g., cumulus clouds). By height, they are referred to as high clouds (e.g., cirrus clouds) when their bases are 6 km from the ground or higher, middle clouds when their bases are between 2 and 8 km, and low clouds when their bases are below 2 km. The low clouds usually include fog, which is composed of small drops (or ice crystals) in suspension mainly less than 1 km from the surface.

Detailed cloud cover characterization from space, including cloud classification, cloud height, cloud opacity and particle size is difficult and still a big challenge. With the advent of advanced multichannel sensors and also multi-sensor and multi-satellite capabilities such as the those associated with the A-Train, as discussed in Chap. 3, the potential for cloud studies is very promising. Knowledge about the cloud cover will undoubtedly improve tremendously with the new data that will come from these systems, but algorithms are still in the process of being developed and validated and it will take some time before the data set is suitable for interannual variability studies.

Among the most useful (and likely most accurate) cloud parameters that can be derived from AVHRR data is cloud frequency, which is generated automatically during the binary cloud identification process as indicated in Chapter 4, section 4.3.1. This parameter will be referred to as cloudiness, which in simple terms refers to the fractional cloud cover that is detected within the satellite footprint during the measurement period. Cloudiness is expressed either as a fraction (less than 1) or as a percentage. The spatial and temporal distribution of cloudiness, as derived from AVHRR data, is an important first step in getting an understanding of the overall distribution of clouds and their general characteristics. The data also provide the means to evaluate the effect of clouds on various surface parameters and enable improved interpretation of these retrieved parameters. Among the surface parameters that can benefit from cloud cover studies are sea ice concentration, sea ice extent, sea surface temperature, ice surface temperature, albedo, and ocean color pigment concentrations, which will be discussed in the next three chapters.

As with the other AVHRR surface temperature and albedo data sets, the cloud data have been generated and stored originally in a polar stereographic grid format with a resolution of 6.25×6.25 km. The resolution is usually degraded to a more manageable standard product, which is in a 12.5×12.5 km grid for large scale studies. Within a grid, our classification scheme provides the binary information of either cloud cover or no clouds. The statistics are compiled within each grid and then used to estimate the fraction of time there was cloud cover during a given period. We thus assume that within each pixel the cloud cover is 100% or cloud free. It would have been useful to estimate the percentage of clouds within each pixel, as we do for sea ice concentration, but because of too many cloud types there is currently no practical way for doing it. There are therefore likely some associated biases in this regard, but such biases would in part be compensated by our inability to identify all the cloud-covered areas. As indicated earlier, the errors in cloud retrieval could be relatively large especially over snow-covered regions. The data were, however, processed in a consistent manner, and can provide very useful information especially since the derived surface measurements, such as surface temperature, are coherent and in good agreement with available in situ data as discussed in Chapter 6. The derived AVHRR albedo also shows temporal and spatial coherence and is in good agreement with snow and sea ice distributions.

5.3.1 Northern Hemisphere

The seasonal variability in the spatial distribution of clouds over the pan-Arctic region is depicted in Fig. 5.17 using averages of all available AVHRR data for each month from January to December. This set of images will be referred to as cloud monthly climatology following similar references to the pressure and wind data. In these maps, high percentages of cloudiness are represented by very light gray or white while the less cloudy regions are represented by darker gray. Also for simplicity in interpretation, land areas are masked (i.e., shown in white) and will not be included in our analysis. The images show a pattern of high cloudiness in the open ocean and relatively low cloudiness over sea ice. This is consistent with less humidity at high latitudes and especially in ice-covered areas. The seasonality of the cloud cover in the Arctic region is, therefore, influenced in part by the seasonality of the sea ice cover. In the Arctic basin, the cloudiness is highest during the summer when the sea ice cover retreats to its minimum extent and the area of open water in the region is most extensive. The contrast in cloudiness over sea ice and ice-free areas may actually be less because of the higher efficiency in detecting clouds in ice-free regions where discrimination of cloud-covered and cloud-free surfaces is less problematic. Over ice-covered areas, the contrast is not as high and hence the error in cloud fraction is higher. Also, the fraction of clouds may be overestimated because of the binary nature of cloud cover identification and the fraction of cloud-free areas within the data elements classified as cloud covered is not considered. Overall, it is apparent that the Northern Hemisphere is generally cloudy at high latitudes and especially over ice-free regions. It also appears that the cloudiness is higher in ice-free regions during the winter months than in summer months.

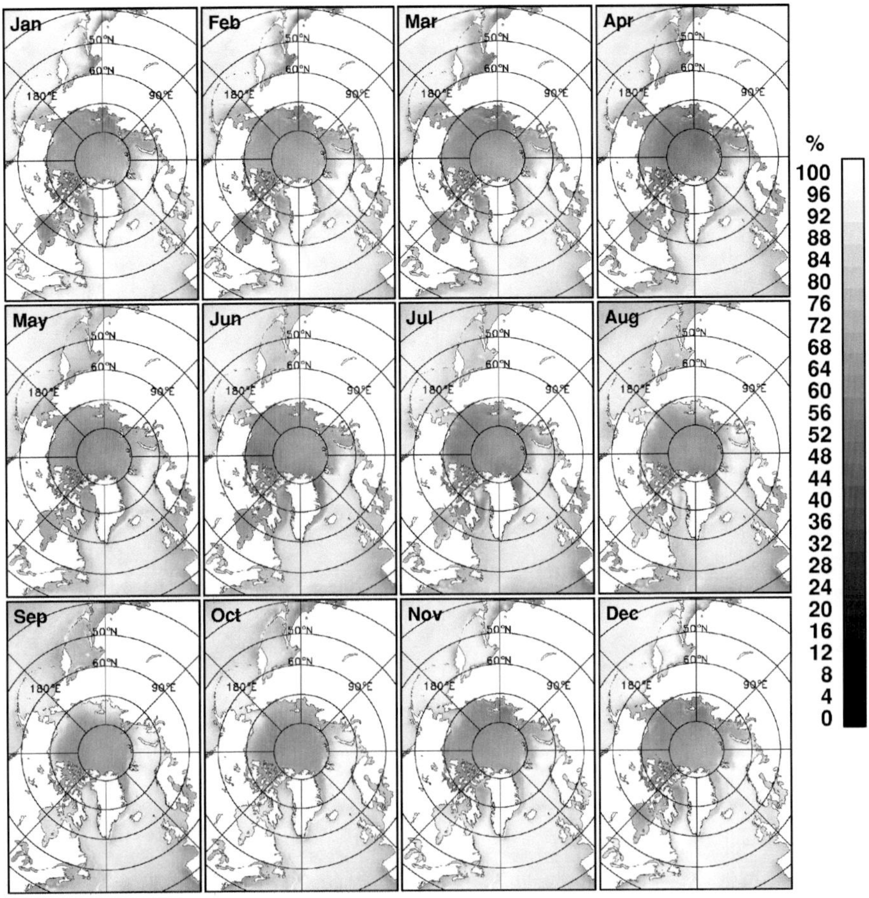

Fig. 5.17 Monthly climatological cloud cover from January to December in the Northern Hemisphere using AVHRR data from 1982 to 2008

The interannual variability of the cloud cover during the winter season is depicted by the February monthly images presented in Fig. 5.18. The maps provide a good general overview of the distribution of the cloud cover in the Northern Hemisphere in winter and allow for easy identification of changes in the cloudiness from one year to another during the 1982–2008 period. The interannual changes appear relatively modest and not as much over open ocean regions as in the sea ice-covered areas. This is caused in part by the persistence of clouds in open ocean regions, and the minor changes from one year to another are not so apparent in the images. For more quantitative comparison, we show in Fig. 5.19 anomaly maps for each February during the 1982–2008 period. The anomaly maps are based on the monthly averages in Fig. 5.18 and the climatology for February presented in Fig. 5.17. The yellows, reds, and purples represent less cloudiness in these maps while the greens, blues, and violets represent increased cloudiness. The anomaly maps show some relatively large positive anomalies as high as 15% in the 1980s. The year with the most

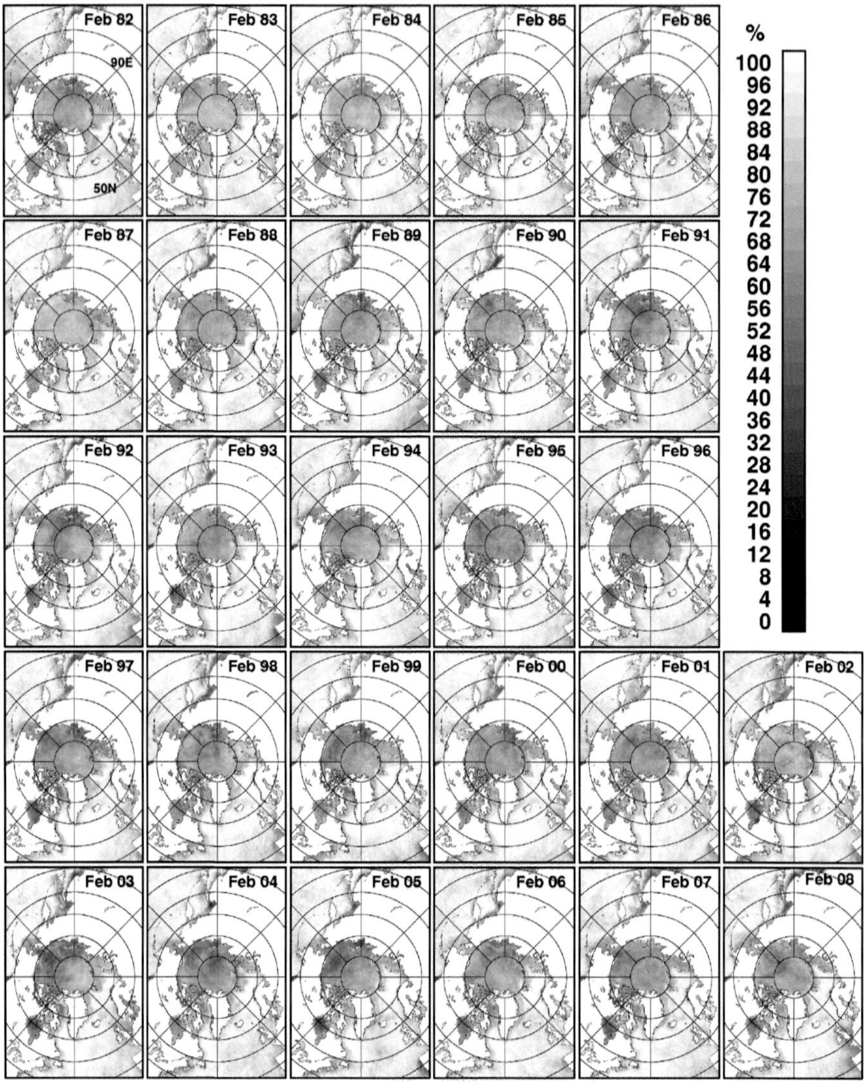

Fig. 5.18 Monthly averages of cloud cover in the Northern Hemisphere as derived from AVHRR data in winter (February) for the years 1982 to 2008

positive anomalies or most persistent clouds in the Arctic region in February is shown to be 1987 followed by 1983, 1984, 1985, and 2002. During the years from 1988 to 2008, a good balance of positive and negative anomalies is apparent except in 2002 when cloudiness was relatively high. From 1996 to 2008, the negative anomalies are relatively more dominant than in previous years except in 2002. Overall, the cloudiness in the pan-Arctic region appears to be on a decline from 1979 to 2008 and primarily over sea ice-covered areas. Interannual changes in the ice-free ocean are relatively more subtle except for a few exceptions as in 1982, 1989, 2005, and 2008.

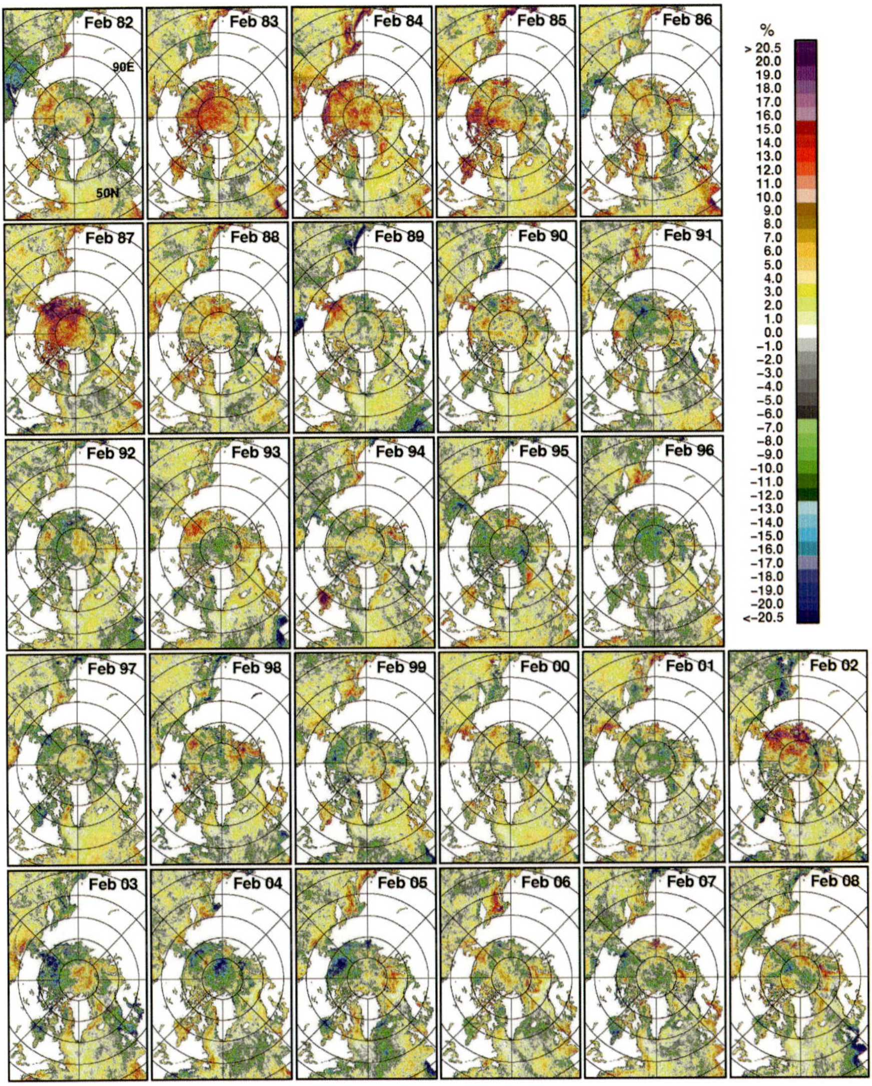

Fig. 5.19 Monthly averages of cloud cover anomalies in the Northern Hemisphere as derived from AVHRR data in winter (February) for the years 1982 to 2008

The corresponding cloud cover interannual variability during the summer is depicted by the set of maps in Fig. 5.20. We use the September monthly maps (for end of summer) because it is the time period when the sea ice cover is least extensive. These monthly maps combined with those in Fig. 5.18 provide the means to assess the yearly cloud distributions during winter and summer months and determine whether clouds are persistent in the same regions during the two periods. The year-by-year variability is again normally associated with the extent and location of the sea ice cover. In ice-covered regions, there are distinct year-to-

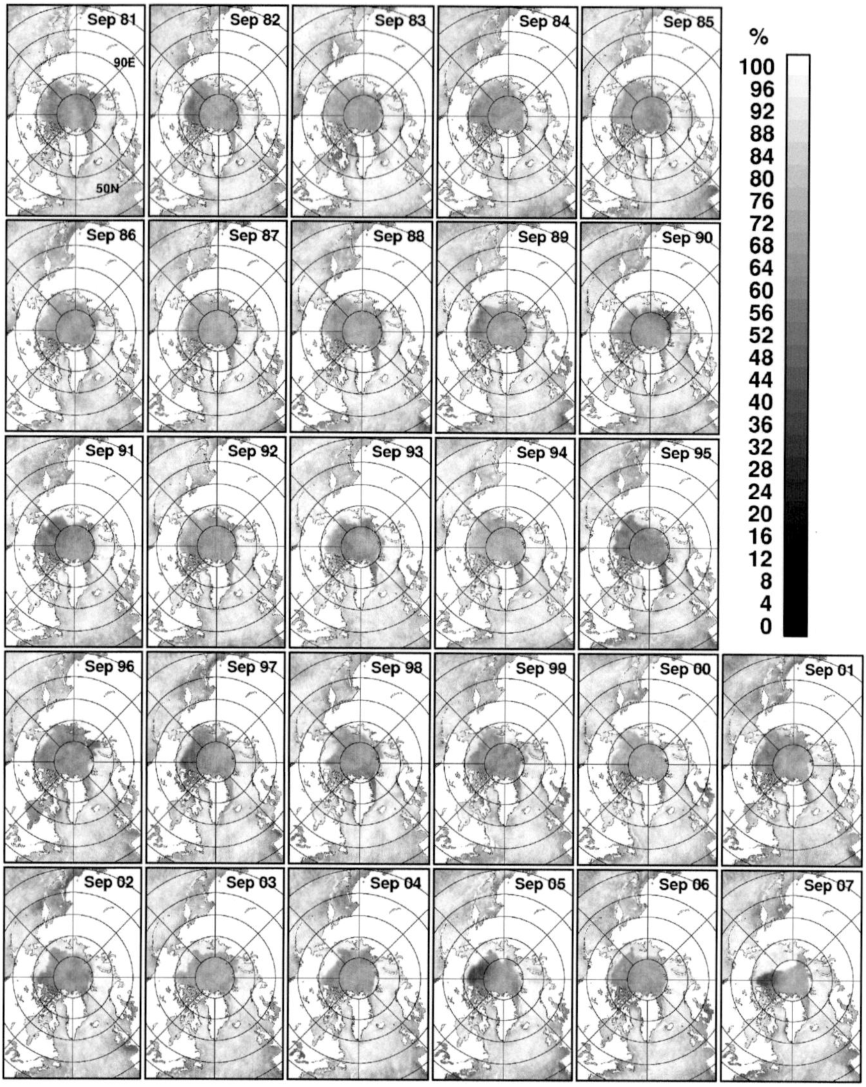

Fig. 5.20 Monthly averages of cloud cover in the Northern Hemisphere as derived from AVHRR data at the end of summer (September) for the years from 1981 to 2007

year variations, and it is apparent that in the Arctic basin, there was less than normal cloudiness in 1990, 1991, 1995, 2005, and 2007. In 2005 and 2007, there was apparently more cloudiness in the open ocean in the Arctic basin, but this is in part because of the considerable retreat in the sea ice cover during the two periods. Since the ice-free areas have relatively more clouds than ice-covered areas, the anomalously large retreat in the sea ice cover is manifested as an anomalously large increase in the cloud cover as well. Quantitatively, the yearly changes are more apparent in the color-coded monthly anomaly maps presented in Fig. 5.20. Overall, the set of images show less cloudiness from September 1981 to September 1994 and relatively more cloudiness from September 1995 to September 2007. The anomaly maps also show that the year-to-year changes are relatively modest during the periods from 1981 to 1994 and to some extent more variable from 1995 to 2007. The cloudiness was unusually high in 2007 when the perennial ice was less extensive than average. Again, the high cloudiness areas were located mainly in open water areas in the Arctic basin. In the sea ice covered part of the Arctic negative anomalies (or less clouds) predominate especially in

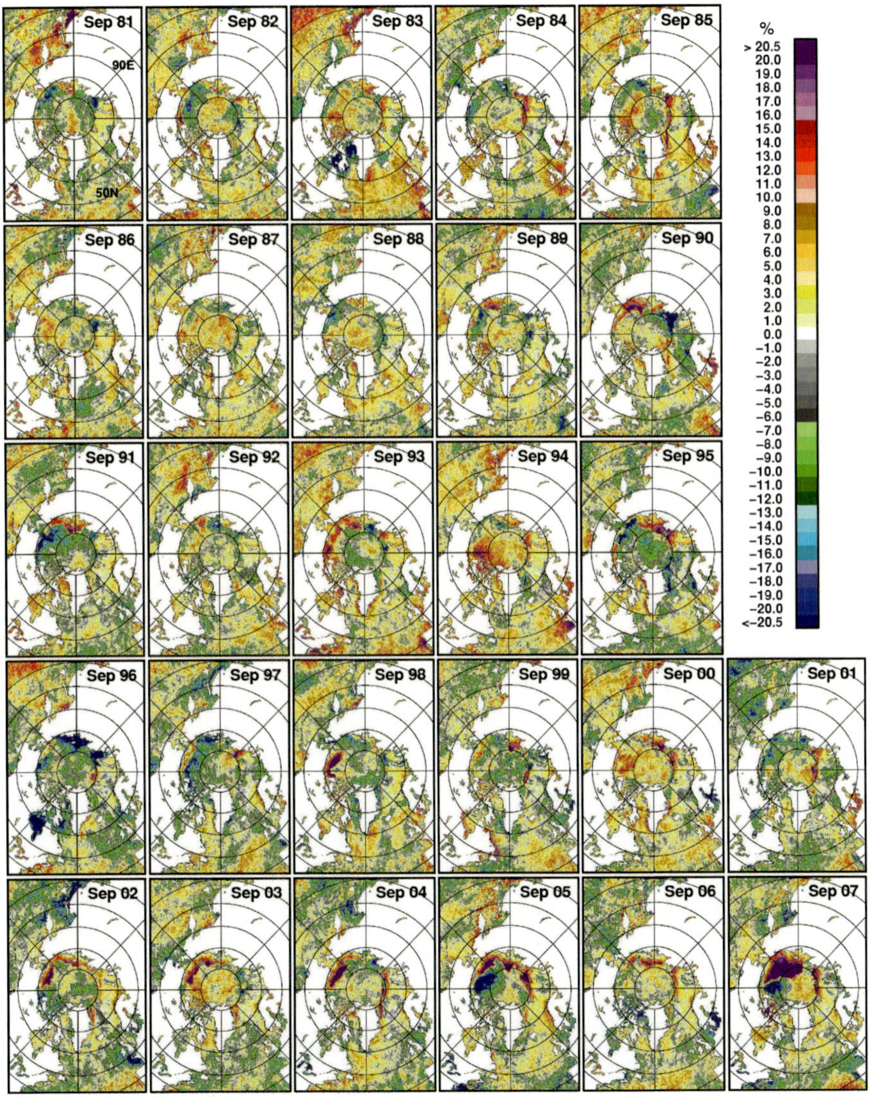

Fig. 5.21 Monthly averages of cloud cover anomalies in the Northern Hemisphere as derived from AVHRR data at the end of summer (September) for the years from 1981 to 2007

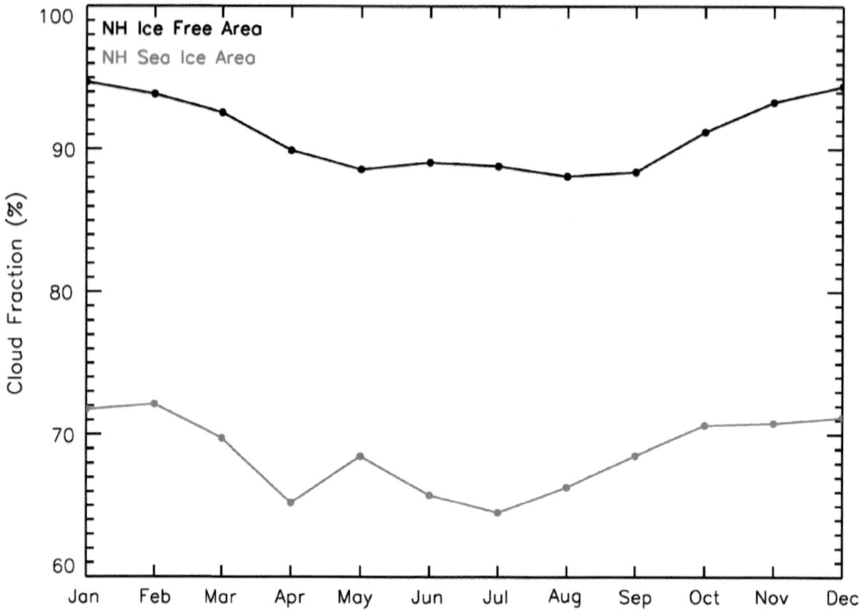

Fig. 5.22 Plots of monthly climatological seasonal distribution of cloudiness over open water (black line) and sea ice (gray line) at latitudes >60°N in the Northern Hemisphere

the Beaufort Sea region where sea ice was retreating the fastest during the period. Similar phenomenon was apparent in the September 2005 data when the perennial ice cover was then a record low. Less clouds means more solar heat reaching the surface, and this may in turn melt more ice and would cause a further retreat in the ice cover. But clouds are also known to have some warming effect since clouds absorb longwave radiation from the Earth's surface and reemit this radiations back to the surface. For clouds to have an overall cooling effect in the Arctic region a large fraction of the cloud cover must be relatively high and thick so that it would reflect a large percentage of solar radiation. At present, we do not know of any set of data we can use to confirm this.

The climatological-monthly averages of cloudiness over sea ice and open water over region >60°N are presented in Fig. 5.22. The plots show that the cloud cover fraction over ice-free ocean is higher than over ice-covered ocean by about 22%. The magnitude of the difference reflects the much higher humidity over ice-free regions than over ice-covered regions. Also, there appears to be a seasonality in the cloud cover with the winter period about 6% higher than the summer period. This is likely associated with greater storminess in the region as indicated by the more persistent lows in SLP pressure during winter than during the summer period.

The interannual variability of the cloudiness in the Northern Hemisphere at latitudes >50°N is illustrated in the plots of monthly anomalies shown in Fig. 5.23. On the average, the cloudiness fluctuates by as much as 2%. In the open ocean the monthly anomalies show significant variability in the 1980s, less interannual variability but a multiyear cycle in the 1990s and significant interannual variability

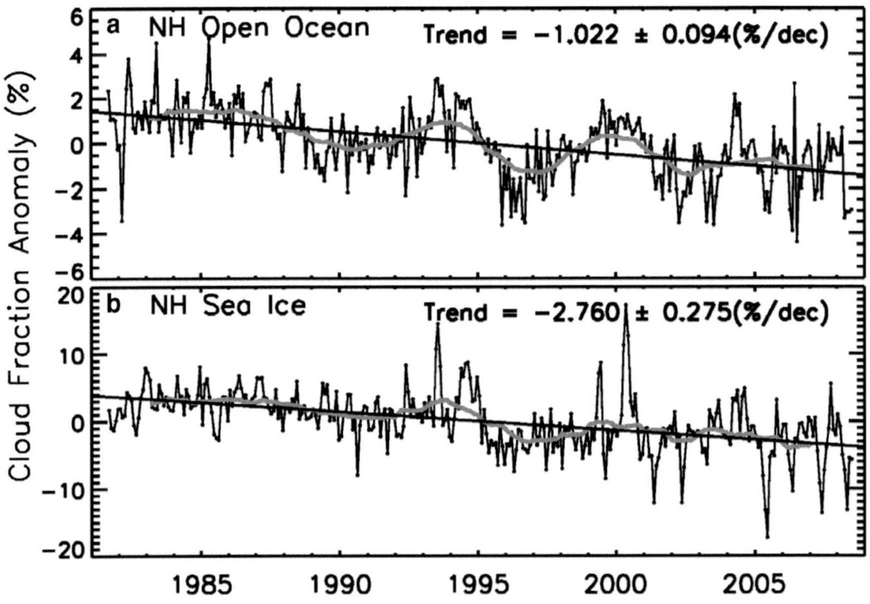

Fig. 5.23 Plots of monthly anomalies of cloud cover and trends in the Northern Hemisphere for the period August 1981 to July 2008 for (**a**) open ocean areas (>55°N) and (**b**) sea ice covered areas. The gray line represents 3-year running averages

in the 2000s. The result of linear regression on the anomalies also indicates a slow decline of the cloud cover of about $-1.02 \pm 0.09\%$ per decade. In the sea ice-covered regions the monthly anomalies indicate a relatively stable cloud cover from 1982 to 1993. After 1993, the cloud cover had much greater interannual variability. The trend in the cloud cover over the sea ice region is more negative than in ice-free regions and estimated to be about $-2.76 \pm 0.27\%$ per decade. Such results are important in light of the relevance of the trend in cloud cover in climate studies. It has been an contentious issue what a warming would do to the cloud cover. The results of a recent study by Clement et al (2009) suggest that a warming ocean would transfer heat to the overlying atmosphere, thinning out the low lying clouds to let in more sunlight that would further warm the ocean. The observed decline in cloud cover could thus be the result of a feedback effect that is associated with observed warming in the Arctic as will be discussed in Chapter 6.

5.3.2 Southern Hemisphere

The monthly climatology of the cloud cover in the Southern Hemisphere for each month from January to December is presented in Fig. 5.24. It is again apparent that the percentage cloudiness is higher over the open ocean regions than over the ice-covered regions. The set of images show the progression in the northward shift of less cloudy areas from the summer months to the midwinter months. Such a shift

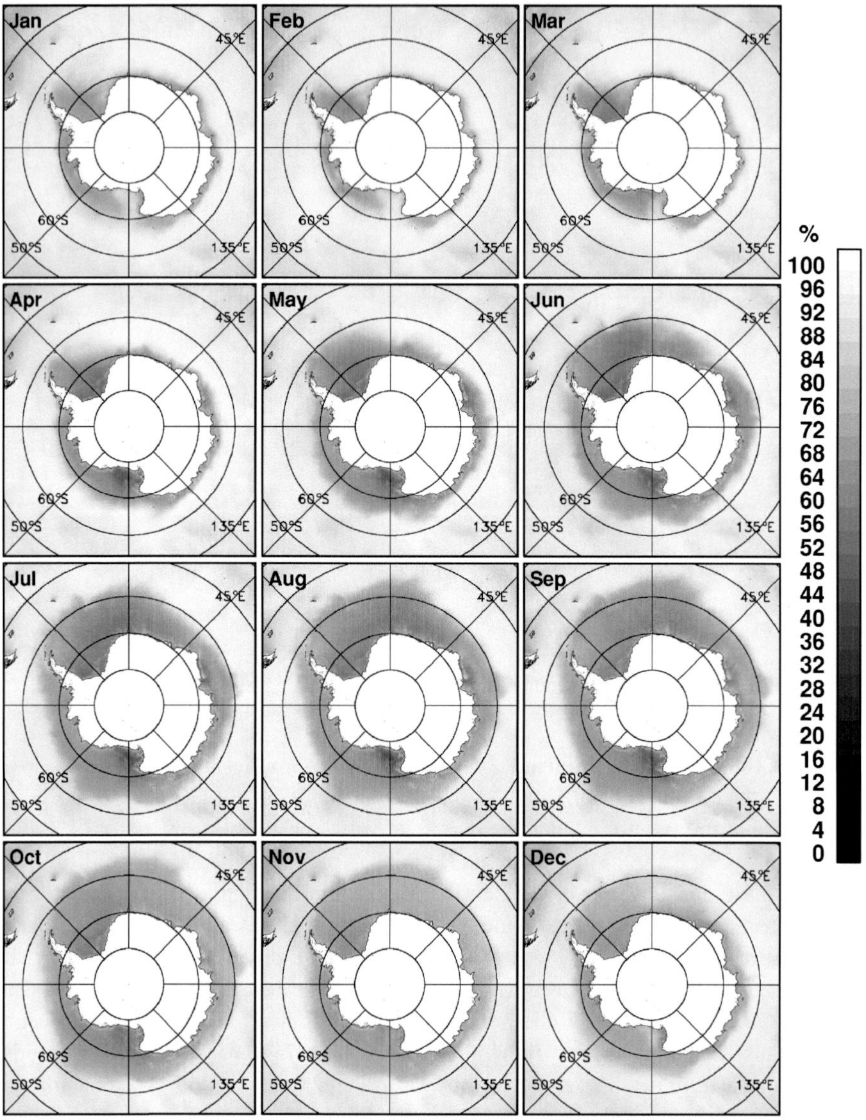

Fig. 5.24 Monthly climatology of cloud cover in the Southern Hemisphere as derived from AVHRR data (1981 to 2008)

is obviously associated with the sea ice growth season and the advance of the sea ice cover to the north. From October to December, a reversal in the seasonal trend is apparent as the sea ice cover declines to its minimum extent. The seasonality of the cloud cover in ice-covered regions is basically the seasonality of the sea ice cover. Also, cloudiness is apparently higher over the sea ice-covered regions dur-

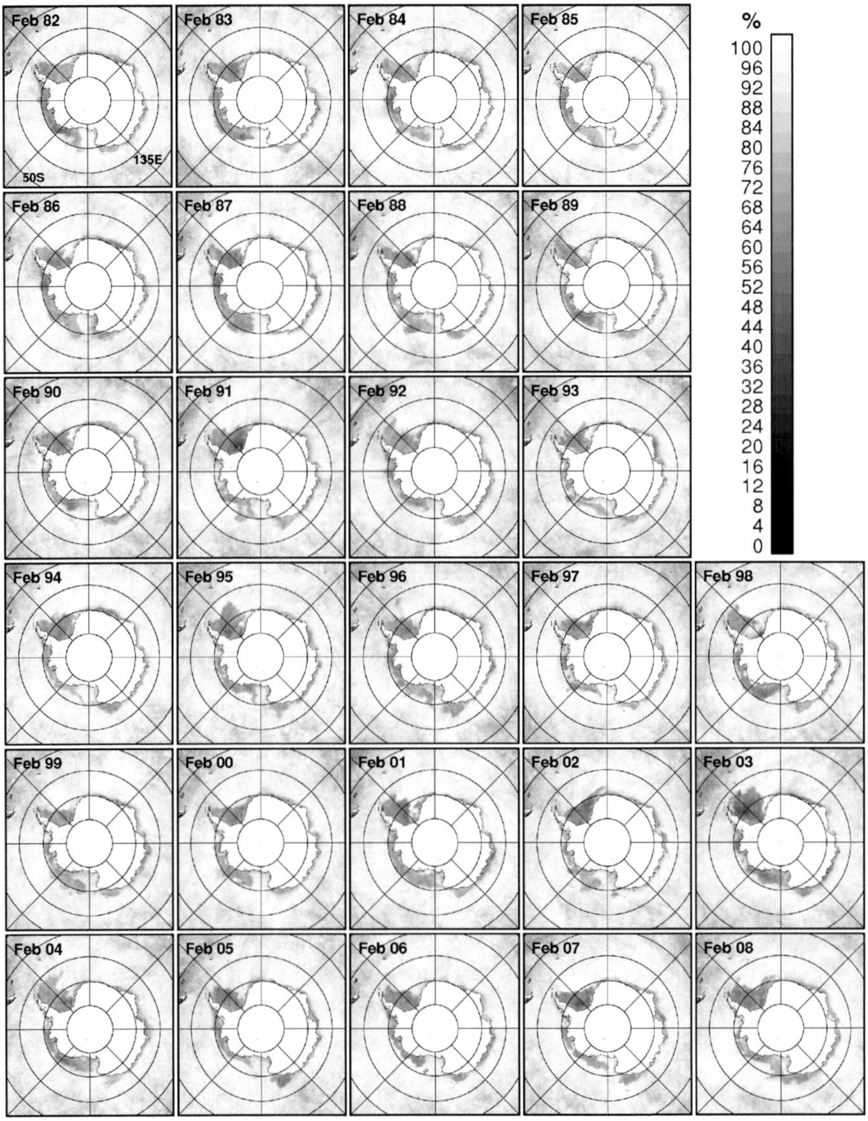

Fig. 5.25 Monthly averages of cloud cover in the Southern Hemisphere as derived from AVHRR data in winter (February) for the period from 1982 to 2008

ing the summer than during the winter in part, because of relatively low ice concentration and extent in the summer and the influence of adjacent open ocean.

During the summer, the sea ice extent is reduced to 20% of its winter extent, and the associated interannual changes in cloudiness is expected to be small. The spatial distribution of the cloud cover each year from 1982 to 2008 during the month of February is illustrated in Fig. 5.25. It is apparent that the cloudiness varies significantly from one year to another. In the ice-free ocean regions, the changes are less apparent with relatively cloudy periods occurring in 1984, 1985, 1994 and 2001, and 2006 while relatively less cloudy periods in 1982, 1989, 1990, 1992, 1996, 1997, and 2003. The interannual changes in the Southern Ocean are greater than expected for this period but mainly because of interannual changes in the location and extent of the sea ice cover and the frequency and location of storms in the open ocean. The year-to-year changes are better quantified in the monthly anomaly maps presented in Fig. 5.26. The anomaly maps provide the means to better identify and quantify the unique changes that happen on a year-to-year basis. In the period from 1982 to 1987, it is interesting to note that over the Bellingshausen Sea it was consistently less cloudy than normal, while over the Weddell Sea it was more cloudy than normal. In later years it was less predictable and was more cloudy than normal in 1989 to 1991, 1993, 1995, 1999, 2000, 2002, 2004, and 2008. This phenomenon is associated with the large variability of the sea ice cover in the region and an apparent decline in the extent starting in 1989 as reported by Jacobs and Comiso (1997). Summer months with more ice cover than normal also show large negative anomalies as in the Weddell Sea in 1991, 1995, 2002, and 2003 and in the Ross Sea in 1987, 1988, 1998, 2003, 2004, and 2008. The maps of cloudiness in the open ocean beyond the ice edge show a general balance of positive and negative anomalies. There were years when the cloud anomalies were mainly positive as in 1984, 1985, 2001 and 2006 and years when the cloudiness were mainly negative as in 1990, 1991, 1992, 1996 and 2003. But generally, there is no indication of a significant trend.

Relatively large interannual variability is also apparent during the winter as depicted by the September monthly averages of cloudiness from 1981 to 2006 as presented in Fig. 5.27. Compared with those in the summer months, the cloud fractions in the Southern Ocean during this time period are significantly less because of winter sea ice cover that blankets much of the region at latitudes greater than 55°S. The relatively lower cloud cover in ice-covered regions is apparent in each of the monthly images. However, there are spatial variations within the ice pack as well. For example, relatively low percentage of cloudiness is apparent in the Ross Sea during some years while high cloudiness is evident in other areas in other years. The year-to-year changes are more apparent in the monthly anomaly maps presented in Fig. 5.28. Overall, it appears that there is a predominance of positive anomalies (or more clouds) in the 1980s and early 1990s while in the late 1990s and 2000s, negative anomalies are more dominant. This suggests that the fraction of clouds in the Antarctic may be declining. The changing location of the winter ice edge is also manifested in the anomaly maps with the ice edge being the site of either high positive or high negative anomaly. Positive anomalies in the region mean that the ice edge is further north than the climatological ice edge while negative anomalies mean that the ice edge has retreated to the south.

The monthly climatology of the cloudiness over the open ocean and sea ice cover in the Southern Ocean is presented in Fig. 5.29. As in the Northern

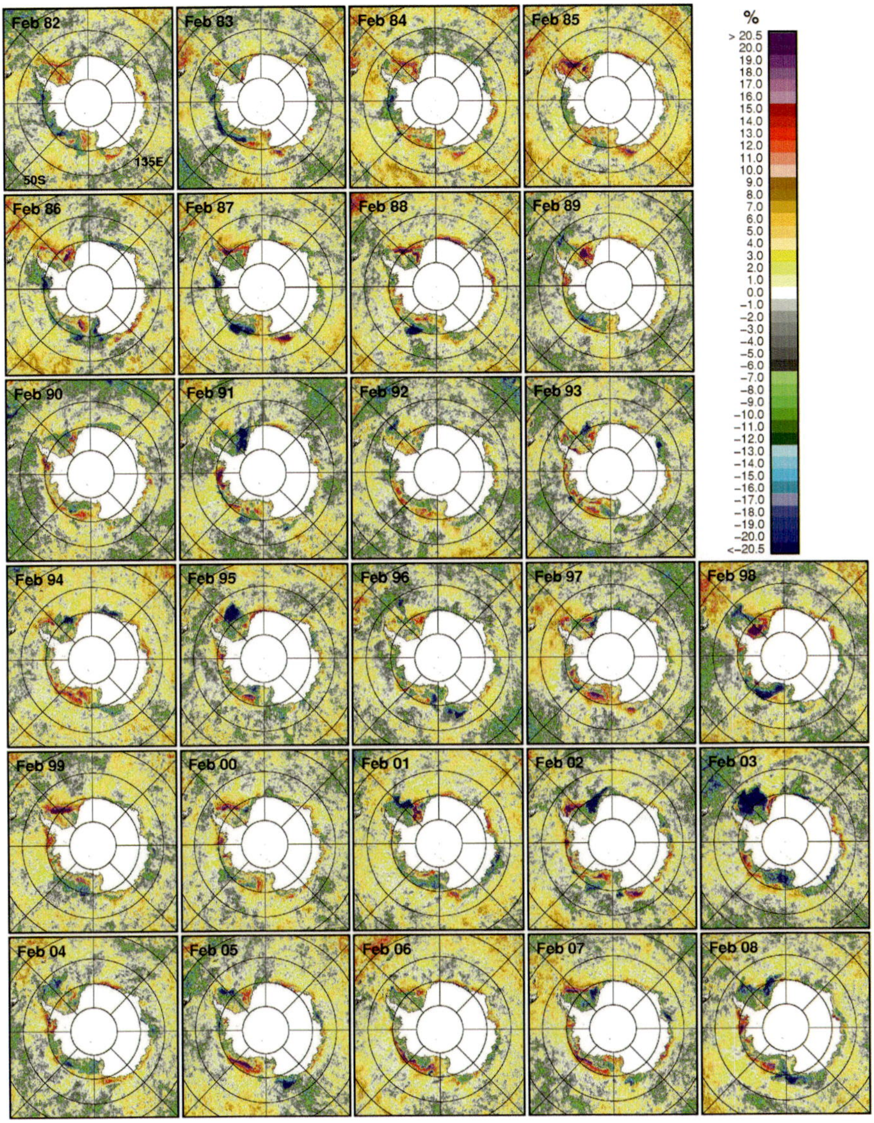

Fig. 5.26 Monthly averages of cloud cover anomalies in the Southern Hemisphere in the summer (February) as derived from AVHRR data for the years1982 to 2008

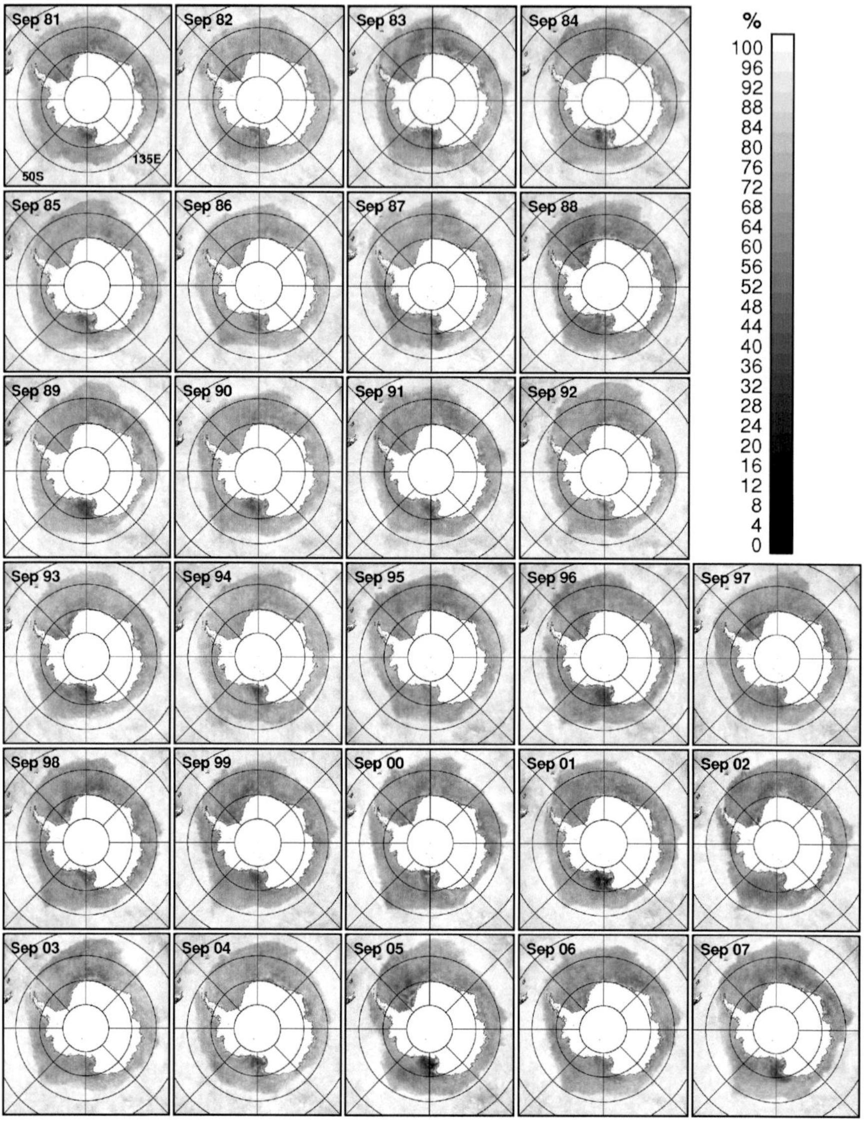

Fig. 5.27 Monthly averages of cloud cover in the Southern Hemisphere in the winter (September) as derived from AVHRR for the years from 1981 to 2007

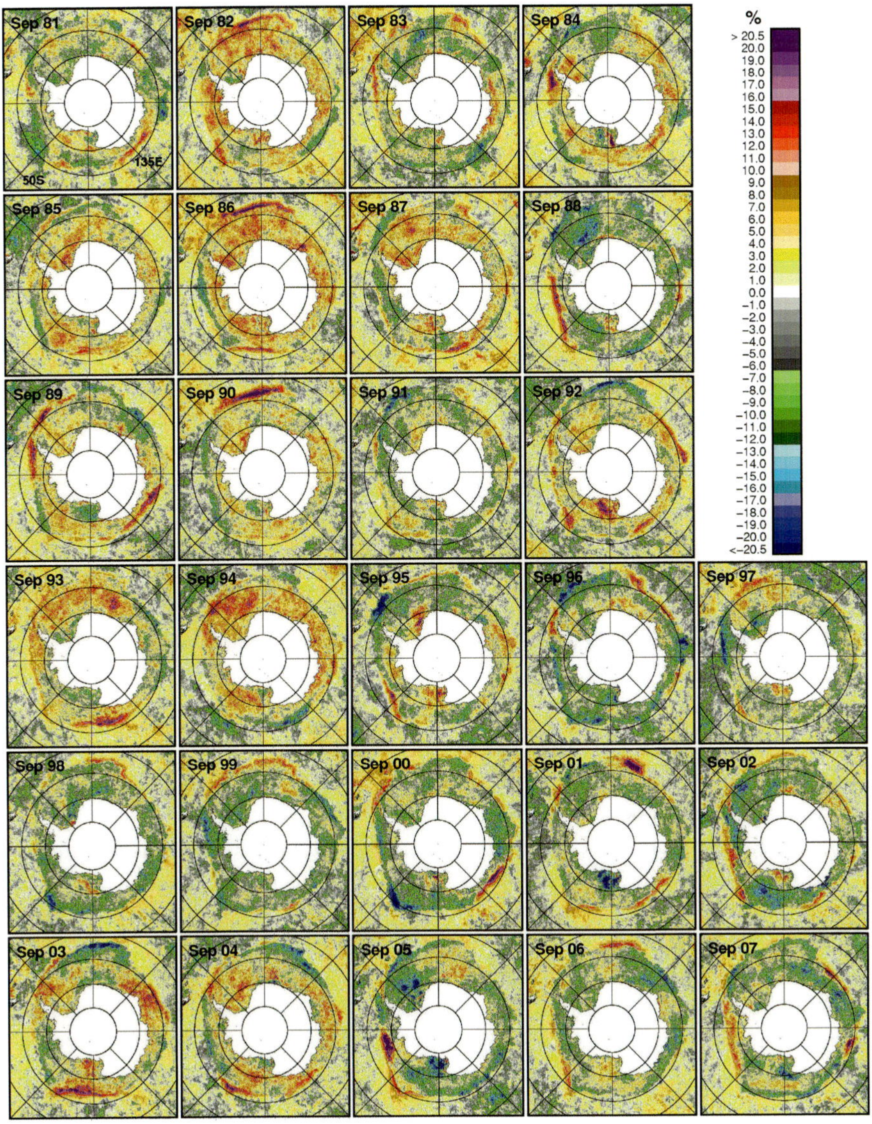

Fig. 5.28 Monthly averages of cloud cover anomalies in the Southern Hemisphere in the winter (September) as derived from AVHRR for the years from 1981 to 2007

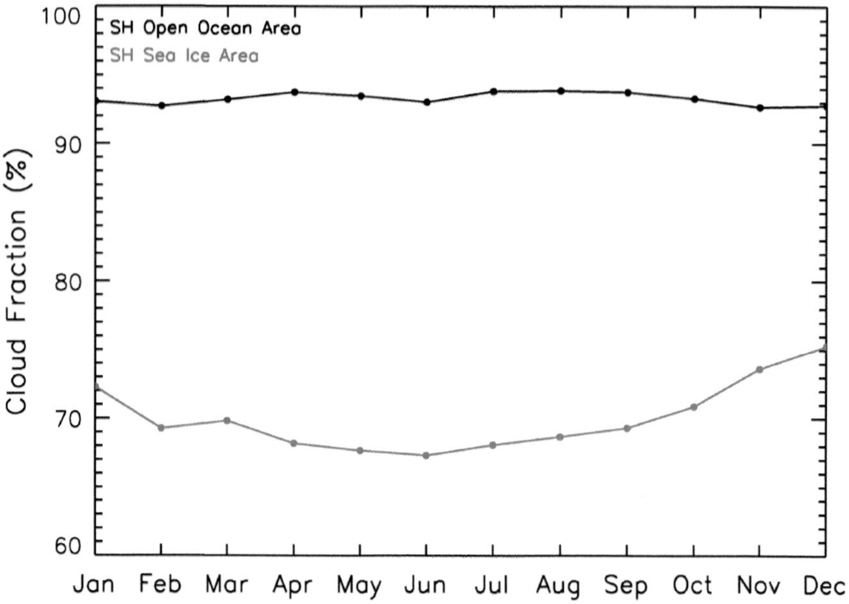

Fig. 5.29 Monthly climatological averages depicting the typical seasonality of cloud fraction in the Southern Ocean (1981 to 2008)

Hemisphere, the difference in the cloudiness of the open ocean and the sea ice covered areas is again quite large at about 21%. The open ocean plot shows a seasonal pattern similar to that of the Arctic in that the winter period is more cloudy than the summer period. The plot, however, shows only a very slight seasonality of only about 1%. The seasonality of clouds in sea ice-covered regions, however, is different from that of the Arctic with the winter period less cloudy than the summer period. This may in part be caused by the much less extensive and more dispersed ice cover in the Antarctic in the summer period. This means more open water within the pack, higher atmospheric humidity and hence higher percentage in cloud cover in the summer.

Monthly anomalies of cloudiness were averaged over open water and ice-covered areas and the results are presented in Fig. 5.30. As in the Northern Hemisphere, the patterns for the open ocean and sea ice-covered areas are similar with both plots showing an increase in cloudiness from 1981 to 1986 and a decrease after that. The two regions also show a peak in 1995 followed by a large interannual variability especially in sea ice-covered areas. Both regions show negative trends in cloudiness with that for the open ocean being very slight and insignificant at $-0.077 \pm 0.068\%$ per decade. The trend in cloudiness over the sea ice-covered areas is also slight but significantly greater at $-1.516 \pm 0.201\%$ per decade.

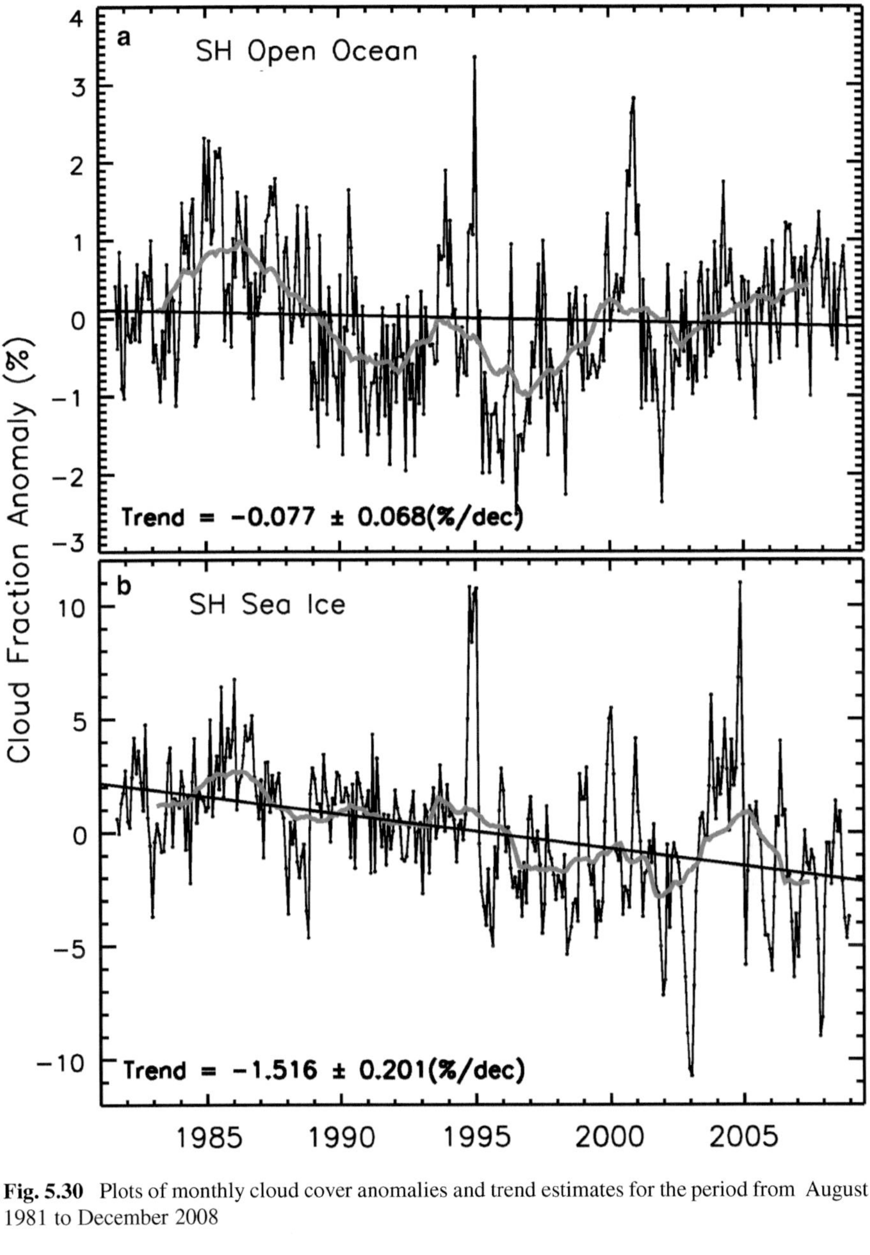

Fig. 5.30 Plots of monthly cloud cover anomalies and trend estimates for the period from August 1981 to December 2008

References

Atlas R et al. (2001) The effects of marine winds from scatterometer data on weather analysis and forecasting. Bull Am. Meteorol. Soc. 82: 1965–1990.

Chapman WL, Walsh JE (1993) Recent variations of sea ice and air temperature in high latitudes. Bull. Am. Meteorol. Soc 74: 33–47.

Chelton DB, Freilich MH (2006) On the use of QuickSCAT Scatterometer Measurement of surface winds for marine weather prediction. Mon. Weather Rev. 134: 2055–2071.

Clement AC, Burgmen R, Norris JR (2009) Observational and model evidence for positive low-level cloud feedback. Science 325: 460–464.

Comiso JC (2000) Variability and trends in Antarctic surface temperatures from in situ and satellite infrared measurements. J. Clim. 13(10): 1674–1696.

Comiso JC (2002) A rapidly declining Arctic perennial ice cover. Geophys Res Letts 29(20):1956, doi:10.1029/2002GL015650.

Comiso JC, Parkinson CL, Gersten R, Stock L (2008) Accelerated decline in the Arctic sea ice cover. Geophys. Res. Lett. 35: L01703, doi:10.1029/2007GL031972.

Deser C (2000) On the teleconnectivity of the "Arctic Oscillation." Geophys. Res. Lett. 27(6): 779–782.

Deser C, Walsh JE, Timlin MS (2000) Arctic sea ice variability in the context of recent wintertime atmosphere circulation trends. J. Clim. 13: 617–633.

Dickson RR, Osborn TJ, Hurrell JW, Meincke J, Blindheim J, Adlandsvik B, Vinje T, Alekseev G, Maslowski W (2000) The Arctic Ocean response to the North Atlantic Oscillation. J. Geophys. Res. 13: 2671–2696.

Freilich MH, Dumbar RS (1999) The accuracy of the NSCAT-1 vector winds: Comparison with National Data Buoy Center buoys. J Geophys Res 104: 11231–11246.

Fyfe JC, Boer GJ, Flato GM (1999) The Arctic and Antarctic Oscillations and their projected changes under global warming. 26(11):1601–1604.

Gong D, Wang S (1999) Definition of Antarctic oscillation index. Geophys Res Lett 26(4): 459–462.

Hurrell JW (1995) Decadal trends in the North Atlantic Oscillation: Regional temperatures and precipitation. Science 269(5224): 676–679.

Hurrell, JW (1996) Influence of variations in extratropical wintertime teleconnections on Northern Hemisphere Temperature. Geophys Res Lett 23: 665–668.

Jacobs SS, Comiso JC (1997) Climate variability in the Amundsen and Bellingshausen Seas. J Climate 10(4): 697–709.

Kalnay E, Kanamitsu R, Krisler R, Collins W, Deaven D, Gandin L, Iredell M, Saha S, White G, Wollen J, Zhsu Y, Chella M, Janowlak J, Eb W, Ropelewski CR, Jenne R (1996) The NCEP/NCAR reanalysis project. Bull. Am. Meteorol. Soc. 77: 437–479.

Krahmann G, Visbeck M (2003) Arctic Ocean sea ice reponse to annular mode-like wind forcing. Geophys. Res. Lett. 30(15): 1793, doi:10.1029/2003GL01754.

Kwok R, Comiso JC (2002) Spatial patterns of variability in Antarctic surface temperature: Connections to the Southern Hemisphere Annular Mode and the Southern Oscillation. Geophys. Res. Lett. 29(14): 1705.

Ledley LS, Huang Z (1997) A possible ENSO signal in the Ross Sea. Geophys. Res. Lett. 24: 3253–3256.

Overland JE, Wang M (2005) The Arctic climate paradox: The recent decrease of Arctic Oscillation. Geophys. Res. Lett. 32: L06701, doi:10.1029/2004GL021752.

Peterson RG, White WB (1998) Slow oceanic teleconnections linking the Antarctic Circumpolar Wave with the tropical El Nino-Southern Oscillation. J. Geophys. Res. 103(C11): 24,573–24,583.

Phillpot HR, Zillman JW (1970) The surface temperature inversion over the Antarctic continent. 75(21): 4161–4169.

Rigor IG, Wallace JM, Colony R (2002) Response of sea ice to the Arctic Oscillation. J. Clim. 15: 2648–2663.
Rossow WB (1989) Measuring cloud properties from space: A review. J. Clim. 2: 201–213.
Schweiger AJ (2004) Changes in seasonal cloud cover over the Arctic seas from satellite and surface observations. Geophys. Res. Lett. 31: L12207, doi:10.1029/2004GL020067.
Serreze MC, Barry RG (1988) Synoptic activity in the Arctic Basin, 1979–1985. J. Clim. 1: 1276–1295.
Simpson JJ, Keller RH (1995) An improved fuzzy logic segmentation of sea ice, clouds, and ocean in remotely sensed Arctic imagery. Remote Sens. Environ. 54: 290–312.
Spinhirne JD, Palm SP, Hart WD, Hlavka DL (2005) Cloud and aerosol measurements from GLAS. Geophys. Res. Lett. 32: L22S03, doi:10.1029/2005GL023507.
Stamerjohn SE, Martinson DG, Smith RC, Yuan X, Rind D (2008) Trends in Antarctic annual sea ice retreat and advance and their relation to El Nino-Southern Oscillation and Southern Annular Mode variability. J Geophys Res 113,C03S90 doi:10.1029/2007JC004269.
Thompson DW, Wallace JM (1998) The Arctic Oscillation signature in the wintertime geopotential height and temperature fields. Geophys. Res. Lett. 25(9): 1297–1300.
Thompson DW, Wallace JM (2000) Annular modes in extratropical circulation. I. Month-to-month variability. J. Clim. 13: 1000–1016.
Thorndike AS, Colony R (1982) Sea ice motion in response to geostrophic winds. J Geophys Res 87(C8): 5845–5852.
Turner J, Comiso JC, Marshall GJ, Connolley WM, Lachlan-Cope TA, Bracegirdle T, Wang Z, Meredith M, Maksym T (2009) Antarctic sea ice extent increases as a result of anthropogenic activity. Geophy Res Lett 36, doi:10.1029/2009GL037524.
Walsh JE, Chapman WL, Shy TL (1996) Recent decrease of sea level pressure in the Central Arctic. J. Clim. 9: 480–486.
Wentz FJ (1992) Measurement of oceanic wind vector using satellite microwave radiometer. IEEE Trans. Geosci. Remote Sens. 30(5): 960–972.
White WB, Peterson RG (1996) An Antarctic circumpolar wave in surface pressure, wind, temperature and sea ice extent. Nature 380: 699–702.
Zhang X, Soteberg A, Zhang J, Gerdes R, Comiso JC (2008) Recent radical shifts of atmospheric circulation and rapid changes in Arctic climate system. Geophys. Res. Lett. 35: L22701, doi:10.1029/2008GL035607.
Yuan X, Li C (2008) Climate modes in southern high latitudes and their impacts on Antarctic sea ice. J Geophys Res 113. doi:10.1029/2006JC004067.

Chapter 6
Variability of Surface Temperature and Albedo

Abstract Results from the analysis of the spatial and temporal variability of surface temperature, including that of liquid water at high latitudes, are presented using mainly NOAA/AVHRR data from 1981 to 2008. The average surface temperature of the region north of the Arctic Circle is shown to be increasing at 0.7°C/decade while that south of the Antarctic circle has been increasing much more moderately at 0.1°C/decade. The trend in the Arctic represents about three times the global trend over the same period and is consistent with the expected effect of ice-albedo feedback. The trend of the Antarctic temperature is less than that inferred using global in situ and meteorological station data since 1981. Data from the same set of AVHRR sensors were used to study the spatial and temporal variability of narrow band albedo in the polar regions and results show the patterns of change that are consistent with warming and the changing snow and sea ice cover.

Keywords Surface temperature • Arctic • Antarctic • Albedo

6.1 Introduction

Surface temperature and albedo are arguably among the most critical parameters in the polar processes and climate change studies. Practically, all processes in the region, and in particular, the cryosphere are directly or indirectly influenced by these two variables. They provide direct information about the state of any type of surface, whether it be sea ice, snow, or liquid ocean. Together, surface temperature and albedo are the key parameters that determine the energy transfer between the surface and the atmosphere, including the net surface radiation flux, surface turbulent sensible and latent heat fluxes, and the transport of energy below the ocean surface (Curry and Webster 1999). Surface temperature and albedo also provide information either directly or indirectly about the state, thickness and distribution of the sea ice cover. Surface temperature would dictate how rapidly sea ice will grow in autumn and winter and how fast it will retreat in spring and summer. It controls the rate of evaporation

J. Comiso, *Polar Oceans from Space*, Atmospheric and Oceanographic Sciences Library 41,
DOI 10.1007/978-0-387-68300-3_6,

that in turn determines the water vapor content of the atmosphere, the cloud cover and indirectly, the amount of precipitation in the region. Moreover, it determines how much melting and meltponding would occur in the spring and summer which at the same time affects albedo and other physical characteristics of the ice cover. Albedo on the other hand, limits the solar energy that is absorbed by the surface especially in ice covered regions where the albedo is relatively high depending on surface dryness, ice type and snow cover (Warren 1982; Allison et al. 1993).

The relatively long record (August 1981 to the present) provided by the AVHRR sensors enables the assessment of the interannual variability and decadal trends in surface temperature and albedo on the Earth's surface. As indicated earlier, the record is actually made up of several AVHRRs with no overlapping data that would have enabled the sensors to be intercalibrated. With no good alternative, surface temperatures derived from the different sensors were made consistent by the use of in situ data as discussed earlier and in Comiso (2003). In particular, initial adjustments were made on a sensor by sensor basis (to get the different AVHRRs consistently calibrated) but because of drifts in the satellite orbit and sensor degradation, yearly and if necessary, seasonal adjustments were made. A similar technique was used to get the albedos from the different AVHRR sensors consistent, but this time because of the lack of real calibration targets we use some target areas with relatively stable albedos as our references. Also, our AVHRR record starts in 1981 instead of late 1978 when the first AVHRR sensor was launched on board the Tiros-N satellite because only a fraction of the continuous stream of data from 1978 to 1981 were stored digitally. As with the cloud frequency data, the AVHRR surface temperature and albedo data have been gridded in a polar stereographic format at a resolution of about 6.25 km by 6.25 km using the GAC data put together by the AVHRR Pathfinder Project. The gridding was done in the same standard format used for SSM/I and AMSR-E data for convenience when doing comparative analyses with the latter. The format provides easy transformation of the relatively high resolution infrared data to coarser resolution data such as the passive microwave data, which are gridded at either 12.5 or 25 km resolution. When compared with other sources of large scale and global surface temperature and albedo data (e.g., NCEP or ECMWF), the satellite infrared and visible data provide improved representation of the spatial distribution of surface temperature and albedo especially in areas where there are no in situ measurements. This is manifested by the stronger coherence of the satellite data with well known surface features like glaciers and elevation gradient regions in the ice sheets of Antarctica and Greenland. Some previously unknown surface features over land and sea ice that are not in other data sources have also been identified.

The amplitude of the seasonal fluctuation of surface temperature and albedo over sea ice covered surfaces is similar to that of snow covered land areas but much higher than that of ice free sea surfaces. For most of the year, there is therefore a much more distinct boundary in surface temperature and albedo between open water and sea ice than between sea ice and land. Since the changes in surface temperature and albedo are influenced mainly by the sun and an atmospheric system that moves around freely over land, sea ice and open ocean, the data that we present and discuss involves all three surfaces. However, the focus of our discussion will be on sea ice covered regions and sea surface temperature (SST). SST will be presented separately because of scale and because of its special role on the distribution of the sea ice cover, the marine biology and the climate of the region.

Surface temperature and albedo are related in many ways. In a climate change scenario in which the surface temperature is going up, the associated decline in the sea ice cover would result in an overall decrease in albedo, which in turn would cause an increase in solar radiation absorbed by the surface and cause further surface warming. Conversely, when the surface temperature is going down, the area covered by sea ice would increase causing further cooling overall. Higher temperatures also cause the formation of more meltponds reduces the albedo of ice covered regions cause the absorption of more solar heat that in turn causes more meltponding and melt of sea ice. Although the relationship between surface temperature and sea ice continues through the winter period, the albedo cannot be measured when the surface is in darkness. During the period of long darkness in the polar regions, long wave radiation is the dominant radiation in the region.

The radiometers were built to very high standards with the radiometer noise reduced to only a small fraction of a degree Celsius and the signal to noise ratios at all channels designed to be as high as possible. There are issues in the retrieval of surface temperature and albedo that need to be reiterated. The main source of error in the retrieval of surface temperature and albedo from AVHRR data is the masking of cloud cover. Clouds are persistent in the polar regions and are also difficult to discriminate from snow/ice covered surfaces. Effective cloud masking especially in snow and ice covered regions has been one of our major objectives. As discussed in Chap. 4, visual inspections of daily maps with and without cloud mask indicate that the automated cloud masking procedures are good but not perfect and they oftentimes provide erroneous classifications. The big errors in cloud masking usually stand out in the daily averages and are deleted using a special filter as described earlier. Furthermore, many of the residual errors are minimized during the averaging procedure and the monthly data used in our study have been shown to be generally in good agreement with available surface measurements. The resulting maps are reasonably accurate and are temporally and spatially coherent. Also, they consistently reproduce some well defined and known surface features in both polar regions and are in good agreement with in situ data.

6.2 Surface Temperature

6.2.1 Northern Hemisphere

6.2.1.1 Surface Temperature over Sea Ice and Land

The spatial details in the distribution of surface temperatures in the entire Arctic was poorly known before the advent of satellite data. Some measurements were made in meteorological and other stations in various locations, but the number of such stations is too small to provide adequate spatial coverage. Good spatial coverage is needed because the atmospheric circulation is constantly changing in the region and with only sparse coverage from a few stations, it is not easy to predict where extreme temperatures occur and how they change from 1 year to another. Monthly distributions of surface temperature in the Arctic have been derived from AVHRR thermal-infrared data for the period August 1981 to July 2008 using the technique explained in Chap. 4. To assess seasonal variability, we took averages of all monthly data available for each month from January to December as in Chap 5. The set of images derived, which we call monthly climatology, is presented in Fig. 6.1. The images represent average measurements on a pixel-by-pixel basis and show distributions of surface temperature in good spatial detail for each month of the year. The northward advance of temperature isotherms from winter to summer and then from summer to the next winter is apparent in the images and it is evident that the isotherms are not spatially symmetrical. This is mainly because the temperature distribution is influenced by the geographic configurations of the land

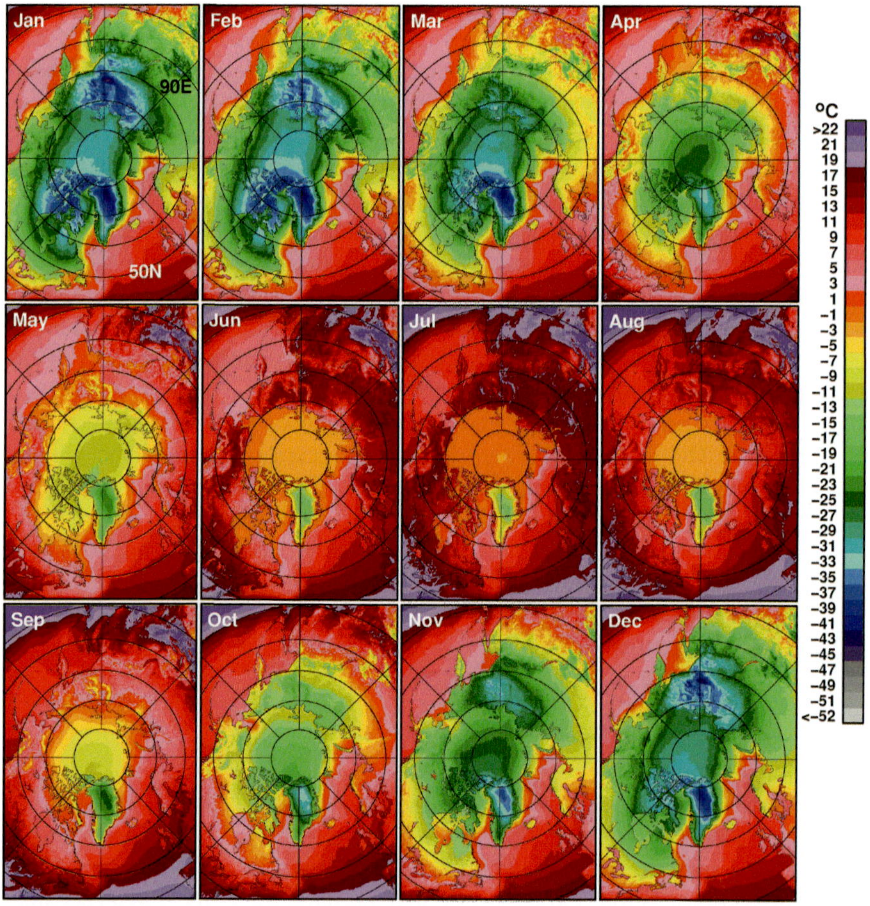

Fig. 6.1 Monthly surface temperature climatology using AVHRR data from August 1981 to July 2008 in the Northern Hemisphere

masses. In these images, subfreezing surface temperatures are usually found in regions where the surface is covered by ice or snow. The coherence of data from 1 month to another is most apparent during the cooling period and to a lesser extent during the melt period. It is remarkable that the contours of low temperatures in East Siberia in December appears very similar in shape and magnitude to those in February after the temperatures reached minimum values in January. In the same general location, the contours of relatively warm temperatures in June are similar to those in August after reaching peak values in July. In the ocean regions, the location of the ice edge can be approximately inferred from the data by noting the location of the −2°C isotherm. As expected, the advance and retreat of the sea ice and snow cover follow this temperature isotherm closely.

It is apparent from the set of images that the coldest month of the year in the Northern Hemisphere is generally January. The January map basically depicts the spatial distribution of the surface temperature that eventually determines the maximum extents of cryospheric parameters, such as sea ice and snow. The maximum extents of these parameters usually occur in February or March because of the lag in the response of the temperature effect. In some regions (e.g., parts of Asia in the top right), the cryosphere goes beyond and further south than what is shown in the map. It is also apparent that the warmest month is July. During this month and subsequent months, the sea ice cover continues to melt and hence to decline in extent as long as the air temperature is above freezing levels. This is basically the reason why the minimum extent of sea ice does not occur until September when the surface temperature has already declined significantly from the peak temperature in the summer. In some regions, the coldest and warmest months may not be January and July, respectively, because of nonuniformity in the impact of atmospheric circulation to different regions in different months as described in Chapter 5. For example, the eastern part of the Canadian Archipelago (e.g., Ellesmere Island) and surrounding areas are usually colder in February than in January. Also, some regions at low latitudes are warmer in August than in July. The coldest area is typically located over the ice sheets of Greenland where the surface temperature stays below freezing point, year round. There are, however, some years when temperatures in other places, like Siberia, are colder than in Greenland as will be shown below. The sea-ice cover can get very cold as well, especially in the Arctic Basin to the north of Greenland and the Canadian Archipelago. This is the same area where statistically, as inferred from buoy data, the oldest and thickest ice floes are expected to be found (Colony and Thorndike 1985). In the summer, the surface temperature over sea ice is basically uniform and close to melt temperatures.

The set of images in Fig. 6.1 illustrates the spatial pulsation of surface temperature in the pan-Arctic region from winter to summer and back to winter. The range of variability in temperature over ice covered areas is quite large, typically from about −41°C to melt temperatures. It is interesting that some relatively well defined features (e.g., yellow areas in Western Siberia north of 60°N at around 135°E) that stand out in May weaken in June, and almost disappear in July, but then reappears again in the same places in August and September. Many other spatial patterns are reproduced from 1 month to another during the various seasons indicating the potential of the data in evaluating regional phenomena. Note that at high latitudes in the Arctic basin (e.g., >85°N), the temperature goes down to near subfreezing temperatures in June then to melt temperatures in July and back down to subfreezing temperatures again in August. This means that the sea ice growth period actually starts in some areas before the ice minimum period is reached in September.

To illustrate the way the surface temperature changes on an interannual basis in winter, monthly averages of surface temperatures in February for each year from 1982 to 2008 are presented in Fig. 6.2. Although February is not the coldest month

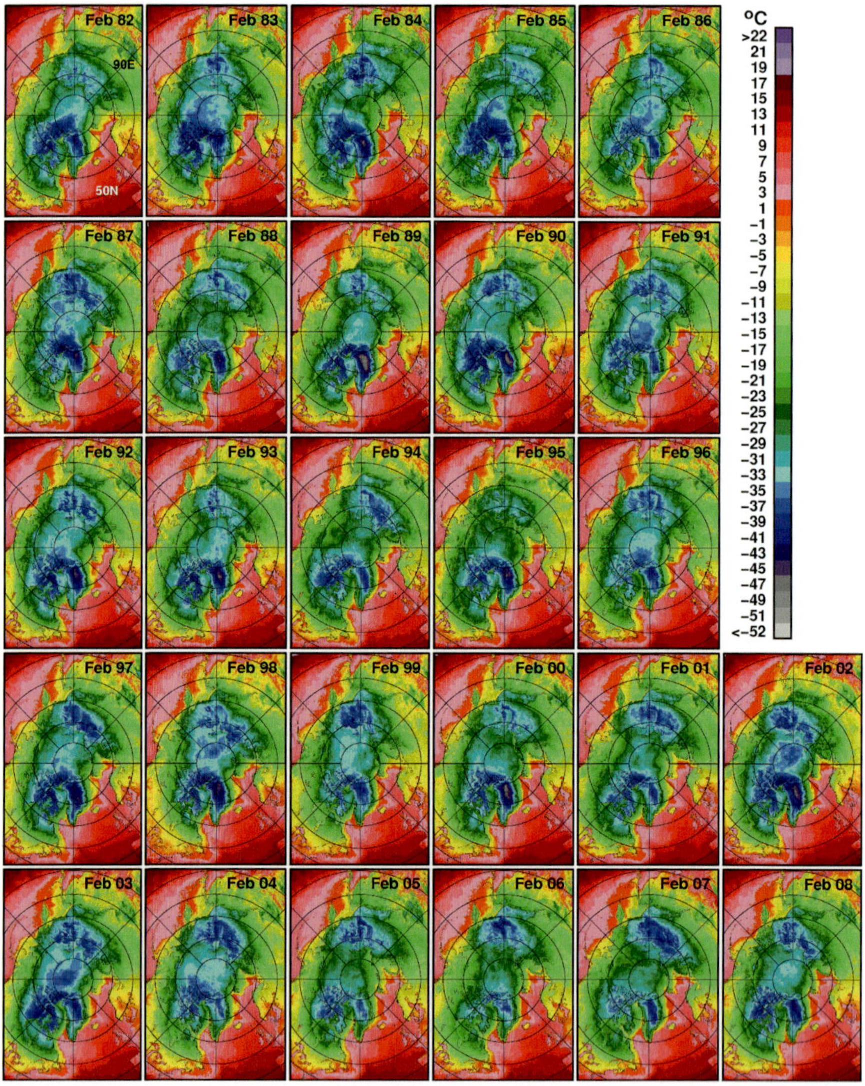

Fig. 6.2 Monthly averages of surface temperature in the Northern Hemisphere in winter (February) as derived from AVHRR data for the period1982 to 2008

everywhere, it is typically the coldest month in the Arctic basin and is sometimes the month when the sea-ice cover reaches its maximum extent. The winter temperature is important to monitor because it is related to the rate of thermodynamic ice growth, and it therefore provides indirect information on how the thickness of undeformed sea ice could vary from 1 year to another. The distribution of surface temperature in the entire pan-Arctic region are generally similar for each year but large interannual variations in the location and shape of some isotherms and distinct features are apparent in many regions. The Arctic Basin is shown to be extremely cold in February with the temperatures being typically about −30°C. However, the pattern of the low temperature values (e.g., areas in blue colors) in the basin changes considerably from 1 year to another. For example, there are years when the Arctic basin was moderately cold (e.g., almost all green) as in 1988, 2005, 2006, and 2007 and years when the temperature is relatively colder than average (e.g., lots of blues) as in 1980, 1986, 1987, 1991, 1992, 1996, 1998, 1999, 2002, 2003, and 2004. Changes in the pattern of cold temperatures are in part associated with changes in wind circulations as described in the previous chapter. Southerly winds, for example, usually carry warm air from low latitude regions and alter the distribution of the isotherms.

The year-to-year changes in the February temperature distribution are better quantified by the anomaly maps presented in Fig. 6.3. The anomaluy map for a certain month is the difference of the average temperature for this month and that of climatology (i.e., Fig. 6.1) for the same month. The color coding for the maps is such that anomalously warm regions are represented by shades of orange, red and purple, while relatively cold regions are represented by shades of gray, green, blue and violet. The locations and extents of relatively warm (or cold) regions in winter for each year show up much better in the anomaly map than in the monthly maps. While some of the patterns look generally similar, none is close to being identical. Relatively warm Februaries in the Arctic basin occured in 1984, 1989, 1994, 1995, 2001, 2005, 2006, and 2007 with the last 3 years being unusually warm. On the other hand, cold Februaries in the same region occurred in 1983, 1987, 1991, 1992, 1998, 2002, 2003, and 2004. During other years, mixtures of positive and negative anomalies in the Arctic basin are observed. Positive anomalies are also shown to be persistent in the Greenland ice sheets in the last few years, especially from 2004 to 2007. During this time period (i.e., 2004–2007), the average temperatures in some areas in Siberia was colder than those in Greenland. Overall, the February anomaly maps provide a unique overview of the spatial and interannual variability of surface temperature during winter conditions.

The surface temperatures vary only moderately in the Arctic basin in the summer because most surfaces covered by snow or ice have values close to that of melt temperature. The interannual variability during this time of the year is depicted in Fig. 6.4 by the monthly maps for September which is the month when both the end of the summer and the beginning of autumn usually occur simultaneously. It is also during this month that the date of summer sea ice minimum occurs and therefore the time when the melting of sea ice ends in most places. The monthly averages in Fig. 6.4 show significant interannual changes during the period 1981 to 2007. Since the dynamic range of surface temperature is not so high at this time of the year, the contours of the isotherms are not as easy to recognize as in February. At this time of the

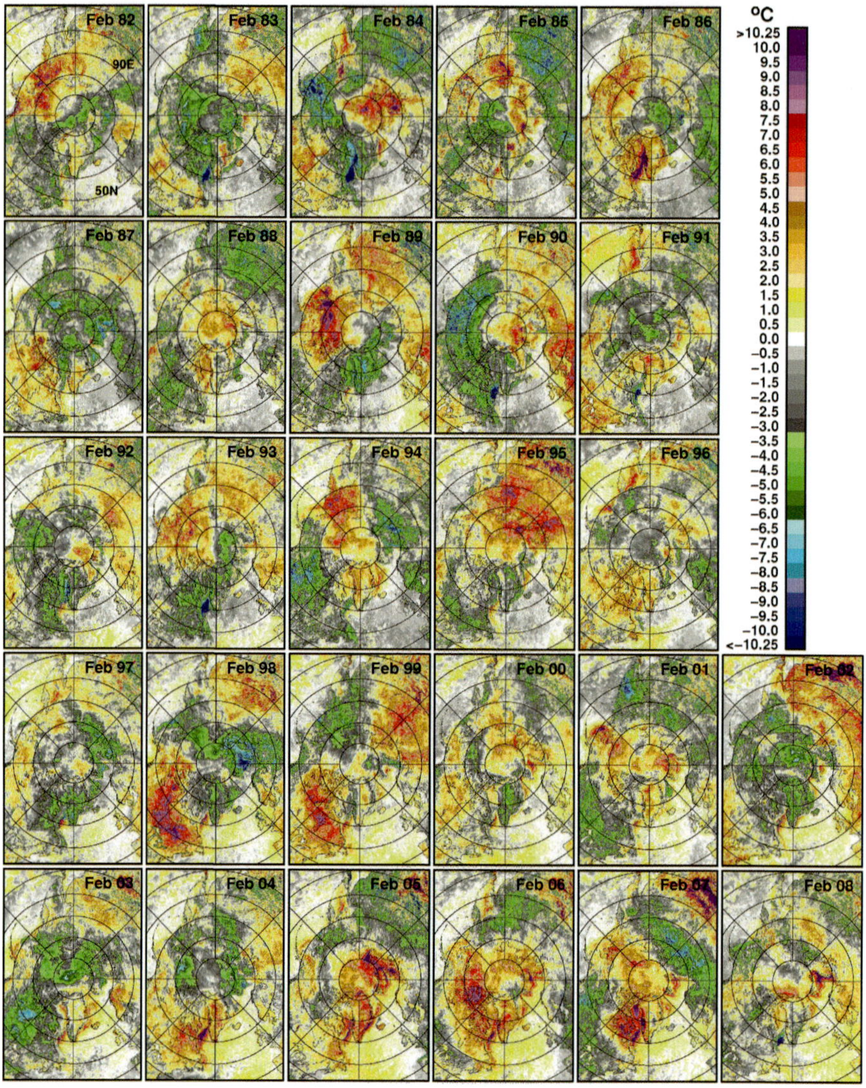

Fig. 6.3 Monthly averages of surface temperature anomalies in the Northern Hemisphere in winter (February) as derived from AVHRR data for the period1982 to 2008

year, except for Greenland and mountain glacier areas, the regions with surface temperature above melt temperatures are outside the Arctic circule.

In the Arctic basin, the average temperatures are primarily subfreezing suggesting that as the extent of the sea-ice cover is reduced to its minimum values in September, growth of sea ice has already started in some parts of the Arctic, especially at high latitudes. Being the month of transition from melt period to freeze-up period, the average monthly temperature in a particular region depends on the exact date of onset of freeze-up. Thus, if freeze-up occurred in the early part of the month, the monthly

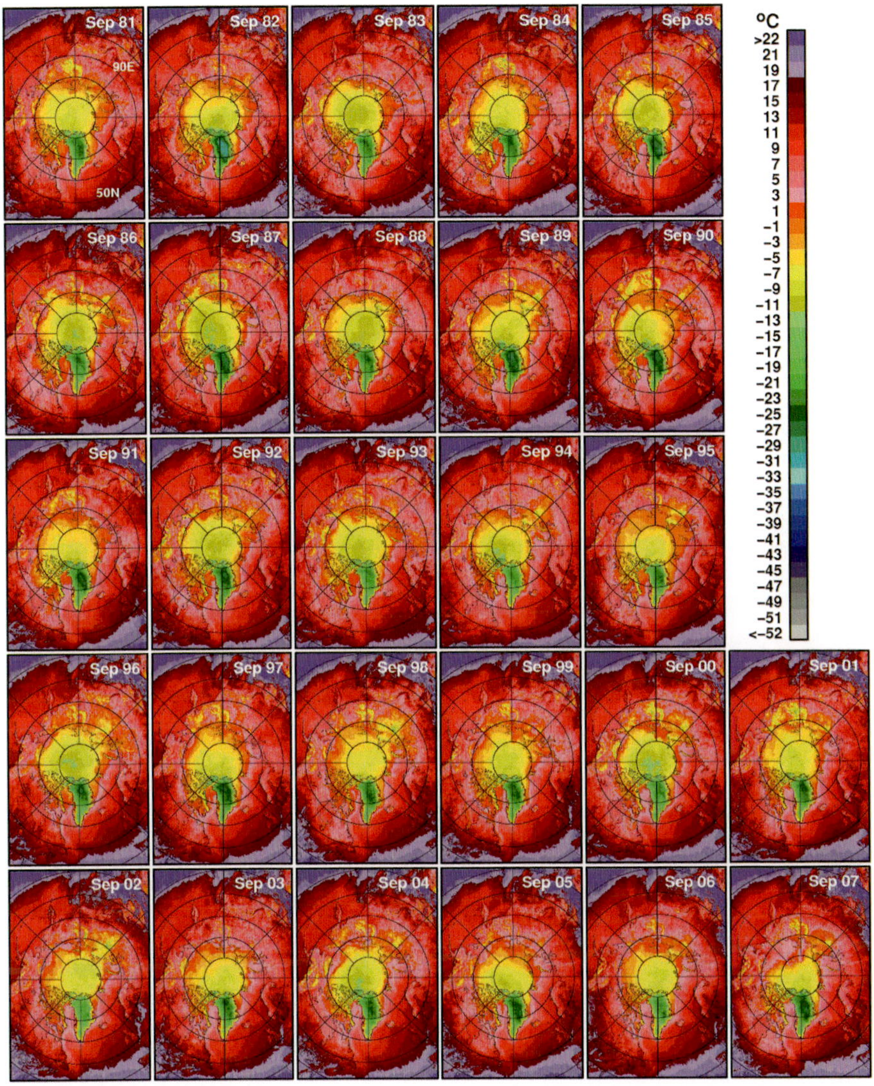

Fig. 6.4 Monthly averages of surface temperature in the Northern Hemisphere at the end of the summer (September) as derived from AVHRR data for the period1981 to 2008

average would be generally colder than when freeze-up occurred in the later part of the month. After freeze-up starts and a sheet of ice has formed on the surface, the ice surface can be much colder than the adjacent liquid surface since the cold air temperatures would start to strongly influence the surface temperature of ice covered surfaces. The images show that the relatively cold areas in the Northern Hemisphere at the end of the summer are the Greenland ice sheet, glaciers, permafrost regions in North America and Russia and the interior of the ice cover. Spatial variations in surface temperatures in these regions are usually associated with topography with the colder temperatures usually found at higher elevations. Interannual changes in the extent and location of the perennial ice cover are the primary factors that affect the temperature distribution in the Arctic basin. The zero-degree isotherms do not stand out as well in the summer as in the winter because the differences in temperature between ice-covered and ice-free areas are only a few degrees.

The interannual variations are better quantified and easier to assess in the monthly anomaly maps presented in Fig. 6.5. The set of images show actually a lot more interannual changes than expected at this time of the year. The most intriguing of

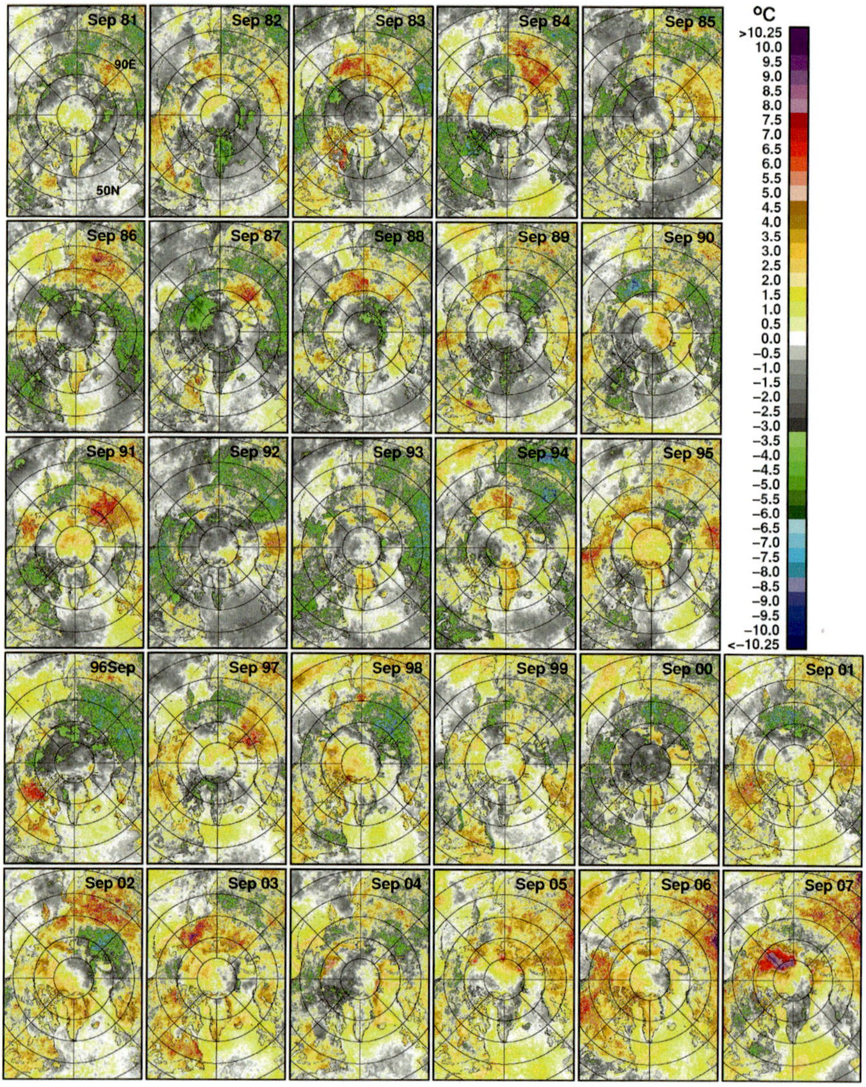

Fig. 6.5 Monthly averages of surface temperature anomalies in the Northern Hemisphere at the end of the summer (September) as derived from AVHRR data for the period1981 to 2007

this is the progressive pattern from generally negative anomalies in the 1980s to primarily positive anomalies in 2000s. This suggests a pattern of warming in the Arctic region. With the exception of 1996 and 2000, the Arctic basin was relatively warm starting in 1995 with the unusually warm years occurring in 2005, 2006, and 2007. The Greenland ice sheet was also consistently on the warm side during the more recent periods. Among the coldest years in the Arctic region are 1992 and 1993 likely caused by cooling because of the spread of aerosols released during the Mount Pinatubo eruption in 1991. In the 1980s, the anomalies are generally negative but there are some regions (e.g., the middle of Siberia between 90°E and 135°E) where strong positive anomalies are evident indicating that these regions were relatively warm especially from 1983 to 1986. Such positive anomalies are remarkable since they occurred during a period of general cooling in the region. Equally surprising was the pattern of negative anomalies basically in the same regions in 1998, 2000, 2001 and 2002 during a period when a general warming in the entire Arctic region was occurring. In 2002, a pattern of positive anomalies occured in Russia at relatively low latitude. The anomaly pattern moved further north in 2003 causing a dominance for this year of positive anomalies at high latitude regions. Such interannual changes can be significant since even a few degrees of positive anomaly in ice-covered region can cause considerable changes in the ice cover.

Plots of monthly averages of surface temperatures at high latitudes from August 1981 to July 2007 as derived from AVHRR are presented in Fig. 6.6. In Fig. 6.6, we show how surface temperature fluctuates in two regions: >60°N (Fig. 6.6a) and

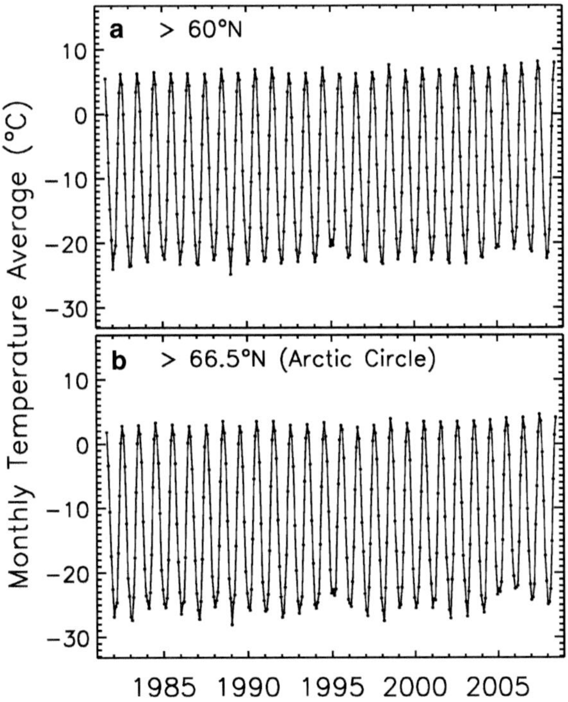

Fig. 6.6 Plots of monthly surface temperature averages in the Northern Hemisphere (**a**) inside 60°N and (**b**) inside the Arctic Circle

>66.5°N (Fig. 6.6b) to demonstrate sensitivity of surface temperature to latitude. The results for >60°N show almost uniform seasonal fluctuations from high values of about 6°C to low values of about −22°C. Inside the Arctic Circle (>66.5°N), the fluctuations are similar, but the values are relatively colder with the range being from about 3 to −26°C. This is consistent with having less of the warmer open water areas at higher latitudes which apparently more than offset the less coverage of the cold areas of Greenland and Sibera. The similarities in the two plots suggest that in mid winter, the surface temperature of snow-covered land areas and snow-covered sea ice areas has basically the same fluctuation characteristics.

The range of temperature variability on a pixel-by-pixel basis is actually considerably larger than what is shown in the monthly averages if daily averages are used since temperature extremes may occur for just a few days. For example, in Siberia, the difference between the minimum and maximum surface temperatures is much greater and approaches 100°C (not shown) comparable to those recorded in meteorological stations. Monthly averaging over large areas tends to suppress the dynamic range of variability and the plots should not be used in comparative analysis with in situ observations during field programs unless similar averaging is done with the latter.

To quantitatively assess the interannual variability and trends, monthly anomalies in temperature in both regions as well as yearly averages are presented in Fig. 6.7.

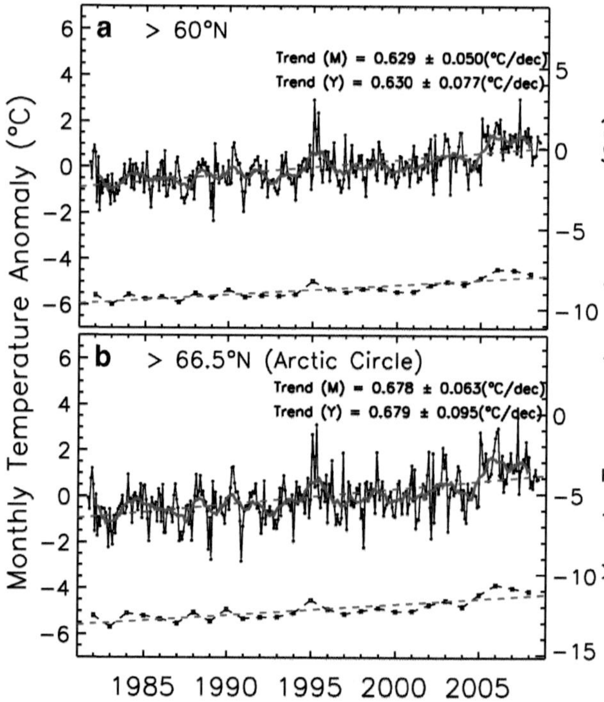

Fig. 6.7 Plots of monthly surface temperature anomalies in the Northern Hemisphere (**a**) inside 60°N and (**b**) inside the Arctic Circle. The gray lines are one-year running averages. Yearly averages are also presented and the dash lines show results of linear regressions that provide the trend values for monthly and yearly data

It is apparent that the monthly fluctuations are significant and can be as high as 3°C as happened in 1995. The time period from 2005 to 2008 also shows relatively more positive anomalies than other years consistently showing significantly warmer temperatures during the period. The warmer temperatures appear to be more significant during the winter period or the cold season as indicated by the minimum values in Fig. 6.6. Twelve-month running averages are shown (in gray line) and yearly averages are also plotted separately to better illustrate interannual variations. The monthly anomaies for the region >60°N yielded a trend of 0.63±0.05°C/decade, while the yearly averages provided an identical trend of 0.63±0.08°C/decade. The error cited is the standard deviation of the slope which is slightly lower for the monthly anomalies than the yearly averages because of statistically more data points used for monthly anomalies than yearly averages in the regression analysis. Inside the Arctic Circle, the anomalies show higher variability and the trends are slightly higher at 0.68±0.06 and 0.68±0.10°C/decade for the monthly anomalies and yearly averages, respectively. For comparison, global temperatures collected from meteorological stations and buoys around the world have on the average increased only by around 0.6–0.7°C from the 1960s to the present yielding a trend of 0.13–0.16°C/decade (Broham et al. 2006). For the satellite period from 1978 to 2008, we estimate a trend of 0.24 +/- 0.02 deg C/decade using a global data set similar to that used by Broham et al. (2006). The Arctic temperature data as inferred from satellite infrared data thus show trends that are about 3.8–4.5 times greater than global average, which is consistent with model predictions (e.g., Holland and Bitz 2003) for polar amplification.

To gain insight into regional variability, similar monthly plots are shown in Fig. 6.8 but separately for sea ice covered regions, Greenland, Eurasia, and North America with the last two study areas confined only to >60°N. The range of values are from −26 to −2°C for sea-ice, −35 to −8°C for Greenland, −26 to +15°C for Eurasia and −25 to +13°C for North America. In the various regions, the temperature minima for each year are very similar for sea ice, Eurasia, and North America and about 10°C lower for Greenland. The temperature maxima are highest in Eurasia, followed by North America, Sea Ice and then Greenland. The variability of surface temperature in Eurasia and North America is very similar with both showing relatively warm summers. The monthly surface temperatures in Greenland show the largest variability with the minimum surface temperatures varying from about −32 to −39°C and the maximum varying from −5 to −10°C.

Monthly anomaly plots are presented in Fig. 6.9 and it is apparent that sea ice has the least fluctuation in surface temperature, followed in order by North America, Eurasia, and Greenland. The relatively high positive anomaly in 1995 identified in Fig. 6.7 is evident in all areas except in Greenland. The enhanced positive anomalies identified in Fig. 6.7 for the period 2005–2008 are also much less in the study areas except over the sea ice region. The regional trends are more varied and estimated to be 0.37±0.07, 0.73±0.12, 0.28±0.09, and 0.67±0.10°C/decade for sea ice, Greenland, Eurasia, and North America, respectively. The highest trend in surface temperature is that for Greenland, where the monthly anomalies go as high as 6.5°C as in 2005. This is an area of concern since the average of maximum temperatures of Greenland is about −8°C and as the surface temperature increases,

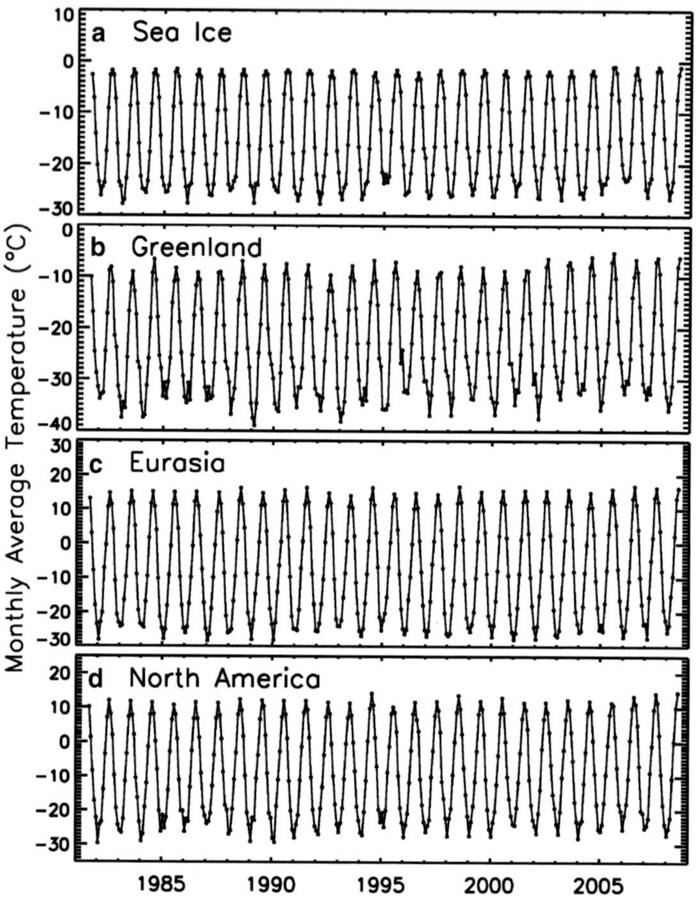

Fig. 6.8 Plots of monthly surface temperature averages over (**a**) sea ice; (**b**) Greenland; (**c**) Eurasia (>60°N); and (**d**) North America (>60°N)

more surface areas get melted causing the percolation of more liquid water to the bottom bedrock. Such water has been postulated to serve as a lubricant that can cause an increase in the velocity of the ice sheet (Zwally et al. 2002b). The Greenland ice sheet has a sea level equivalence of about 7 m and any phenomenon that might facilitate the melt process should be better understood. The next highest trend in temperature is that for North America which is the site of hundreds of glaciers located at the Queen Elizabeth Islands and Alaska. Many of these glaciers have been reported to have declined substantially in volume in recent years (Abdalati 2006). The trend in temperature over the sea ice cover is more moderate at 0.37°C/decade, while that for Eurasia is only about 0.28°C/decade. In Eurasia, the relatively modest trend is in part because of the positive anomalies in the region in the 1980s as described earlier. The trend is moderate over sea ice mainly because in summer, the sea ice temperature is close to melt temperature and does not change much from one year to another.

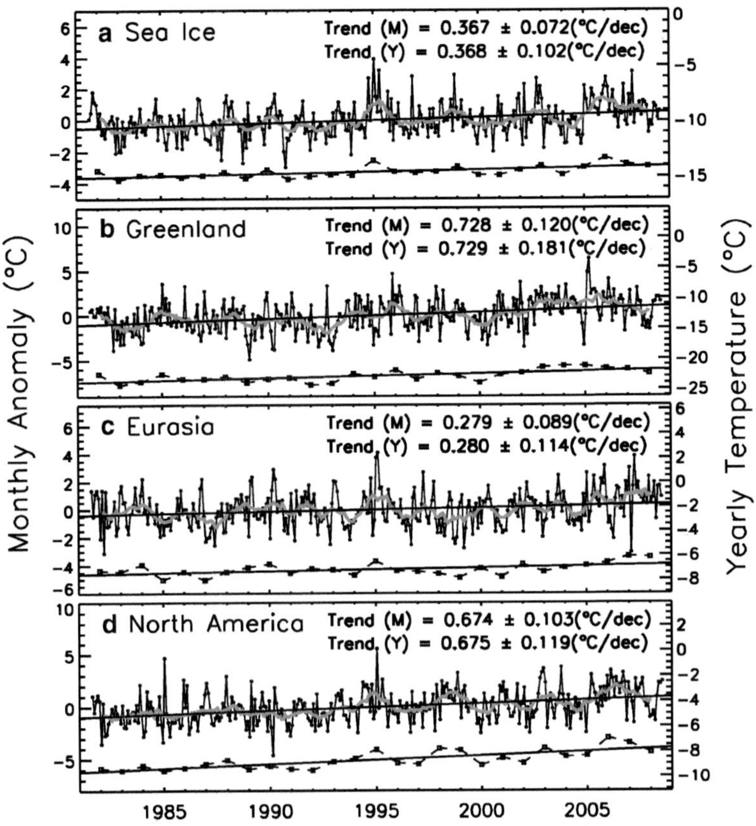

Fig. 6.9 Plots of monthly surface temperature anomalies from 1981 to 2008 over (**a**) sea ice; (**b**) Greenland; (**c**) Eurasia (>60°N); and (**d**) North America (>60°N). The gray lines are one year running averages. Yearly averages are also presented and the lines represent results of linear regressions that provide the trend values for monthly and yearly data

Although the trend of monthly surface temperature for the entire sea ice cover is relatively low, the impact is likely the most discernible since even just slight changes can significantly affect the sea ice extent and thickness. To gain insights into regional variability and trends, we present anomaly plots of average monthly surface temperature of sea ice in various regions of the Arctic in Fig. 6.10. The various regions are similar to that used in Fig. 4.2 of Parkinson et al. (1987) with slight modifications. In the modified version, the Eastern and Western areas of the Arctic are separate study regions to enable the study of processes in the North American side separately from those in the Eurasian side of the Arctic Basin. Also, the Kara Sea becomes part of the Eastern Arctic while Barents Sea becomes a distinct sector. The new study areas were also used for ocean color studies as presented graphically in Fig. 9.9. The anomaly plots in Fig. 6.10a and b show high variability and linear regression analysis of the data yielded trends in the Eastern and Western Arctic of 0.6 ± 0.11 and 0.65 ± 0.12°C/decade, respectively. These are significantly higher than for the entire sea ice cover and comparable to that for North America. The sectors

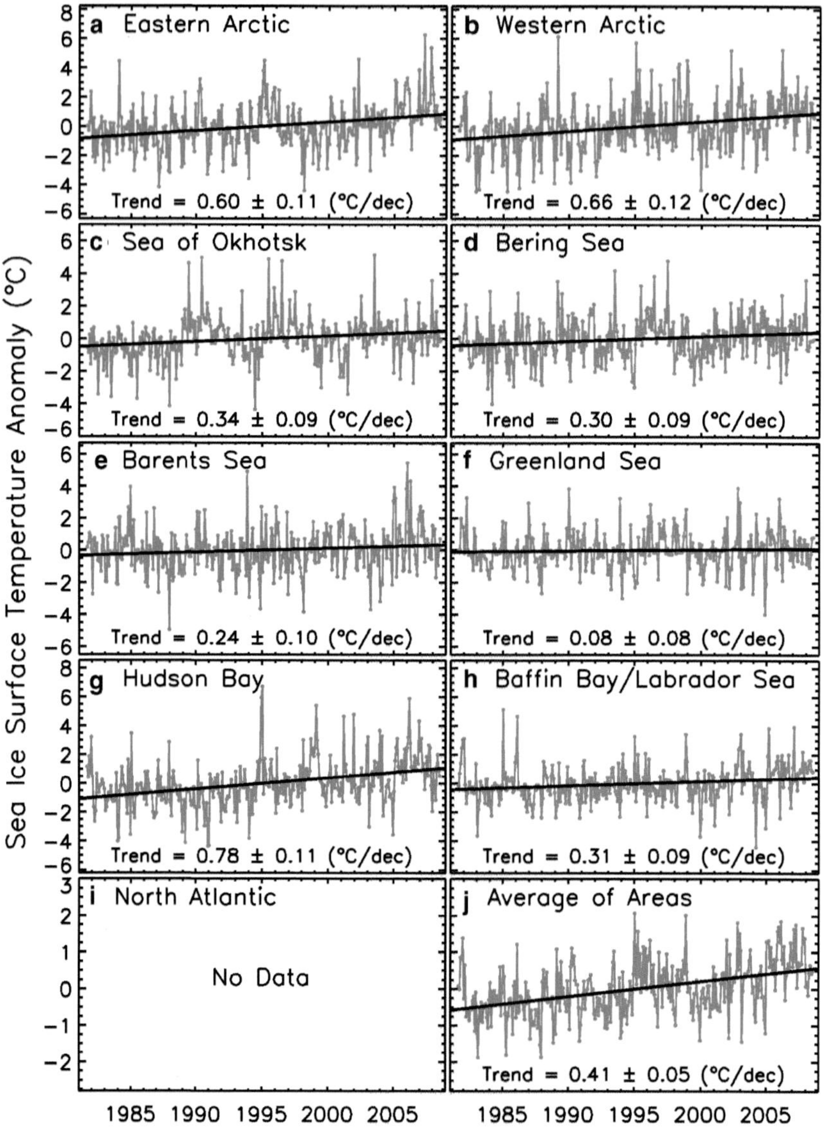

Fig. 6.10 Plots of monthly surface temperature anomalies from 1981 to 2008 in the ice covered regions of (**a**) the Eastern Region; (**b**) Western Region; (**c**-**h**) other sectors (see Fig. 9.10) and (**j**) average over all areas. The trend lines derived from linear regression analysis and trend values are also provided

are part of the Arctic Basin where the sea ice in summer has been declining rapidly and such trends make ice in the region and most especially, the perennial ice cover very vulnerable to volume loss (Hakkinen and Mellor 1990). The Sea of Okhotsk and the Bering Sea show comparable trends at 0.34 ± 0.09 and 0.30 ± 0.09 °C/decade, respectively, which are more similar to that for the entire sea ice cover. The trends at the Barents Sea and the Greenland Sea are more moderate at 0.24±0.10 and 0.08±0.08 °C/decade, respectively. The trend of sea ice surface temperature in the Greenland Sea is surprisingly much less than that for the Greenland ice sheet. This trend is affected in part by changes in the rate of advection of sea ice from the Arctic Basin to the Greenland Sea. The trend at Hudson Bay is the largest at 0.78±0.11 °C/decade, while that for Baffin/Labrador Seas is 0.31±0.09 °C/year which is closer to that for the entire sea ice cover. The average trend from these areas is 0.41±0.05 °C/decade which is close to that of the entire sea ice cover.

For completeness, yearly averages of surface temperatures in the Northern Hemisphere were generated and are presented in Fig. 6.11. The yearly averages were

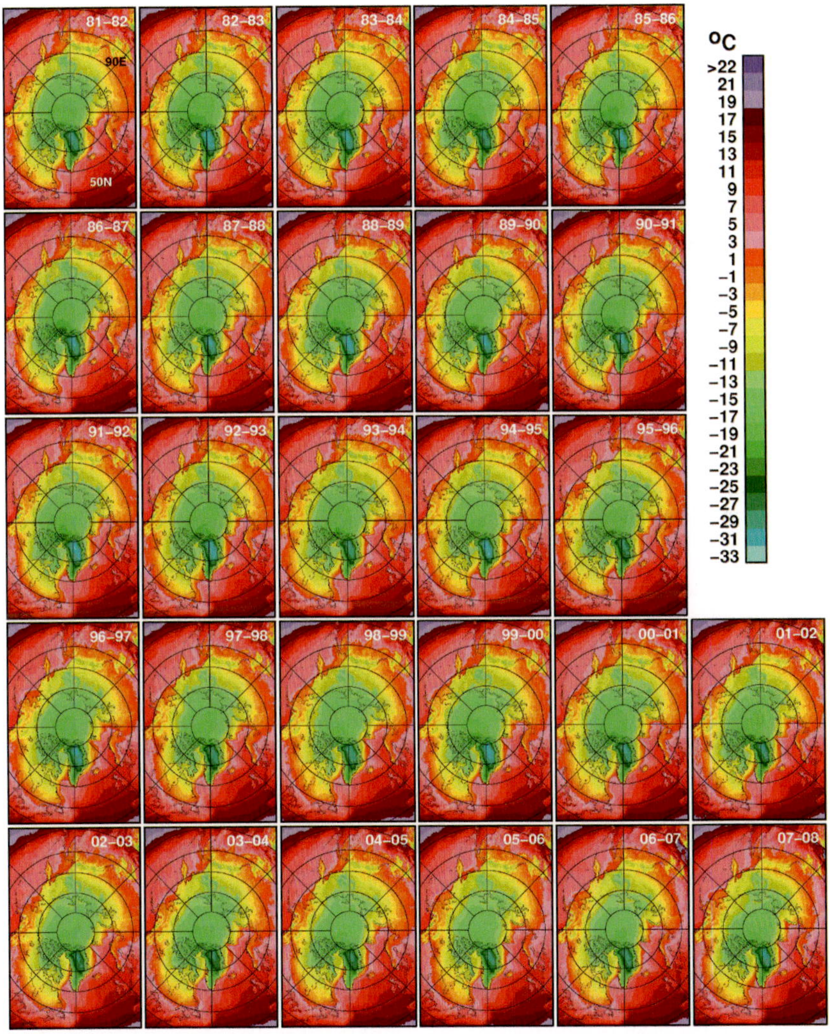

Fig. 6.11 Yearly surface temperature in the Northern Hemisphere for the period from August 1981 to July 2008 as derived from AVHRR data

generated such that each year starts on 1 August and ends on 31 July the following year. Such averaging was done for the Northern Hemisphere to enable comparisons of the average temperatures during each ice season. It also avoids having to end a yearly average in December which is the time when surface temperature is rapidly declining to minimum values in the middle of the ice growth season. It also enables direct comparisons with yearly averages similarly generated for the Southern Hemisphere, where the averaging is also from austral summer (1 January) to the following austral summer (31 December). The yearly temperature maps shown in Fig. 6.11 are quite different from those in February and September and look very much like the April climatological data. The images generally look very similar with many surface features being consistently depicted from 1 year to another. Some of the distinct features that are faithfully reproduced include those associated with glaciers and high elevation areas. The data can thus be used to monitor interannual changes in surface temperature in these features. When compared with the winter averages, the yearly averages show much less fluctuation in part because the averaging is over a longer time period and biases that may be associated with an early or late winter are not relevant with the data. A closer analysis of the images, however, shows much more variations than is apparent. For example, the distribution of temperatures in Greenland looks quite colder some years (e.g., 1982–1983, 1991–1992, 1992–1993, and 1999–2000) than other years. The region north of Greenland and the Ellesmere Island which is consistently the coldest in the sea ice covered regions also show significant interannual variability and was especially cold in 1982–1983, 1984–1985, 1986–1987, 1988–1989, 1990–1991, 1991–1992, and 1992–1993. This region is usually covered by thick perennial ice and is the general location of the oldest ice types (Colony and Thorndike 1985). The region is also an area of heavy ridging and had been the site of extremely thick ice (>10 m) as observed by upward looking sonars from submarines (Wadhams 2000).

Some of the subtle interannual changes which are difficult to resolve from the images in Fig. 6.11 are better quantified in the anomaly maps shown in Fig. 6.12. It is apparent that the transitional patterns that depict significant changes from one time period to another in the summer anomalies (Fig. 6.5) shows up even much stronger in these yearly anomaly maps. In the yearly anomalies presented in Fig. 6.12, the negative anomalies are clearly very prominent in the 1980s and from 1991 to 1994 while the other maps show predominantly positive anomalies especially from 2004 to 2008. Some warming is evident in 1989-1990 and 1990-1991 but this was followed by three years of cooling which is likely associated with the aerosol released during the Mount Pinatubo volcanic eruption in 1991. It is also apparent that the Eurasian side experienced some episodes of cooling from 1997 to 2001. Some cooling in this region persisted in the 2000s but from 2006 to 2008, warming became more prevalent. Anomalies were highly positive in the Central Arctic in the last three years ending in 2008. It is likely that the surface conditioning associated with the warming events in 2005-2006 and 2006-2007 led to the drastic decline in the perennial ice cover in 2007 which will be discussed in the next chapter.

The yearly anomalies presented in Fig. 6.12 show patterns of warming in the Arctic that are possible to obtain only through the use of satellite data. A strong manifestation of an apparent warming during the 1981 to 2008 period is revealed and it is also evident that the warming is not spatially uniform. Comparing with the anomaly maps in

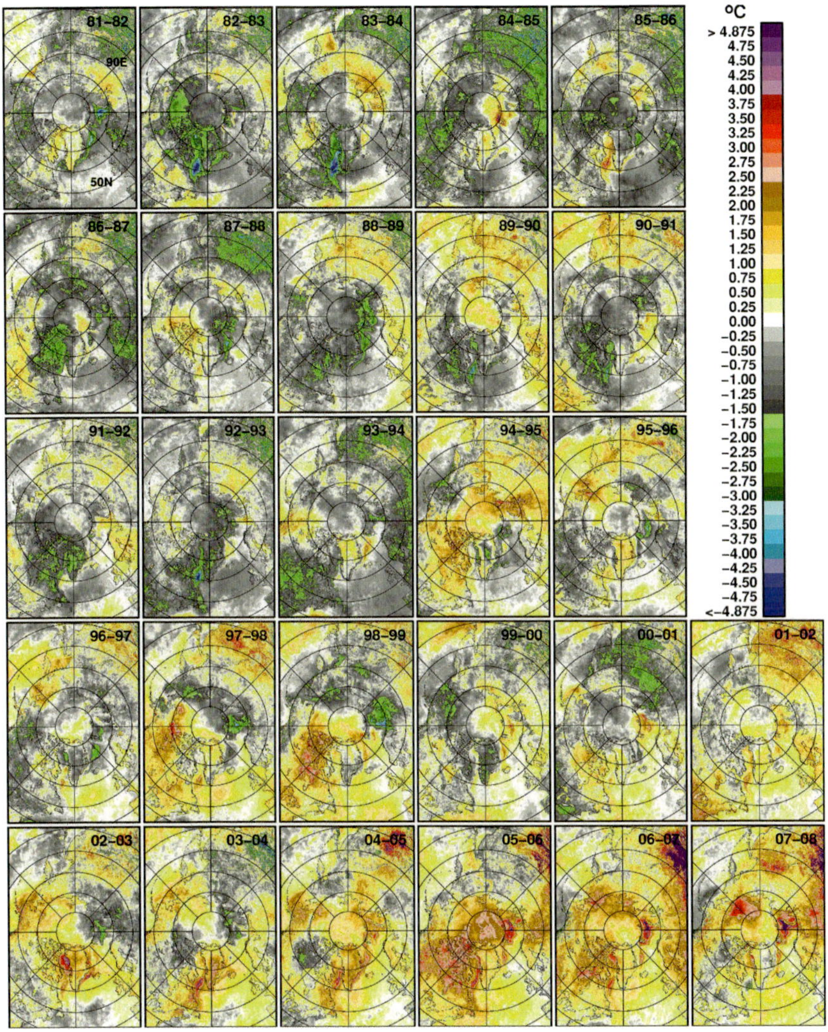

Fig. 6.12 Yearly surface temperature anomalies in the Northern Hemisphere for the period from August 1981 to July 2008 as derived from AVHRR data

Fig. 6.5, the images also show that warming is not just a summer phenomenon but more of a year round event. The persistence of enhanced warming anomalies from 2001 to 2008 also suggests a need for continued monitoring of the region.

6.2.1.2 Sea Surface Temperature in the Northern Hemisphere

The interannual variations in Sea Surface Temperature (SST) on a global scale have been the subject of many investigations (e.g., Kilpatrick et al., 2001) because of their relevance to climate change. Studies of the variability of SST is not trivial because it fluctuates only by small amounts since the ocean is such a huge reservoir of energy that suppresses changes at the surface. The accuracy requirements are thus more demanding that those of other surfaces. Many studies have taken advantage of what has been referred to as the Reynold's SST data set (Reynolds et al. 2002) which is based on thousands of ship and buoy measurements around the globe. In recent years, Reynold's SST also incorporates satellite data, but mainly to fill in gaps not covered by buoy and ship data. Also, this data set has been gridded at a resolution of only 110 km because in situ measurements do not provide the spatial details that would allow for a better resolution. True global data sets have been possible to generate with the advent of satellite systems and in recent years, there have been many options including the use of thermal infrared data and more recently, microwave data such as those provided by Aqua/AMSR-E as described by Shibata et al (2010). Microwave sensors provide day/night almost all weather measurements and global coverage at a good time resolution. Since it provides synoptic and good temporal coverage, the sensor provides true averages of surface temperatures. While ideally suited for large scale SST studies, the current data record is just about 6 years long and the spatial resolution is quite low, the field-of-view being around 43 by 73 km. Also, because of sidelobe and contamination effects, data near shorelines and sea ice are also deleted. We use the AVHRR data for our time series studies not only because of the relatively long time record but also because they provide distributions of SST in the polar regions that are consistent with those from AMSR-E and Reynolds.

To illustrate how compatible these three different sources of data are, sea surface temperature maps for a winter (i.e., February) and a summer (i.e., September) month as derived from AVHRR, AMSR-E and Reynolds data are presented in Fig. 6.13. The AMSR-E SST is derived using a technique described by Shibata et al (2010). It is apparent that the three data sets provide strikingly similar retrievals of SST for the two periods. The temperature isotherms are located at approximately the same latitudinal locations in the three images and basically, many of the distinct features are similarly represented. There are, however, subtle discrepancies some of which may be due to differences in resolution and data sampling. The AVHRR and AMSR-E SSTs are the most similar of the three data sets with the Reynolds data showing some discrepancies likely at areas where buoy or ship measurements are lacking. It is interesting to note that detailed undulations of the isotherms some of which are associated with ocean eddies and fronts are depicted in a similar fashion in both AVHRR and AMSR-E data. Because of the lack of resolution, the same features are not evident in the Reynolds data. As pointed out earlier, data near the sea ice cover and land are not retrieved in the AMSR-E data but otherwise, it probably provides the best representation for monthly average data. The key concern with AVHRR has been the availability of the data only in cloud free conditions. However, the good agreement of AVHRR and AMSR-E data in both winter and summer months indicates that the bias due to clouds is likely small. The results from AVHRR also agree in magnitude with the Reynold's data except near the ice edges where buoy data are scarces and in areas where resolution makes a difference.

Despite the differences in technique and type of instrumentation used for the retrieval of SST, it is encouraging that the AVHRR, AMSR-E and Reynolds data provide very similar SST distributions. The temperature isotherms in the Pacific

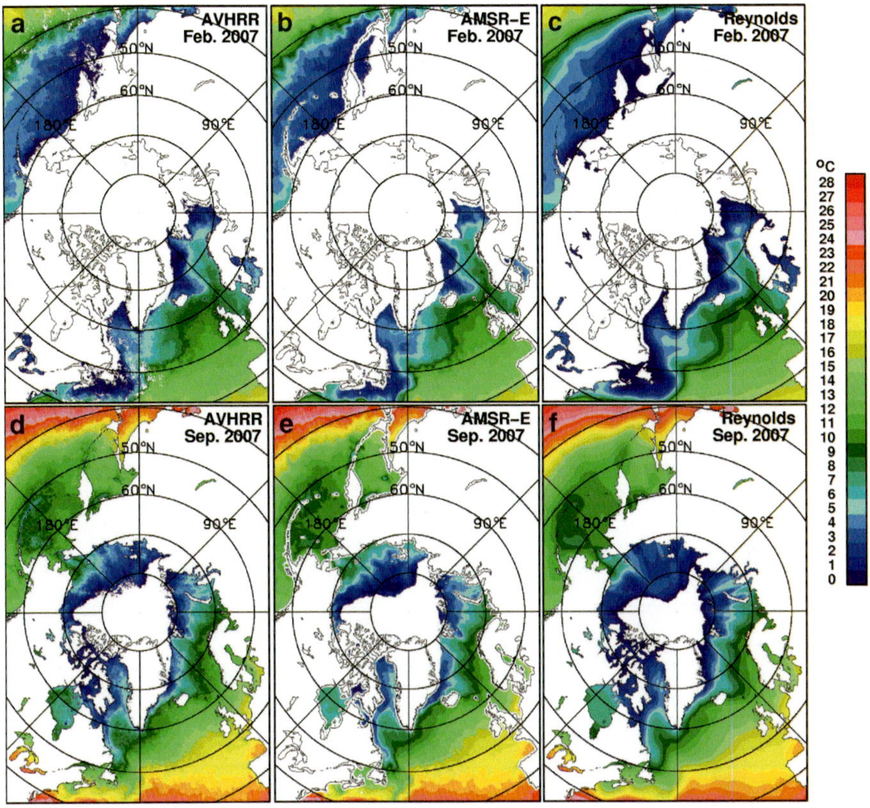

Fig. 6.13 Comparison of monthly average SSTs using (a) AVHRR; (b) AMSR-E; and (c) Reynold's data during a mid-winter month (February 2007); and (d) AVHRR; (e) AMSR-E; and (c) Reynold's data during an end of the summer month (September 2007)

and Atlantic Oceans as depicted in the February and September maps for 2007 in Fig. 6.13 generally follow the latitudinal lines at the lower latitudes. At about 45–50°N, it is apparent that the Pacific Ocean is relatively colder than the Atlantic Ocean in both periods, a manifestation of the effects of the Gulf Stream in the Atlantic and sea ice in the Bering and Okhotsk Seas. Between 50°N and 60°N, the SST becomes more unpredictable, but the contours of near freezing temperatures are close to the locations of the ice edges making these contours unsymmetric with respect to the North Pole. The relatively warm temperatures at high latitudes in the Barents Sea are likely caused by the transport of warm surface water by the North Atlantic Current as indicated in Fig. 2.2.

To complement the climatological images in Fig. 6.1, we use the 6-year AMSR-E data to calculate multiyear-monthly averages and the results are presented in Fig. 6.14. The scale has been enhanced as in Fig. 6.13 to better illustrate the seasonal variability of SST in the Northern Hemisphere. The data basically provide the contemporary version of the seasonal variability of SST and at the same time, the means

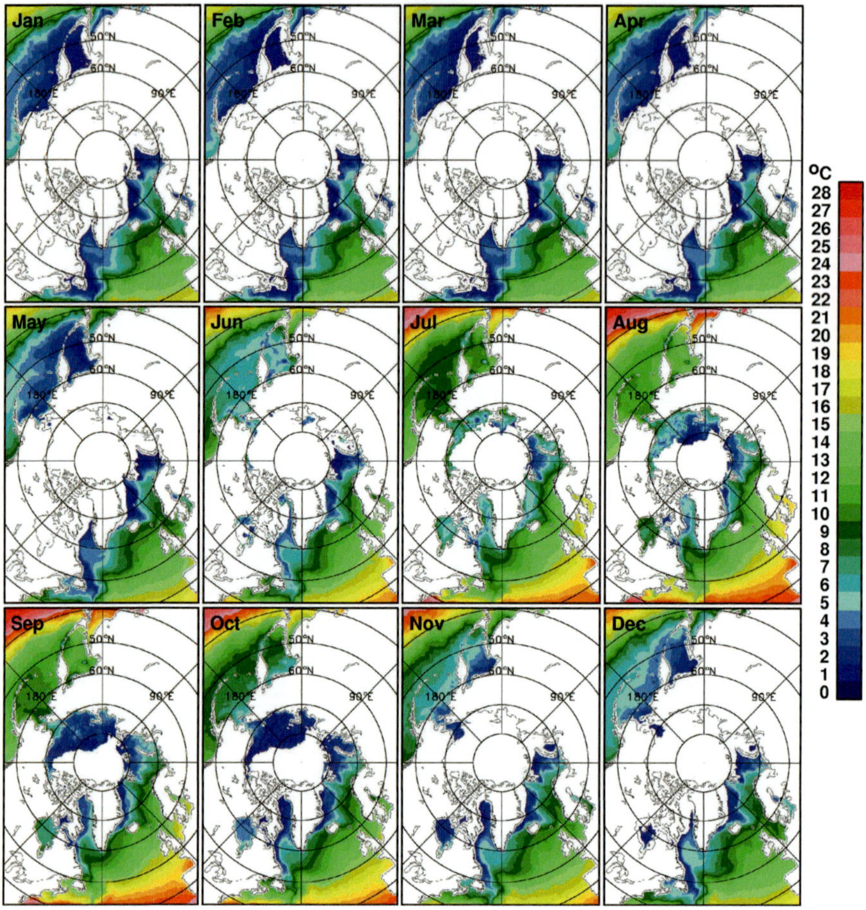

Fig. 6.14 Averages of surface temperatures from January to December in the Northern Hemisphere using AMSR-E data for the period from May 2002 to April 2008

to assess SST distributions using a different type of instrument. In these multiyear averages, much of the detailed undulations of isotherms observed in the monthly images in Fig. 6.13 have been averaged out and were replaced by smoother contours. The set of images in Fig. 6.14 shows generally how contours of the same temperature advance to the north from winter to summer and retreat in the opposite direction from summer to winter. It is interesting that some of the distinct features of an isotherm are preserved from 1 month to another. An example is the 5°C isotherm (light blue) in the Pacific Ocean (south of the Bering Sea) where the feature is basically maintained from January to April. Another example is the 5°C isotherm in the Greenland and Barents Seas the shape of which is mainly preserved from October to December. The persistence of these features indicates that they are mainly oceanographic characteristics as opposed to that caused by the more variable atmospheric forcing. The set of images also shows that the maximum SST values do not occur in the same month for the different regions. For example, the SSTs in the Arctic basin are shown to have the highest values in August, but in the North Pacific Ocean, the SSTs are higher in some places in September than in August. Using weekly averages (not shown), the peak value for SST usually occurs toward the end of August.

The set of multi-year monthly images also shows some interesting patterns in the various seas. For example, although the Sea of Okhotsk is located at lower latitude than the Bering Sea, it is generally colder than the latter except in August and September. The coldest part of the Okhotsk Sea region in June and July is actually located in the central portion of the sea, while in November, December, and the rest of the winter period, the coldest region is closer to land. The influence of cold air from Siberia which as shown in Fig. 5.1 is generally very cold during the winter season is likely a factor. In the early summer months (i.e., June and July), the SSTs in the Barents and Laptev Sea are generally lower than those in the Bering Strait and Beaufort Sea but by August and up to October, the reverse is true especially at high latitudes.

It is also apparent that as indicated in the previous section the Atlantic Ocean is warmer than the Pacific Ocean in every month of the year. This is an important phenomenon since it provides insights into the difference in the salinity distribution and deep water production in the Atlantic and Pacific Oceans as discussed by Broecker et al (1985). In the Atlantic Ocean, warm surface water is advected to the North (see Fig. 2.5) where they cool and downwell into the deeper part of the ocean. On the other hand, relatively cool deep water upwells to the surface in the northern Pacific before it gets advected to the south. Such difference in the circulation patterns is what makes the surface temperature in the Atlantic Ocean genrally at a higher value than that of the Pacific Ocean. Such temperature difference also makes the Atlantic Ocean relatively more salty and in some regions the excess salt leads to deep-water formation.

To assess the interannual variability of SST, Fig. 6.15 shows anomaly maps for each February from 1982 to 2008 using AVHRR data. The climatology used in the estimate of these anomaly maps is the average of all data in February from 1982 to 2008 as in the previous section (see Fig. 6.1). The color scale for the anomaly maps are binned at 0.2°C interval to provide more information (than in Fig. 6.3) about the subtle variations in SST from 1 year to the next. It is apparent that SSTs were relatively low in February in 1986, 1987, and 1988 in both Pacific and Atlantic Oceans and also low in the Atlantic Ocean in 1983, 1985, 1990, 1991, and 1994. They were also relatively cold in the Pacific Ocean in 2000, 2001, and 2008. On the other hand, SSTs were relatively high in both Pacific and Atlantic Oceans in 1998, 1999, 2003, 2004, 2005, and 2006, while they were high in the Pacific Ocean only in 1991 and 1997. The SSTs were also relatively high in the Atlantic Ocean in 2000, 2001, 2002, 2007, and 2008. Note that there were years when it was unusually high in the Pacific Ocean and at the same time unusually cold in the Atlantic Ocean as in 1991, while a reversal can happen as in 2001 and 2008. Furthermore, there were years when it was unusually warm at low latitudes and colder than normal at higher latitudes. Generally, the anomalies are relatively modest and within ±2.0°C but even relatively small changes in SST are important to identify. Overall, it is apparent that SSTs were generally warmer in recent years when compared with those of earlier years, especially in the Atlantic Ocean.

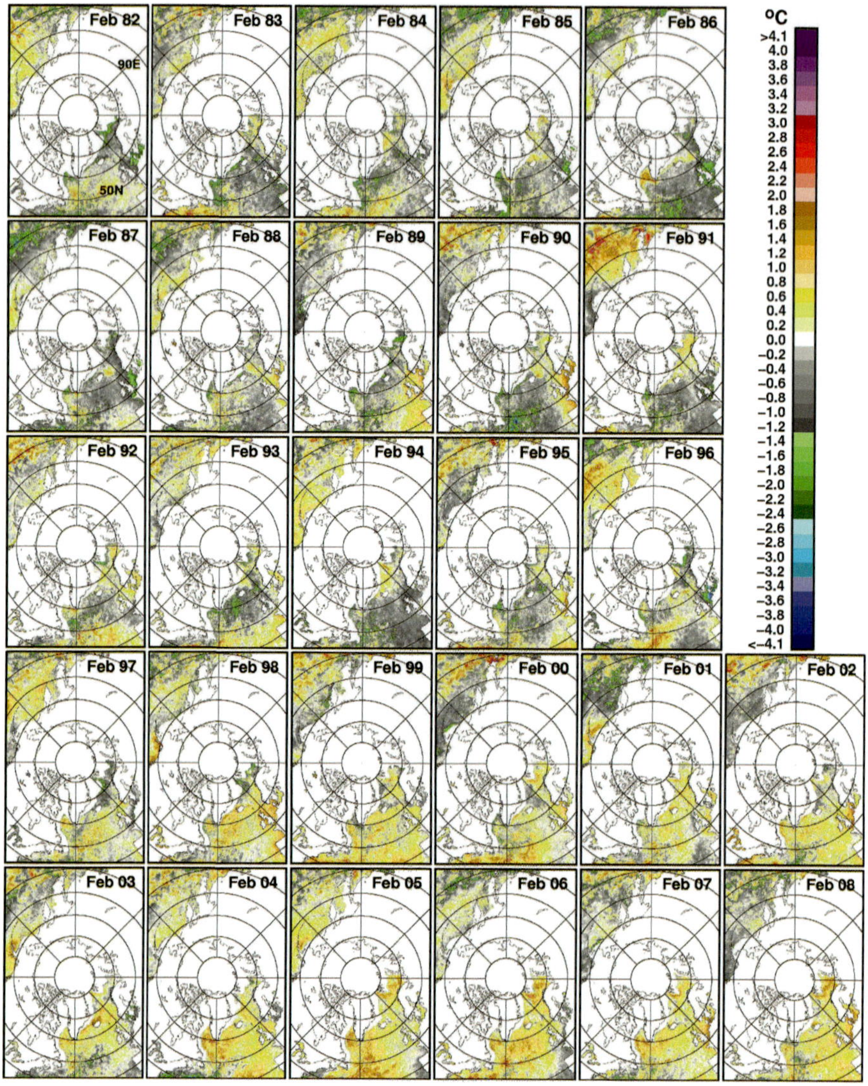

Fig. 6.15 Monthly average SST anomalies in the Northern Hemisphere for the period from 1982 to 2008 during a winter month (February)

A similar set of monthly anomaly images but for September is presented in Fig. 6.16. It is apparent that the anomalies in September have higher values than those in February. Negative anomalies are clearly prominent for the years 1981–1995, while for the later years the anomalies are primarily positive. There were some exceptions like the positive anomalies in the Pacific Ocean in 1986, 1988, 1989, and 1994 and negative anomalies in the Pacific Ocean in 1997 and 2002. Strong positive anomalies in SST occurred in the Pacific Ocean in 1994, 1998, 2000, and 2006. It is also interesting to note that a strong positive anomaly occurred in the Pacific Ocean at around 40°N in 1989, 1994, 1998, 2000, and 2006. A relationship with the ENSOs in 1994, 1998,

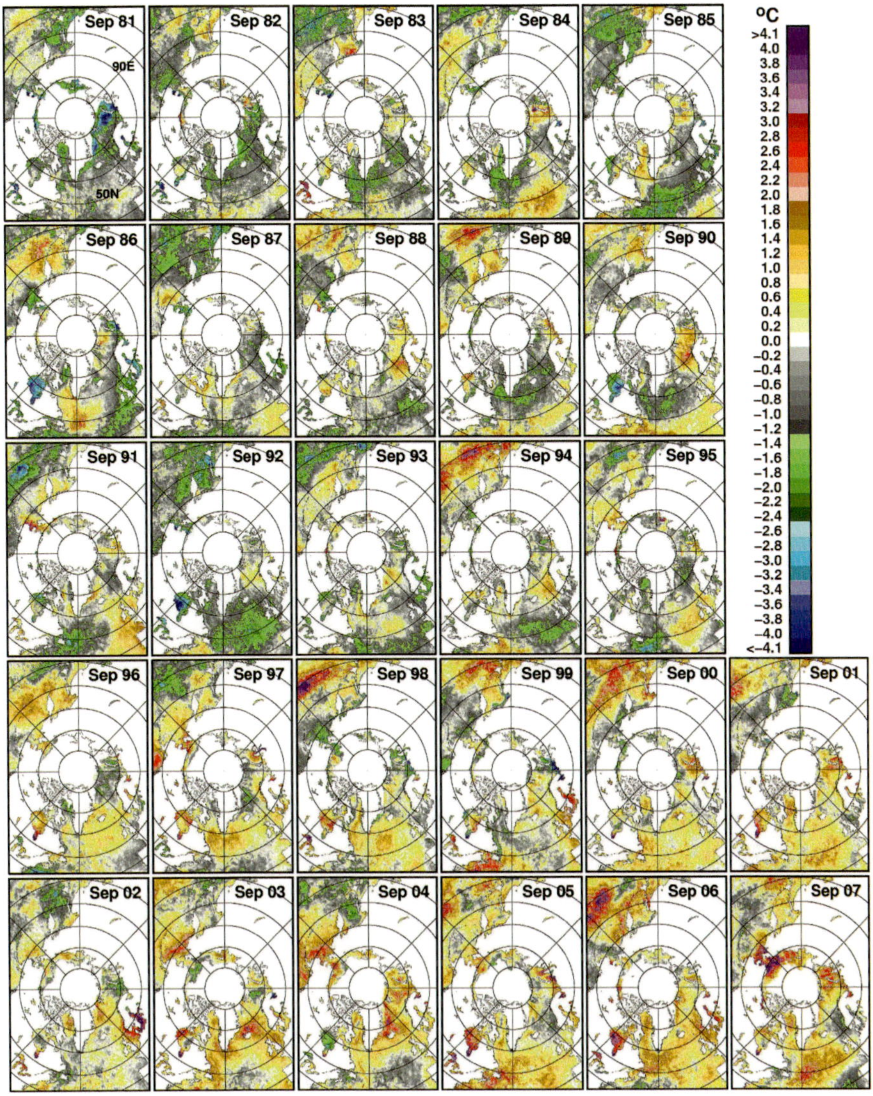

Fig. 6.16 Monthly average SST anomalies in the Northern Hemisphere for the period from 1981 to 2007 during an end of the summer month (September)

and 2006 is possible, but no such effect was observed during the strong ENSOs in 1982–1983 and 1991–1992. An unusually big positive anomaly at the Bering Strait/Chukchi Sea area is apparent in 2007 coinciding with a cyclone in approximately the same location time, as indicated in Fig. 5.5i. It was in 2007 when the sea ice cover retreated very fast in July, August, and early September 2007 (Comiso et al. 2008). During this period, there were also reports that cloud cover in the region was less prominent than normal (Schweiger and Lindsay 2008). A similar but smaller anomaly occurred in the Bering Strait in 2003 and 2004. Possible intrusion of warm water from the Bering Sea to the Arctic may have contributed to the reduced ice cover in the Beaufort Sea region during these years.

The average SST in the Arctic Basin for >65°N is of interest in light of rapid changes in the Arctic sea ice cover in the region. The average SST and also the maximum value for each year is presented in Fig. 6.17. It is apparent that there is a large interannual variability and that the values in recent years are higher than those

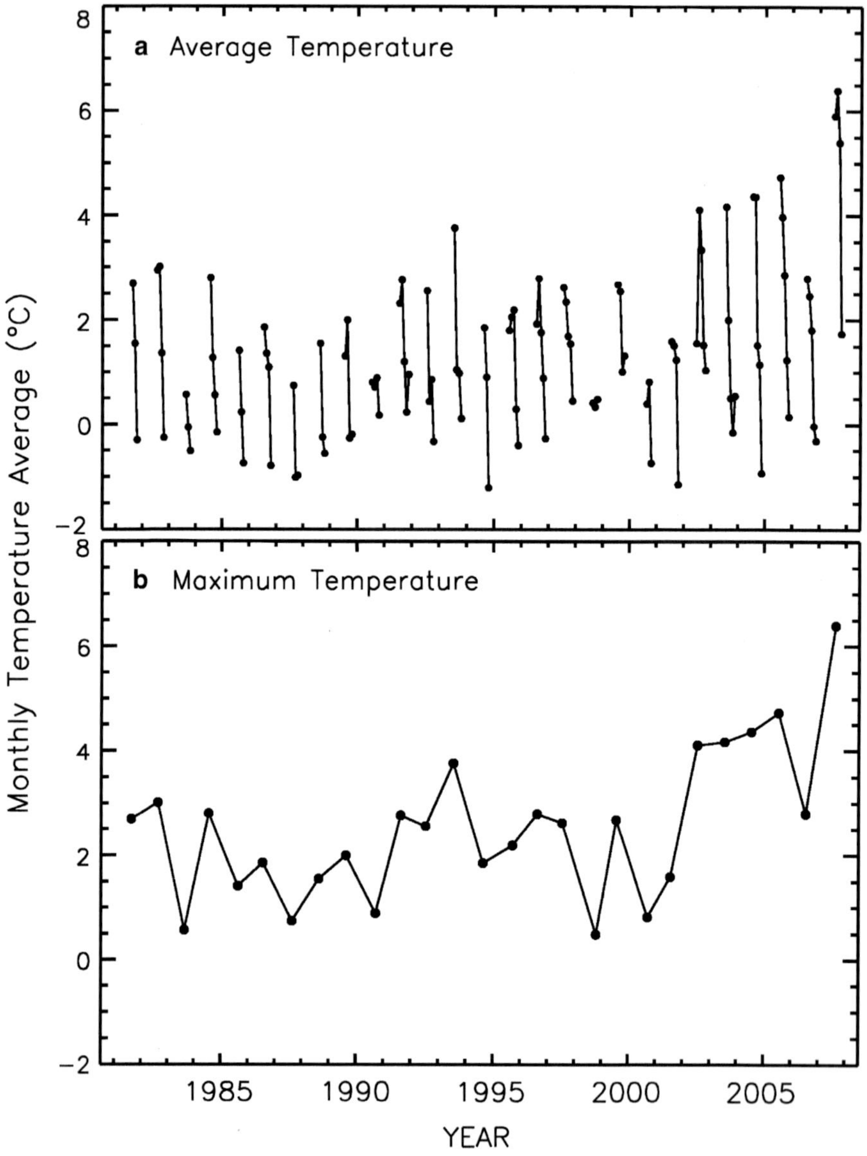

Fig. 6.17 Plots of (**a**) monthly averages and (**b**) maximum values of SST from 1982 to 2008 at latitude regions >65° N in the Northern Hemisphere

in earlier years. The trend in the anomalies of the average temperature is estimated to be 0.71±0.12°C/decade, while that of the maximum temperature is 0.77±0.26°C/decade. It appears that the average SST and maximum SST were unusually high in 2007. A record high SST that lasted for a few weeks in 2007 in the Chukchi Sea region was also reported by Shibata et al. (2010).

6.2.2 *Southern Hemisphere*

6.2.2.1 Surface Temperature over Sea Ice and Land

In the Southern Hemisphere, the spatial distributions of surface temperature are very different from those in the Northern Hemisphere because of environmental and geographical differences as discussed earlier. Climatological monthly surface temperatures similar to those shown for the Northern Hemisphere as derived from AVHRR data from 1981 to 2008 are presented as color-coded maps in Fig. 6.18. The set of images shows the progression in the distribution of surface temperature from summer to winter and from winter back to summer in the Antarctic region. The seasonality of surface temperature for the entire region is quite large considering that the area covered by sea ice changes in extent from a minimum of about 3 million km^2 to a maximum of about–19 million km^2. The key surface features of interest are SST in ice free ocean surfaces which fluctuates very little with seasons, sea ice covered surfaces which change moderately, and ice sheet or land surfaces which change substantially during an annual cycle. The −2-degree isotherm usually defines the approximate boundary between ice free ocean and sea ice. A large pulsation in the latitudinal location of the −2-degree isotherm is evident as the sea ice cover advances to the north from February to September and retreats back to the south from October to February. Although SST and surface ice temperature are primarily influenced by surface air temperature, the atmospheric temperature is also influenced by the sea ice cover and the open ocean through exchanges in fluxes during ice-ocean-atmosphere interactions (Ackley 1981; Hibler 1979).

Although the surface temperatures are usually coldest in July or August and warmest in December or January, there is a lag in the temperature effect and the sea ice cover is most expansive in September and least expansive in February (or March). The surface temperature contours over the open ocean and sea ice is asymmetric in January and February with the coldest areas located in the Western Weddell Sea and the Bellingshausen/ Amundsen Seas. These are also the areas where sea ice is most likely to be found at the end of the summer. In March through July, the northward advance in temperature isotherm is apparent with the −7°C isotherm becoming more and more circular and symmetric with respect to the continent. In March through May, there is a discontinuity of the temperature contours at the Antarctic Peninsula. This suggests that the climate on eastern side of the Antarctic Peninsula is different from that on the western side. From July to October, the difference is minimized and the −5°C contour is almost like continuous rings around the Antarctic continent. At the lower temperatures, the discontinuity is still evident and this is primarily because the surface

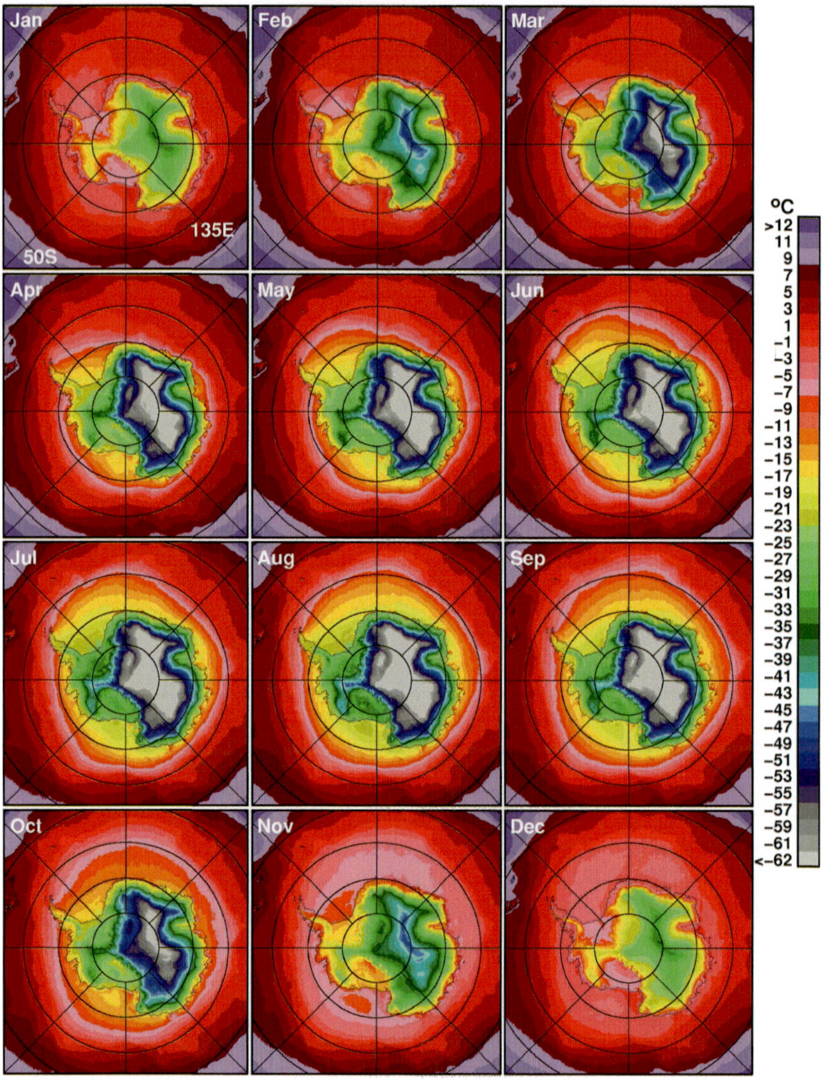

Fig. 6.18 Climatological monthly surface temperatures from January to December in the Southern Hemisphere using AVHRR data from August 1981 to July 2008

temperature in the Antarctic Peninsula is generally much lower than that of the sea ice cover. Generally, the symmetry is preserved until ice breakup in November. In the continent, the coldest month appears to be August and the coldest spot is in the Antarctic Plateau. In the sea ice covered regions, the surface temperature is lowest in the Western Weddell Sea followed by the Ross Sea. Both of these regions are sites of winter coastal polynyas as will be discussed in Chap. 8. The higher frequency of these polynyas in the Ross Sea makes the coastal region generally warmer than sea ice covered areas at lower latitudes. The coastal region is also the scene of an early spring polynya. The highest rate of advance for the zero degree isotherms also occurs in the Weddell Sea with the advance generally toward the northeast following the Weddell gyre and the atmospheric pattern shown in Fig. 5.12.

To gain insight into interannual variability, monthly averages of surface temperatures during an austral summer month (February) from 1982 to 2008 are presented in Fig. 6.19. The variations from 1 year to another are relatively subtle and not so easy to identify in the open ocean but are much more apparent over sea ice covered areas

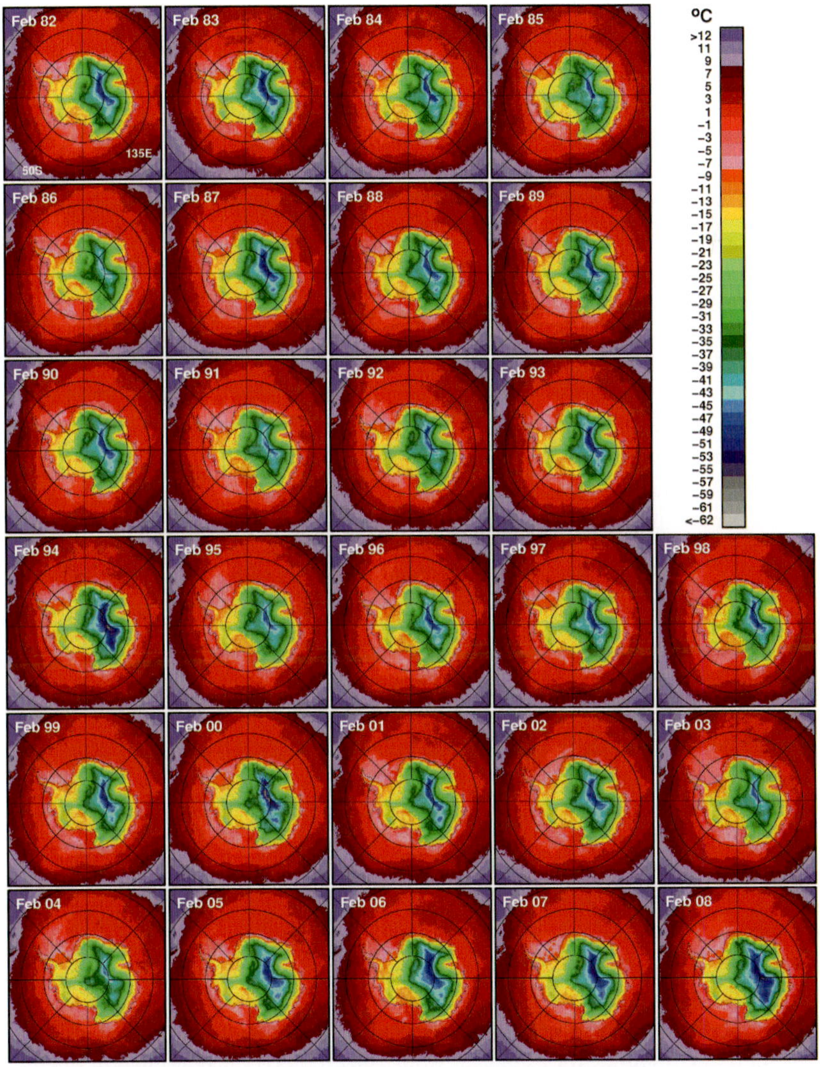

Fig. 6.19 Monthly averages of surface temperatures in the Southern Hemisphere during a summer month (February) as derived from AVHRR data for the period1982 to 2008

and the continent. In the ocean region, the year-to-year changes are mainly associated with the location of the sea ice cover which determines where the zero-degree isotherms are located. The lack of symmetry is also apparent to the north and farther from the ice edge mainly because the melt of sea ice is not spatially uniform and the distribution of melt water which affects the distribution of surface temperature depends on the history of the ice cover in the region. In the Antarctic continent, interannual variations in surface temperature for a certain area can be relatively large but the spatial distributions for the different years are similar since the contours of temperature isotherms follow closely the topographic contours. Strong year-to-year fluctuations are evident in many areas especially in East Antarctica (e.g., from February 1993 to February 1994, and from February 1999 to February 2000). The surface temperatures in the continent are usually correlated with the surface elevation and in some cases with those of surrounding areas, especially those in the sea ice covered region. For example, the surface temperatures in the Antarctic continent were relatively cold in 1994 and 2000, as manifested by the extremely low values in the East Antarctic plateau and at the same time surface temperatures in the Western Weddell Sea and other areas were also unusually cold. There are however exceptions and the temperature patterns are normally influenced by atmospheric circulation which is dictated by the Southern Annular Mode as discussed in Chap. 5.

The year-to-year variations of the February surface temperatures are better quantified in the anomaly maps presented in Fig. 6.20. It is apparent that there are no temporally coherent patterns of surface anomalies in the open ocean, sea ice and continental regions. The surface temperature anomalies were mainly positive in the open ocean and sea ice covered regions in 1983, 1992, 1993, 1997, 1998, 2003, and 2006, while they were mainly negative anomalies in 1986, 1989, 1999, 2000, 2001, 2002, 2007 and 2008. In the continent, temperature anomalies were strongly positive in 1986, 1991, 1996, 1998, 2003, and 2004, while they were strongly negative in 1994 and 2008 and also mainly negative in 2000, 2005, 2006, and 2007. Some of the negative anomalies in the Western Weddell Sea region are actually associated with the changing location of the perennial ice cover in the region. The presence of sea ice makes a significant difference although the contrast of surface temperature over sea ice and open ocean is not as much in the summer as in winter. The negative anomalies in the continent in 2008 were also likely linked to the relatively strong La Niña during the year. It is interesting to note that there were years as in 1983 when relatively low temperatures in the continent were accompanied by relatively high temperatures in the Southern Ocean. The reverse is true in 1986 and 1996. Concurrent occurrences of negative anomalies in both continent and the Southern Ocean occurred in 1987 and 2000, 2007 and 2008, while concurrent occurrences of positive anomalies happened in 1995, 1997, 1998, and 2003. The set of images shows some distinct patterns of unusual warming or unusual cooling but overall, there is no suggestion of a trend or a pattern of change.

The monthly averages of surface temperatures shown in Fig. 6.18 indicate that winter is much colder than summer in the Antarctic region. The September monthly temperature distributions for each year from 1981 to 2007, as presented in Fig. 6.21, show similar patterns but much lower values than those of Feburary. The contours of zero degree isotherms are much further to the north and much more symmetric than

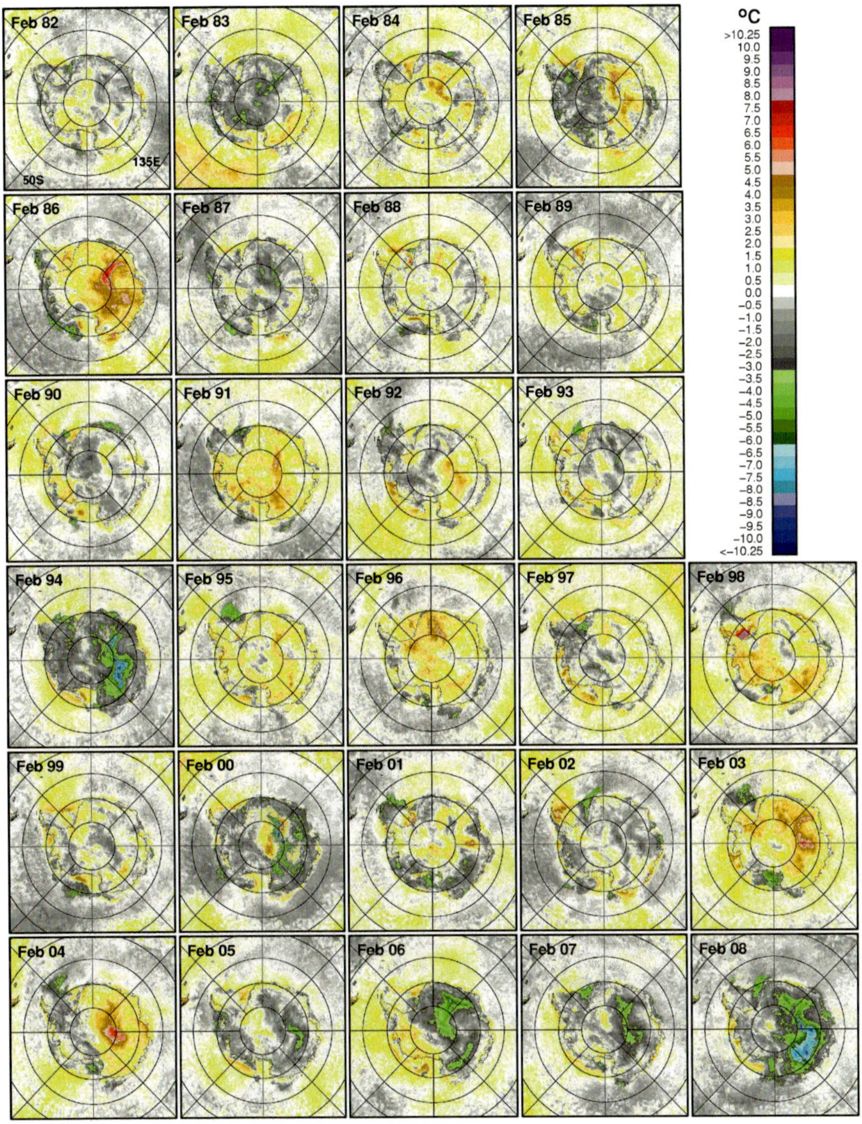

Fig. 6.20 Monthly averages of surface temperature anomalies in the Southern Hemisphere during a summer (February) as derived from AVHRR data for the period1982 to 2008

those in February. Extremely cold temperatures over the Antarctic continent are apparent especially in the Antarctic plateau. The coldest region in the maps is consistently at the high elevations of the Antarctic plateau where the lowest temperature on the Earth's surface has been observed. The coldest surface temperatures in the sea ice and open ocean regions are not easy to identify but occur during this month (September) as well. Significant interannual changes are apparent with the shape and

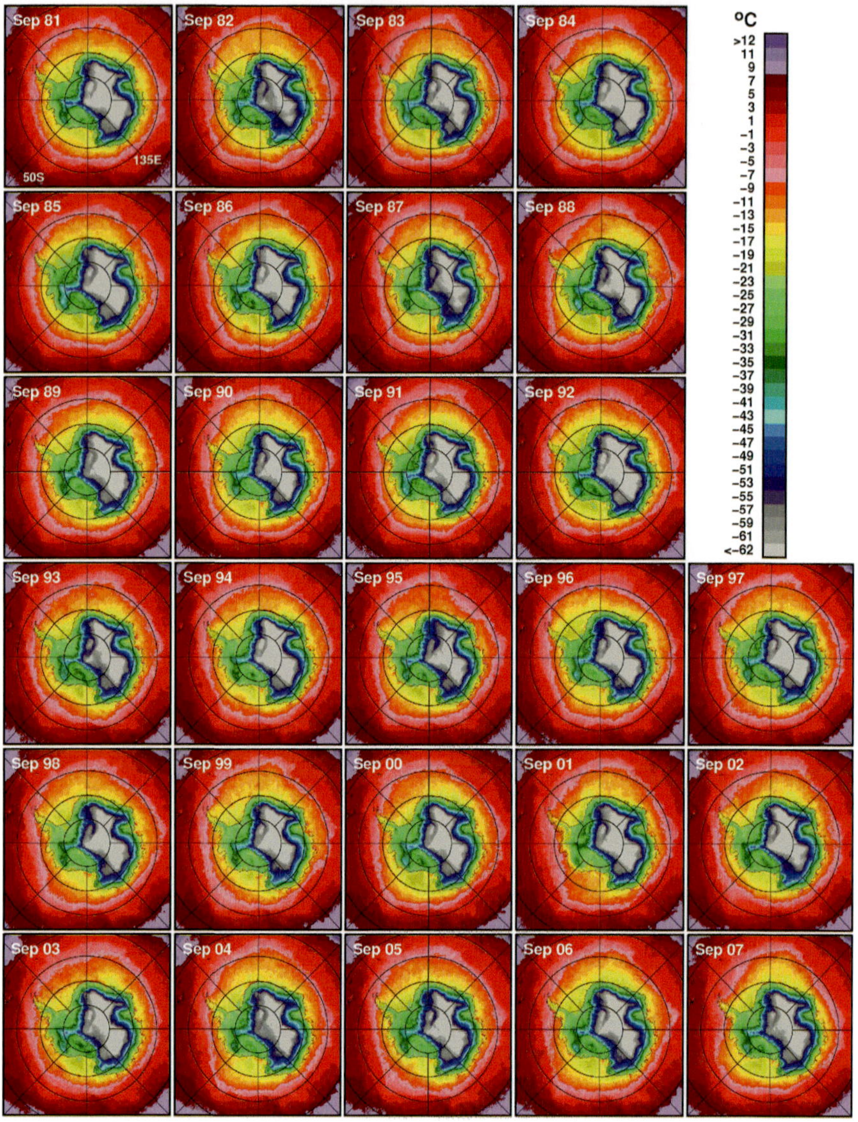

Fig. 6.21 Monthly averages of surface temperatures in the Southern Hemisphere during a winter month (September) as derived from AVHRR data for the period1982 to 2008

location of the surface temperature zero-degree isotherm, usually located between 55°S and 65°S, changing from 1 year to another as the ice cover changed in response to changes in temperature and atmospheric circulation. There were years as in 1981, 1985, 1989, 1993, 1997, and 2003, when the −7° contour was almost circular, but there were also years like in 1983, 1995, 2001, and 2007 when the patterns were less symmetrical. The spatial patterns of the surface temperature in the sea ice covered regions are strongly influenced by ice concentration and atmospheric circulation.

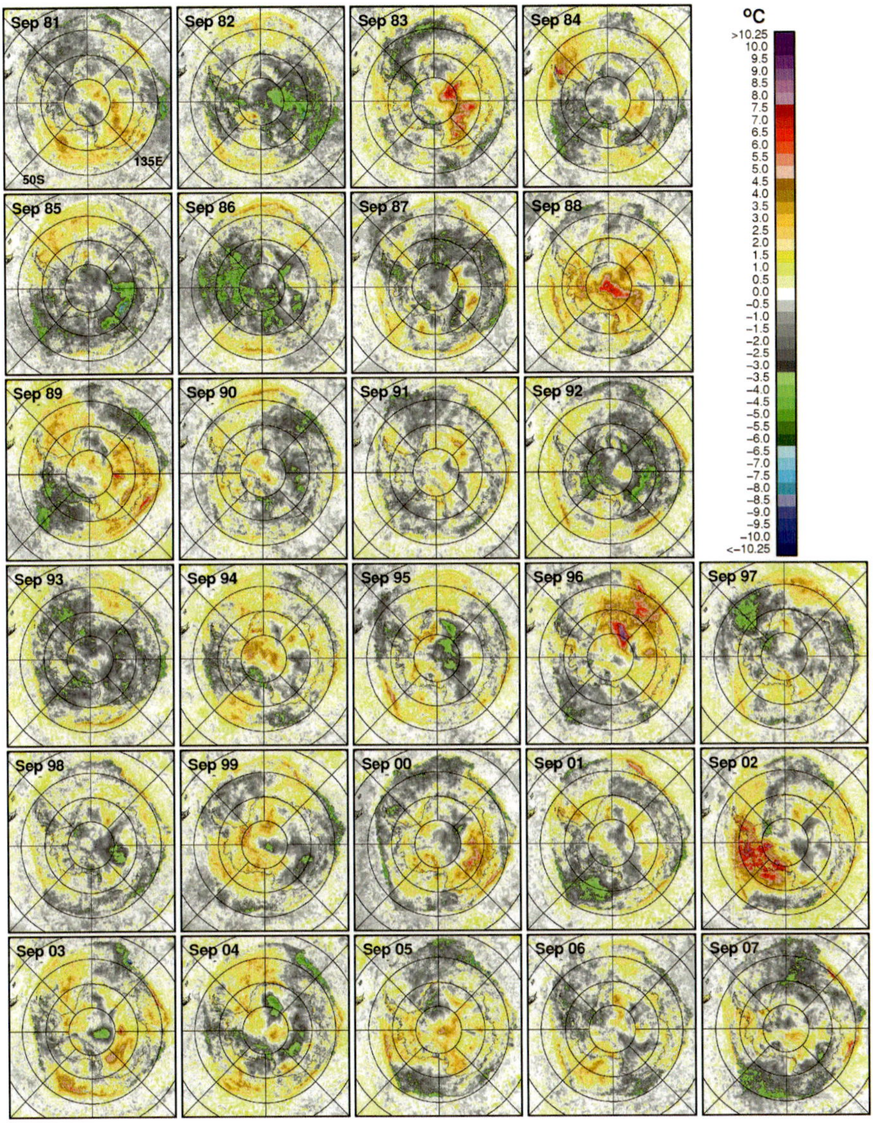

Fig. 6.22 Monthly averages of surface temperature anomalies in the Southern Hemisphere during a winter month (September) as derived from AVHRR data for the period1982 to 2008

Areas that are constantly subject to divergence are more likely to have higher temperatures than those in consolidated ice regions.

The year-to-year variations in the September surface temperatures are better quantified in the anomaly maps presented in Fig. 6.22. The set of images shows patterns of anomalously cold or warm regions which are in different locations in different years but no decadal changes are apparent. Relatively strong positive anomalies are apparent in the eastern side of the Antarctic plateau in 1983, the Antarctic continent and surrounding areas in 1988, the western side of the Antarctic plateau and surrounding ice cover in 1996 and the West Antarctic and the Antarctic Peninsula in 2002. Conversely, relatively strong negative anomalies also occurred in the Antarctic Plateau in 1982 and 1985, in the entire Antarctic continent in 1986, 1987, 1992, and 1993, in the western Weddell Sea in 1997 and in the Ross Sea in 2001 and 2007. From 2003 to 2006, the temperature anomalies show no preferred patterns while some cooling in sea ice covered regions is apparent in 2007. Such cooling was likely associated with abnormal lows in approximatedly the same region as depicted in Fig. 5.14.

Although clear cut warming or cooling patterns are not discernible from the set of images in Fig. 6.22, there is actually some kind of order in the winter anomaly distribution that is associated with the Antarctic Circumpolar Wave (ACW) as described by White and Peterson (1996) and discussed in Chap. 5 (Sec. 5.2.2). The ACW was described as a mode-2 system propagating around the continent along the sea ice margin with a period of about 8–9 years. Alternating patterns of warm and cold anomalies along the ice edge are postulated and for a mode-2 system this is repeated twice around the continent. At a particular longitude, warm (or cold) anomalies should occur at the ice edge location every 4 years. For example, the warm anomaly near the Antarctic Peninsula in 1984 propagated to the east in 1985 and 1986 and then in 1989 (after 5 years) a warm anomaly in the region occurred again. This was again repeated in 1994, but in 1998 the warming anomaly was very weak. It reappeared in a strong way in 2002 and then in 2005. The pattern is depicted graphically to demonstrate the propagation of the wave in Chap. 8 (Sec. 8.4). Such atmospheric patterns complicate the interpretation of the interannual variability of the temperature data making it difficult sometimes to discern what may be a relatively weak climate change signal from the data set.

Plots of monthly averages of temperatures from 1981 to 2008 for the region > 60°S and inside the Antarctic Circle (>66.5°S) are presented in Fig. 6.23 for comparison with similar plots in the Arctic. It is apparent that the Antarctic region is much more seasonal than the Arctic region. This is mainly because of environmental factors as discussed earlier in this section with the high elevation regions in the Antarctic plateau so much colder in winter than sea ice covered regions in the Arctic. The maximum and minimum values are also shown to be more variable inside 60°S than inside the Antarctic Circle. The averaging includes ice free ocean surfaces and the variability is in part because of the interannual variability in the location of the ice edge and the extent of the sea ice cover within the study area. Inside the Antarctic Circle, there is a lot less influence of the warmer open water areas and the range of values is generally from −12 to −35°C when compared with −8 to −25°C for the monthly averages for the region > 60°S.

Interannual changes and trends are better quantified when the monthly anomalies are used as presented in Fig. 6.24. The monthly anomalies for the averages of surface temperature >60°S are presented in Fig. 6.24a and it is apparent that that interannual variability is of the order of ±1°C and there is also a cyclical pattern of around 7 years. The 3-year running average (in gray) is also shown and it appears that the pattern is very similar to the yearly distribution (dash line). Trend analysis indicates an overall warming of about 0.081 ± 0.039°C/decade using the monthly anomalies

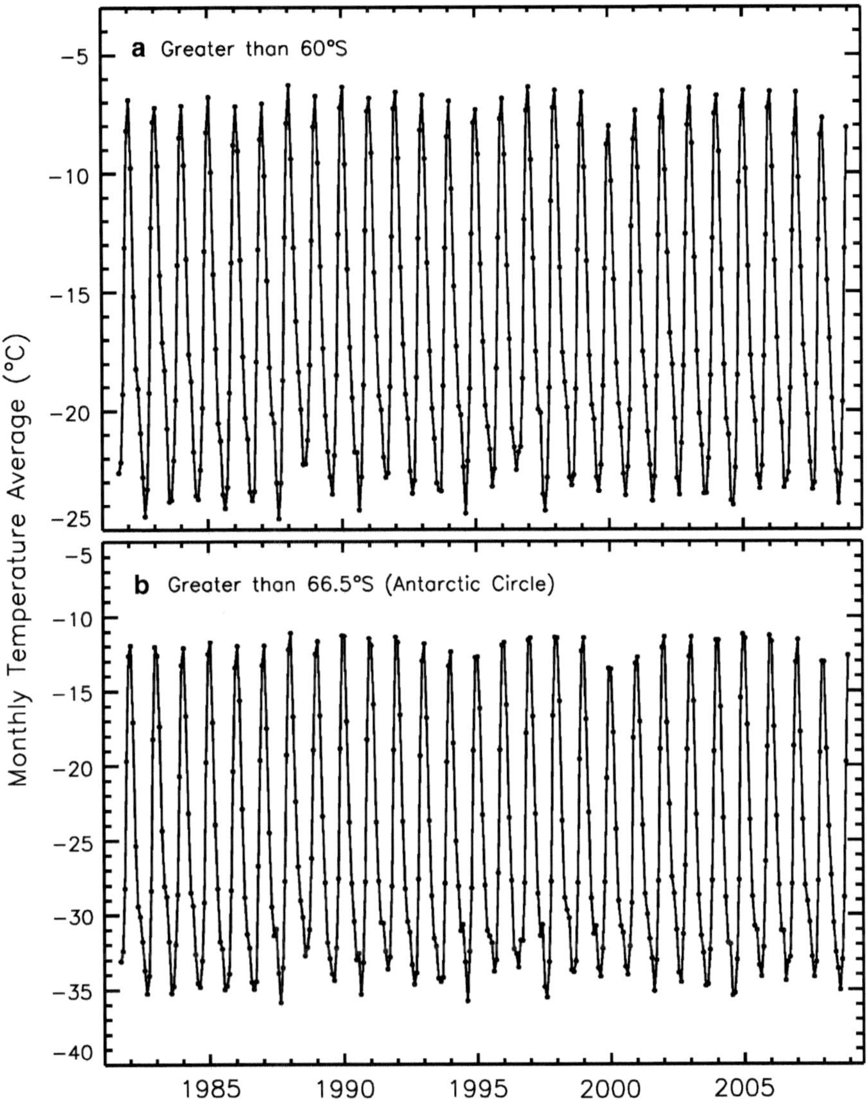

Fig. 6.23 Plots of monthly averages of surface temperatures in the Southern Hemisphere from 1981 to 2008 for the regions (a) > 60°S and (b) >65.5°S

and 0.099 ± 0.062°C/decade using the yearly averages. The trend is much more moderate than that for the Arctic, but the sign is opposite to what has been reported earlier (Doran et al. 2002). For comparison, the monthly anomalies for average temperatures inside the Antarctic Circle are presented in 6.24b and it appears that the variability in the data is very similar to that of the >60°S. The trends are also comparable but slightly higher with the monthly anomalies yielding 0.107 ± 0.057°C/decade, while that derived from the yearly averages is 0.132 ± 0.80°C/decade Plots of monthly

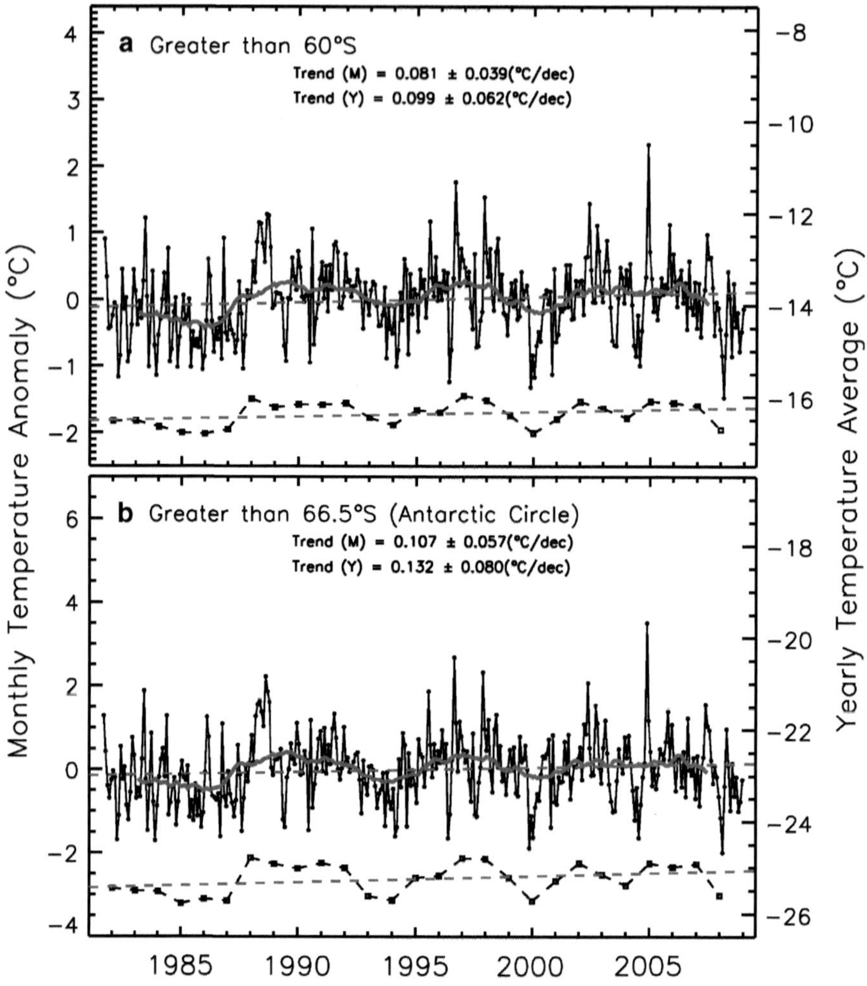

Fig. 6.24 Plots of monthly averages of surface temperature anomalies in the Southern Hemisphere from 1981 to 2008 for the regions (**a**) > 60°S and (**b**) >65.5°S. Gray line represents one year running averages. Yearly averages are also included and the dash lines represent results of linear regressions that are used to estimate the trends as indicated

averages over the Antarctic continent, sea ice cover and ice free ocean at regions >55°S are presented in Fig. 6.25. The range of variability for these different surfaces are quite different with the surface temperature at the continental ice sheet ranging from about −45°C to −15°C, that for sea ice is from −12°C to −3°C while that for open ocean is from 1°C to 3°C. Since these are monthly averages for large regions the range of values in some locations and at a higher temporal resolution can be very different. The plots indicate that even on a monthly basis, the average ice sheet temperatures are much colder than that of sea ice and open ocean. It is also apparent that the range of seasonal variability varies from one year to another. At a scale of a

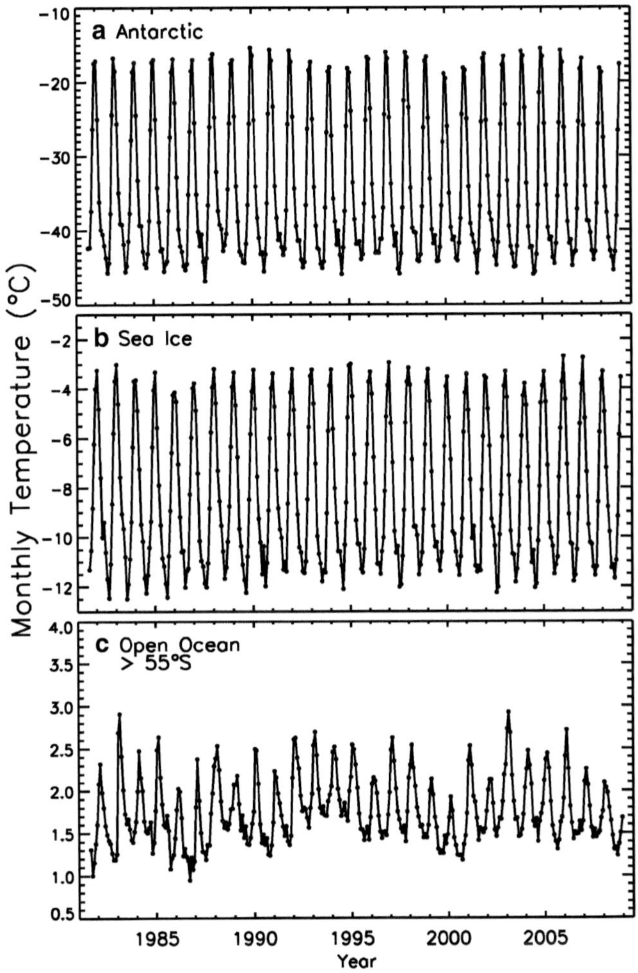

Fig. 6.25 Plots of monthly averages of surface temperatures from 1981 to 2008 in (**a**) the Antarctic continent, (**b**) sea ice covered areas; and (**c**) open oceans in the region >55°S

fraction of a degree, the interannual changes in the temperature of open ocean areas is quite significant. The yearly variability for the different surfaces is better quantified in the anomaly plots shown in Fig. 6.26. With the scale adjusted for the different surfaces to fit the dynamic range the interannual variations are quite large but no obvious strong trend is discernible. Linear regression results, however, indicate consistently positive but very modest trend of 0.098 + 0.074°C/decade for the continental ice sheet, 0.114 + 0.028°C/decade for sea ice covered regions and 0.39 + 0.015°C/decade for open ocean areas. Trends inferred from yearly averages are slightly higher but generally consistent. These results show that the trends in surface temperature are much lower in the Antarctic than those in the Arctic region.

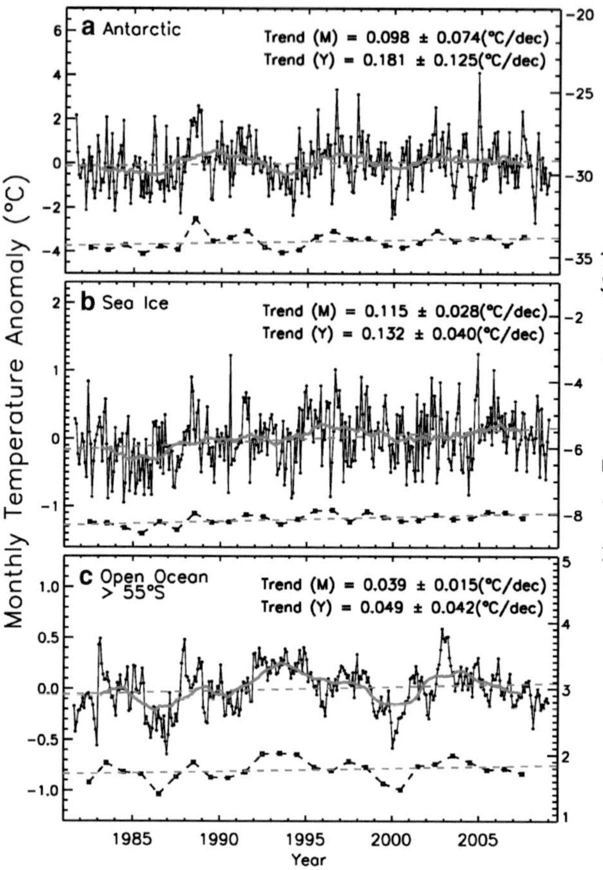

Fig. 6.26 Plots of monthly surface temperature anomalies from 1981 to 2008 in (**a**) the Antarctic continent; (**b**) sea ice covered areas; and (**c**) open oceans in the region > 55°S. Gray line represents one year running averages. Yearly averages over the same period are represented by the black dash line

Color-coded maps of yearly averages of surface temperatures in the Antarctic for the years 1982–2008 are presented in Fig. 6.27. The distributions of temperature isotherms are shown to be very similar indicative of relatively minor changes in surface temperature during the satellite period. Changes are most apparent on the year to year basis in the sea ice regions and especially near the marginal ice zones, indicative of slightly different locations of the sea ice cover during different years. In particular, significant changes in the Western Weddell Sea and the Ross Sea occurs from one year to another because of different atmospheric circulation pattertns and associated changes in the sea ice cover. The yearly variations are better quantified in the yearly surface temperature anomalies shown in Fig. 6.28. It is apparent that there are some years when positive anomalies are dominant in the entire region as in 1988, 1995, 1996, 1997, 1998, 2001, 2002, 2003, 2005 and 2006 and some years when negative anomalies are similarly dominant as in 1985, 1986,

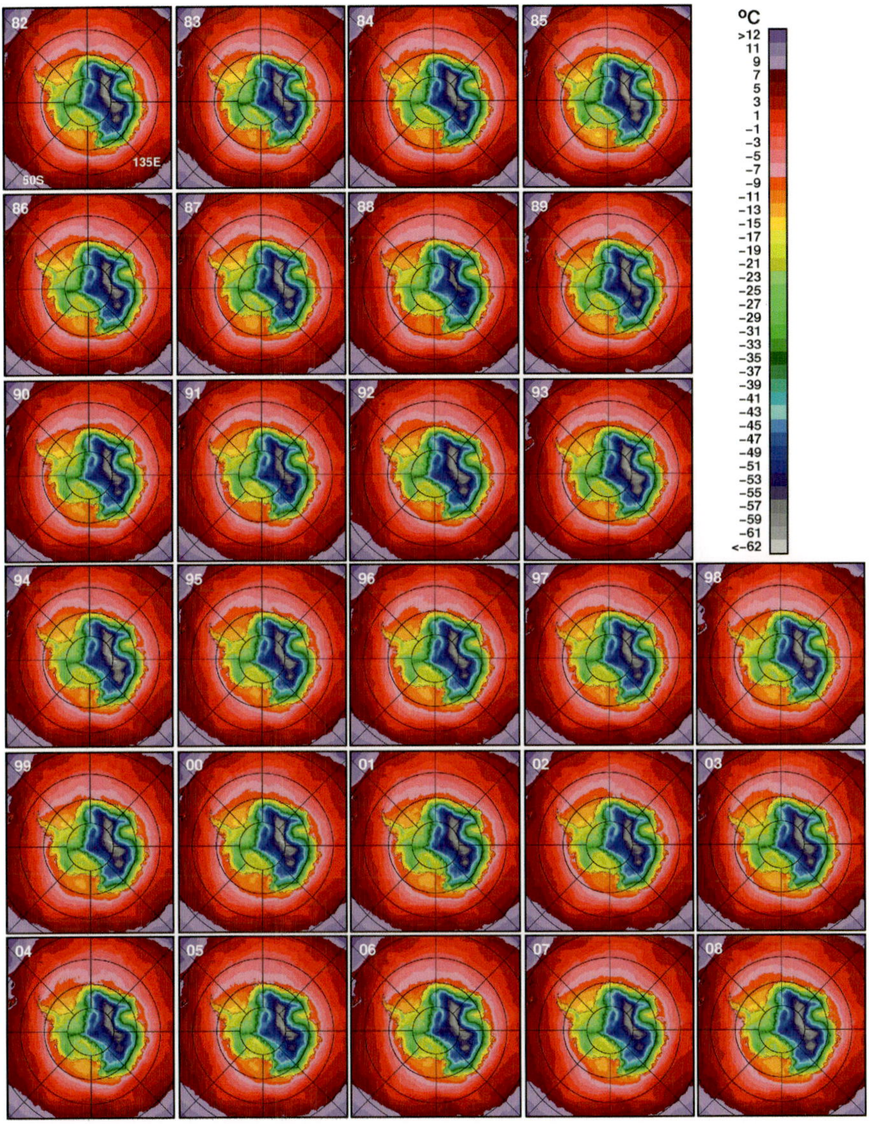

Fig. 6.27 Yearly averages of surface temperatures in the Southern Hemisphere from 1982 to 2008

1987, 2000 and 2008. There are also times when the ocean regions are relatively warm when compared with the continent (e.g., 1982, 1983, 1992, 1993, 1994) and times when it is relatively cold when compared with the continent (e.g., 1988).

In the sea ice covered areas, plots of monthly surface temperature anomalies are shown separately in Figure 6.29 for the five different Antarctic sectors as defined in Zwally et al. (1983). The variability of the peak values occurring in the summer as inferred from the monthly average plots (not shown) is relatively moderate while

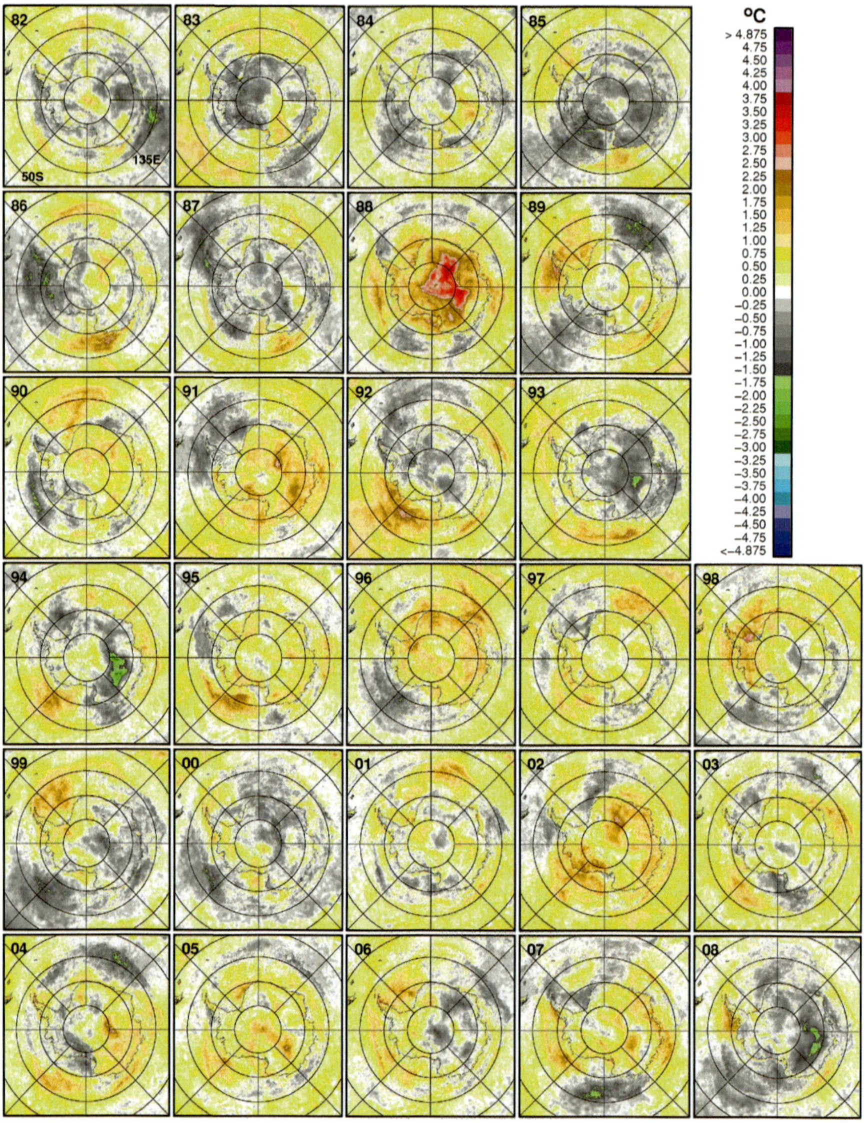

Fig. 6.28 Yearly surface temperatures anomalies in the Southern Hemisphere from 1982 to 2008

those of the low values occurring during the winter is larger. The latter is likely associated in part with the effects of divergence during the winter period that can cause temporary increases in average surface temperature as more open water is exposed within the ice pack. The range of variability of the surface ice temperature for the various sectors is surprisingly similar the ranges being approximately: −3.5 to −13°C, −2.5 to −12°C, −3 to −10°C, −3 to −13°C and −2.5 to −10.5°C for the Weddell Sea, Indian Ocean, West Pacific Ocean, Ross Sea and Bellingshausen/

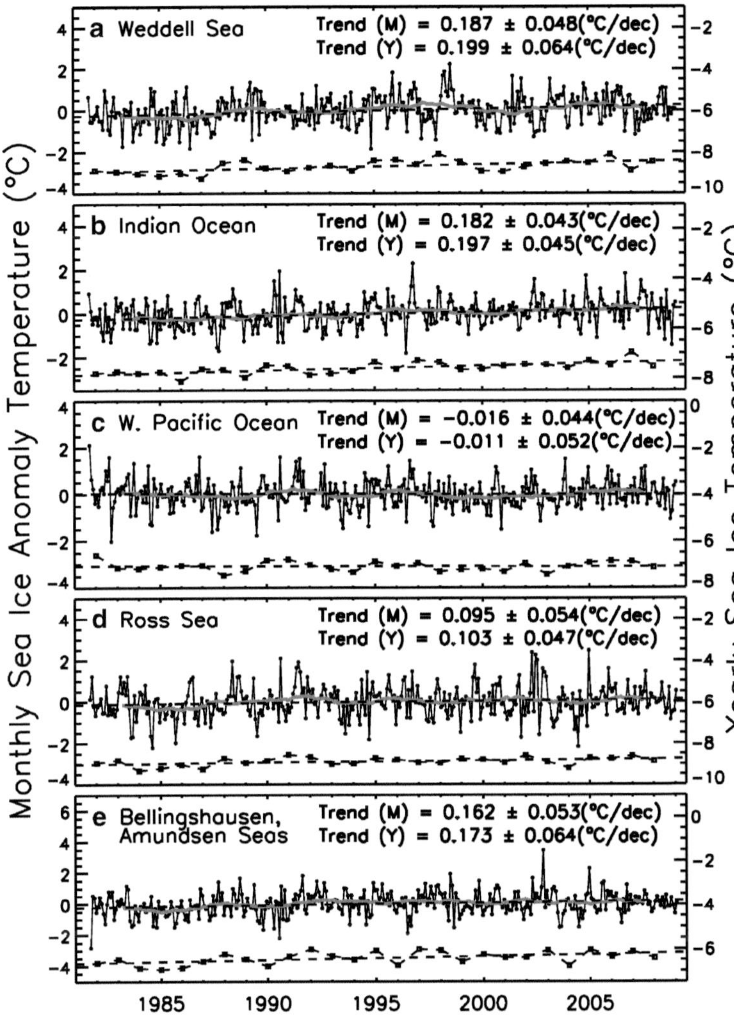

Fig. 6.29 Plots of monthly averages of surface temperatures in ice covered areas in the following regions: (**a**) Weddell Sea; (**b**) Indian Ocean; (**c**) West Pacific Ocean; (**d**) Ross Sea; and (**e**) Bellingshausen/Amundsen Seas

Amundsen Seas sectors, respectively, as inferred from the monthly average data (not shown). The winter values are lowest in Weddell and Ross Seas in part because the land/ocean boundaries in these sectors are closest to the South Pole. The monthly anomaly plots presented in Figure 6.29 indicate that the year-to-year changes are relatively subtle. The fluctuation of the monthly anomalies appear similar for all sectors with the changes confined basically within ±2. Trends in surface temperature over the period 1981 to 2008 were derived using linear regressions on the monthly anomaly data and except for the Western Pacific Ocean region where the trend is close to zero, the trends are 0.20 ± 0.05, 0.18 ± 0.04, -0.02 ± 0.04, 0.10 ± 0.05 and

0.16 ± 0.05°C per decade in the Weddell Sea, Indian Ocean, Western Pacific Ocean, Ross Sea and Bellingshausen/Amundsen Seas, respectively. Yearly averages of sea ice surface temperature for the various sectors are also presented (dash line) and the yearly fluctuations are also shown to be relatively small. The trends inferred from the data are also small and consistent with those from the monthly anomalies.

Overall, satellite historical data provide an excellent means to examine seasonal, interannual and decadal changes in the surface temperature of high latitude regions. The seasonal fluctuations of surface temperature are large, especially over the Antarctic continent and the Greenland Ice Sheet. The fluctuations of monthly anomalies are relatively similar but the trends in surface temperature are quite different in the two regions with the trend for the region north of the Arctic Circle being about 7 times larger than that of the region south of the Antarctic Circle. The reason for such a large discrepancy is currently unknown but the data show that climate change as reflected in different parts of the globe can be very different.

6.2.2.2 Sea Surface Temperatures

The typical seasonality of SST in the Southern Ocean is presented in Figure 6.30, using multi-year monthly averages as determined from AMSR-E data from 2002 to 2008 and at a more appropriate scale than in Figure 6.18. The area covered by the monthly maps is the extended region of the Southern Ocean that includes areas near 30°S. This enables the ability to make assessments of the location of polar fronts in the various oceans and how they may change from one month to another. The maps thus fully cover the Antarctic Circumpolar Current in its entirety and show meandering paths as they are influenced by topographic features, winds and other effects. The temperature isotherms are quite symmetrical with respect to the continent. Advances of these isotherms to the north are apparent from January to August after which they go back to locations during the summer period. Some of the unique features include a sharp diversion to the north from the circumpolar isotherm in an area to the east of Argentina which becomes very prominent in winter causing abnormally low SSTs in the region (compared to other regions at the same latitude) during this time period. A similar but more subtle northward advance at the Ross Sea near 180°E is also apparent and most prominent in the winter period as well.

To assess how SST in the Southern Ocean changes on an interannual basis, monthly anomalies in March and December, representing austral autumn and austral summer periods are presented in Figures 6.31 and 6.32, respectively. These months were chosen to complement those presented in Figures 6.20 and 6.22 and also to provide information about SST in the extended region during the peak of phytoplankton blooms as discussed in Chapter 9. These are also periods when the ice cover are close to minimum values and the impact of the location of the ice edge is not as much of a factor as during a winter period. In these images, the Antarctic continent and sea ice are shown in white while the other land areas are in black. The set of images show that yearly variability in early austral autumn can be considerable in some regions. For example, in the Amundsen Sea at around 135° W, and 55°S, SST was abnormally

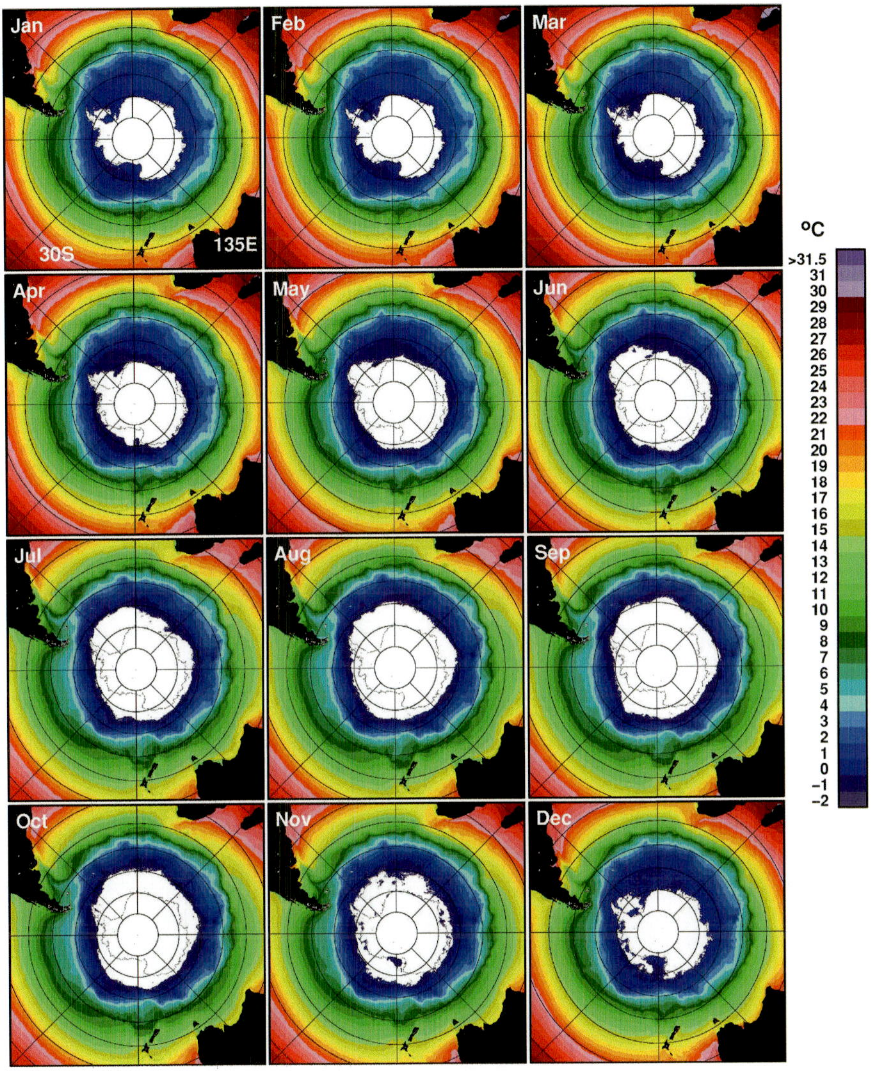

Fig. 6.30 Monthly averages of sea surface temperatures from January to December in the Antarctic using AMSR-E data from May 2002 to April 2008. The maps show an extended study area for the Southern Hemisphere that includes 30°S

high in March 1983, 1992, 2001, and 2003 and abnormally low in March 1989, 1999, 2000, and 2008. The high SST area in this region sometimes migrate to lower latitudes as in March 2001, 2004, and 2006. Also, in the Indian Ocean near 45° E, and 50°S, the SST was abnormally high in 1992, 1997, 2000 and 2001 and abnormally low in 1982, 1983, 1988, 1995, 2006 and 2008. The interannual variability of the SST anomalies in early austral summer (i.e., December) is similar to those in early autumn. The magnitudes of the anomalies are quite similar and in some cases, the

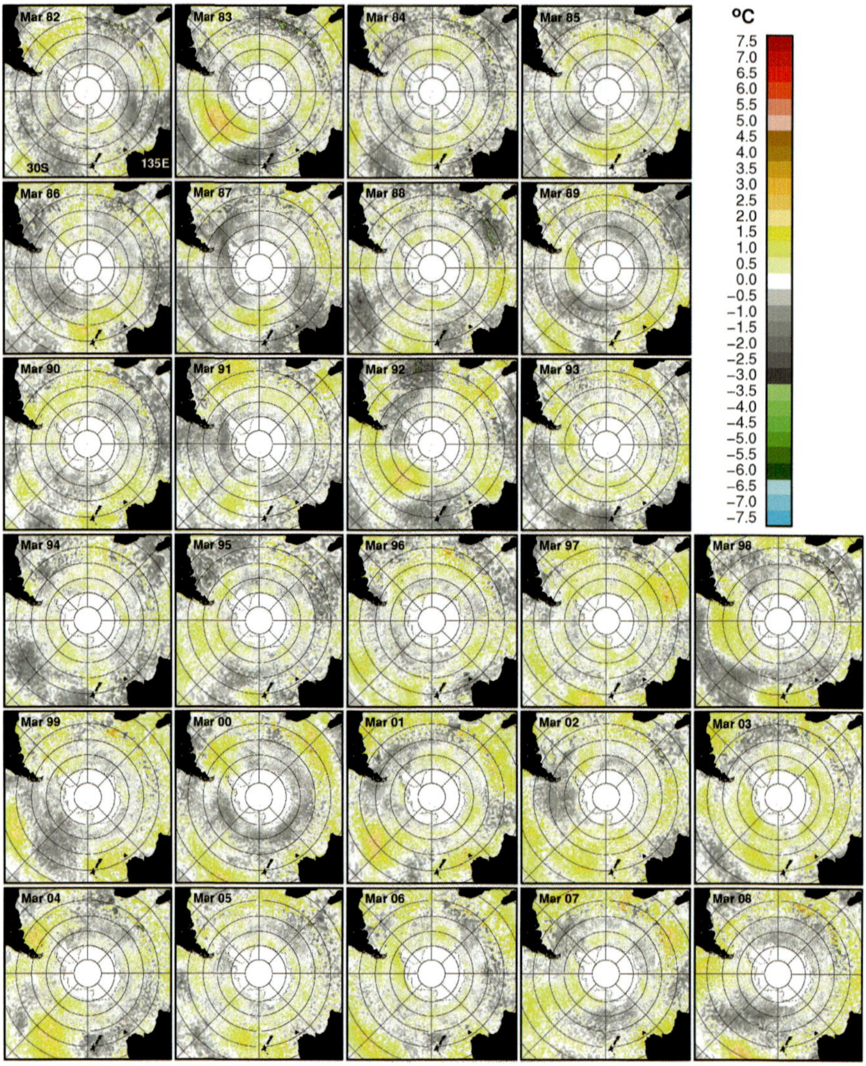

Fig. 6.31 Monthly sea surface temperature anomalies in the Southern Hemisphere during an early autumn month (March) for the years 1982 to 2008. The maps show an extended study area for the Southern Hemisphere that includes 30°S

anomalies appear to be persistent from early summer to early autumn. For example, the abnormally high SSTs in the Amundsen Sea in December 1991 is likely the same anomaly that is apparent in the March 1992 data suggesting that the anomaly was persistent for at least three months. Also, an extensive area of low SST anomalies at the Bellingshausen Sea in December 2001 is apparent in the March 2002 data. In some areas there are apparent influences of the decay of sea ice. With the sea ice cover declining rapidly in December, the negative anomalies at high latitudes are

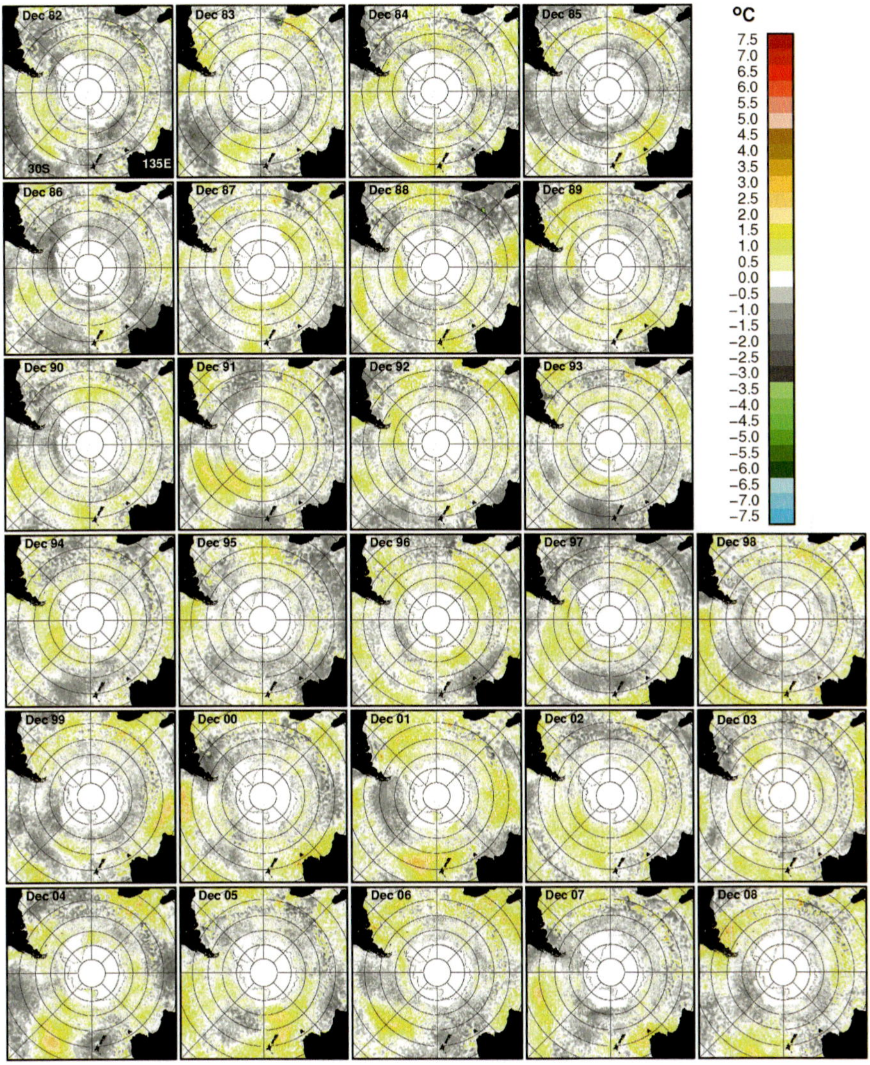

Fig. 6.32 Monthly sea surface temperature anomalies in the Southern Hemisphere during a summer month (December) for the years 1982 to 2008. The maps show an extended study area for the Southern Hemisphere that includes 30°S

likely influenced by the presence of meltwater in the region. The multiyear-monthly averages shown in Fig. 6.30 actually show that the 3°C contour does not change much throughout the year and basically show the average location of the edge of the sea ice cover during the winter period.

The presence of relatively high negative anomalies even at mid-latitudes, indicates that the interannual changes in the location of the sea ice cover may not be an important factor affecting the interannual variability of SST in these regions. Wind induced upwelling can be an important mechanism that leads to relatively high SST values. Large interannual variations in wind and pressure patterns in the region have been shown in Section 5.2.2, and expected to be a key factor. The associated influence of the Antarctic Circumpolar Current and the Antarctic Circumpolar Wave needs to be considered as well. Overall, patterns of change are not apparent but it appears that regions of relatively high SST anomalies are slightly more extensive in the latter years compared with earlier years suggesting a modest warming in the Southern Ocean area. Yearly anomalies (not shown) provide confirmation but such results have to be interpreted in the context of many interacting atmospheric and oceanic processes that affects the SST.

Plots of monthly averages and monthly anomalies are presented in Figs. 6.33 and 6.34. The monthly plots show the distinct seasonality of SST in the Southern Ocean in the five different sectors. The amplitudes of interannual variability are less than 2°C with those in the Bellingshausen/Amundsen Seas being the most variable. The monthly anomalies show that SST in the Southern Ocean has been relatively stable with the trends being quite small and mixtures of positive and negative values. The most positive trend occurred in the West Pacific Ocean sector at 0.07 ± 0.02°C per decade while the most negative trend occurred in the Bellingshausen/Amundsen Sea sector at -0.05 ± 0.02°C per decade. The trend of SST in the Ross Sea is 0.03 ± 0.03°C per decade while those of the Weddell Sea and Indian Ocean are both 0°C per decade. These results indicate that the average SST in the Southern Ocean (for the areas >55°S) has not changed much since 1981.

6.3 Albedo

The albedo of ice and snow has been shown to have large variability due to spatial and seasonal variations in the physical characteristics of the surface (Warren 1982; Allison et al. 1993). The reflected radiation includes contributions from scattering and absorption within the snow and ice which varies with the dielectric properties of the material, the thickness of snow and ice, the size of scatterers (e.g., grain size of snow, size of air pockets and brine inclusions) and wetness of the surface. The albedo of sea ice can change considerably depending on the stage it is in during the growth and decay process. During early formation stages the sea ice cover is relatively thin and hence the albedo is relatively low, but as the ice thickens the albedo becomes relatively high especially as the surface acquires a dry snow cover of about 10 cm thick. During the autumn and winter period, the albedo remains high and altered only by changes in snow thickness and grain size and layering within the snow or at the snow/ice interface. Fresh snow usually has high albedo, while old grainy snow has lower albedo. Albedo over snow and ice is also affected by contaminants such as soot. The transport of these contaminants from the south to the north in the Arctic has been reported in several studies and may alter the albedo of the surface (e.g., Grenfell et al. 2002; Warren and Wiscombe 1980; Warren and Clarke 1990). In the spring and summer, the albedo of sea ice declines as its snow cover

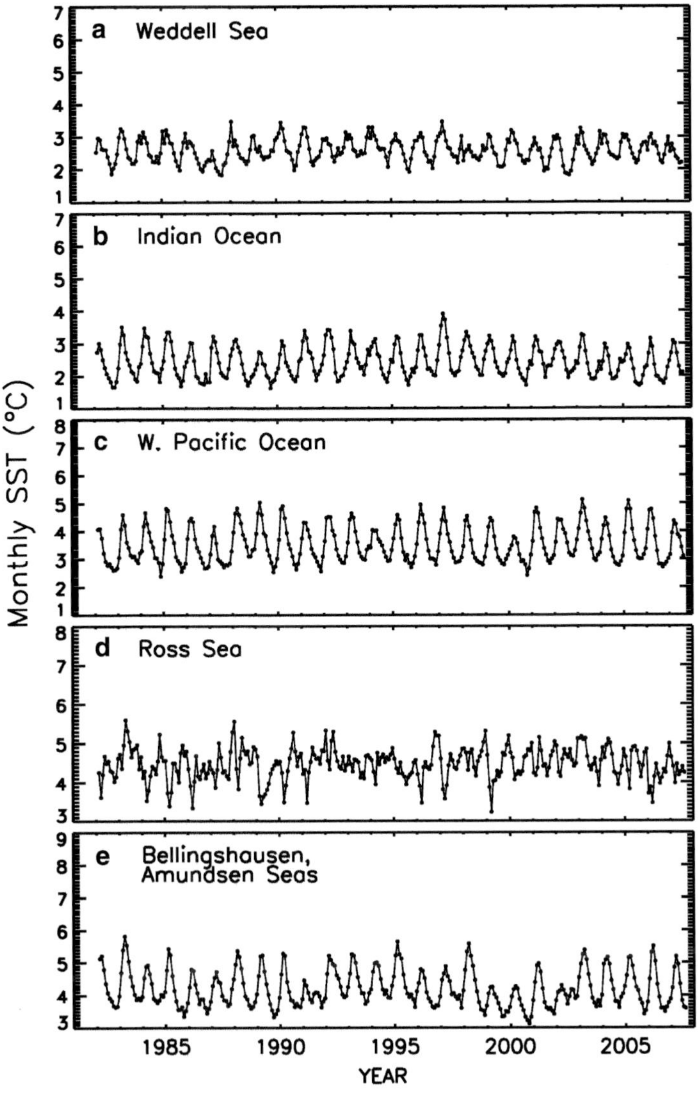

Fig. 6.33 Plots of monthly averages of SST from 1981 to 2008 for the regions >55°S in the (**a**) Weddell Sea; (**b**) Indian Ocean; (**c**) West Pacific Ocean; (**d**) Ross Sea and (**e**) Bellingshausen/Amundsen Seas

starts to melt and turns into slush. The albedo declines further as meltponds are formed and the snow cover gets totally melted.

In addition to its important role in heat exchange, knowledge of the albedo of a surface can also be useful for assessing the physical characteristics of the surface including the surface type, thickness, and the condition of the snow cover. While more work needs to be done, albedo has been used to obtain crude estimates of the thickness of sea ice (Martin et al. 2005; Drucker et al. 2003). This only applies to new

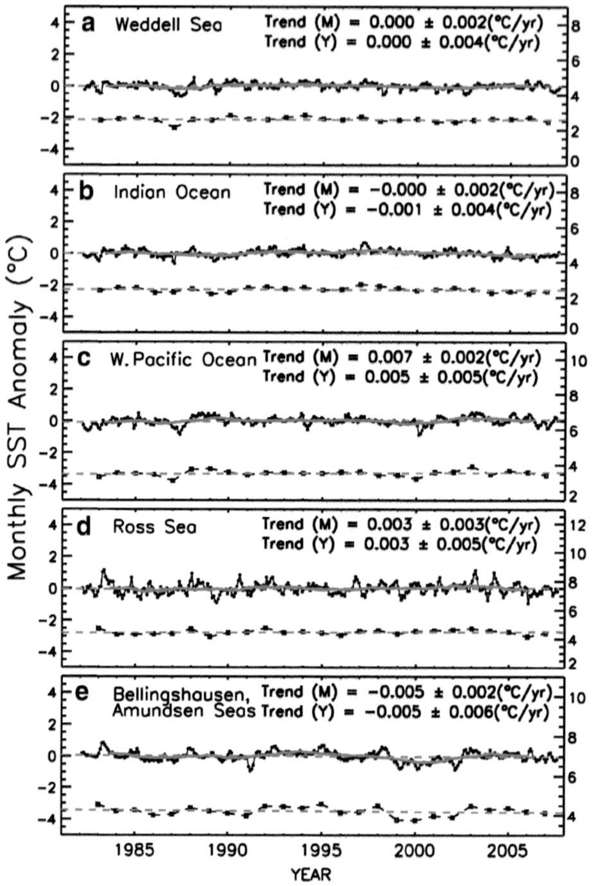

Fig. 6.34 Plots of monthly anomalies of SST from 1981 to 2008 for the regions >55°S in the (a) Weddell Sea; (b) Indian Ocean; (c) West Pacific Ocean; (d) Ross Sea and (e) Bellingshausen/ Amundsen Seas

ice types especially during the transition period from grease ice to nilas that in turn becomes gray ice and then young ice as described in Chap. 2, Sec 2.3.2. As the sea ice increases in thickness, there are more scatterers within the ice causing an increase in reflectivity or albedo. This makes the albedo data useful in assessing the age of the ice and in the classification of new, young and first year ice types. The main drawback for this application is that new ice are usually formed in leads and divergence areas the dimensions of which are usually smaller than the resolution of satellite data. Also, once snow cover accumulates, much of the information about thickness disappears since the reflectivity of snow is quite high and dominates that of snow-free sea ice. Furthermore, rafting in new ice areas and raised edges in relatively new pancakes makes estimates of thickness from albedo difficult. Nevertheless, such applications have been implemented with reasonable accuracies in areas of persistent ice formation such as in coastal polynya regions (Kwok et al., 2007)

6.3.1 Northern Hemisphere

The albedo of the snow and ice cover in the Arctic is known to vary regionally and from one season to another (Grenfell and Maykut 1977; Perovich et al. 2002a, b). The factors that cause regional variability have been discussed earlier, but these factors may not take effect at the same time and the seasonality of the albedo in various regions of the Arctic may be different. Furthermore, the extent of the snow and ice cover in the Northern Hemisphere is very seasonal. The albedo of the entire Arctic region is thus expected to be very seasonal but not necessarily predictable. The only practical way to quantify the regional and seasonal variability of the Arctic region is through the use of satellite data. As indicated in Section 4.3.2 estimates of the broadband albedo from AVHRR data have some complications that are difficult to overcome and cloud contamination can be a serious problem. The data that will be presented are narrow-band albedos (at 0.6 μm) during clear skies conditions but with no atmospheric and anisotrophic reflectance corrections as discussed in Chapter 4. For optimal utilization, the data has also been normalized such that the albedo over Greenland is consistent for each month with MODIS narrow-band albedo data. While not the ideal product for surface albedo studies, this data is spatially coherent and temporally consistent providing the means to assess how albedo has been changing in the polar regions.

The seasonality of the narrow-band albedo of the Arctic region as observed by the AVHRR sensor from 1981 to 2008 is depicted by the monthly climatologies presented in Fig. 6.35. The monthly climatologies are in this case the averages of all AVHRR albedo data available for each month. The conspicuous hole (in black) in the middle of the image for the month of October represents missing data not covered by the sensor because the surface is in darkness. The black hole gets bigger as the period of darkness gets longer in the Arctic region and therefore data for November, December, January and February are not included. The monthly images provide the means to assess the way the albedo in the Northern Hemisphere generally changes from 1 month to another starting in March. The high albedo areas are those covered by sea ice, glaciers, ice sheets (i.e., Greenland), and other areas covered by snow. Low albedos are mainly those in ice-free oceans or snow-free land and ice surfaces that are meltponded or in the process of decay. The set of images shows the progression in the areal extent of high albedo areas from a relatively large extent in March to minimum extent in August. The relatively low albedo above 85°N in March is caused in part by low sun angle during the month. Areas with high albedo over land are areas that are continuously covered by snow during the month or are covered by glaciers or ice sheets. Sporadic snow cover contributes to intermediate values of albedo over land areas. The relatively low albedo in the summer months is associated mainly with the loss of snow cover, growth of vegetation, reduced sea ice concentration and the formation of meltponds. Some distinct features can be identified such as the high albedo sea ice covered areas in the Sea of Okhotsk in March. In this sea and other seasonal ice regions, the albedo is shown to decrease from the highly consolidated thick ice region to the outer region dominated by new ice and/

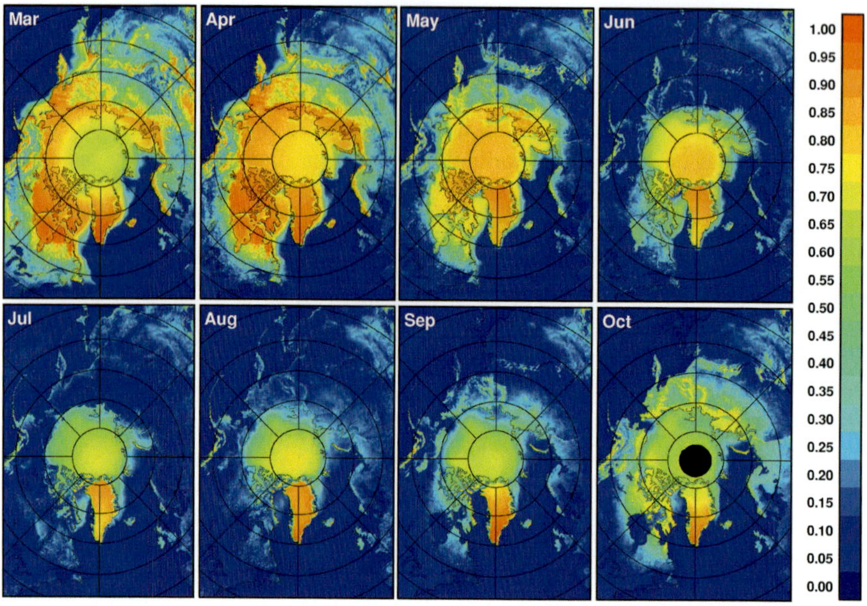

Fig. 6.35 Monthly climatologies of narrow-band albedos (0.6μm) in the Northern Hemisphere as derived from AVHRR data from 1981 to 2008

or low concentration ice. Also, over the Stanovoy Range in Eastern Russia is a high albedo feature immediately south of 60°N (at around 135°E) which persists from March to May, has basically disappeared from June to September and becomes visible again in October. Similarly, the well-defined high albedo feature at the western side of Alaska is shown to be maintained throughout the year. The consistent presence of these distinct features in the data demonstrates that despite aforementioned difficulty of masking the persistent clouds in the polar regions, the sensor is able to provide good surface data. Overall, the climatology shows that albedo at high latitudes in the Northern Hemisphere is highly seasonal with the seasonality driven primarily by the seasonality of the snow and sea ice cover.

The monthly changes in albedo are coherent with those of surface temperatures as discussed in the previous section and sea ice concentration as presented in Chap. 7. Albedo is high in snow and sea ice covered areas where the surface is equal or below freezing temperatures. The spatial variation in albedo thus provides the means to assess the seasonal changes in the spatial extent and characteristics of the snow and sea ice cover. During the autumn and winter periods, high albedos are caused by the presence of snow over land and sea ice surfaces. Variations in the albedo over snow covered surfaces may be caused by variations in grain size, liquid content and depth (Warren 1982; Warren, 1984) as discussed earlier. Over snow free regions in land areas, the seasonal variations in surface albedo also occur because of seasonal changes in the state of the vegetation and forest. During summer, the only regions with high albedo are Greenland, glaciers (e.g., parts of Ellsmere Island and Alaska) and the Arctic sea ice cover.

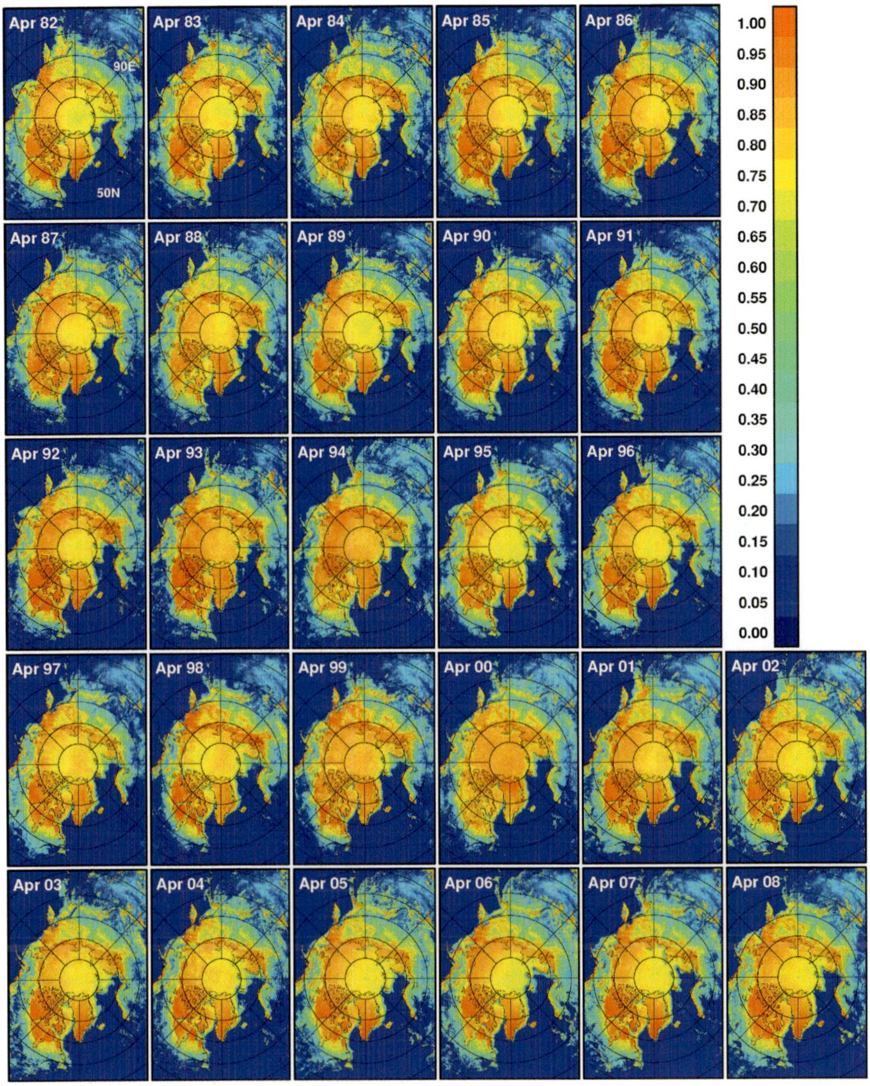

Fig. 6.36 Monthly averages of narrow-band albedo (0.6 μm) in the Northern Hemisphere for each month of April from 1982 to 2008

To illustrate how the albedo changes from 1 year to another during spring, monthly averages of albedo in April for each year from 1982 to 2007 are presented in Fig. 6.36. The distribution of surface albedo for the different years at this time of the year is shown to be generally very similar suggesting no drastic changes during the period. The extents of high albedo areas are also very similar. The modest changes over land are consistent with the observation that the extent of the snow cover has not changed much during the period (Frei et al. 1999; Armstrong and

Brodzik 2001). Over sea ice, the biggest differences occur at the peripheral seas like the Sea of Okhotsk where the average albedo is relatively high as in 1982, 1985, 1991, 1999, 2000, 2001, 2002, and 2003 and relatively low in other years. Similar interannual variability associated with the sea ice cover is also apparent at the Barents Sea.

The monthly averages for each year provide the means to assess spatially how and where the albedo at high latitudes changes from 1 year to another at a time when the overall albedo of the region is near its highest value. The year-to-year variations during April are better quantified in the anomaly maps presented in Fig. 6.37. Overall, there appears to be significant variability from 1 year to another, especially in the seasonal sea ice regions and more specifically near the sea ice margins. Positive anomalies are quite prominent, for example, in the Okhotsk Sea in April 1982, 1985, and 2001 and in the Bering Sea in 1985, 1988, 1991, 1995, and 2008. Relatively thick ice cover with intact snow cover and a delay in the ice melt and breakup in spring would cause such an anomaly. In the interior, there were years when the albedo was anomalously low, as in Northern Canada in 1984, 1987, and 1993, Western Russia in 1995, and the Barents Sea in 2006, 2007, and 2008. Over land, this usually means an early melt of snow in the region and over the sea ice covered areas, it means an early retreat of the ice edge due to melt or wind compaction. The albedo was also anomalously high in the Arctic basin in 1993, 1994, 1999, and 2000 and in Russia in 1998, 1999, and 2000. This is likely associated with a late melt period for the snow cover. In some years, there is a balance of negative and positive anomalies, while in other years, either negative or positive anomalies would dominate. Overall, however, there appears to be more positives in later years (i.e., from 1998 to 2008) than in earlier years.

A similar set of images but using monthly averages in August is presented in Fig. 6.38 to assess the interannual variability of the summer albedo from 1981 to 2007. At this time, the average albedo of the region is near its lowest value as indicated in Fig. 6.35. The albedo of the Greenland ice sheet is shown to be consistently very high which is expected on account of its snow cover. It was also used as the basis for ensuring consistency of the albedo provided by AVHRR data as discussed in Chap. 4. In the Arctic basin, the interannual variability of high albedo areas is highly influenced by the location, extent and surface characteristics of the perennial ice cover. At this time, the sea ice cover declines quite rapidly as it approaches minimum values which usually occur in September but freeze-up at high latitudes appears to have already started in August. High albedo values in the areas surrounding the North Pole in 1997, 1998, 1999, and 2000 in the perennial ice region are quite apparent but adjacent regions did not show similarly high values. The high albedo suggests that melting has ceased and the surface was dry and consolidated during the month. On the other hand, the albedo was relatively low near the North Pole from 2002 to 2007 while, adjacent areas show similarly low albedo values. During these years surface temperatures were on the average higher than previous years as indicated previously (see Fig. 6.12) and more surface melting than normal might have been going on.

Over land, the presence of intermediate albedo values in some places in August suggests the presence of snow cover in these areas either from the previous winter

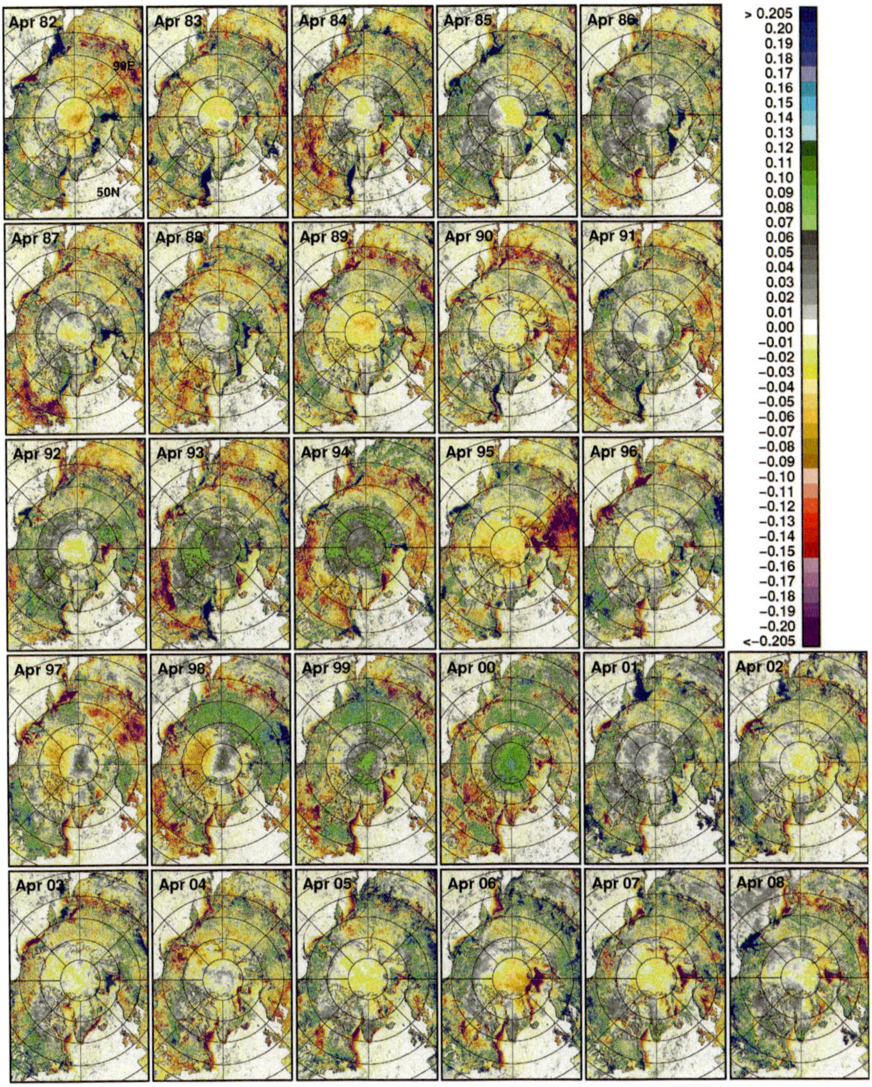

Fig. 6.37 Monthly anomalies of narrow band albedo (6 μm) in the Northern Hemisphere for each month of April from 1982 to 2008

or from new snow fall. An example is the area in the top right portions of the images (around 40°E, 120°W) where two adjacent mountains in Mongolia, called the Hangay and Altay mountains, are located. The albedo feature is sometimes split into two, which means that there is snow cover in these mountains only and not in between, but sometimes there is no spit as in 1993 which suggests that an unusual snow event may have occurred causing the region between these mountains to be snow covered as well. A decline in albedo from the 1990s to the 2000s is also

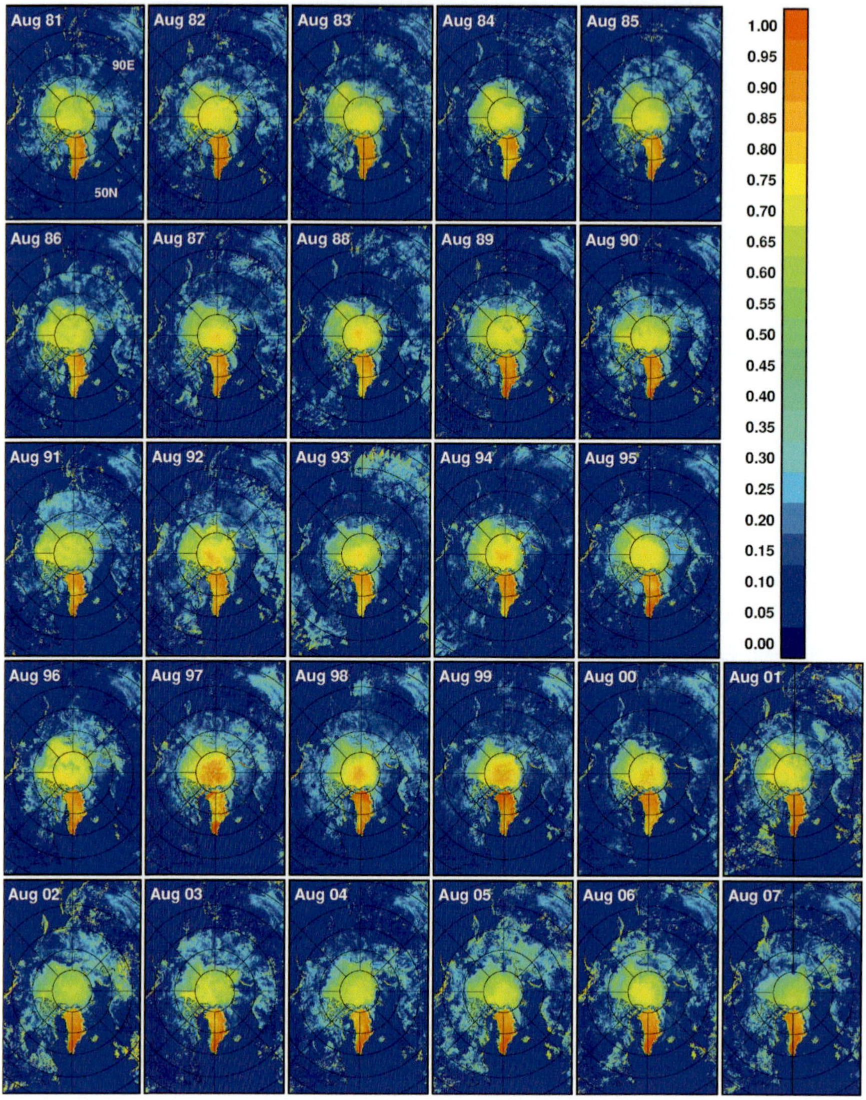

Fig. 6.38 Monthly averages of narrow-band albedo (0.6 μm) in the Northern Hemisphere for each month of August from 1981 to 2007

apparent in the Arctic basin likely associated with declining perennial ice cover as has been reported (Comiso 2002; Stroeve et al. 2007; Comiso et al. 2008) and will be discussed further in Chap. 7. The perennial ice extent was highly variable but basically stable until the late 1990s when it started to show significant decline. Consistently low values were observed since 1998 culminating in a dramatic decline in 2007 (Comiso et al. 2008) followed by a modest recovery in 2008. The monthly anomaly maps presented in Fig. 6.39 are consistent with this phenomenon and

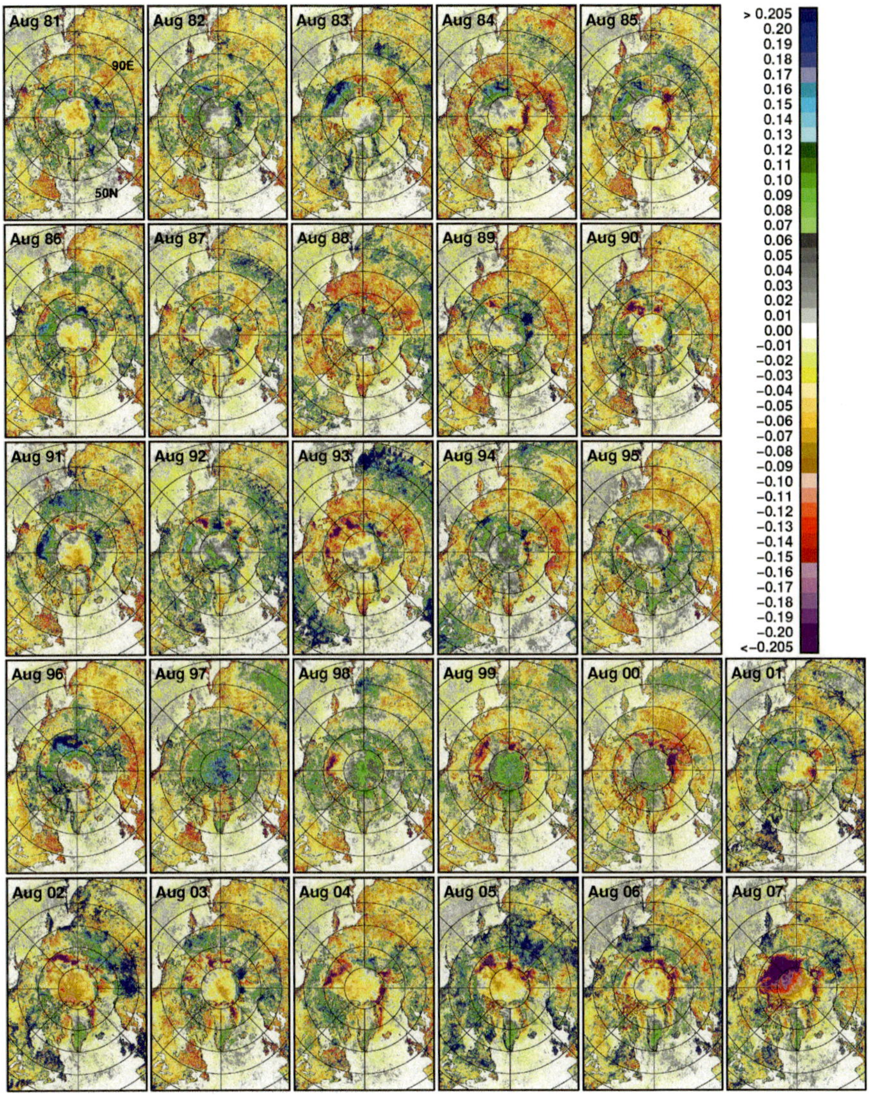

Fig. 6.39 Monthly anomalies of narrow-band albedo (0.6 μm) in the Northern Hemisphere for each month of August from 1981 to 2007

provide insights into how the Arctic albedo changed on an interannual basis from 1981 to 2007 and how the sea ice retreat is reflected in the data. Although high albedo values are confined to Greenland and the sea ice covered regions, the monthly anomaly maps for August show significant interannual changes all over the Arctic region. It is evident that August is a transition period for the albedo of many surfaces. Positive anomalies in albedo over land as in Northern Russia in August 1991, 1997, 2002 and 2005 can be due to early freeze up that allows snow

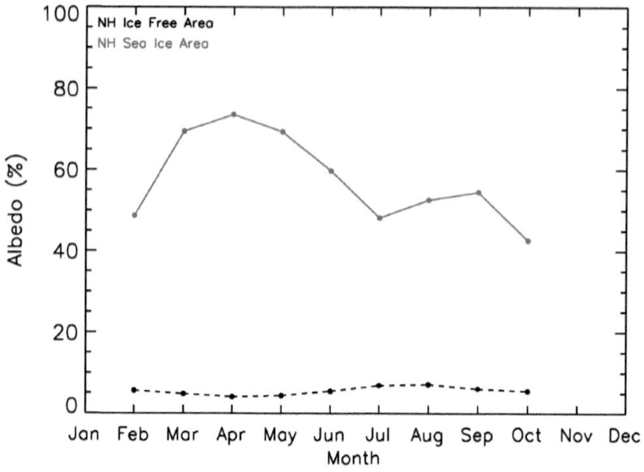

Fig. 6.40 Plots of monthly averages showing the seasonality of narrow-band albedo in the Northern Hemisphere for sea ice covered and ice free areas

cover to accumulate and survive for a relatively long time. Negative anomalies as happened in the same region in August 1988, 1993 and 2000 could be caused by the melt of existing snow cover. Similar phenomenon apparently happened in other regions as well. In the Arctic Ocean, August is the time period when the ice cover is on a rapid decline to its minimum value. The albedos in some areas have intermediate values because the average in these areas represents surfaces that changed in type (e.g., sea ice to ice free surface) during the month. In the Arctic basin, positive anomalies are more dominant around the coastline (or near the ice edges) in the 1980s than in the 1990s and 2000s. An exception is the anomaly map in 1996 which shows large areas of positive anomalies in the Beaufort Sea region reflecting the relatively extensive perennial ice cover during the year. It is interesting to note that in 1998, 1999 and 2000, high negative anomalies are apparent at the sea ice edges but inside the pack the anomalies are mainly positive. This may mean a more compact sea ice cover or less melting in the inner zone during the period. During the years from 2001 to 2007, strong negative anomalies in August are apparent at the ice edges and also at the inner pack. There was apparently a transition, as mentioned earlier, in the surface characteristics from the late 1990s to the 2000s. The positive anomalies in the earlier period suggests that the ice cover was highly consolidated and dry while the negative anomalies in the latter period indicate relatively wet or broken up ice cover that may be caused by warmer temperature and more storms in the region. Such scenario is plausible since the monthly temperature anomalies shown in Fig. 6.5 indicate warmer temperatures in later years while the pressure and wind maps in Figs. 5.5 and 5.6 suggest more storminess.

The average albedo for each month for ice free and ice covered regions in the Arctic are depicted in the plots shown in Fig. 6.40. The average albedo for sea ice (solid line) is shown to increase from February to April and then to decline after that up to July with the average being almost constant from July to October. The dip in

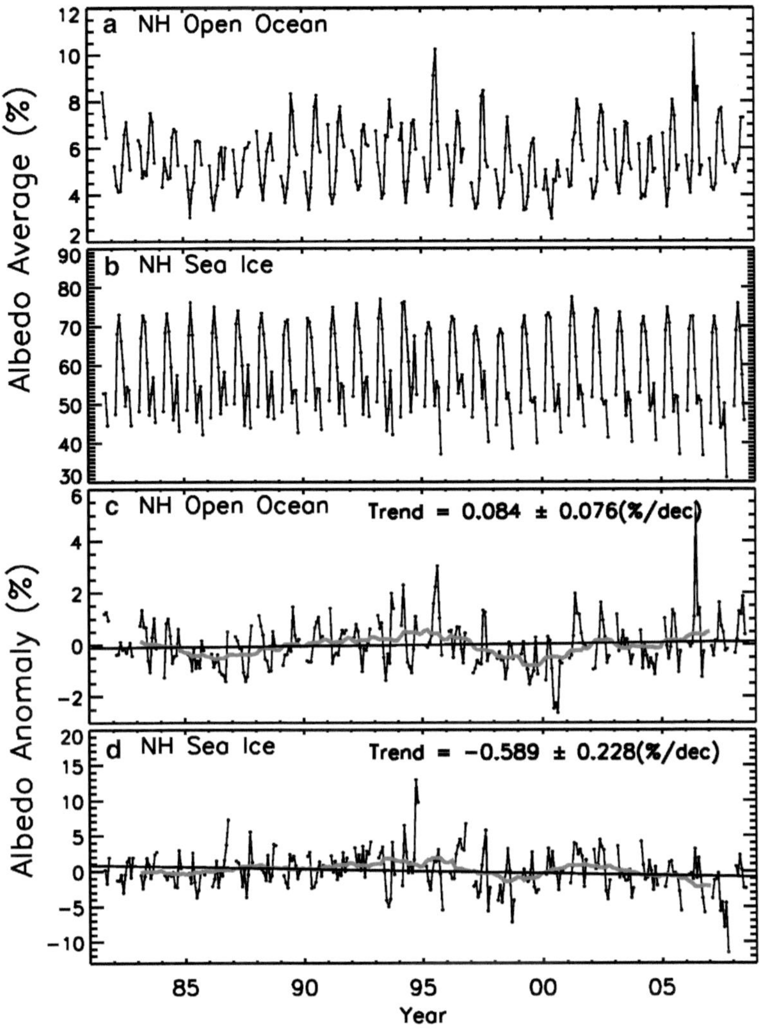

Fig. 6.41 Plots of monthly averages of narrow-band albedo from August 1981 to September 2008 for (a) NH open water and (b) NH sea ice regions. Plots of monthly anomalies for the same period are also shown in (**c**) for NH open water and (**d**) for NH sea ice

albedo in July may be associated with melponding during the period. Plots of the monthly averages for each year over sea ice covered regions and over open water are presented in Fig. 6.41a and b and it is apparent that there is a significant interannual variability especially over sea ice covered regions. Note that the scale in the open water plot is different from that used in the sea ice plot and the interannual variability for open water albedo is relatively small (about 2%). The monthly anomalies of the average albedo for open water and sea ice are presented in Fig. 6.41c and d. The interannual variabilities are only a few percents and the trend lines derived from linear regressions of the data points indicate opposite trends for open water and sea ice. The open water albedo is shown to have a negative trend of 0.084±0.076%/decade, while the sea ice albedo has a positive trend of −0.589±0.228%/decade. The trends are relatively modest and the errors reflect only statistical error. As indicated in Chap. 4, we used MODIS albedo data in Greenland to renormalize the AVHRR data for improved temporal consistency. The observed decrease in the monthly albedo over the sea ice covered region does not reflect the overall decline in albedo associated with the rapid decline of the Arctic sea ice cover. The changes in the monthly averages reflect mainly changes in the surface optical characteristics of the sea ice in the region that may include changes in the extent of meltponding, the length of the melt season, the granularity of the snow surface, and other parameters.

The yearly average maps of albedo presented in Fig. 6.42 provide a means to evaluate interannual variability of narrow band albedo during the daylight months of the year. The averaging was done only for the period from March to October of each year since albedo is not measured in a large fraction of the Arctic during the winter season. Although the monthly albedo maps for April and August provide unique information about interannual changes, the yearly averages provide a more complete assessment of the yearly variability. It is apparent that the images for each year are quite similar. The region of variability is usually in the Arctic Basin and basically on account of different locations and extent of the perennial ice cover as revealed during the summer period with the most significant changes occurring in the last decade. In the land areas, the interannual variability is associated mainly with the interannual changes in the snow cover. It is apparent that many of the distinct features over land and sea ice are reproduced each year and altered only by interannual changes in surface characteristics. The changes in albedo from one year to another are more quantitatively documented in the yearly anomaly maps presented in Fig. 6.43. The yearly anomalies show variabilities similar to those in the April and August anomalies presented in Fig. 6.37 and 6.39, respectively, but the magnitude of the anomalies are much smaller and the scale has to be adjusted to better depict some of the noteworthy features. Over land areas, the yearly changes are relatively minor from 1982 to 2000 and also for 2003 and 2004. Positive anomalies are shown to be more dominant in 2001, 2002 and 2005 to 2008 suggesting more snow cover in land areas during these years. In the Arctic basin, positive anomalies are prevalent inside the ice pack in 1994, 1997, 1998, 1999, and 2000. They are also prevalent in the seasonal ice regions of the basin in 1983, 1984, 1085, 1986, 1988, 1992, 1994 and 1996. Negative anomalies are observed in a large fraction of the Arctic basin in 1995, 1998, 2000, 2002, 2003, 2004, 2005, 2006, 2007, and 2008. As indicated earlier, the record low extent in the summer sea ice cover occurred in 2007 while the second lowest occurred in 2008. It is interesting to note that when negative anomalies were dominant in the Arctic basin during the 2005 to 2008 period, positive anomalies over land were concurrently dominant. This suggests the occurrence of more snow precipitation and more snow cover over land that may in part be caused by more evaporation because of more open water in the Arctic Basin during the summer. However, there are exceptions to this phenomena as in 2001 and 2002.

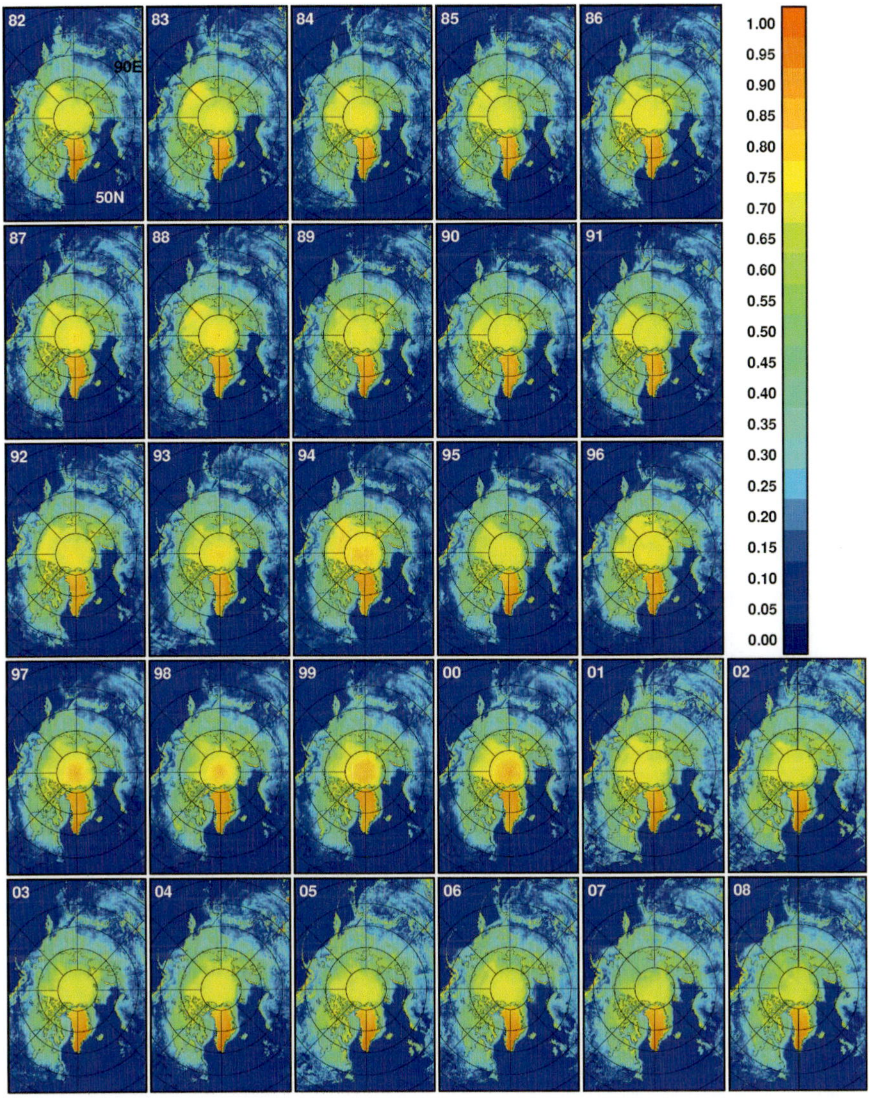

Fig. 6.42 Yearly averages (i.e., March to October) of narrow-band albedo (0.6 μm) from 1982 to 2008

6.3.2 Southern Hemisphere

The seasonal variation of the albedo of the Southern Hemisphere is depicted by the monthly albedo climatology presented in Fig. 6.44. The images were created the same way as in the Northern Hemisphere but with the austral autumn and winter months of May, June, July and August missing. As in the Northern Hemisphere, a

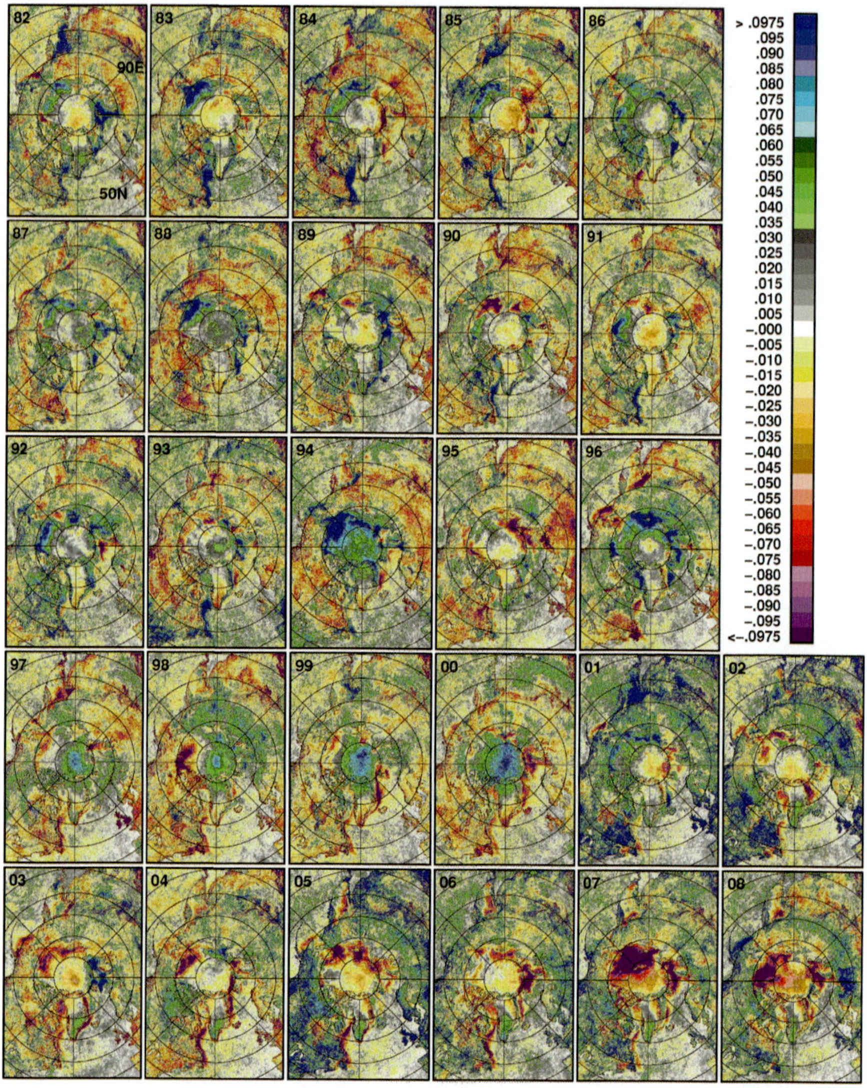

Fig. 6.43 Yearly Anomalies (i.e., March to October) of narrow-band albedo at 0.6μm from 1982 to 2008

black circular area in the middle of the April map where no data was acquired because of darkness. The month-to-changes in albedo are associated mainly with the growth and decay of the sea ice cover since the Antarctic continent is covered by snow year round and the albedo of the entire region is not expected to change much. The minor changes in the albedo of the Antarctic continent are shown to be mainly near the periphery where such changes may be caused by surface melting or the introduction of new snow at the surface. Conversely, the albedo of the sea ice region is far from uniform. The distribution in January shows high albedo

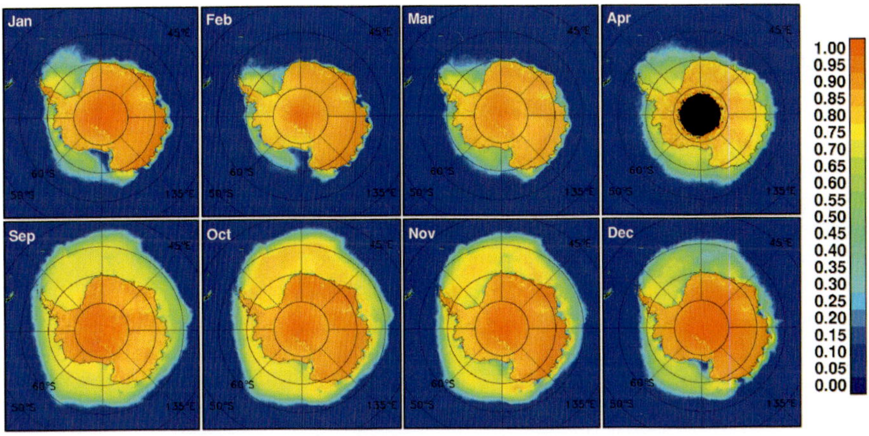

Fig. 6.44 Monthly climatologies of narrow-band albedo in the Southern Hemisphere (0.6 μm) from January to December derived from AVHRR data from August 1981 to July 2008

in some regions but large areas of intermediate values likely on account of surface melt and the rapid retreat of sea ice during this period. The melt of sea ice and associated ice retreat continues on in February where the highest albedo corresponds to the core region covered by the perennial ice. In March and April, rapid ice growth is manifested by the progressively increasing albedo from the ice edge to the inner pack. The images in September and October show the albedo of the Southern Hemisphere when the sea ice cover is most expansive. Inside the pack, changes in ice concentration, snow cover and ice thickness lead to changes in albedo. In October, a reduction in albedo in the Ross Sea along the coastal area is apparent indicating that the spring polynya has started to show up. This region widened in November as the sea ice cover starts to melt and retreats to the north. By December, the albedo over sea ice is again predominantly in the intermediate values because of ice melt, relatively low ice concentration and rapid ice retreat.

The interannual variabily of the surface albedo at the end of the summer period is illustrated in the March monthly albedo maps of each year from 1982 to 2008 as presented in Fig. 6.45. The albedo of the Southern Hemisphere at high latitudes is at this time period dominated by that of the Antarctic continent which has an area of about $14 \times 10^6\,km^2$. During summer minimum ice extent, the average area of sea ice is only about $3 \times 10^6\,km^2$ or 21% of the area of the continent. Only minor variations in the albedo of the continent are apparent and they occur mainly near the periphery where the surface melts during the summer and fresh snow may have been introduced.

The yearly variations are better quantified in the March anomaly maps shown in Fig. 6.46 and it is evident that indeed, many of the changes occur at the sea ice covered regions and near the continental boundaries. Yearly changes in the Weddell, Bellingshausen, Amundsen and Ross Sea regions are associated with yearly changes in the location of the summer ice cover in the region as discussed in Chap. 7. In the

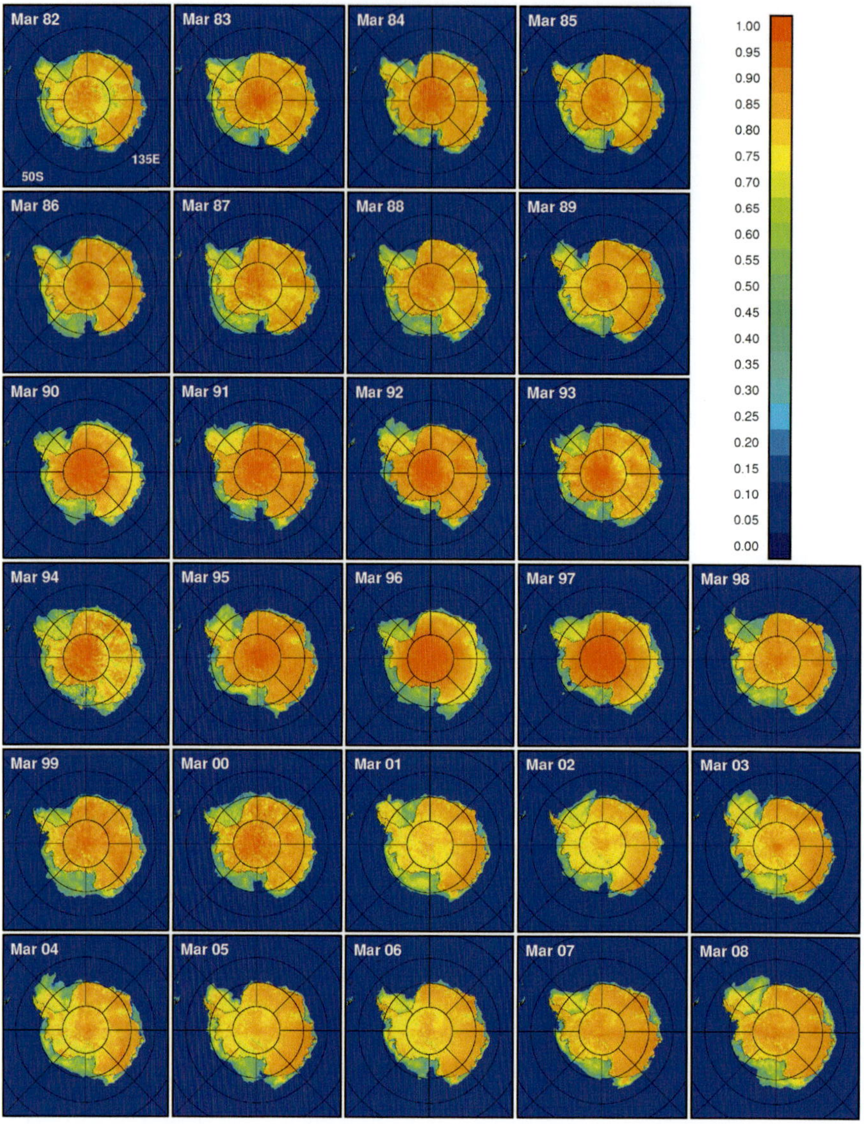

Fig. 6.45 Monthly averages of narrow-band albedo (0.6 μm) in the Southern Hemisphere for an early spring month (March) from 1982 to 2008

Weddell Sea region, the changes in albedo are associated with the large variability of the extent and location of the summer ice in the eastern side of the Antarctic Peninsula. The positive anomalies in the region as observed in 1991, 1992, 1994, 1995, 2001, 2003, 2004, and 2008 are also indicative of either anomalously extensive summer ice cover or early freeze-up date. During the other years the images usually show a mixture of negative and positive anomalies. The Bellingshausen and Amundsen Seas also show positive anomalies which are more persistent in the 1980s

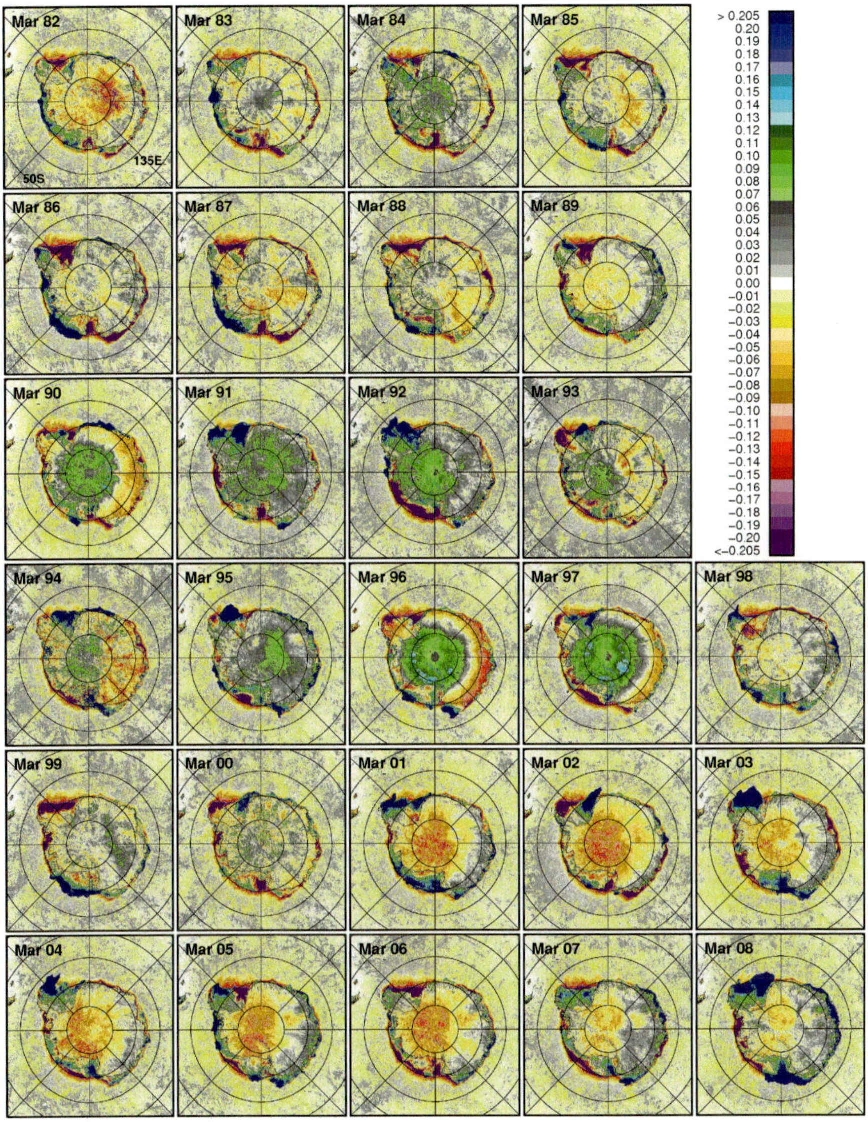

Fig. 6.46 Monthly anomalies of narrow-band albedo (0.6 μm) in the Southern Hemisphere for an early spring month (March) from 1982 to 2008

than in later years. In the Ross Sea mixtures of positive and negative anomalies are apparent but there were cases as in March 1992 and 1997 when negative anomalies are dominant while in March 1999, 2001, 2003 and 2008, positive anomalies are dominant in the region. Over Antarctica, positive anomalies are prevalent in 1990, 1991, 1992, 1996 and 1997 while negative anomalies are dominant in 2001, 2002, 2004, 2005 and 2006. The cause for such interannual variations is not known but they may be associated with interannual changes in snow precipitation patterns over the continent.

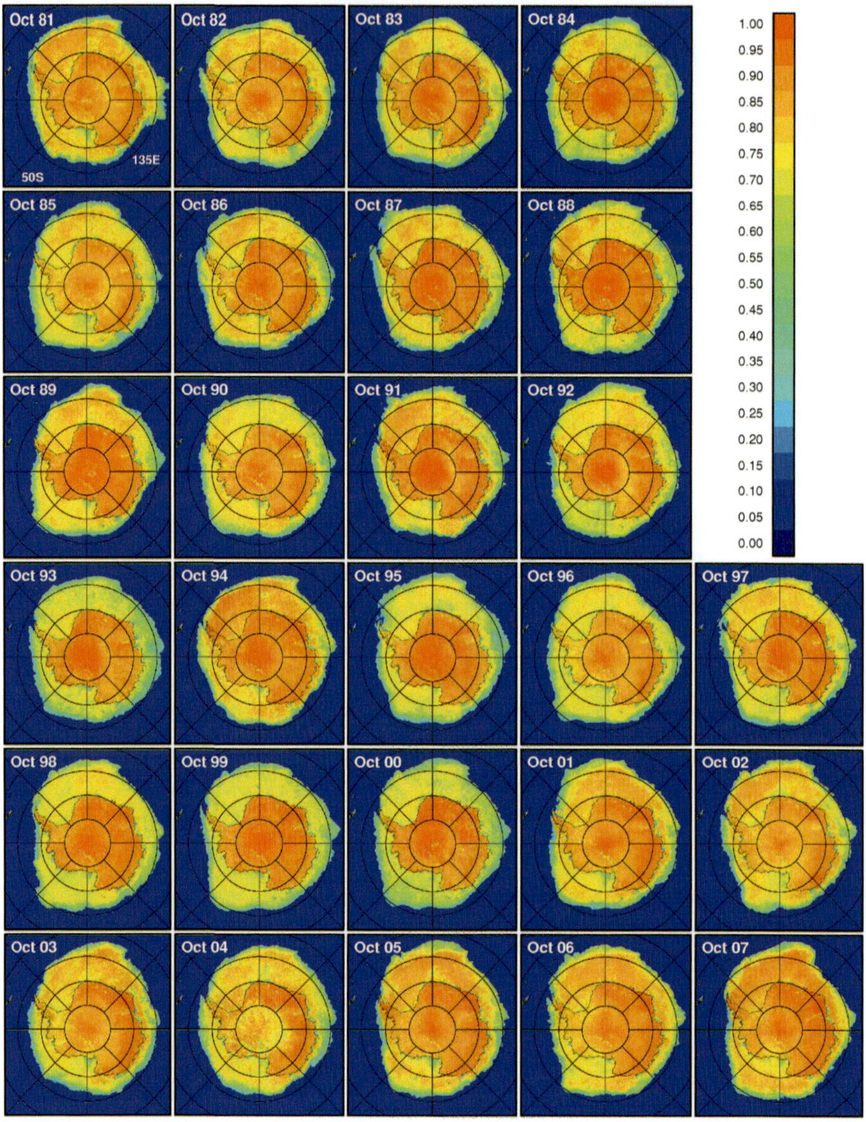

Fig. 6.47 Monthly averages of narrow-band albedo (0.6 μm) in the Southern Hemisphere for an autumn month (October) from 1981 to 2007

Monthly maps of albedo for the different years in October, which is about the beginning of the melt period, are presented in Fig. 6.47. At this time, the extent of the sea ice cover is greater than that of the continent by a factor of 1.4 and therefore has a bigger impact than the latter on the total albedo of the region. The albedo of Antarctica is generally higher than that of the sea ice cover in part because the latter usually has significant components of liquid water surfaces such as those found in leads and polynyas. Coastal polynyas are apparent and represented mainly by intermediate values of albedo (greens) and are most prominent at the Ross Sea. Deep ocean polynyas are also apparent especially near the Maud Rise as in October 1994, 2000 and 2004. The inner zone of the sea ice cover are sometimes lower than average as in October 1993, 1995, 1999 and 2000. We don't exactly know why but the lower albedo could be caused by a relatively stormy year that would cause the formation of a lot of leads or by an early melt season that would alter the albedo of the surface.

The interannual variability of the extent and distribution of the winter sea ice cover are the primary cause of the yearly variability of the albedo in the region. This is apparent in the October monthly anomaly maps shown in Figure 6.48 for the different years. The anomaly maps show that the significant yearly changes are located mainly at the ice edges, with the greens and blues representing advances at the ice edge from average values while the purples represent retreats at the ice edges. Along the circumpolar ice edge, alternating advances and retreats are apparent which is again a manifestation of the Antarctic Circumpolar Wave mentioned earlier in Sec. 6.2.2. Note that northerly winds can cause compaction of the ice that could enhance albedo as may have happened in October 1994 but they could also cause ice breakup and melt as may have happened in October 1990 and 1999 in the Weddell Sea. Southerly winds on the other hand may cause cooling, more consolidation and an expanded ice cover as may have happened in October 1981, 1991 and 2005. It is also apparent that the phenomenon is not strictly Mode 2 and can be Mode 3 as indicated in the October 1994 and October 2003 images. The pressure maps presented in the Chap. 5 Sec. 5.2.2 also indicate that pressure patterns can be unpredictable.

Plots of climatological averages of albedo over sea ice and ice free areas for the months of continuous daylight that starts with August through summer and up to April are presented in Fig. 6.49. The albedo for sea ice is shown to be less seasonal than that of the Arctic likely because the Antarctic ice cover is more divergent and there is more new ice and open water inside the ice pack. Also, the peak value occur in October which is the month when the sea ice cover is thick and highly consolidated in most regions. The average albedo of open water area is also less seasonal than that of the Arctic region with the albedo being close to 0.06 (or 6%) for most of the year.

The monthly averages of albedo over sea ice and open water for each month except the winter months from August 1981 to July 2008 are presented in Fig. 6.50a and b. Again, the scales for the plots for open water and sea ice are different and the interannual variability of sea ice albedo is significantly higher than that of open water. The monthly albedo over open water is shown to be relatively variable up to 1995 after which the albedo was relatively constant. Over sea ice, the monthly albedo was relatively constant up to 1995, declined slightly from 1995 to 2001 and from 2001 it increased moderately until 2008. The monthly anomalies of the average albedo for open water and sea ice along with running averages are presented in Fig. 6.40c and d. The interannual variabilities reflect those of the monthly variabilities and the trend lines derived from linear regressions of the data points

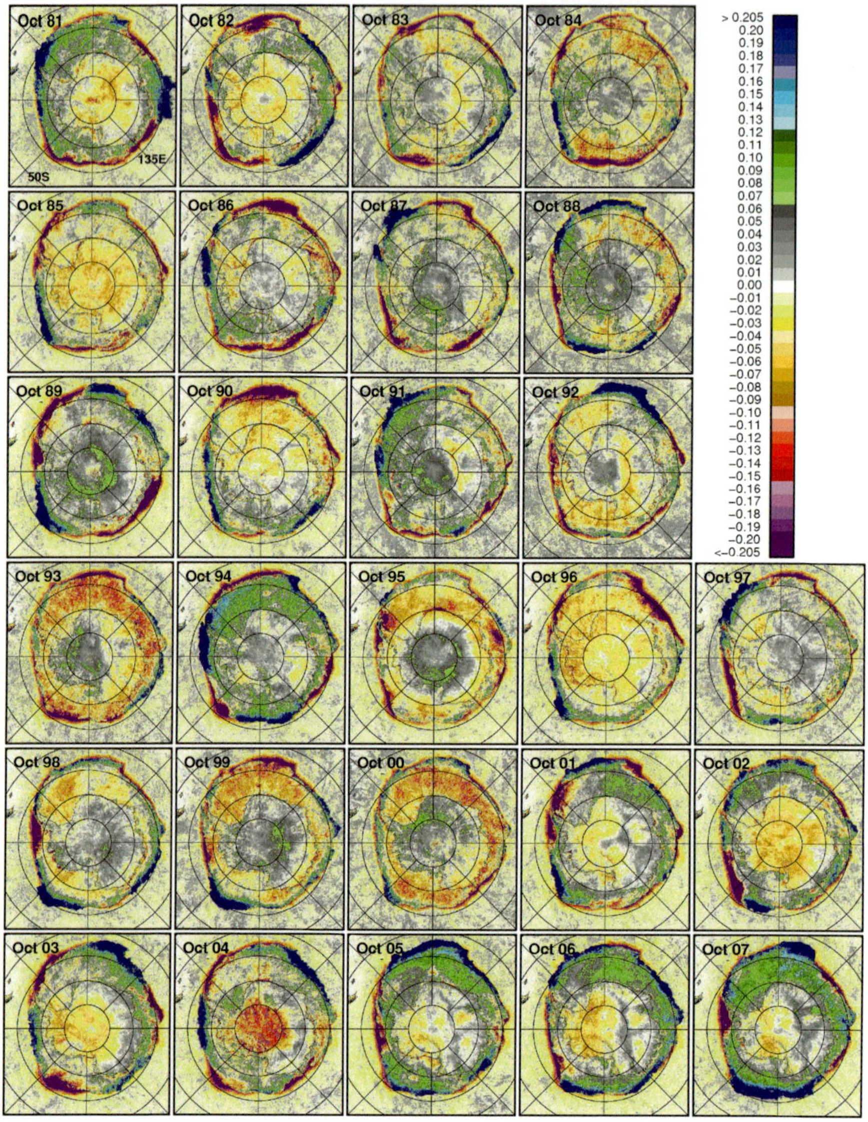

Fig. 6.48 Monthly anomalies of narrow-band albedo (0.6 μm) in the Southern Hemisphere for an autumn month (October) from 1981 to 2007

indicate opposite trends for open water and sea ice. The open water albedo is shown to have a negative trend of -0.31 ± 0.09%/decade while the sea ice albedo has a positive trend of 1.34 ± 0.35%/decade. The trend for sea ice albedo in the Antarctic has the same sign as the trend for sea ice extent which is also positive as is shown in Chap. 7. The trend is also opposite in sign to that of the Arctic where the sea ice

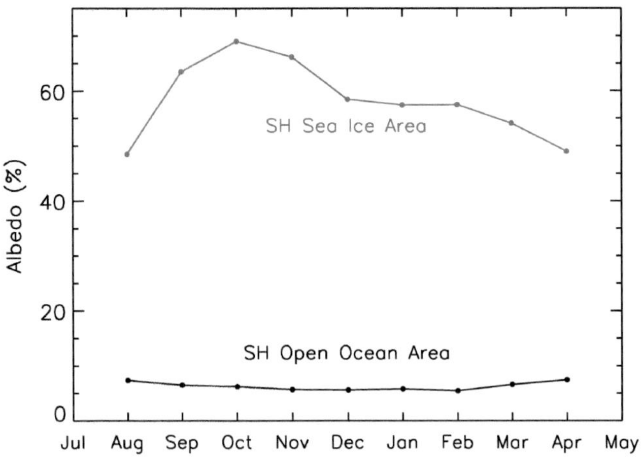

Fig. 6.49 Plots of monthly climatologies of narrow-band albedo (0.6 um) in the Southern Hemisphere for sea ice and ice free regions

cover has been declining significantly. Although the results show that the trends in albedo in the Southern Hemisphere has been relatively modest, such information is valuable and they could provide insights into the observed changes of other parameters in the polar regions.

The yearly averages in albedo are presented in Fig. 6.51. The averaging goes from September of one year to April of the following year. As in the Arctic, this allows continuous coverage through the peak of solar insolation in the summer season. The high albedo areas in the yearly average are obviously located in the Antarctic continent where the albedo is consistently high. The albedo over sea ice covered regions is relatively low because of the large seasonality of the ice cover. In most of the sea ice covered areas, the average albedo is unexpectedly low because it represents observations of the albedo of both open water and sea ice. The maps also depict a relatively high albedo in the perennial ice regions in the Western Weddell and the Bellingshausen/Amundsen Seas region and the relatively low albedo in the Ross Sea spring polynya region. It is apparent that the high albedo regions were more extensive in the 2002-2003 and 2007-2008 maps the other periods in the Weddell Sea. A low albedo region north of the Ross Ice shelf is also apparent every year but the extent significantly declined in 1997-1998, 2000-2001, 2002-2003 and 2003-2004 periods. The variability of the albedo in the region is influenced by the size and persistence of the spring polynya that in turn had been affected by the calving of large icebergs in the region.

The color-coded monthly anomaly maps presented in Fig. 6.52 show relatively small interannual changes in the albedo of the Antarctic continent but large interannual changes in the albedo of the sea ice cover. Large negative anomalies, especially near the ice edges, are apparent in the 1989–1990, 1996–1997, and 1999–2000, while large positive anomalies are apparent in 2002–2003,

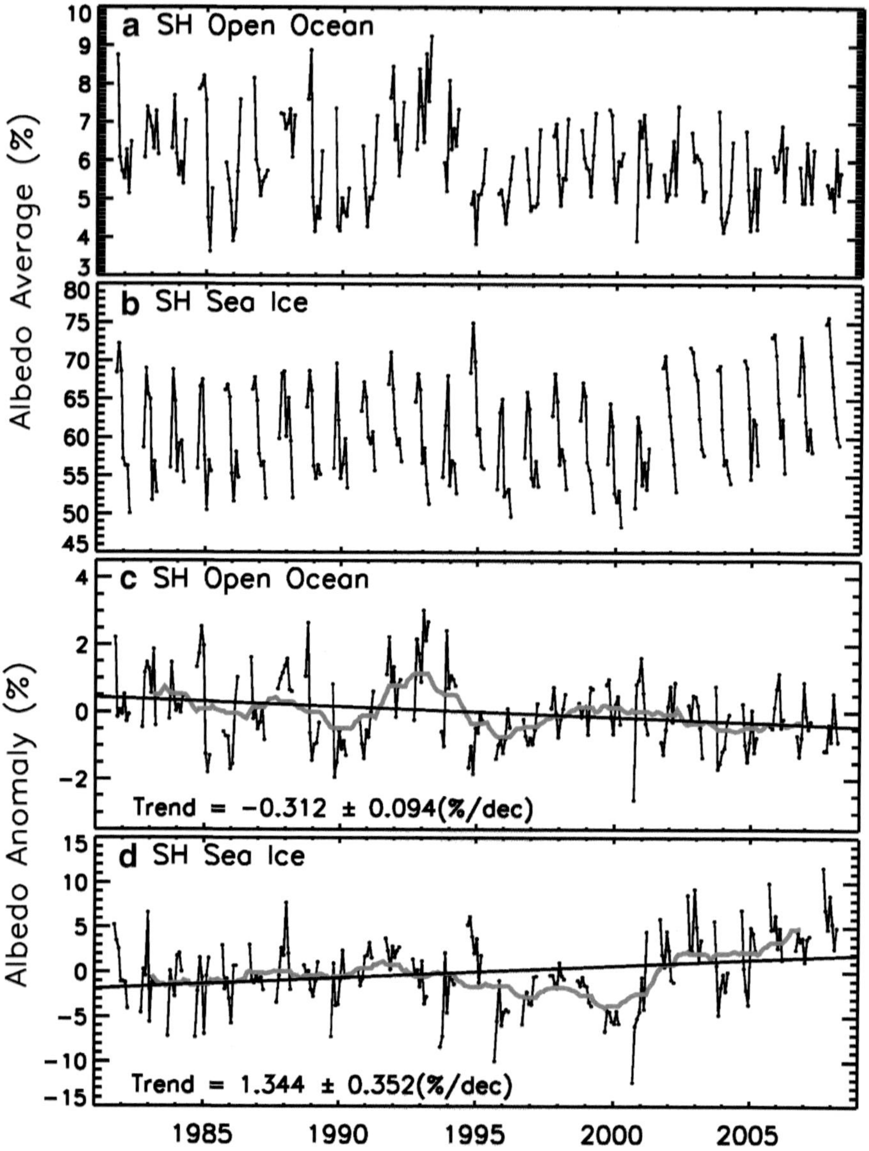

Fig. 6.50 Plots of monthly average and monthly anomalies for (a, c) open ocean and (b, d) sea ice covered regions

2005–2006, and 2007–2008. It is interesting that the negative anomalies were so dominant on opposite sides of the continent in 1989–1990, while the same is true in 2007–2008 but with opposite signs. The phenomenon is a strong manifestation of a mode-2 system which is apparent as well in the September pressure maps in

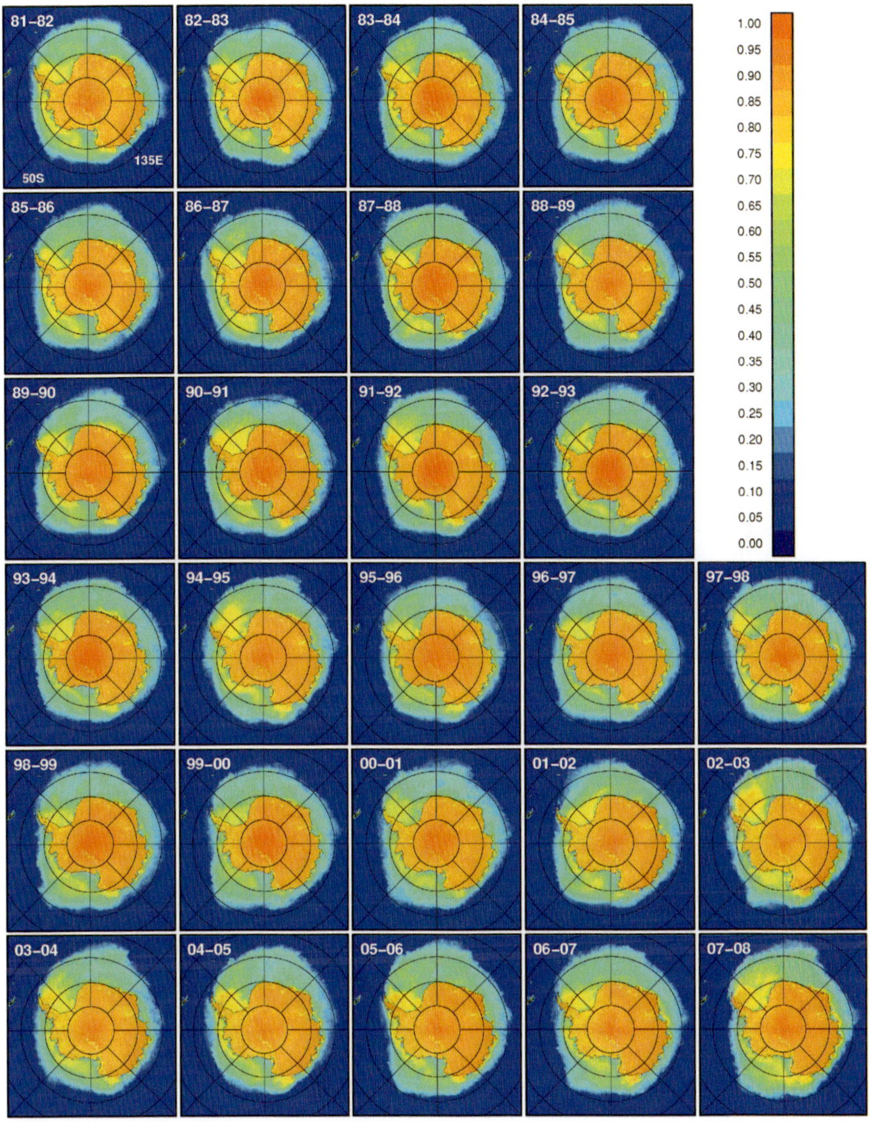

Fig. 6.51 Yearly Yearly averages (i.e., from September of one year to April the following year) of narrow-band albedo (0.6 um) in the Southern Hemisphere from September 1981 to April 2008)

1989 and 2007 shown in Figs. 5.14 and 5.15 and the September temperature anomaly maps shown in Figure 6.22. They are among the few occurrences of the mode-2 system in a generally complex atmospheric circulation system in the Southern Hemisphere.

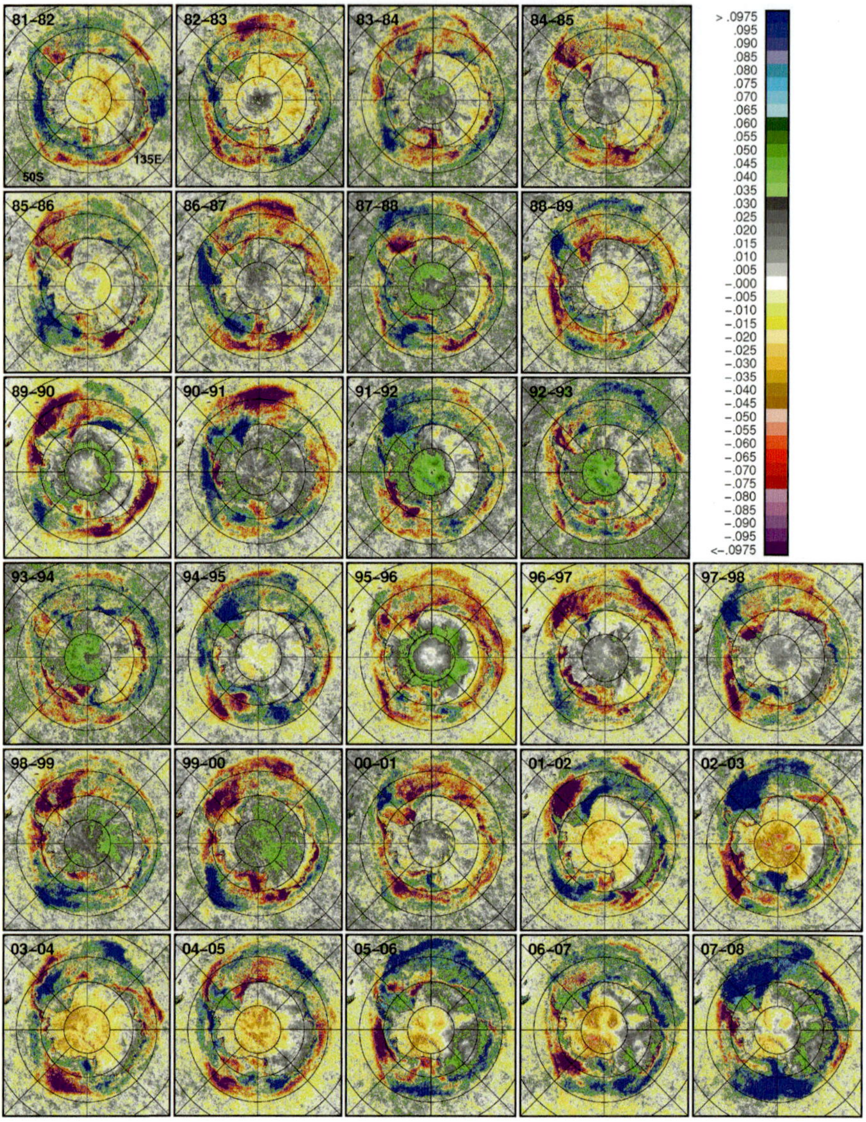

Fig. 6.52 Yearly anomalies (i.e., from September of one year to April the following year) of narrow-band albedo (0.6 um) in the Southern Hemisphere from September 1981 to April 2008)

References

Abdalati W (2006) Recent changes in high-latitude glaciers, ice caps and ice sheets. Weather 61(4):95–101

Ackley SF (1981) Sea-ice atmosphere interactions in the Weddell Sea using drifting buoys. In: Sea Level, Ice, and Climatic Change (Proc. Canberra Symp., December 1979), IAHS 131: 177–191

Allison I, Brandt RE, Warren SG (1993) East Antarctic sea ice: albedo, thickness distribution and snow cover. J Geophys Res 98(C7):12417–12429

Armstrong RL, Brodzik MJ (2001) Recent Northern Hemisphere snow extent: A comparison of data derived from visible and microwave satellite sensors. Geophys Res Lett 28 (19):3673–3676

Brohan P, Kennedy JJ, Harris I, Tett SFB, Jones PD (2006) Uncertainty estimates in regional and global observed temperature changes: A new data set from 1850. J Geophys Res 111, D12106, doi:10.1029/2005JD006548

Colony R, Thorndike A (1985) Sea ice motion as a drunkard's walk. J Geophys Res 90:965–974

Comiso JC (2000) Variability and trends in Antarctic surface temperatures from in situ and satellite infrared measurements. J Climate 13:1674–1696

Comiso JC (2002) A rapidly declining Arctic perennial ice cover, Geophys Res Lett 29(20):1956, doi:10.1029/2002GL015650

Comiso JC (2003) Warming trends in the Arctic. J Clim 16(21):3498–3510

Comiso JC, Parkinson CL, Gersten R, Stock L (2008) Accelerated decline in the Arctic sea ice cover. Geophy Res Lett 35:L01703, doi:10.1029/2007GL031972

Curry JA, Webster PJ (1999) Thermodynamics of atmospheres and oceans. Academic, San Diego, CA

Doran PT, Priscu JC, Lyons WB, Walsh JE, Fountain AG, McKnight DM, Moorhead DL, Virginia, RA, Wall DH, Clow GD, Fritsen CH, McKay CP, Parsons AN (2002) Antarctic climate cooling and terrestrial ecosystem response, Nature doi:10.1038/nature710

Drucker R, Marin S, Moritz R (2003) Observations of ice thickness and Frazil ice in the St,. Lawrence Island Polyya from satellite imagery, upward looking sonar and salinity/temperature moorings. J Geophys Res 108, doi:10.1029/2001JC9001213

Frei, A, Robinson DA, Hughes MG (1999) North American Snow Extent: 1900-1994. Int. J. Climatol. 19:1517–1534

Grenfell TC, Maykut GA (1977) The optical properties of ice and snow in the Arctic Basin. J Glaciol 18:445–463

Grenfell TC, Light B, Sturm M (2002) Spatial distribution and radiative effects of soot in the snow and sea ice during the SHEBA experiment. J Geophyus Res 107(C10):8032, 1029/2000JC000414

Hakkinen S, Mellor GL (1990) One hundred years of Arctic ice cover variations as simulated by one-dimensional, ice-ocean model. J Geophys Res 15:15959–15969

Hibler III WD (1979) A dynamic-thermodynamic sea ice model. J Phys Oceanogr 9(4):815–846

Holland MM, Bitz CM (2003) Polar amplification of climate change in the coupled model intercomparison project. Clim Dyn 21:221–232

Kilpatrick KA, Podesta GP, Evans R (2001) Overview of NOAA/NASA advanced very high resolution radiometer Pathfinder algorithm for sea surface temperature and associated matchup database. J Geophys Res 106(C5):9179–9197

Kwok R, Comiso JC, Martin S, Drucker R (2007) Ross Sea Polynyas: Response of ice concentration retrievals to large areas of thin ice. J Geophys Res 112, C12012,doi:10.1029/2006JC003967

Martin S, Drucker R, Kwok R, Holt B (2005) Improvements in the estimates of ice thickness and production in the Chukchi Sea polynyas derived from AMSR-E. J Geophys. Res Lett 31, L005505, doi:10.1029/2004GL0022013

Martin, S., R.S. Drucker and R. Kwok, 2007. The areas and ice production of the western and central Ross Sea polynyas, 1992-2002, and their relation to the B-15 and C-19 iceberg events of 2000 and 2002, J. Marine Systems, 68:201–214

Parkinson CL, Comiso JC, Zwally HJ, Cavalieri DJ, Gloersen P, Campbell WJ (1987) Arctic Sea Ice 1973-1976 from Satellite Passive Microwave Observations, NASA Spec. Publ. 489, Washington DC

Perovich DK, Tucker III WB, Ligett KA (2002a) Aerial observations of the evolution of ice surface conditions during summer, J Geophys Res 107(C10):8048, doi:10.1029/2000JC000449

Perovich DK, Grenfell TC, Light B, Hobbs PV (2002b) Seasonal evolution of the albedo of multiyear Arctic sea ice. J Geophys Res 107(C10):8044, doi:10.1029/2000/JC000438

Reynolds RW, Rayner NA, Smith TM, Stokes DC, Wang W (2002) An improved in situ and satellite SST analysis of climate. J Clim 15:1609–1625

Schweiger AJ, Lindsay RW (2008) Relationships between Arctic sea ice and clouds during autumn. J Clim 21:4799–4810

Shibata A, MUrakami H, Comiso JC (2010) Sea surface temperature in Arctic Ocean from 2002 to 2009 observed by AMSR-E. Japan J. Remote Sensing (In Press)

Stroeve JC, Holland MM, Meier W, Scambos T, Serreze M (2007) Arctic sea ice decline: faster than forecast. Geophys Res Lett 34:L09501, doi:10.1029/2007/GL029703

Wadhams P (2000) Ice in the ocean. Gordon and Breach Science Publishers, London, UK

Warren SG (1982) Optical properties of snow. Rev Geophys Space Phys 20:67–82

Warren SG, Clarke AD (1990) Soot in the Atmosphere and snow surface in Antarctica. J Geophys Res 95(D2):1811–1816

Warren SG, Wiscombe WJ (1980) Model for the spectral albedo of snow. J Atmosheric Sciences 37:2734–2745

White WB, Peterson RG (1996) An Antarctic circumpolar wave in surface pressure, wind, temperature and sea ice extent. Nature 380:699–702

Zwally HJ, Comiso JC, Parkinson CL, Campbell WJ, Carsey FD, and Gloersen P (1983) Antarctic sea ice 1973-1976 from satellite passive microwave observations. NASA Spec Publ 459, Greenbelt, MD

Zwally HJ, Comiso JC, Gordon AL (1985) Antarctic offshore leads and polynyas and oceanographic effects, in Oceanology of the Antarctic Continental Shelf, ed. by S. Jacobs, Antarctic Research Volume 43:203–226

Zwally HJ, Comiso JC, Parkinson CL, Cavalieri DJ, Gloersen P (2002a) Variability of the Antarctic sea ice cover. J Geophys Res 107(C5):1029–1047

Zwally HJ, Abdalati W, Herring T, Larson K, Saba J, Steffen K (2002b) Surface melt-induced acceleration of Greenland ice-sheet flow. Science 297:218–222

Chapter 7
Characteristics and Variability of the Sea Ice Cover

Abstract Mesoscale and large scale characteristics and interannual variability of the Arctic and Antarctic sea ice cover are described and evaluated using geophysical data derived from visible, infrared, and microwave sensors. Thirty years of satellite data were used to assess global, hemispherical, and regional trends and to gain insights into the changing sea ice cover in the two hemispheres. The extent of the sea ice cover in the Arctic is shown to be declining at –4%/decade, whereas in the Antarctic, it has been increasing unexpectedly at 1%/decade. The most intriguing result, however, is the rapid decline in the Arctic perennial ice cover area of about −13%/decade and possible connection with greenhouse gas warming. Equally intriguing is the relative stability of the Antarctic sea ice cover and associated effects of the ozone hole, the Southern Oscillation and the Antarctic Circumpolar Wave.

Keywords Mesoscale characteristics • Ice types • Ice dynamics

7.1 Introduction

The sea ice cover is the frozen portion of the polar oceans and is one of the most seasonal geophysical parameters on the Earth's surface, second only to the more variable but less predictable snow cover. At any given time, the extent of the global sea ice cover ranges from 17.5×10^6 to $28.5 \times 10^6 \mathrm{km}^2$. This corresponds to a significant fraction (3.4–5.6%) of the total surface area of the Earth. In the Arctic region, the ice cover doubles its size from summer to winter, while in the Antarctic, the corresponding seasonal change is about fivefold (Zwally et al. 1983; Parkinson et al. 1987; Gloersen et al. 1992). The large expanse and seasonality have made sea ice a subject of strong scientific interest, especially in the last few decades when climate change became a societal and political issue. Recent studies show that the influence of sea ice is indeed large and has gone beyond original expectations. We now know that sea ice is a key component of the Earth's climate system and an important part of the ecosystem in the polar regions. In particular, sea ice is a good insulator that is effective in keeping ocean heat energy at high latitudes from leaking to the atmosphere during

J. Comiso, *Polar Oceans from Space*, Atmospheric and Oceanographic Sciences Library 41,
DOI 10.1007/978-0-387-68300-3_7,

winter. At the same time it has a relatively high albedo and allows only a small fraction of solar energy to be absorbed by the surface during spring and summer. The sea ice cover also redistributes salt in the ocean since it enhances surface salinity where it forms while introducing low salinity water where it melts. Moreover, in some special regions, as in coastal polynyas in the Antarctic, the rate of growth of sea ice is so high in winter that it actually causes the formation of high density bottom water that becomes part of the global thermohaline circulation (Zwally et al. 1985; Smith et al. 1990; Markus et al. 1998; Comiso and Gordon 1998). On the other hand, the melt of sea ice causes the formation of a stratified surface layer at the marginal ice zone that gives rise to increased growth of phytoplankton (Alexander and Niebauer 1981; Niebauer et al. 1995; Smith and Nelson 1985; Sullivan et al. 1993) and high productivity in the region. Through these and other mechanisms, sea ice impacts both physical and biological characteristics of the oceans.

Because of general inaccessibility and harsh weather conditions, our knowledge about the extent and distribution of the ice cover was very limited until the advent of satellite passive microwave data. Although there were earlier passive microwave systems, it was the ESMR, launched in December 1972, that revealed for the first time relatively accurate spatial distribution and extents of the global sea ice cover (Zwally et al. 1983; Parkinson et al. 1987). The Russian satellite, Cosmos 243, which was launched in 1968, provided good information about the location of the ice edges, but it was a nonscanning nadir viewing system, and it took several days to obtain a complete coverage of a highly variable sea ice cover. Because of the strong contrast of the brightness temperature of open water and sea ice, ESMR data at 19 GHz were successfully used to quantify the location of the ice edge, the approximate fraction of open water within the ice pack and the seasonal variability of the ice cover (Zwally et al. 1983; Parkinson et al. 1987). The introduction of the dual polarized SMMR data at five frequencies, from 6 to 37 GHz, provided the means to derive sea ice concentration at significantly better accuracy (Gloersen et al. 1992; Comiso 1986), as discussed in Chap. 4. Passive microwave sensors provide day/night, almost all-weather observation of the sea ice cover and, therefore, synoptic coverage at a very good temporal resolution. The ice cover is usually depicted in the data in terms of ice concentration maps gridded at a resolution of 25 km by 25 km. Because of this relatively coarse resolution, many of the spatial details in the distribution, mesoscale features, and surface properties are not very well represented in the passive microwave data. However, since the mixing algorithm converts the brightness temperature data to ice concentration (or open water fraction) for each data element, the data can be used for various applications, including mass balance, heat flux, and polynya studies.

Sea ice is constantly changing with time at many scales, from formation to melt stages. In this chapter, we examine the mesoscale and large scale variability of the sea ice cover in both hemispheres using high, medium, and low resolution satellite data. The mesoscale characteristics of sea ice correspond to processes that occur at the meter to kilometer scales, including the effects of many environmental factors such as temperature, wind, ocean current, eddies, tides, and the rotation of the Earth. The structure and spatial distributions of the sea-ice cover change constantly through thermodynamic, dynamic, and other processes. Many small scale spatial features such as regions of divergences in which leads are formed, and regions of convergences in which the average thickness gets enhanced through rafting and ridging are of scientific interest especially in heat flux and mass balance studies. Wind induced ice advection and the upwelling of warm water contribute to changes in spatial distribution that includes the formation of polynyas and unusual ice features such as the Odden, as will be described in Chap. 8. The mesoscale characteristics are important to know, in part, because their integrated effects determine the large-scale characteristics of the sea ice cover. The large-scale characteristics are usually described in terms of total ice extent and area, ice thickness, snow cover, temperature, and other features that represent the combined effect of mesoscale processes as the sea ice cover goes through the seasonal growth and decay stages.

Mesoscale studies usually require the use of high resolution satellite data because they usually involve analysis of spatial features that have dimensions of a few meters to a few kilometers. The resolution requirements depend on applications. For applications that require about 5 km resolution and a relatively long time series, the most suitable data available are the AVHRR visible and infrared GAC data during clear skies conditions, as discussed in Chap. 6, which are available from 1981 to the present. AVHRR LAC and HRPT data have resolutions of about 1 km at nadir, but available data are limited in spatial and temporal coverage. The global and continuous data from the Aqua/MODIS and Terra/MODIS sensors are excellent source of mesoscale data during cloud free conditions since they provide resolution of as good as 250 m and continuous global coverage, starting from 2000. When available, data from Landsat and SPOT, which have resolutions from 10 to 30 m as described in Chap. 3, provide even more useful information, but such data in the polar regions are again spotty. For day/night all-weather coverage, the Synthetic Aperture Radar (SAR) data, which have a resolution of about 35 m, can be very useful (Tsatsoulis and Kwok 1998). Ice floes and ridges within the ice pack are recognizable at this resolution, but the interpretation of the data is difficult, especially in the MIZ, and time series of images (or ancillary data) is needed to resolve some of the ambiguities. Mesoscale studies can also benefit from retrieved SAR data of the drift of sea ice and the approximate thickness of seasonal ice, especially that of thin ice (Liu et al. 1998; Kwok et al. 1998). Global coverage by SAR at a reasonably good time resolution would be highly desirable, but the acquisition of such data has not been possible because of instrumental and other limitations. A relatively comprehensive set of SAR data over a large fraction of the Central Arctic has been put together by the University of Alaska SAR Facility/ NASA DAAC. The collection consists primarily of data acquired by ERS-1, ERS-2, and Radarsat SAR systems in the Western Arctic Basin, including the North Pole, Chuckchi Sea, and Beaufort Sea. Other active microwave systems that can be useful but provide lower resolution are the scatterometers and radar altimeters. Scatterometers were carried by ERS-1, ERS-2, Adeos-1, QuickSCAT, and Adeos-2, and they provide more comprehensive coverage than SAR and have strong potential for many polar process studies (e.g., Drinkwater and Lytle 1997; Liu et al. 1998). The data coverage and resolution are similar to those from passive microwave, but more research is needed to accurately interpret the data, and algorithms have to be developed to retrieve geophysical parameters. The radar altimeter provides good complementary information, such as height of the ice freeboard, which can be used to assess sea ice thickness and surface roughness (Fetterer et al. 1992; Laxon et al. 2003). The use of laser altimeters, such as that provided by ICESat, also enables measurement of ice freeboard height that, in turn, can be used to obtain sea ice thickness (Kwok et al. 2007; Zwally et al. 2008).

Large-scale characterization and variability studies of the sea ice cover have been done using, primarily, passive microwave data (Gloersen et al. 1992; Parkinson et al. 1999; Zwally et al. 2002). The basic geophysical parameter derived is sea ice concentration, as described in Chap. 4, which provides the important information required to fully characterized the spatial distribution of sea ice and open water within the ice pack. The parameter is not meant for detailed mesoscale studies that can be done only with high resolution data, but it provides the ability to study the large scale characteristics of the global sea ice cover at a fairly good time resolution. Among the most basic information that can be derived from the ice concentration data are ice extent and ice area. Ice extent is defined as the integral sum of the areas of all data elements that have at least 15% ice concentration, while ice area is defined as the sum of the products of the area of each data element and the ice concentration within this data element. In simple terms, ice extent is the total area of the ice covered regions (i.e., regions with at least 15% ice cover), while ice area is the total area that is actually covered by sea ice. As indicated in Chapter 2, the ice cover can be described in terms of an inner zone, where the ice cover appears like a continuous ice sheet with occasional breaks and polynyas, and an outer zone, where the ice cover consists of individual (not consolidate) components, including ice bands, pancake ice, and loose ice floes. The outer region is often referred to as the marginal ice zone (MIZ), which varies in width, depending on wave conditions and wind velocities. Ship observations have indicated that the width of the MIZ is highly variable and ranges from a few kilometers to several tens of kilometers. Although it is effective in detecting the location of the MIZ, the standard ice concentration maps, which are generated at 25 km resolution, cannot be used to fully characterize the region (Worby and Comiso 2004). However, they provide temporally consistent estimates of ice extent and ice area that are invaluable for mass balance studies. Moreover, some useful characterization of the outer zone can be inferred from AMSR-E data which provide ice concentrations at resolutions as high as 6 km.

The observed growth and decay of sea ice is primarily a response of the ocean surface to changes in surface temperature, winds, and other external forcing. The growth and decay periods are out of phase in the two hemispheres by about six months. Thus, when it is a growth season in one hemisphere it is concurrently a melt season in the other hemisphere. The typical magnitudes of the seasonal growth and decay in both hemispheres are depicted in Fig. 7.1. The plots show monthly averages of ice extents from January to December. Each data point represents monthly averages, using about 30 years of passive microwave satellite data and will be referred to as the monthly sea ice climatology. The seasonal distributions for the two regions are clearly out of phase, and the amplitude and shape are also different. In the NH (Fig. 7.1a), the plot looks approximately symmetrical, with growth period being about as long as the decay period. In the SH (Fig. 7.1b), the distribution looks

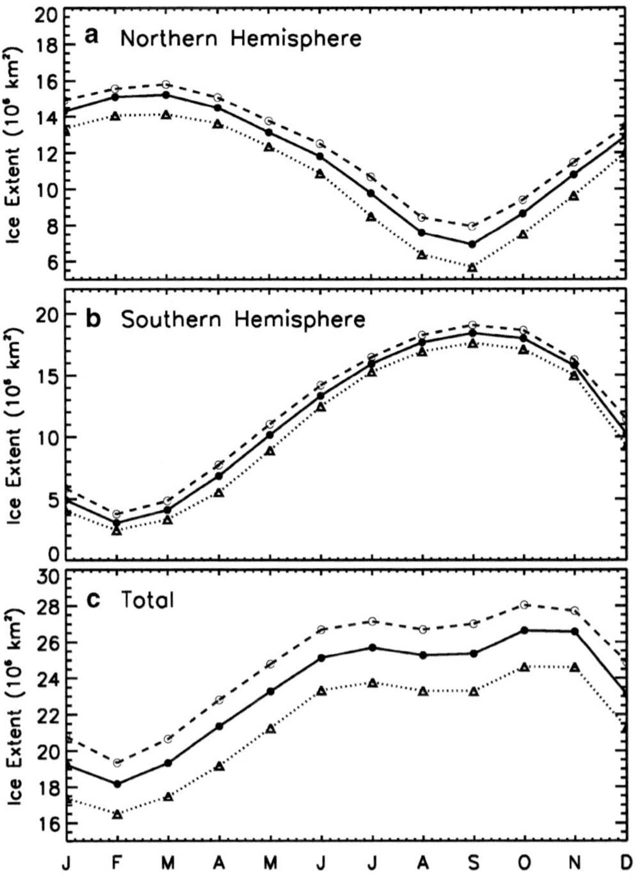

Fig. 7.1 Sea ice extents as derived from passive microwave data (1979 to 2008) in (**a**) the Northern Hemisphere; (**b**) the Southern Hemisphere; and (**c**) the entire globe

less symmetric with the growth period being about seven months while the decay period is shorter, at only five months. The difference in the length of the seasons is likely caused mainly by the difference in geographical and environmental conditions, as discussed earlier. In particular, the rate of retreat of sea ice in the Arctic is relatively moderate because of the presence of land at the southern boundaries, while in the Antarctic, the ice retreat is accelerated by direct intrusions of large ocean waves and warm water from lower latitudes. The relatively warm Southern Ocean at the end of the summer also acts to dampen the growth of sea ice, while cold air temperatures from the Antarctic continent in mid winter contribute to the lengthening of the growth season. In the Arctic, the growth and decay of sea ice follows the same seasonal pattern as surface temperatures observed from surrounding land boundaries. Thus, the presence of land at mid latitudes that rapidly cools down to subfreezing temperatures in autumn and winter enables ice to form faster in the NH (than in the SH) during the period. During spring and summer, their presence also inhibits big waves and the incursion of warm water from the south, preventing the rapid decay of ice. The presence of relatively cold land areas also allows sea ice to form at relatively low latitudes.

In the NH, the mean monthly extent, represented by a solid line in Fig. 7.1a, is shown to vary from a minimum of $6.8 \times 10^6 km^2$ in the late summer (September) to about 15.2 km^2 in the winter (March). The corresponding values in the SH (Fig. 7.1b, solid line) are $3.1 \times 10^6 km^2$ (in February) and $18.4 \times 10^6 km^2$ (in September), respectively. Although the minimum values for sea ice in the NH in the summer are greater than the corresponding values in the SH, it should be noted that the average albedo in the SH is much greater than that of the NH because of the presence of the snow covered Antarctic continent, which has a total area of about $14 \times 10^6 km^2$. In the NH, except for glaciers and the Greenland ice sheet, many of the snow covered areas over land are no longer covered by snow in late spring and summer. Thus, during the summer, when the solar insolation is at its peak, changes in the sea ice cover in the NH is likely to have a greater impact on climate change (e.g., due to ice albedo feedback) than the sea ice cover in the SH.

Climatological global sea ice extent, which is simply the sum of data in Fig. 7.1a and b, is presented in Fig. 7.1c. The plot shows that at any one time, the total ice cover ranges, on the average, from 19.0 to $27.5 \times 10^6 km^2$ but the extent can be as low as $17.5 \times 10^6 km^2$ and as high as $28.5 \times 10^6 km^2$. The lowest value in global extent, as depicted Fig. 7.1c, occurs in February, which is the month of lowest extent in the Antarctic climatology and the month of maximum extent in the Arctic climatology. The occurrence of peak values in November is due to the seasonal asymmetry in the Antarctic ice cover. It is also interesting to note that despite large hemispherical differences in the ice cover, the total area is almost constant from June through November, averaging at about $27 \times 10^6 km^2$. There is a slight dip during this period, when the Arctic and the Antarctic approaches minimum and maximum extents, respectively. It is interesting to note that, overall, the seasonal distribution of the global sea ice cover is more akin to that of the Antarctic ice cover than to that of the Arctic, but with reduced amplitude.

The averages of the sea ice extents and ice areas for the different seasons and for each year from 1979 to 2008 are depicted in Fig. 7.2. Seasonal averages of data from the Northern and Southern Hemispheres are presented in Figs. 7.2a and b, respectively, while the sums of the Northern and Southern Hemisphere averages for the same period are shown in Fig. 7.2c. The seasons are defined, as in previous chapters, with the NH winter represented by the average of December, January, and February data, while the austral winter is the average of June, July, and August data. The relative magnitudes of the averages for each season of the ice extent and ice areas are actually different from expectations. For example, the ice cover is expected to be most extensive in winter, but data in the NH show that the extents and areas in winter and spring (Fig. 7.2a and b) are very similar. Also, the minimum values for these parameter do not occur in the summer but actually in autumn. This is, in part, caused by the delay in the response of the ice cover to the seasonal changes in surface temperature. In particular, the maximum extent in winter, as shown by the climatology, usually occurs in March, which is defined as a spring month, while the minimum values usually occur in September, which is an autumn month. The similar values for winter and spring are consistent with the afore-

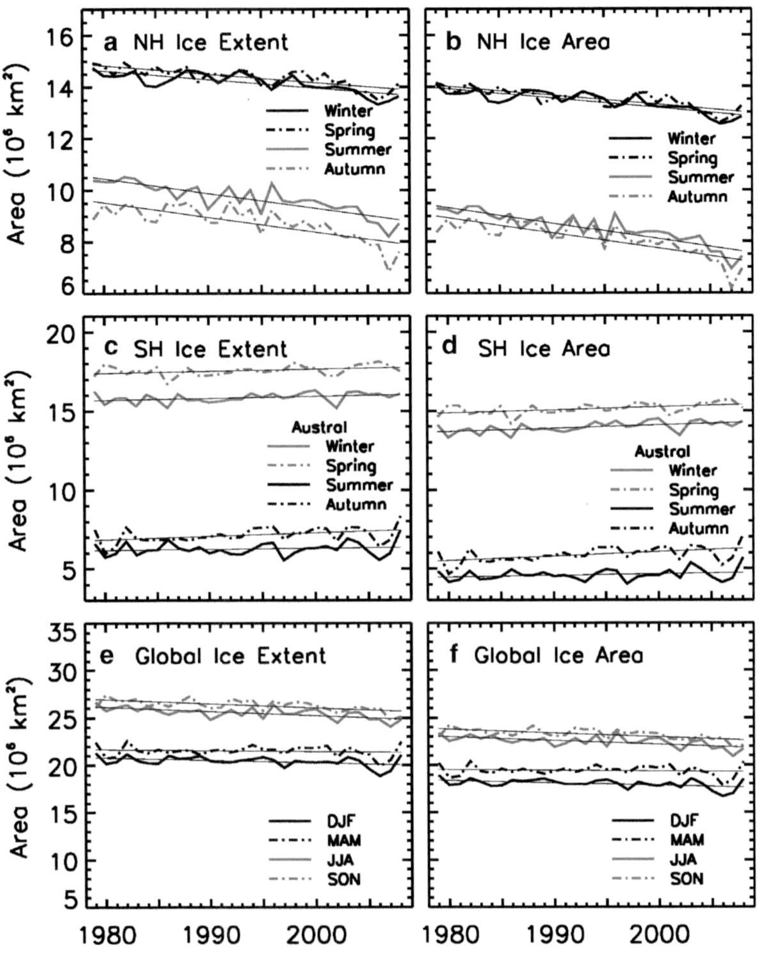

Fig. 7.2 Plots of averages of sea ice extents and ice areas for each of the seasons from 1979 to 2008 in the (**a–b**) Northern Hemisphere; (**c–d**) Southern Hemisphere and (**e–f**) Global Oceans

mentioned symmetry of the seasonal growth and decay cycle of sea ice in the NH. In the summer and spring, the values are more distinct because the difference between the August and September values are greater than those in February and March.

The seasonality of the extents and areas in the Southern Hemispheres (Fig. 7.2c and d) are somewhat different than that of the NH because the winter and spring data are more distinct with the spring values significantly higher than the winter values, while the summer and autumn extents are more similar in magnitude. The spring values are higher than those of winter because the maximum extent occurs in September, which is a spring month, and the October values are only slightly lower than the September values. But unlike those in the Arctic, the autumn values are higher than the summer values, as expected, since the minimum in the Antarctic usually occurs in February, which is a summer month.

In the NH, yearly fluctuations in the seasonal average values are relatively modest in winter and spring, while larger fluctuations occur in summer and autumn, especially in the 1990s. On the other hand, in the SH, the seasonal values fluctuate more in the austral winter and spring, especially in the 1980s, than in the austral summer and autumn. It is interesting that the large yearly fluctuations occur in the two hemispheres during the same months (i.e., Arctic summer and Antarctic Austral winter). Except for a few cases (e.g., spring peaks in 1985 and 1989), the fluctuations in the two hemispheres during the different seasons are generally uncorrelated.

The seasonal averages provide a means to assess the interannual trends in the ice cover for the different seasons. The dash lines in the plots represent trends in the ice cover as inferred from linear regression analysis of the data. In the NH, the trends in ice extent are -2.32 ± 0.37, -2.28 ± 0.35, -5.87 ± 0.69, and $-6.50\pm1.00\%$ per decade in winter, spring, summer, and autumn, respectively, while the corresponding trends for ice area are -2.81 ± 0.35, -2.68 ± 0.36, -7.20 ± 0.79, and $-7.37\pm1.02\%$ per decade. The trends in ice extent and ice area are comparable except in the summer, when the trend in ice area is significantly more than that in ice extent. This implies increasing open water areas within the pack that may be caused by more frequent storms and hence more divergence in the summer. The results show that the most rapid decline occurs in the summer and followed in order by autumn, spring, and winter. In the SH, the trends in extent are -0.77 ± 0.39, 0.73 ± 0.36, 1.03 ± 1.41, and $3.08\pm1.33\%$ per decade in winter, spring, summer and autumn, respectively. The corresponding trends in ice area are 1.43 ± 0.42, 1.22 ± 0.42, 2.10 ± 1.64, and $4.66\pm1.54\%$ per decade. The values are all positive except for a negative but insignificant trend in extent in winter. The trends are consistently higher for ice area than for ice extent, suggesting less open water (or less divergence and storms) within the pack in later years.

The interannual variability of the global sea ice cover (i.e., combined NH and SH data) during different seasons (as defined for the NH) is depicted by the plots of yearly averages as presented in Fig. 7.2e and f. It is interesting that the global averages show much less variability than the averages in the two hemispheres. This suggests no connectivity in the ice cover in the two hemispheres, and that the less variability is mainly the result of the averaging process. However, the results show overall stability in the global sea ice cover during the satellite era. The summer and autumn seasons have very similar values and are significantly higher than those of winter and spring but the contrast is not as big as for the individual hemispheres. This is mainly because this time period also includes the ice cover during the austral winter and spring, which are more extensive than the corresponding values in the NH. The trends in ice extent are relatively low and negative at -1.29 ± 0.50, -0.50 ± 0.56, -1.75 ± 0.38, and $-1.67\pm0.37\%$ per decade for winter, spring, summer, and autumn, respectively, while the corresponding trends in ice area are -1.56 ± 0.52, -0.45 ± 0.61, -1.83 ± 0.43, and $-1.78\pm0.40\%$ per decade. These results indicate that, overall, the trends in the global sea ice cover for the different seasons are consistently negative but relatively small.

These results provide basic information about the state and large scale characteristics of the global sea ice cover. The quantitative estimates of the typical extent of the sea ice cover over an entire annual cycle in the Northern and Southern Hemispheres provides a quick assessment of the magnitude of the coverage and the seasonality of the extents. They also provide estimates of seasonal averages of the ice extent and area in the two hemispheres and how they have been changing during the 30 year period of continuous satellite coverage. The results of regression analysis of these averages indicate an unexpected trends with the sea ice cover declining significantly in the Northern Hemisphere while increasing moderately in the Southern Hemisphere. The results also show that the total ice cover is vast with the extent ranging from 17 to 28 million km^2 at any one time and comparable to the total area of North America (i.e., 24.7 million km^2). Such an extensive area on the surface of the Earth covered by an important climate parameter must be studied. In the following sections we will present detailed information about the mesoscale and large scale characteristics and variability of the sea ice cover as observed from space.

7.2 Sea Ice in the Northern Hemisphere

The distributions of the sea ice cover in the Northern Hemisphere in both winter and summer as depicted by two different sensors are presented in Fig. 7.3. The set of images show weekly data in the visible and microwave frequencies in April and September in 2007. The images indicates how consistently the sea ice cover is represented by medium resolution (6.25 km) AVHRR visible channel-1 data at 0.6 μm, and the relatively coarse resolution (12.5 km) AMSR-E ice concentration data. The AVHRR data (Fig. 7.3a and b) show many of the mesoscale features of the ice cover that are usually apparent to an ice observer in the NH but not discernible from the standard passive microwave data (Fig. 7.3c and d). Having data at a relatively high resolution such as that provided by the AVHRR sensor is desirable, but the set of images provided in Fig. 7.3 indicates that despite the coarse resolution the passive microwave data provide good sea ice cover information. The high contrast in the albedo of open water and sea ice is apparent in April (Fig. 7.3a), when the surface is practically all covered by snow and also in September (Fig. 7.3b), but to a lesser degree because of less snow cover and more bare ice. The contrast in the microwave brightness temperature data at some frequencies is also high and hence the ability to estimate the ice concentration quite effectively. Comparing the two sets of images, it is apparent that both the visible and microwave data provide very similar information about the sea ice cover, with the spatial distribution, especially in the locations of the ice edges and open water areas within the ice cover being basically the same. For example, the low albedo areas in the visible image in April (e.g., in the Bering Sea and Greenland Sea) are also areas of relatively low concentrations in the corresponding passive microwave data. In the September images, relatively low albedo areas (e.g., north of Alaska) are also shown as relatively low concentration areas in the passive microwave image. Moderately low albedo, as depicted in the AVHRR data, may represent new ice with no snow cover, or areas of meltponding (Allison et al. 1993; Perovich et al. 2002). Such surfaces also have relatively low emissivity (Grenfell et al. 1994; Eppler et al. 1992) at microwave frequencies and hence the relatively low ice concentration as discussed in Chap. 4, Sec. 4.3.6. It is apparent, however, that there are details in the ice cover that are depicted so much better in the medium resolution visible image than the

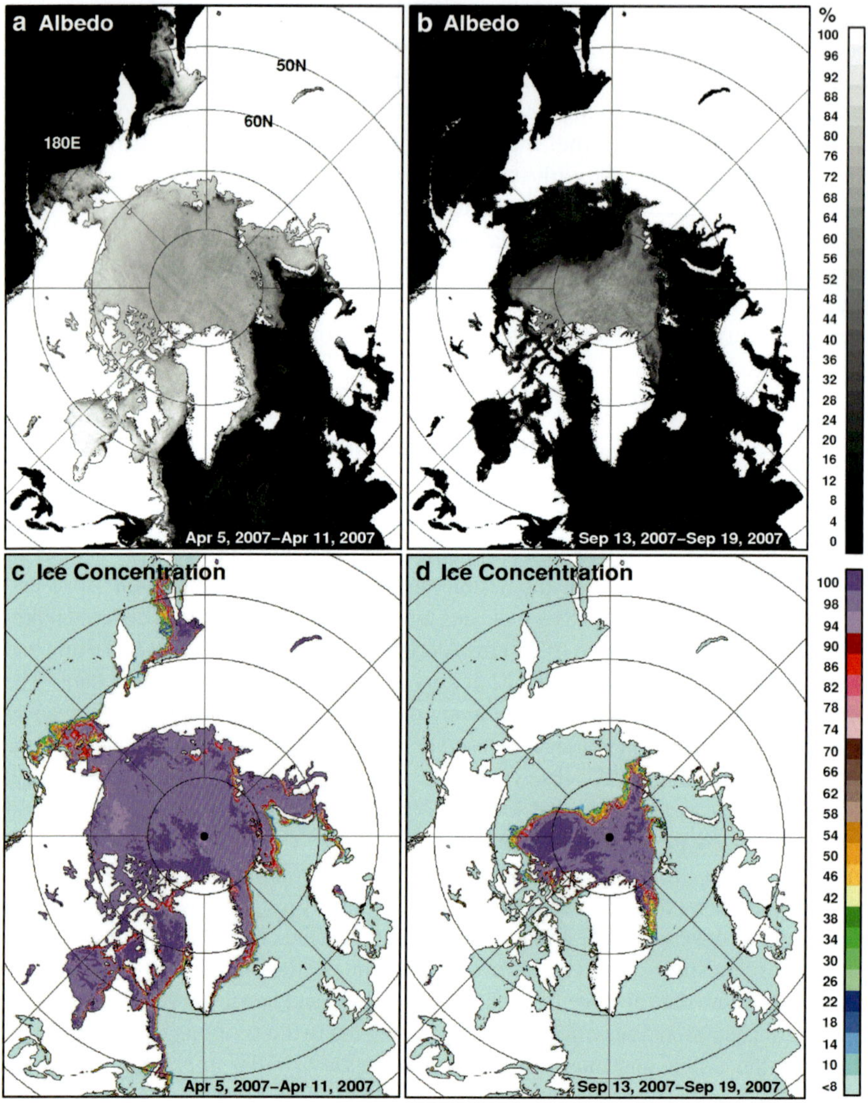

Fig. 7.3 Weekly averages of sea ice cover using AVHRR visible images (0.6um) on (**a**) 7-11 April 2007 and1 April 2007 and (**b**) 13-19 September 2007 and AMSR-E ice concentration maps on (**c**) 5-7 April 2007and (**d**) 13-19 September 2007

low resolution passive microwave image. In the Arctic, the special region of interest would be the MIZ, where polar lows, eddies, and the Odden occur as discussed in Chap. 8. In the inner zone, the ice cover is usually consolidated except in polynya regions such as the North Sea water, the Bering Sea, and the Chukchi Sea polynyas. There is thus a need for sea ice to be studied at different scales in order to get a full understanding of its overall role and impact on the ocean and climate system.

7.2.1 Mesoscale Characteristics

There are many interesting and important mesoscale processes in the polar regions that are depicted by and can be studied using satellite data. The resolution requirements depend on the geophysical processes being considered and could range from a few meters to several kilometers. Among the regions that have been of primary interest are the marginal ice zones, divergence regions, and polynya regions. These are regions noted for strong ice–ocean–atmosphere interactions and enhanced heat and salinity fluxes. Ability to identify relevant features and measure associated parameters in these regions is important because they allow for a better understanding of the variability of the fluxes and associated impacts on the ocean, sea ice cover and the atmosphere. The key drivers of these mesoscale processes are winds, currents, temperature, upwelling of warm water, tidal effects, and strong storm events. The mesoscale characteristics detectable using satellite data are leads, rafted and ridged ice, redistribution of snow and ice, eddy formation, oddens, coastal polynyas, and deep ocean polynyas. Rafting is most prevalent over predominantly thin ice regions where sheets of ice can easily go on top of each other. Ridging happens in the inner zone and, mainly, in thick ice regions and occurs when large ice floes collide, causing the breakup and file-up of thinner ice types caught in between. Rafting and ridging contributes to changes in the thickness distribution of the sea ice cover. The impacts of polynyas, Oddens, eddies, cold outbreaks and other features will be discussed in the next chapter.

High resolution satellite data are excellent tools for mesoscale studies but they are not as readily available as passive microwave data and the coverage in time and space over the polar regions is relatively sparse. When available, they provide the ability to examine the spatial distribution of the ice cover in great detail and also the distribution of different ice types, including new ice, first year ice, and older types. However, interpretation of the data is sometimes not trivial. For example, Fig. 7.4a and b show an ice covered region in the inner zone of the Arctic in winter as viewed, almost simultaneously, by SAR and Landsat sensors, respectively. At first glance, it may seem apparent that the two sensors provide different representation of the ice cover but after more careful examination, ice cover features that are common to both of them start to emerge. To illustrate, some distinct features have been marked with 1, 2, and 3 in both images. The SAR image show features of large ice floes with relatively high backscatter (near white) values, regions of intermediate backscatter (light gray), and areas of low backscatter (dark gray). On the other hand, the Landsat image show mainly different characterization of the same surface with the snow covered surfaces in white, refrozen leads or new ice in light gray, and unfrozen leads in dark gray or black. The high backscatter, as detected by SAR, has been associated with multiyear ice, while the low backscatter usually represents undeformed first year ice (Onstott 1992). The reason for the difference is associated with different salinity and structure of the two ice types as discussed in Chap. 2. In contrast, the albedo of the surface, as depicted by the Landsat image, is basically the same for both, first-year and multiyear ice once the sea ice surface is covered by snow. This phenomenon is apparent near the general area marked #3 in Fig 7.4a and b, where the Landsat image shows only one type of surface while the SAR

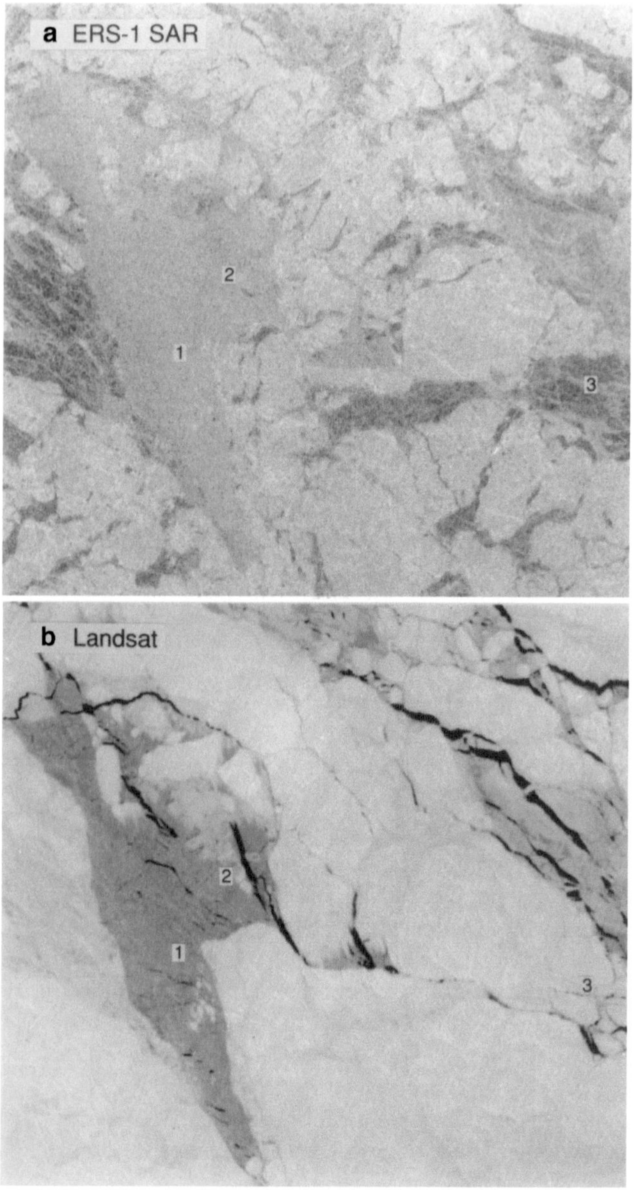

Fig. 7.4 Sea ice cover as observed almost concurrently at the Beaufort Sea (71.77°N, 145.28°W) on 16 April 1992 by (**a**) ERS-1 SAR) and (**b**) Landsat-5 thematic mapper Courtesy of Koni Steffen of the University of Colorado and ASF (SAR image, ESA copyright, 1992). From Comiso (1995) with permission from the American Geophysical Union

image shows a good delineation of first-year ice covered areas (dark gray) from the multiyear ice areas (light gray). Without the SAR image, it would not have been possible to assess, from the Landsat image alone, that the large floe in #3 consists of both, first-year and multiyear ice. Alternatively, the Landsat image provides useful information in other areas that cannot be inferred from the SAR image. For example, the thin ice region shown in area #1, is an almost featureless and homogeneous surface in the SAR data, but it is depicted as an active ice region with many cracks or leads (dark area) in the Landsat data. The big lead to the right of the area marked #2 in the Landsat image is totally undetectable in the SAR image. The areas with intermediate value in the SAR image (e.g., area marked #1) represent new ice, the signature of which can be different from that of first year ice because of the presence of salt flowers on the surface (Onstott 1992).

The ice cover as shown in Fig. 7.4 basically depicts the mesoscale characteristics of a typical sea ice cover in the inner zone of the Arctic basin. In the winter, the Arctic basin is almost fully consolidated with the continuity of the ice sheet being disrupted only by occasional leads formed by winds and tides that are refrozen within a few hours. There are also some polynyas formed, especially near islands, as will be discussed in the next chapter. Although such leads or refrozen leads shows up very well in the visible images, they may not be easy to identify in the SAR image. The refrozen leads become young ice and then first year ice which are relatively thin and during strong wind conditions, they are first to give in to the dynamic energy that causes the formation of ridges. In some areas, the ridges built up over time and they become the thick component of the perennial ice cover. The SAR data enable identification of the distributions of multiyear ice, first-year ice, and new ice. Rafting in first-year ice regions and ridging in multiyear ice areas are also detectable with the SAR data. On the other hand, the information is not as complete until the comparative analysis with Landsat data is done. The two sensors provide images that complement each other in that the more recent mesoscale processes are revealed by the Landsat image, while the ones associated with longer time history of the surface are provided by the SAR image. As for mesoscale characteristics, the SAR data alone provide useful information about the distribution of multiyear and first-year ice, while Landsat data provide locations and distributions of leads and new ice. The synergy of measurements from these two different sensors and the similarity in the resolution makes them ideal for many mesoscale process studies. This kind of data set, however, is not readily available because of persistent cloud cover, which largely limits the availability of a joint Landsat coverage and SAR coverage.

The distribution of the ice cover in the outer zone is much more complex than that in the inner zone because the outer zone is a highly dynamic region that is easily affected by slight shifts in surface temperature. The region is frequently altered by the changing direction and strength of winds and waves, making the location of the ice edge somewhat unpredictable. Since the average surface temperature in the region is close to freezing, slight cooling or warming may mean a big change in the character of the ice cover. In the NH, many interesting mesoscale processes occur at the ice margins in the Arctic and peripheral seas. To illustrate, the ice cover in the Sea of Okhotsk, as observed by Envisat/ASAR in mid-February 2003, is presented in Fig. 7.5. The image depicts some of the unique features and interesting processes that can occur in the outer region of the ice pack, usually referred to as the marginal ice zone (MIZ). The portion of the image near the top left corner has a relatively low backscatter (dark gray), which, as mentioned in Chap. 4, is likely covered by near

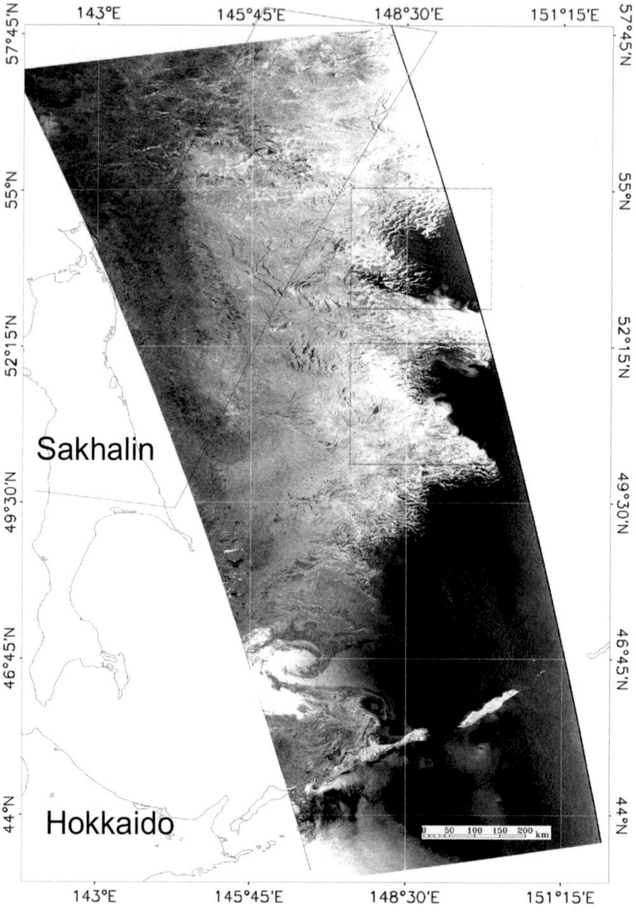

Fig. 7.5 Sea ice cover at the Sea of Okhotsk on 8 February 2003 as observed by Envisat/ASAR. ASAR image, ESA copyright (2003), was provided by Leonid Mitnik, 2004)

consolidated first year ice. Other satellite data (e.g., AMSR-E) indicate that the region has close to 100% ice cover. The middle part has higher backscatter (i.e., light gray) but does not represent multiyear ice, since the Okhotsk Sea is covered only by seasonal ice. It is evident that the region is consolidated except for the presence of narrow leads, oriented in different directions. The relatively high backscatter is likely caused by uneven surface topography, primarily due wind-induced dynamics that cause rafting and ridging. Also, the ice cover may consist mainly of pancake ice floes, which usually have relatively high backscatter because of rough edges (Onstott 1992) that may be frozen together. It is also apparent that the ice edges are dominated by ice bands, some oriented perpendicular to the MIZ, while others are parallel to the MIZ. The different orientations reflect the different wind patterns and directions in different locations. A cyclone is also apparent near the bottom (high backscatter area), reflecting the presence of a polar low that can significantly impact the distribution of the sea ice cover, as will be discussed in the Chap. 8. The event may be associated with cold

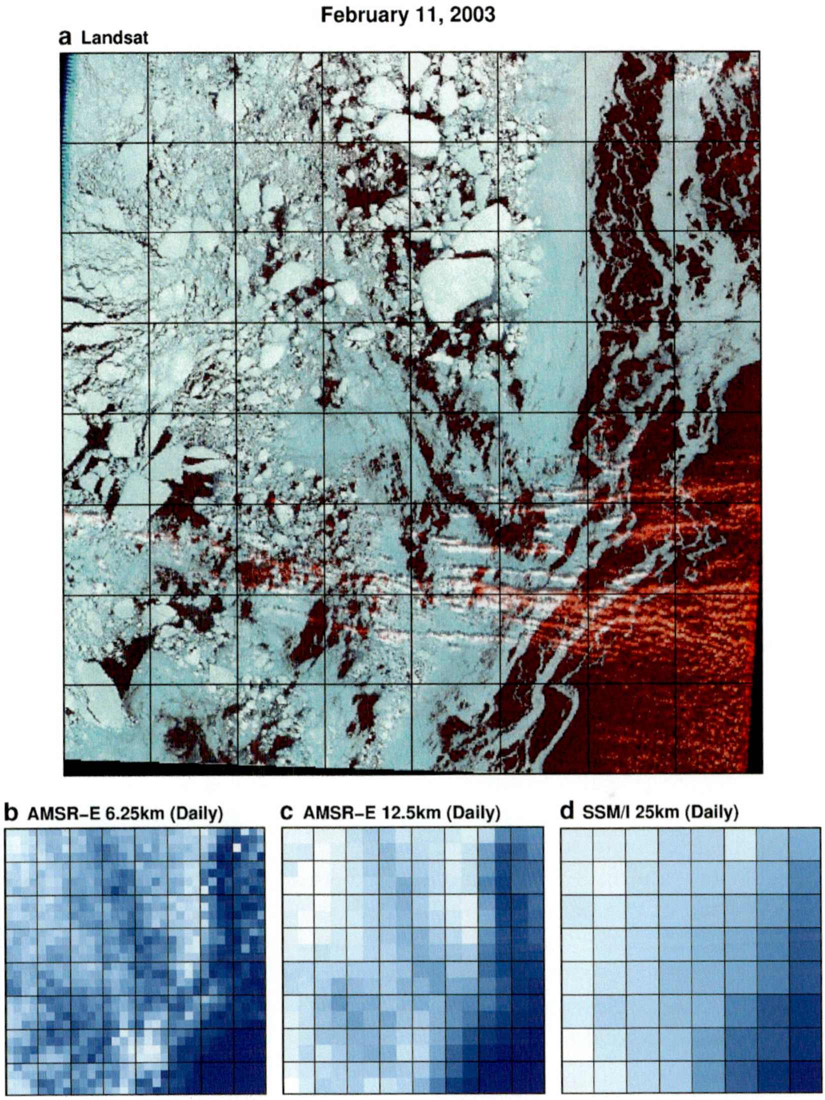

Fig. 7.6 Landsat image of sea ice in the sea of Okhotsk and near simultaneous EOS-Aqua/AMSR-E images over the same region at (**a**) Landsat image of sea ice in the sea of Okhotsk and near simultaneous EOS-Aqua/AMSR-E images over the same region at (**b**) 6.25 km resolution using the 89 GHz channel only, (**c**) 12.5 km resolution using the Bootstrap Algorithm and (**d**) 25 km resolution using the Bootstrap Algorithm

air outbreaks that lead to the extraction of huge amounts of energy from the ocean and extremely severe weather conditions.

A Landsat image over the same general, but smaller region in the Okhotsk Sea on 11 February 2003, which is approximately the same period, is presented in Fig. 7.6a. A similar image was shown in Chap. 4 (Fig. 4.24), and we show another one to emphasize the value of such images taken under clear skies conditions for mesoscale studies. The image provides a good example of an ice cover in the marginal ice zone, which has been subject to a strong dynamic force and is in a relatively rapid growth stage. The presence of big floes (top left) indicates that there was originally a relatively thick consolidated ice cover in the region but was broken up by winds, tides, waves and other environmental forcing. The ice cover shown in the image is dominated by an almost continuous distribution of thinner ice types (light blue) consisting primarily of pancake ice. Since the image was taken three days after the ASAR image in Fig. 7.5 was taken, some mismatches in the location of the floes are possible. The Landsat resolution is not good enough to detect individual floes but the pancakes have raised edges and hence the relatively high backscatter seen in the SAR image in Fig. 7.5. The thinner ice type gets easily advected by wind at the ice edge (see right hand side of image), forming bands of ice, which, in this case, are oriented almost parallel to the ice edge. Because the surface temperature in the region was generally sub-freezing as observed from a ship in the region (Nakayama, personal communication, 2003) during the period, the dark regions between bands were mainly covered by grease ice.

The mesoscale characteristics of the ice cover are reproduced to a limited extent by passive microwave data. Ice concentration maps derived from EOS-Aqua/AMSR-E and SSM/I data are presented in Fig. 7.6b–d and it is apparent that sensor resolution can make a big difference on the ability to evaluate the mesoscale characteristics of the ice cover. Figure 7.6b shows an AMSR-E ice concentration map, using the 89 GHz data gridded at 6 km, while Fig. 7.6c shows ice concentrations from the standard sea ice algorithm that utilizes AMSR-E 18 and 36-GHz data and gridded at 12.5 km resolution. The image shows compatibility of microwave with the Landsat data in that the major features are generally depicted in the AMSR-E data with the 89-GHz channel providing a closer match. It is apparent that if detailed spatial distribution is desired, including the detection of large floes and divergence regions, high resolution data are required. In this case, the 89 GHz channels could provide useful information for such studies. A word of caution is, however, needed because such a good matchup was obtained when the weather was good (i.e., clear skies condition), and the 89 GHz data can provide misleading information when weather conditions are not favorable, as discussed in Chap. 4. Nevertheless, the 89 GHz data can provide a good alternative, when appropriate, to Landsat, SPOT, and other high resolution data since they provide synoptic and global coverage. For studies that do not require the 6.25 km resolution, the 12.5 km data is currently the better option since they provide improved accuracy through the use of frequency channels that are not as vulnerable to atmospheric effects as the 89-GHz channel. The standard historical product with a resolution of 25 km, based on SSM/I data, is shown in the Fig. 7.6d and it is apparent that much of the mesoscale details are washed out by the coarse resolution. It may seem evident that the historical data is not providing a good representation of the ice cover but the results from direct correlation analyses between the Landsat data with each of the passive microwave data sets indicate that the 25 km data provides the highest correlation. This is, in part, due to the difference in the time of observation by the two sensors, and a slight mismatch due to dynamics would have a stronger impact on the correlation analysis when high resolution data are used. It is, however, useful to know that despite

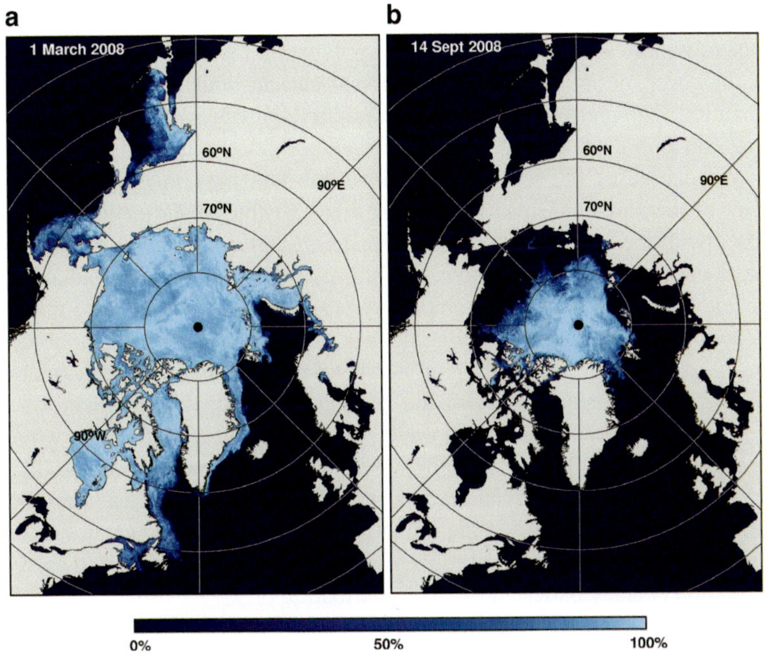

Fig. 7.7 Passive microwave high resolution (6.24 km) ice concentration images during (**a**) maximum extent on 1 March 2008 and (**b**) minimum extent on 14 September 2008 in the Northern Hemisphere using AMSR-E 89 GHz data

the coarse resolution, the standard passive microwave data provide accurate and consistent representation of the ice cover on a large scale.

Examples of a daily sea ice concentration maps of the entire Arctic region, as depicted by the 89 GHz data, are shown in Fig. 7.7. The images represent the Arctic sea ice cover during maximum and minimum extents in 2008 and it is apparent that mesoscale details about the sea ice cover is captured with greater fidelity than the standard version of the passive microwave data. Such data can be acquired on a twice-a-day basis, with the ascending orbits separated from the descending orbits, thereby, providing information about diurnal changes in the surface characteristics and heat fluxes in ice-covered regions. The data also provide some useful information along the ice edges and in the ice pack that are not available in the standard data set. In the winter image (Fig. 7.7a), the structure of the ice edges is represented in good enough resolution that sometimes the presence of ice banding and the orientation of the bands in the region is detectable. Areas of divergences are also apparent and it is usually possible to detect from the data the locations of big leads and relatively small polynyas. Large floes are also detectable. During the end of summer in September, the data are also useful in evaluating the state of the ice cover during this period. For example, the data at the ice edges (e.g., in the Beaufort Sea) show large areas of low concentration ice, indicating that the region will become mixtures of old ice and new ice during the onset of freeze-up. Low concentration ice cover is also apparent within the ice pack north of Ellesmere Island, suggesting the presence of a large area of divergence or a polynya in the region. The 89 GHz data should, however, be interpreted, as mentioned earlier, with atmospheric effects in consideration. The presence of a weather system (e.g., storm) is usually depicted as high ice concentration areas. Since such weather systems normally cause divergence that decreases the ice concentration, the data may misrepresent the true character of the surface. However, in most areas, distinct features that can only be associated with the ice cover are usually apparent, and in these regions where high resolution data may not be available, the information could be very valuable.

Accurate and consistent characterization of the ice edge, using passive microwave data is important because the ice edge defines the location of the marginal ice zone and the ice concentration in this region is critical in the estimate of the extent of the sea ice cover. The characteristics of the ice edge is seasonal and depend on many environmental factors such as winds, waves, tides, and temperature. In the winter, the ice edge region is usually dominated by pancake ice that grows in size and thickness through thermodynamic and dynamic processes. The accumulation and consolidation of frazil ice at the surface leads to the formation of pancakes, which are noted for their circular features and distinctly raised edges caused by the floes bumping each other. The formation characteristics and spatial distribution of sea ice as inferred from ship observations have been reported previously (e.g., Lange et al. 1989). In a special case where the wind is calm and the advance of sea ice is dictated mainly by thermodynamics, the structure of the marginal ice zone is almost predictable. During ice growth stages, the ice edge in this scenario would be represented by an almost continuous distribution of small pancakes. These pancakes usually grow in size by sticking to each other, about five or more of them at a time. The marginal ice zone would, thus, normally look like a progression of pancakes of different sizes with the smallest ones near the open water region and the biggest ones near the consolidated ice region. In most cases, the distribution of sea ice at the MIZ would be strongly influenced by winds and waves. In spring and summer, the ice edge usually consists of ice bands that broke up from consolidated ice regions and advected toward the open seas. The breakup can be caused by tidal effects, waves, and/or wind. The bands would be from hundreds of meters to several kilometers apart, depending on location, wind strength and direction, wave amplitude, and sea surface temperature.

To illustrate how the MIZ is represented in the visible and passive microwave data, Fig. 7.8 shows sea ice cover images from MODIS, AMSR-E, and SSM/I on 29 June 2005. During clear skies conditions, MODIS narrow-band albedo data at 0.6 μm with a spatial resolution of 250 m provide a good representation of the sea ice cover and the location of the ice edge because of the large contrast in the albedo of sea ice and open water (Allison et al. 1993). The MODIS data show detailed characteristics of sea ice cover and its mesoscale distributions both near the ice edges and the inner pack. Comparing this image with the AMSR-E and SSM/I passive microwave images, it is apparent that all three images provide similar representation of the sea ice cover. The AMSR-E data at 12.5 km resolution provide better representation of many of the features of the ice cover observed by MODIS than SSM/I, indicating the advantage of high resolution data in identifying ice edges, leads,

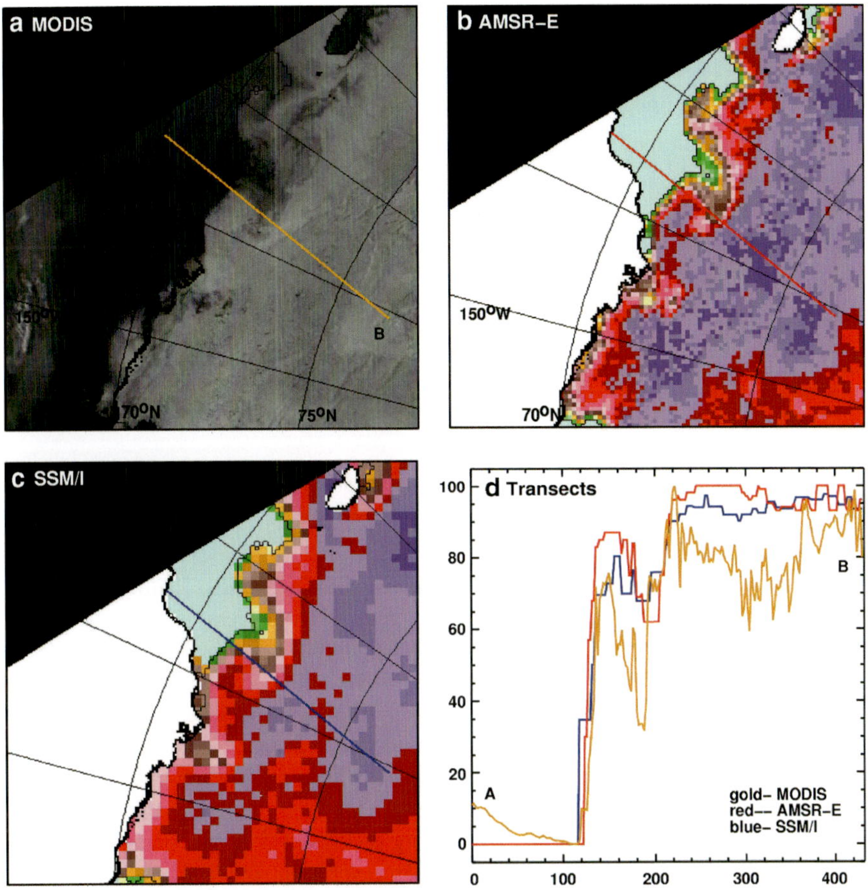

Fig. 7.8 Sea ice cover in the Beaufort Sea on 29 June 2004 as observed by (**a**) MODIS; (**b**) AMSR-E; and (**c**) SSM-I data. Radiance and ice concentration data along a transect line with end points A and B across the Marginal Ice Zone are presented in (**d**). The different sensors provide approximately the same location of the sea ice edge. From Comiso and Nishio (2008) with permission from the American Geophysical Union

polynyas, and other characteristics of the sea ice cover. For a more quantitative comparison, the values from the three images along a transect, defined by a line from A to B (see MODIS image), are presented in Fig. 7.8d. The MODIS radiances (in gold) are shown to be relatively low in the open water area (at A) and actually goes down because of calmer water near the ice edge and then increase substantially to much higher values within the pack because of the high albedo of sea ice. Some fluctuations within the pack are apparent likely due to the variability of the albedo of sea ice surfaces during this time on account of ice melt, meltponding, ice breakup and ice decay. The corresponding plot for ice concentrations from AMSR-E is shown in red, while that from SSM/I is shown in blue. It is apparent that the ice edge location (i.e., where the ice concentrations increase rapidly), as detected by AMSR-E, is closer to that identified by MODIS than the SSM/I data. The 15% ice edge detected by the SSM/I is further south than that detected by AMSR-E by about 12.5 km, while the AMSR-E is further south of MODIS value by less than 6 km. The plots thus basically provide quantitative information about the width of the marginal ice zone and the location of the ice edge. Although the interpretation of MODIS data at the ice edge is complicated by the presence of many new ice types that have different albedos, the location of the ice edge is discernible and appears to be in agreement with those from passive microwave data. The SSM/I usually finds the ice edge farther from the pack because of coarser resolution, side-lobes and smearing effects as discussed in Comiso and Nishio (2008). For time series studies of ice parameters such as ice extent and ice area, it is important to take this into consideration, making sure that biases associated with resolution are removed when combining data from different sensors such as SSM/I and AMSR-E.

7.2.2 Large-Scale Seasonal and Interannual Variability

During the last few decades we have observed unusual changes in the large scale characteristics of the sea ice cover (Comiso et al., 2008; Stroeve et al., 2007). It is the passive microwave satellite data that have provided the unique capability to study the long term variability of the sea ice cover. The advent of ESMR data in 1973 enabled the assessment and analysis of the large scale characteristics for 4-years (Zwally et al., 1983; Parkinson et al., 1987) but there were several gaps in the data and the accuracy was limited because it was only a one-channel sensor. The spatial extent and seasonality of the sea ice cover in the NH have been documented almost continuously since 1978 using SMMR and SSM/I data. Since June 2002, AMSR-E data, which provide higher resolution and improved accuracy, became available as well. The best data set that can be used to illustrate the seasonality of the sea ice cover as observed in recent years is likely the monthly averages that can be obtained from the 12.5 km AMSR-E data from June 2002 to the present. Although the data record is not as long as that of the SSM/I data, the AMSR-E data set represents contemporary sea ice cover at improved accuracy, especially at the ice margins. The spatial distribution of the ice cover for each month from January to December, as derived from AMSR-E data, is presented in Fig. 7.9. This set of images depicts the monthly evolution of the sea ice cover from winter months through spring, summer, and autumn, and back to winter. Similar monthly averages from 30 years of SMMR and SSM/I data show generally similar patterns, but are a little more extensive because of the decline in the ice cover since the 1980s, especially in the summer.

The set of maps in Fig. 7.9 shows how far south the sea ice typically advances in winter and how far north the sea ice retreats in summer in recent years. The peak ice coverage in any region occurs in January, February, or March, depending on location and atmospheric circulation. It is apparent that sea ice reaches its southernmost coverage at 44°N in the Sea of Okhotsk in the Pacific Ocean region and at 46°N in the

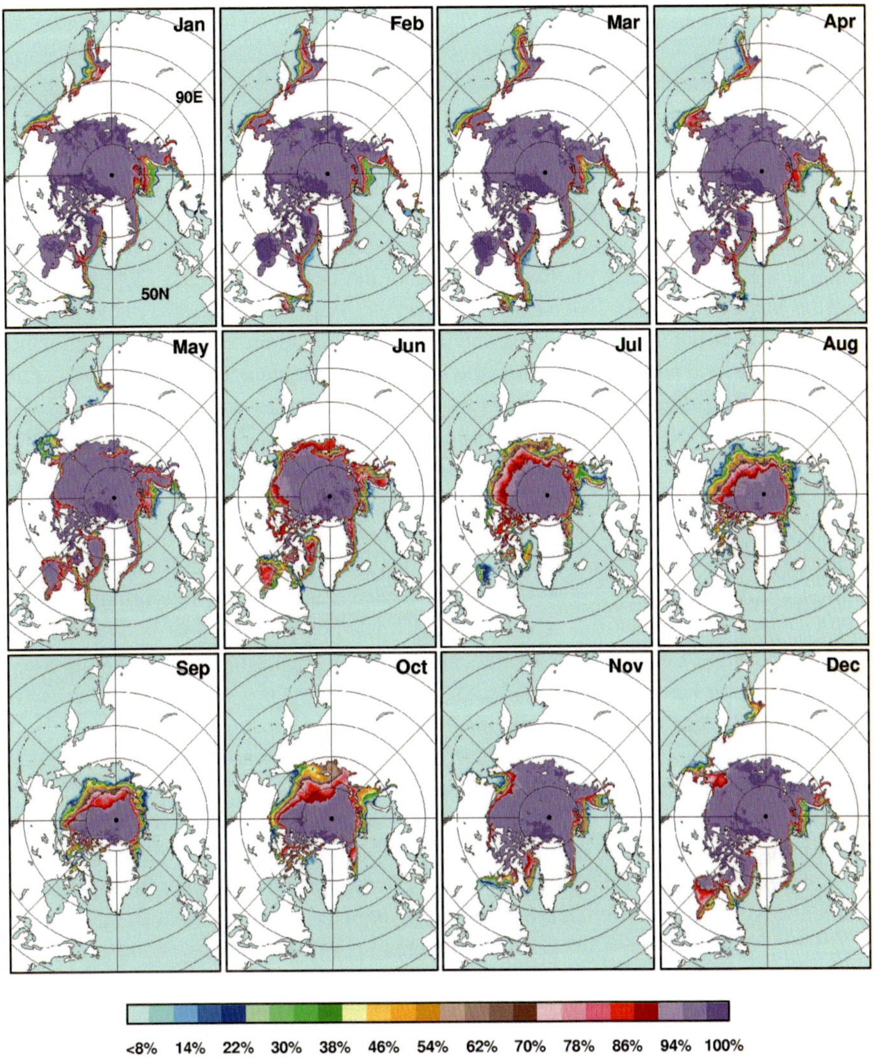

Fig. 7.9 Monthly averages (climatology) of sea ice concentration from January to December in the Northern Hemisphere using EOS-Aqua/AMSR-E data from June 2002 to May 2008

Labrador Sea in the Atlantic Ocean. The regions beyond the Arctic Basin in the Eastern (North Atlantic Ocean) sector have ice coverage confined mainly to bays and seas close to land because of the influence of warm surface water from the Gulf Stream in the North Atlantic Ocean. Similarly, a relatively warm Pacific Ocean at high latitudes keeps the ocean from getting frozen beyond the Bering Sea and the Sea of Okhotsk. During spring breakup, sea ice retreats from its maximum extent in the peripheral seas, starting from the southern region and into the Arctic Basin. The retreat rate is fastest in the Bering and Okhotsk Seas in April, in Hudson Bay in June, and in the Beaufort,

Chukchi, and Siberian Seas in July. By the end of summer, sea ice has melted in much of the bays and seas in the NH and what is left is confined in the Arctic Basin and parts of the Canadian Archipelago. Around the middle of September, only ice floes that are thick enough to survive the summer melt season remains. September is usually the month when the melt period ends and the ice freeze-up period starts. The monthly images from September to December show where ice growth is most prominent in the Arctic Basin as well as in the peripheral seas. The rate of growth varies depending on location and is shown to be relatively high in the Siberian Sea in October, in the Chukchi and Beaufort Sea in November, and in the Bering Sea, Sea of Okhotsk, and Hudson Bay in December. Sea ice also forms at the Baltic Sea, as shown in the January to April images, but in a narrow inlet, and because of uncertainties associated with land contamination and antenna side lobes, the ice cover is not included in the analysis. The series of monthly maps also shows that even before ice has completely covered the Arctic Basin, ice cover already starts to accumulate in bays (e.g., Hudson Bay) and peripheral seas. The Arctic Basin becomes fully ice covered in December and remains as such until June when ice at the peripheral seas is almost all gone.

Although the location of the ice cover has large interannual variability, the multiyear maps in Fig. 7.9 provide the means to identify the patterns of monthly changes in the ice cover during an ice season. The use of averages for each month during the last 6 years, as derived from AMSR-E data, instead of the 30-year averages from SMMR and SSM/I data, has the distinct advantage of representing the contemporary ice cover, which has changed considerably from earlier years, especially in the summer. The maps are generally useful in identifying the location of rapid growth in autumn and winter and ice retreat patterns in spring and summer. In addition, they also provide the means to identify the location of areas where ice concentrations are consistently very high and those that are consistently very low. This includes identification of typical locations of the ice edge, and the approximate size of the marginal ice zone during different months of the year. Because of the multiyear averaging, the gradient of ice concentration at the ice edge, as depicted in these climatology maps, is usually high when the ice edge does not change much from year-to-year but relatively low when the interannual variability of the ice edge location is high. Thus, the multiyear maps may not provide the typical ice edge gradient but provide information about interannual variations of the ice edge in various regions. The gradients at the ice edges are normally much higher when daily averages of ice concentration (e.g., see Fig. 7.11) are used nstead of the multiyear averages. Generally, however, the maps depict typical seasonal patterns that can be useful in scientific and commercial navigation and field program projects.

Plots of monthly averages of ice extent, ice area, and ice concentration for each month for the entire NH, from November 1978 to December 2008, as derived from SMMR and SSM/I data, are presented in Fig. 7.10a–c, respectively. The SMMR and SSM/I ice concentration data have been processed to be compatible and consistent with AMSR-E data during the period of overlap as described by Comiso and Nishio (2008). It is also the same data set that was used to generate the climatological averages shown in Fig. 7.1. The distributions in Fig. 7.10 for both ice extent and actual ice area clearly show the large seasonality of the sea ice cover and how they fluctuate from one year to another. The peak values are shown to have much less interannual variability

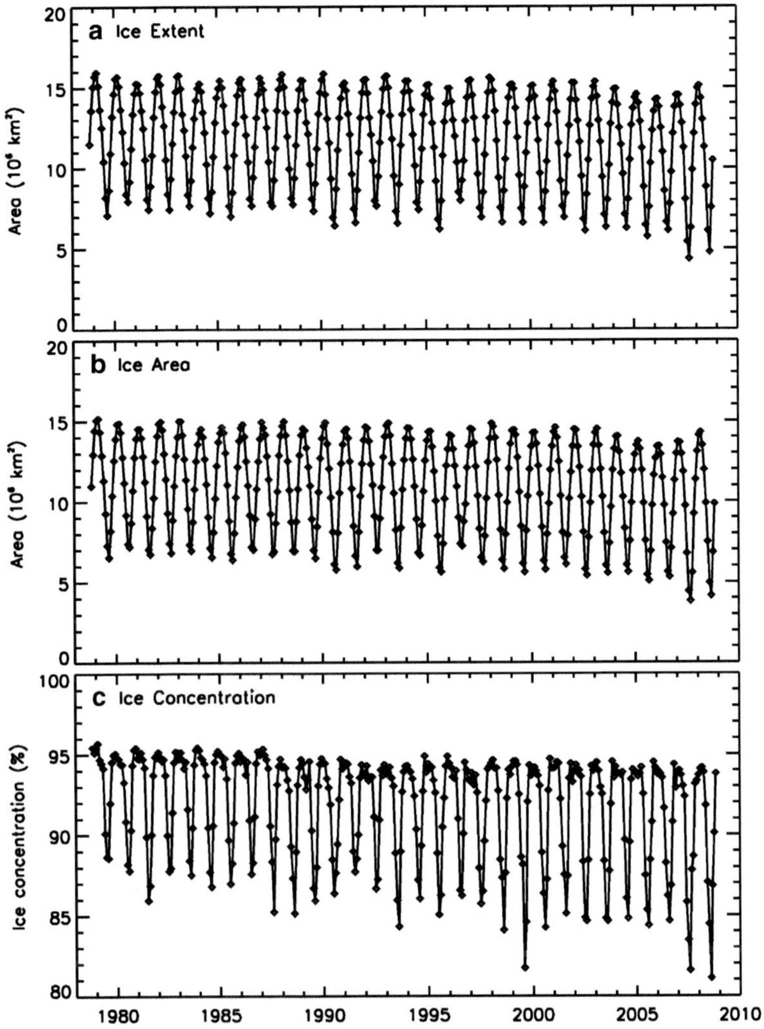

Fig. 7.10 Plots of monthly averages (**a**) sea ice extent; (**b**) sea ice area; and (**c**) sea ice concentration for each month from November 1978 to December 2008 in the NH using SMMR and SSMI data

than the minimum values, especially in the 1990s. Relatively low values in the maximum extent occurred in 1996, 2005, 2006, and 2007, compared to previous years, while low values in minimum ice extent are apparent in 1990, 1991, 1993, 1995, 1998, 1999, 2000, 2002, 2003, 2004, 2005, 2006, 2007, and 2008 (Fig. 7.10a). It is apparent that there are a lot more low values in the late 1990s and in the 2000s than in previous years, especially for minimum extent. In the ice area plot (Fig. 7.10b), the same phenomenon occurs with the minimum values during the more recent years appearing even relatively lower. The monthly averages of sea ice concentration are basically uniform in winter at about 95% (Fig. 7.10c), while the low values averaging about 86% have much larger variability. Unusually low average ice concentrations occurred in the

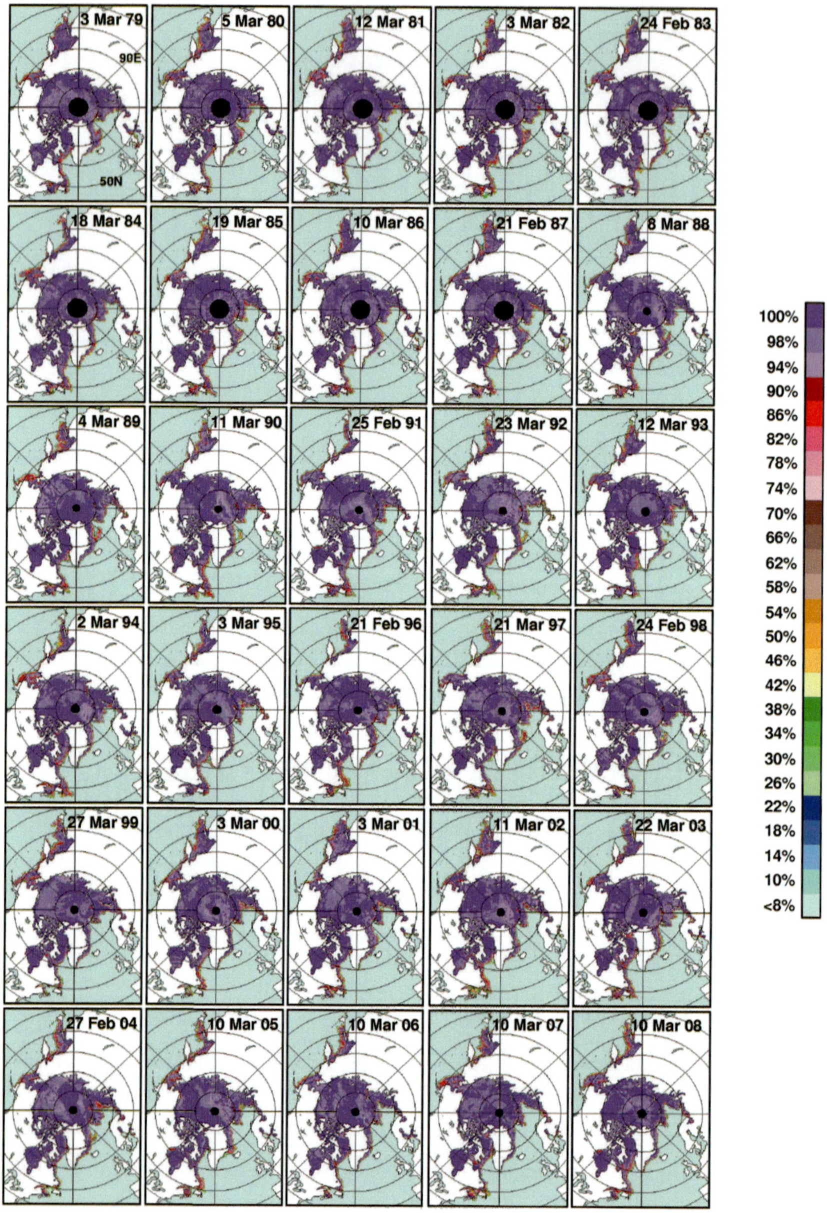

Fig. 7.11 Daily averages of sea ice concentration in the Northern Hemisphere during maximum extent for each year from 1979 to 2008 as derived from SMMR and SSM/I data

summers of 1999, 2007 and 2008. It is apparent that the ice concentrations are generally lower in the recent years than in earlier years suggesting stronger winds and more divergence during the more recent years. The declining trend in ice concentration during the summer period is actually consistent with more storms that may be associated with a declining surface level pressure as reported by Walsh et al. (1996) and Yang et al. (2001). It is also apparent that high values in winter do not necessarily mean high values in the summer and viceversa. For example, high extents and areas in the winters of 1979 and 1990 were followed by abnormally low values in the subsequent summers, in part, because relatively cold winters are not necessarily followed by cold summers. This appears counter-intuitive since unusually cold winter months are usually associated with enhanced ice thickness that makes the ice cover less likely to melt completely in the subsequent summer. However, the summer may be unusually warm and changes in the atmospheric circulation and ice dynamics might cause more ice to be advected our of the Arctic region. Also, low summer values in 1980, 1990, 1991, 1993, and 1995 were followed by a recovery in the subsequent winter, which again may be a case of a warm summer being followed by a relatively cold winter. Note that the low ice extent in summer of 2002 was followed by a relatively low ice extent in the subsequent winter which is an opposite scenario. The apparent lack of correlation of the ice cover in one season with that of another season is an indication of the complexity of the system that affects the ice extent variability.

Atmospheric patterns in the Arctic have been studied extensively and have been postulated to be controlled primarily by the Arctic Oscillation (Thompson and Wallace 1998), as discussed in Chap. 5. The pattern of the Arctic circulation has also been reported to have periodic shifts in directions (Proshutinsky and Johnson 1997), and sometimes may cause the drift of the ice cover toward Greenland through Fram Strait to be accelerated (Kwok and Rothrock 1999). The AO has been shown to be correlated with the sea ice cover, with relatively high sea ice extents occurring when the AO-indices were negative while relatively low sea ice extents occur when AO-indices were positive. However, the AO-indices have been near neutral during the last decade despite a declining sea ice cover, and this has been interpreted as a shift in the phase of the AO (Overland and Wang 2005). Such phenomenon makes it very difficult to assess the specific role of the atmospheric circulation on the observed variability of the sea ice cover.

To gain insights into how the sea ice cover in the NH has been changing, color-coded ice concentration maps during ice maximum and minimum extents for each year from 1979 to 2007 are presented in Figs. 7.11 and 7.12, respectively. The corresponding images for 2008 have been presented previously in a different format and resolution in Fig. 7.7. In these maps, the circular area centered at the North Pole (shown in black) is an area not covered by the satellite sensor because of the inclination of the satellite polar orbit as described in Chap. 3. This is, in part, limited by the swath width of the imaging sensor, which varies from one sensor to another. The area of missing data is shown to be significantly bigger in, SMMR than in SSM/I data and is almost negligible in AMSR-E data (see Fig. 7.7). The set of images during maximum extents shows how far south the seasonal ice cover gets during the various winters and how the areal coverage in different regions changes from 1 year to the other during the winter period. The images show very similar ice cover patterns for the different years, but significant interannual variabilities are apparent, reflecting interannual changes in atmospheric and oceanic conditions as described in Chaps. 5 and 6.

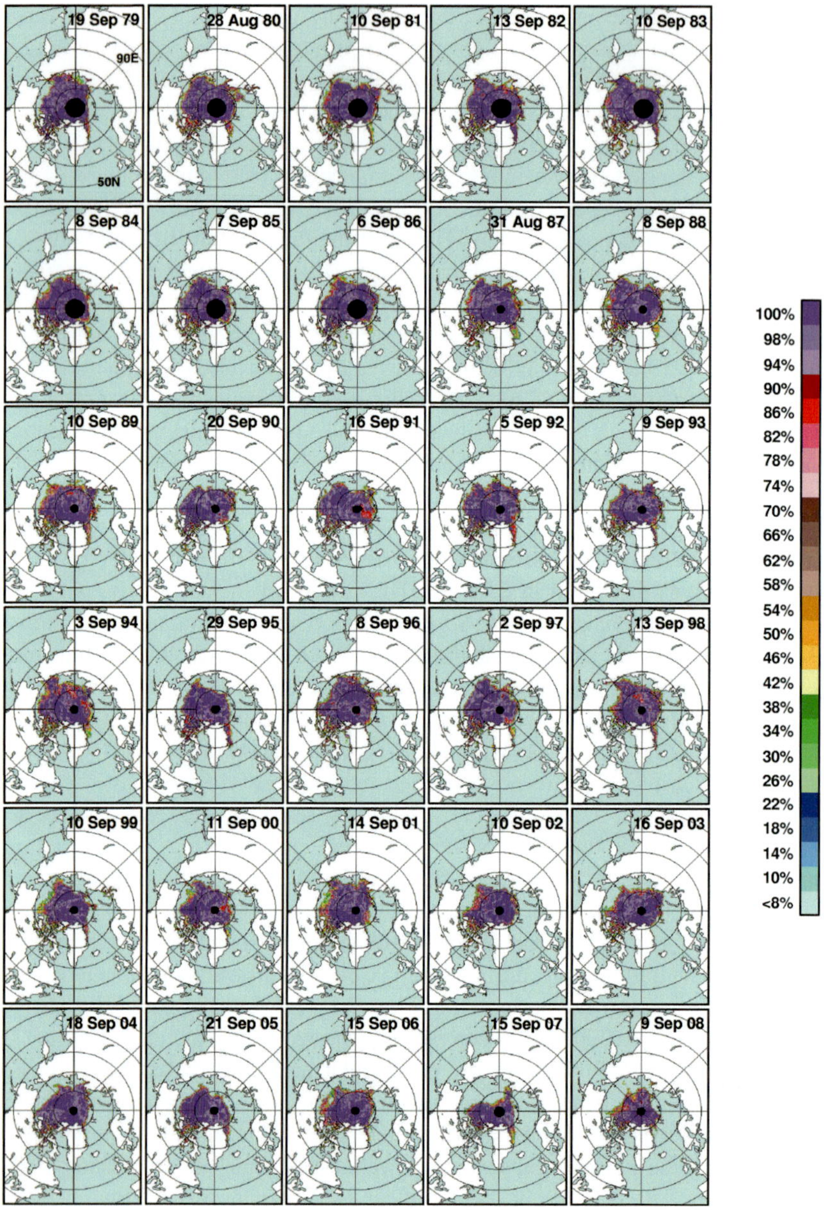

Fig. 7.12 Daily averages of sea ice concentration in the Northern Hemisphere during minimum extent for each year from 1979 to 2008 as derived from SMMR and SSM/I data

The ice-covered regions at relatively low latitudes are of special interest because they may provide more information about a changing climate than those at higher latitudes. For example, it is apparent that the most extensive ice cover in the Okhotsk Sea (top left) occurred in 1979, gradually decreasing to an anomalously low value in

1984, followed by a recovery up to 1990 and then a decrease to another low in 1991. The AVHRR temperature anomaly record, discussed in Chap. 6, shows anomalously high temperatures in the region in February 1984 and 1991, suggesting a strong influence of surface temperature on the sea ice cover during the two periods. A similar sea ice pattern occurred in subsequent years with a strong recovery in 2001, and then low values in subsequent years followed by a mild recovery in 2007. The temperature anomaly maps in Chap. 6 (i.e., Fig. 6.3) also show some cooling in 1987, 1988, and 2001, indicating temperature influence on the Okhotsk Sea ice cover as well. Comparing the sea ice cover in the region in the 1980s with those in the 1990s and 2000s, it is apparent that the ice cover in the region is slowly declining. While the trend is likely caused, in part, by the influence of surface temperature, there are also extra-polar phenomena, such as the ENSO and the Pacific Decadal Oscillation (PDO), that may contribute to the decline. Note that during low ice years, the sea ice cover is concentrated more along the coastal regions instead of the northern regions suggesting significant oceanographic and bathymetric effects on the distribution of sea ice. A similar phenomenon occurs at the Gulf of St. Lawrence (bottom left), but the latter is out of phase, with consistently more ice cover from 1982 to 1997 than the subsequent period except in 2003. Ice areas in the Greenland Sea also vary substantially from year-to-year with the variability of the ice extents usually associated with the formation of the Odden, as will be described in Chap. 8. Relatively low extent of sea ice in the winters of 2005, 2006, and 2007 has been reported previously (Comiso 2007), and mainly in the Bering Sea, Baffin Bay, and the Barents Sea. A slight recovery in these regions is apparent in 2008, which may, in part, be caused by a global cooling associated with a relatively strong *La Niña* during the time period. The maximum ice cover occurred mainly in early March, but there were other years when it occurred in February (i.e., in 1983, 1991, 1996, 1998, and 2004) and some years (i.e., 1997 and 2003) when it occurred in late March. Although the exact date of maximum extent varies significantly from one year to another, a trend toward an earlier date to indicate warming is not apparent.

The ice cover at the end of the summer is of interest because it is more vulnerable to climate change than the winter ice cover. In the summer, the surface temperatures are near or at melting temperatures, and even a slight warming during this period could cause a significant change in ice extent. In contrast, winter surface temperatures in the Arctic are tens of degrees lower than freezing temperatures and a slight warming may have a much smaller impact on the extent of the ice cover. The ice that survives the summer is usually called the perennial ice cover, which consists mainly of thick multiyear ice floes that are the mainstay of the Arctic sea ice cover. The multiyear ice floes in the Arctic are historically very thick, averaging about 3 m or more, as inferred from submarine data (Wadhams and Comiso 1992). The perennial ice has had a strong survival rate because of its thickness, and statistical studies have shown that the residence times of the floes can be as long as seven years (Colony and Thorndike 1985). The perennial ice cover has also been relatively extensive historically, with the extent being usually about half of the maximum extent of sea ice and resides mainly in deeper parts of the Arctic Ocean (Parkinson et al. 1987; Gloersen et al. 1992; Johannessen et al. 1994). The ice survives in part because it has been protected from the upwelling of warm water below the mixed layer by a cold and stable ocean layer, called halocline, as described in Chap. 2. The state of the perennial ice cover is important to monitor because, in addition to Greenland, it is the key contributor to the ice-albedo feedback, which causes the amplification of climate signals in the region. It also dictates the average ice thickness distribution of the ice in the Arctic basin, which determines the ability of the perennial ice to survive the summer. A drastic reduction of the perennial ice can affect the Arctic Ocean circulation and can also profoundly alter the climate, primary productivity, and ecology of the region.

The spatial distribution of the perennial ice is approximately represented by the ice concentration data during the minimum extent, as discussed in Comiso (2002). To illustrate how the spatial distribution of the perennial ice cover has been changing, we present daily ice concentration maps during the ice minimum for each year, from 1979 to 2008 (Fig. 7.12). The maps show large year-to-year changes in the location and extent of the ice cover before the ice growth season gets started. Note that not all the ice cover in these maps are perennial ice since the end of the melt period does not occur simultaneously in all areas of the Arctic. Some new ice may have started to form in the inner zone while ice may still melt in the outer zone. The perennial ice cover, as defined, is not really "perennial," since the ice floes do not last that long because of ice dynamics and they can be flushed out of the Arctic basin. Every year, a large fraction of these floes gets advected through Fram Strait to the Greenland Sea and the Atlantic Ocean where they melt. Large variability in drift velocities has been observed and the loss of in ice mass through Fram Strait has been quantified to be about $1\times10^6\,km^2$ (Vinje and Finnekasa 1986; Kwok and Rothrock 1999). The perennial ice can increase through unusually large production of second-year ice in the Arctic basin, which is usually preceded by the formation of thick first-year ice during the previous winter/growth period. Thick first-year ice may be the result of abnormally cold conditions during growth period or abnormally stormy weather conditions that lead to thicker ice through rafting and ridging. Ability to survive the summer melt depends on many factors, including thickness, sea surface temperature, and environmental forcing. Interannual changes in the ice cover are apparent in Fig. 7.12, starting with large open water areas in the Chukchi, Siberian, and Laptev Seas in 1989 and 1990. Also, after an extensive ice cover in 1996, the open water area in the Arctic basin started to increase in the summer of 1997 and got considerable attention with the advent of anomalously large open water area in the Beaufort and Chukchi Seas in 1998. Except for a slight recovery in 2001, the areal coverage of the perennial ice has been consistently low during the last decade. It is also apparent that the sea ice cover in 2007 and 2008 is much less extensive than in other years as also indicated in the plot in Fig. 7.10.

The changes in the ice extent and ice area during the yearly ice maximum and minimum are quantified in the plots presented in Fig. 7.13. The yearly fluctuations for the maximum ice extents and areas are relatively large, but negative trends are apparent, with the trends being 2.2 ± 0.4%/decade and 2.5 ± 0.4%/decade for ice extent and ice area, respectively. The ice area is about 4.4% lower than the ice extent on the average, reflecting the changing fraction of open water or new ice in the pack. In contrast, the trends for minimum ice extent or perennial ice are 11.3 ± 1.7%/decade and 12.5 ± 1.6%/decade for ice extent and ice area, respectively, representing almost five times the trends in the maximum ice cover. A rapid decline

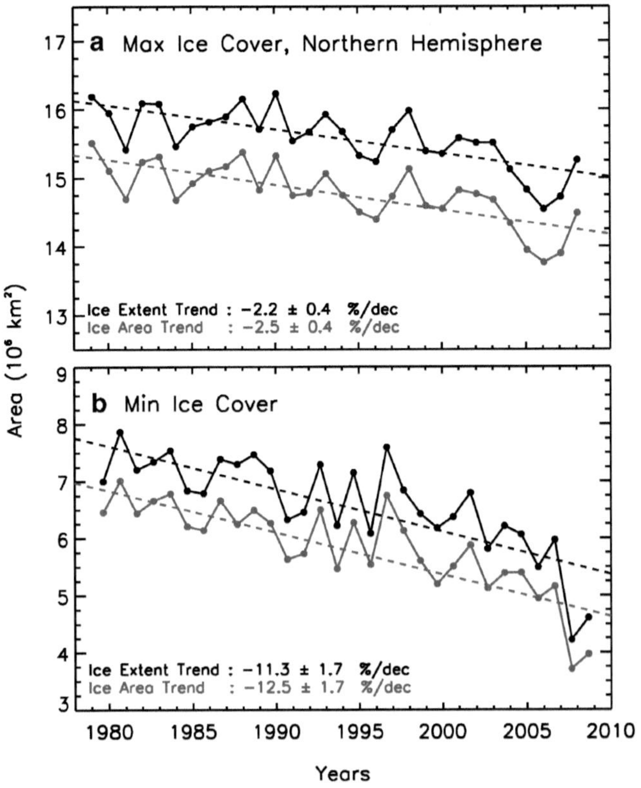

Fig. 7.13 Plots of sea ice extent and ice area during (**a**) maximum and (**b**) minimum ice extents. Dash lines represent the results of linear regression that provide the trend values

in the perennial ice has been previously noted by Comiso (2002) but not quite as high as these values, indicating an acceleration in the decline as described by Comiso et al. (2008). The trend has been shown by Stroeve et al. (2007) to be three times higher than predicted by numerical models. A continuation of such a decline is expected, especially because of observed warming in the region and ice albedo feedback effects. A modeling study has indicated that a thinning of the Arctic ice of about 1.3 m may have occurred from 1988 to 2003 primarily on account of ice albedo-feedback (Lindsay and Zhang, 2005). Such effect may have also been responsible for the precipitous drop in the perennial ice extent and area in 2007.

7.2.3 Monthly Sea Ice Anomalies and Trends

The satellite passive microwave data provide statistically consistent and relatively accurate ice concentration maps of the ice cover on a daily basis. However, estimating the anomalies for the ice maximum or minimum for each year is problematic because the dates for maximum or minimum are different for each year and there

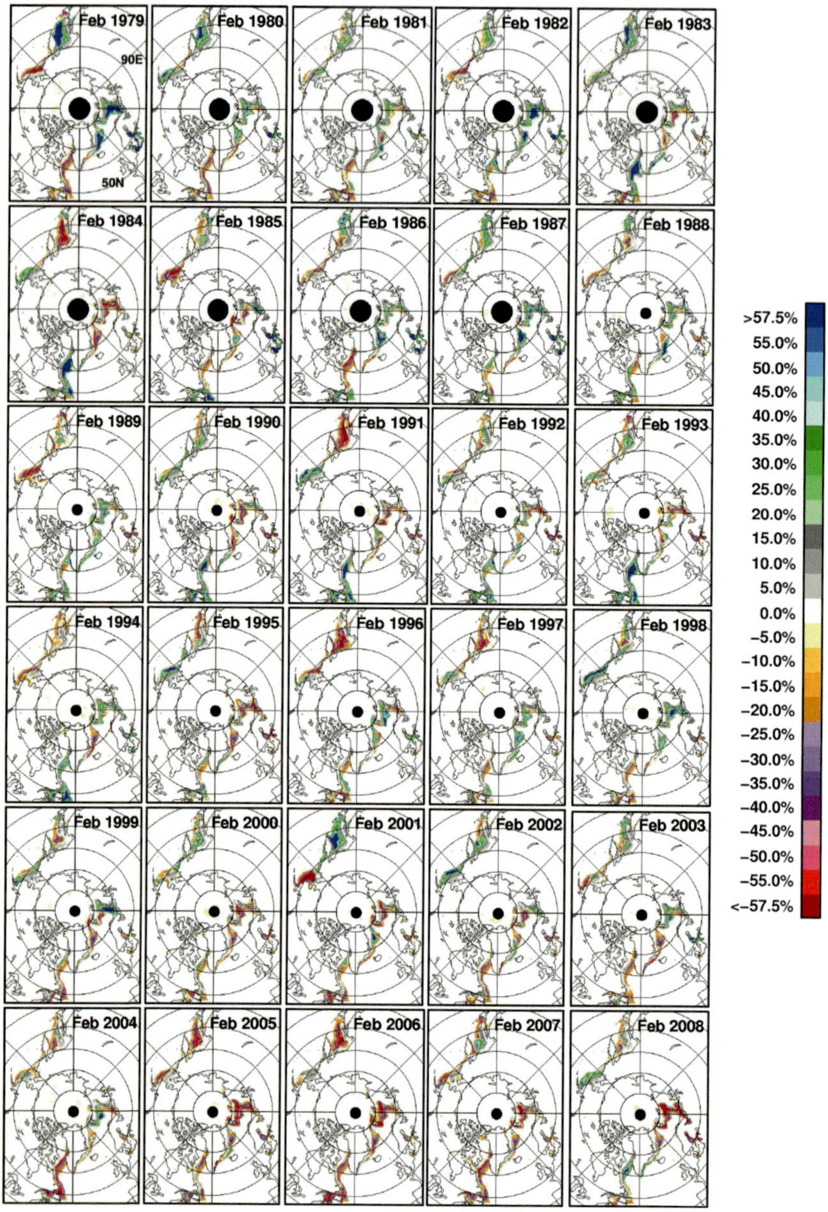

Fig. 7.14 Monthly anomalies of sea ice concentration in the Northern Hemisphere as derived from SMMR and SSM/I data for each winter month (February) from 1979 to 2008

are large interannual fluctuations in the daily data. Therefore, to quantitatively assess the magnitude of the yearly changes during winter and summer, we chose to use monthly instead of daily anomalies. In particular, we use monthly anomalies during the peak of winter (i.e., February) and also during the end of summer (i.e., September).

The monthly data also provide insights into the variability of the ice cover during maximum and minimum extents, as indicated in Fig. 7.13.

Color-coded monthly anomaly maps of ice concentration in a winter month (i.e., February) of each year, from 1979 to 2008, are presented in Fig. 7.14. Each anomaly map is generated by taking the difference in every data element of the monthly ice concentration and the corresponding monthly climatological ice concentration using data from 1979 to 2008. Positive anomalies (indicating more ice cover) are shown in the maps as grays, greens, or blues, while negative anomalies (indicating less ice cover) are in oranges, purples, and reds. During winter, the changes in the Central Arctic basin are only minor because it is generally all frozen up with near 100% ice cover. The largest interannual variabilities are, therefore, most apparent in the seasonal ice regions. The large year-to-year changes depicted in the eastern side (e.g., Barents Sea) are likely caused by interannual changes in wind circulation or ice dynamics that makes the general location of the perennial ice cover (and therefore, that of the seasonal ice) different from 1 year to another. The region is shown to have positive anomalies when the perennial ice shifts to the eastern side (e.g., 1979, 1982, 1998) and negative otherwise (e.g., 1984, 1990, 1995, 2000, 2005, 2006, 2007). The ice cover anomaly in the region was persistently negative from 2005 to 2007, in part, because of the observed decline in the Arctic winter sea ice cover (Comiso 2006). In the peripheral seas, positive anomalies in a certain area are oftentimes accompanied by negative anomalies in adjacent areas and vice versa. This phenomenon is apparent between the Okhotsk and Bering Seas in Feb. 1979, 1982, 1984, 1991, and 2001 and between the Baffin Bay and the Greenland Sea in Feb. 1979, 1984, 1986, 1992, 1994, 1997, and 1998. Similar patterns are also apparent between the Greenland Sea and the Barents Sea. It is interesting that in some years the signs of the anomalies are similar between these regions. However, since 2004, they are practically all negatives except for some positives in 2008, reflecting a slight recovery as described earlier. While the interannual increase or decrease in ice extent and area can be easily quantified, as presented in Fig. 7.13a, the locations of the anomalies that caused such increase or decrease are shown to be quite unpredictable and could happen anywhere in the Arctic region. This suggests that the anomalies in winter are mainly the effect of changes in atmospheric condition and wind circulation.

The sea ice cover in the peripheral seas apparently goes through different phases of change during different years. There are some years when changes in the ice cover in some seas such as the Sea of Okhotsk and the Bering Sea are in phase with each other while there are other years when they are out of phase with each other. The sea-saw and interconnectivity are, in part, influenced by some oscillatory patterns such as the Arctic Oscillation (Thompson and Wallace 1998) and the Pacific Decadal Oscillation. The atmospheric circulation in the Arctic basin also go through some phases, and although it is generally cyclonic it becomes anti-cyclonic, in other years as described by Proshutinsky and Johnson (1997). The impact of the increasing greenhouse gases in the atmosphere has been expected to be most apparent in the winter period when much of the region is in darkness and infrared radiation dominates. During the 2004 to 2007 period the anomalies were practically all negatives, which meant a consistent decline. This pattern is suggestive of a direct influence of warming due to greenhouse gases as cited in Comiso (2006). A slight recovery is shown in the 2008 anomaly map,

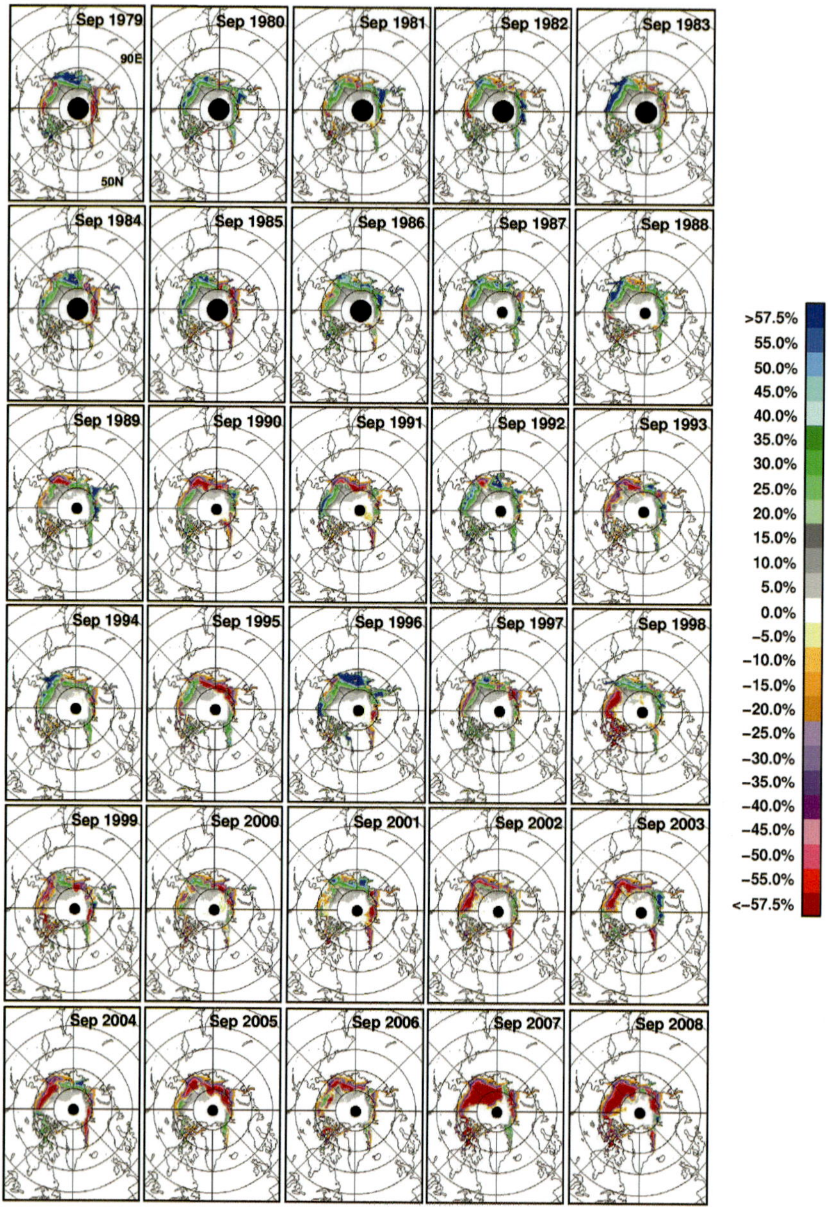

Fig. 7.15 Monthly anomalies of sea ice concentration in the Northern Hemisphere as derived from SMMR and SSM/I data for each end of summer month (September) from 1979 to 2008

with the Bering Sea and Baffin Bay showing some positive anomalies. The recovery is likely associated with a global cooling due to a persistent La Niña during the year. Changes in climatic condition are oftentimes unpredictable and the interpretation of sea ice cover data, especially during the winter period, can be very difficult.

Anomaly maps for September, depicting changes in the ice cover at the end of the melt period, are presented in Fig. 7.15. Sea ice, during this period, is confined to the Arctic basin, and the anomalies are significant mainly near the edges of the perennial ice cover. A normal ice year would be represented by approximately equal areas of negative and positive anomalies. In 1979, an alternating sequence of negative and positive anomalies around the Arctic is apparent, starting with the negatives near Alaska, positives near Siberia, negatives in the Barents Sea, and finally positives in the Greenland Sea and the Canadian Archipelago. It turns out that such configuration is not so typical since an examination of the other maps indicates a more random distribution with the possible exception of those in 1996 and 2001. The series of images shows a dominance of positive anomalies in the 1980s, some large negative anomalies in the 1990s, and a dominance of negative anomalies in the late 1990s and 2000s. The negatives are mainly in the East Siberian Sea and Laptev Sea regions in 1989, 1990, 1991, 1993, and 1995, but starting in 1997, large negative anomalies are apparent in the Western region (i.e., Chukchi and Beaufort Seas) except in 2001. During the period from 2002 to 2007, much of the western and eastern sides show large negative anomalies, culminating in a huge anomaly, representing a precipitous decline in 2007 that was almost duplicated in 2008.

The September anomaly maps clearly show a changing sea ice cover the trend of which is most evident during the last decade of data. They also provide insights into the interannual variability of the perennial ice cover from 1979 to 1997 which may be considered as normal years for the perennial ice and compare them with those of the more recent decade. Although the maps during this period are slightly biased on the positive side because of strong negative anomalies in the 2000s, some interesting patterns are apparent. In the Barents Sea, the negative anomalies in the region in 1979 were followed by positive anomalies in 1980, 1981 and 1982, which, in turn, were followed by negative anomalies in 1983, 1984 and 1985 and again back to positive anomalies after that. Such periodic changes may be associated with changes in atmospheric circulation that is basically controlled by the Arctic Oscillation in the region. Thus for some years, the perennial ice may be advected toward the western regions, and for some years, the advection goes in the opposite direction. Similar patterns also occurred in the Laptev/Siberian Seas region with the positive anomalies in 1979 being followed by slightly negative anomalies from 1980 to 1983, and then positive anomalies from 1984 to 1987, and again by negative anomalies from 1988 to 1991. Such periodic patterns are no longer evident in the anomaly maps from 1992 to 2001 except for a big shift from highly negative in 1995 to highly positive in 1996. From 2002 to 2008 the anomalies were overwhelmingly negative everywhere except for some positives in 2003 and 2004. It is apparent that the interannual variability of the perennial ice in the last decade is quite different from the earlier years of satellite observation.

To quantitatively evaluate interannual changes, averages of the monthly anomalies in extent for the entire hemisphere and various sectors in the Arctic region (following previous analysis by Gloersen et al. 1992; Parkinson et al. 1999; Comiso and Nishio, 2008) were estimated for every month from 1978 to 2008. Plots of the results are presented in Fig. 7.16. Each data point in the anomaly plots is the difference of the average ice extent for the month (see Fig. 7.10) and that of the

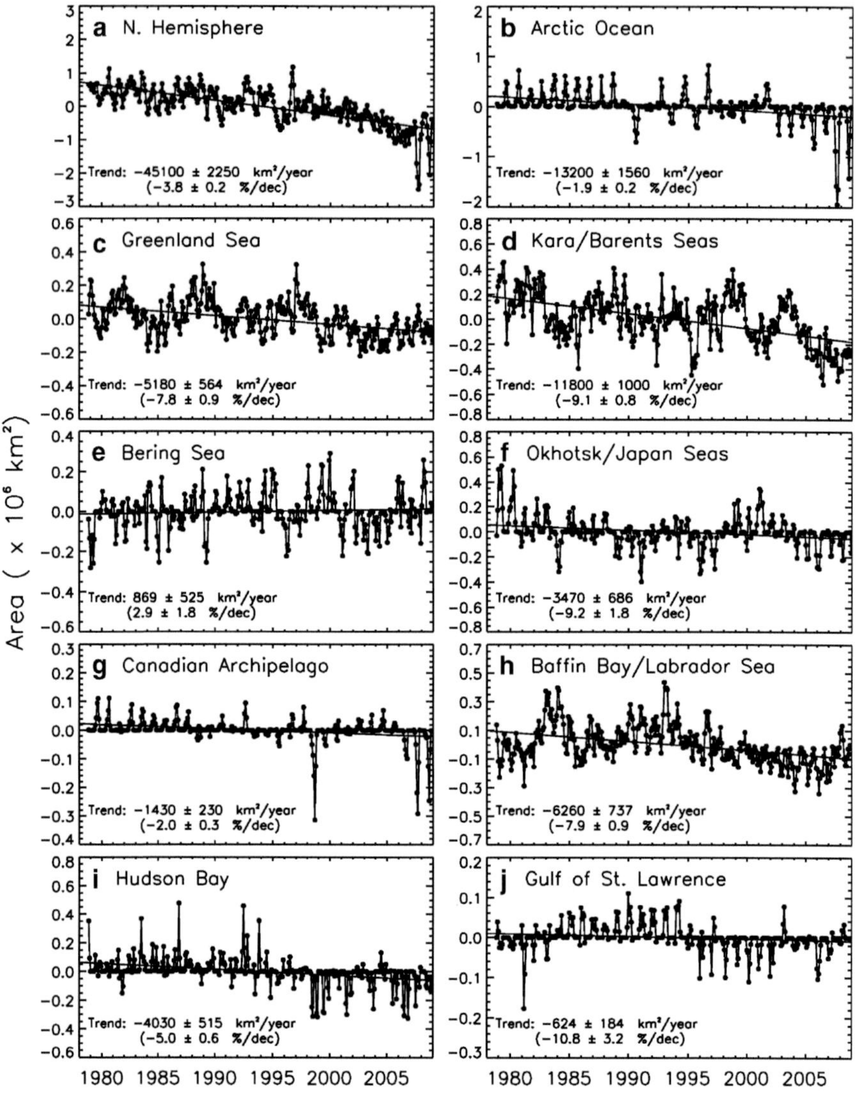

Fig. 7.16 Plots of monthly anomalies in the sea ice extents and trend lines for (**a**) entire NH and (**b**–**j**) the different Arctic sectors (see Fig. 9.10)

climatological average of the same month, based on the 30-year period presented in Fig. 7.1. The monthly anomaly data provide a means to assess more accurately (than the monthly data), the interannual variability and trends, since the large seasonality is subtracted. Statistically, they also provide higher statistical accuracy for trend analysis than yearly averages. The anomalies in ice extent and ice area again show coherence and reflect very similar characteristics. It is also apparent that the abnormally low and high values stand out better in these plots than in the monthly plots shown in Fig. 7.10.

The plot in Fig. 7.16a represents the overall interannual variability of the sea ice cover in the entire NH. The year-to-year changes are substantial but the trend is relatively moderate from 1978 to 1996 with some significant dips in 1990–1993 and a big recovery in 1996. However, from 1996 to the present, it is almost a linear decline. The overall trend is 2.2%/decade, but from the 1996 to the present, the trend is 8%/decade, suggesting that an acceleration in the decline of the ice cover has been occurring (Comiso et al. 2008).

Since the Arctic basin is basically covered entirely by sea ice most of the year, the anomalies in the Arctic sector (Fig. 7.16b) are mainly zero and become significantly different from zero only during the summer melt period. Note that the anomalies are generally positive in the 1980s and are mainly negative in the 1990s, except in 1992, 1994, and 1996. This is consistent with previous discussions on minimum extent that indicates declines in the summer ice cover. The plot, however, provide quantitative estimates about the magnitude of the changes in the perennial ice cover in the region from 1 year to another. The anomalies in the Greenland Sea sector (Fig. 7.16c) show more variability through the 30-year period. A cyclical pattern is apparent in the distribution for this sector, mainly as a result of a similar variability in the extent of the Odden, as discussed in Chap. 8 and as described previously by Comiso et al. (2001). In the Kara/Barents Seas the ice extent variability (Fig. 7.16d) is similar to that of the Greenland Sea but a cyclical pattern is not as apparent. In this sector, significant decreases are apparent in 1983, 1985, 1992, 1995, and 2000, while increases occurred in most other years. The sea ice anomalies in the Bering and Okhotsk Seas (Fig. 7.16e and f) have similar variability but with opposite trends, with the former having negative anomalies in the 1980s and mainly positive anomalies in the 1990s. In the Okhotsk Sea, the anomalies were on the average positive in the 1980s, except in 1984, and negative in the 1990s, indicating a rapid decline of ice in the region (Parkinson et al. 1999). There was a recovery, with positive anomalies being prevalent from 1998 to 2002, but after that, it has been again a consistently negative anomaly. In the Canadian Archipelago, the anomalies in extent were generally small except for the big drop in extent in 1998 and again in 2007 and 2008. These were the years when the region was basically ice free in the summer and when a record open water area was observed in the Beaufort Sea. In the Baffin Bay/Labrador Sea region, periodic fluctuations in the anomalies (Fig. 7.16h) were apparent in the 1980s, but there was a steady decline from 1993 to 2006 followed by a mild recovery in 2007 and 2008. In the Hudson Bay, there is not much variability except for the generally positive anomalies from 1978 to 1995 and the generally negative anomalies from 1995 to 2008. At the Gulf of St. Lawrence, there is a suggestion of periodicity with the anomalies being all negative up to 1983, followed by practically all positive anomalies up to 1995 and then almost all negative anomalies up to the 2008. It is apparent that there is a lack of coherence in the variability of the anomaly distributions at the different sectors. The peaks and the dips generally do not occur at the same time in the different regions, and periodicities, where they occur, are usually out of phase from each other. The lack of coherence is partly the result of a complex atmospheric system in the Arctic that produces different ice distributions in the different regions during different periods. Overall, negative trends of varying degree of significance are observed in all sectors, except at the Bering Sea sector, in which the ice cover have been increasing at the rate of 2.90±1.8% per decade. The regions where the ice cover has been declining the most are the Gulf of St. Lawrence, Kara/Barents Sea, Okhotsk/Japan Seas, Baffin Bay/Labrador Seas, and Greenland Sea with the trends being −10.8±3.2, −9.1±0.8, −9.2±1.8, −7.9±0.9 and −7.8±0.9% per decade, respectively.

To illustrate how the trends are spatially distributed on a seasonal basis, we present the results of trend analysis of ice concentration for the different seasons in Fig. 7.17. In winter, negative trends are very prominent in all areas except the Bering Sea where a cooling has been detected in the surface temperature data. The highest decline in ice concentration occurred in the Barents Sea, and Greenland Sea

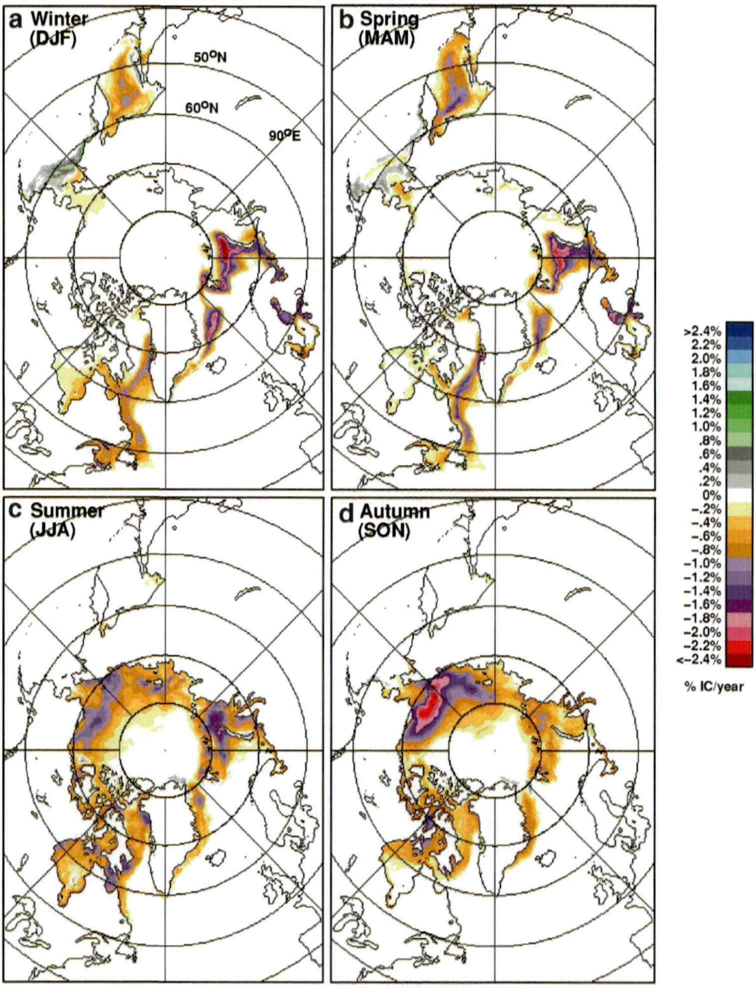

Fig. 7.17 Trends in sea ice concentration in the Northern Hemisphere as derived from SMMR and SSM/I monthly ice concentration data (November 1978 to December 2008) during (**a**) winter; (**b**) spring; (**c**) summer and (**d**) autumn

during this season. In the spring, the trends are very similar to that of winter, with the trend at the Okhotsk Sea being more negative, indicating that sea ice in the region has been melting earlier in the more recent period. In the summer, it is practically all negative trends in the pan Arctic region with the Beaufort Sea, Chukchi Sea, Kara Sea, and Barents Sea having the largest negative trends. In autumn, it is also negative trends throughout the Arctic Basin, with the Beaufort and Chukchi Seas, where the perennial ice has been rapidly declining, being the region of highest negative trends. Overall, the images in Fig. 7.17 indicate that the trends in the pan-Arctic region are overwhelmingly negative in all seasons.

7.2.4 Variability and Trends in Ice Thickness

Satellite data have been used successfully to assess the variability and trends in the extent and area of the sea ice cover as described above but to be able to do mass balance studies effectively, knowledge of the thickness distribution and how it changes with time is also needed. Much of our knowledge about the large scale thickness distribution of the sea ice cover in the NH has been inferred from submarine upward-looking sonar data (Wadhams 1997; Wadhams and Comiso 1992). The submarine data constitute tens of thousand of miles of data but because of the vastness of the Arctic region, they provide sampling of just a small fraction of the region. Comparative analysis of submarine sea ice thickness data in the 1960s with those of more recent years have indicated large negative changes in the average thickness of the Arctic ice cover (Rothrock et al. 1999; Wadhams and Davis 2000). Because the ice is so dynamic, interpretation of these measurements and trends are difficult especially in terms of temporal changes in ice thickness. However, the measurements serve to provide quantitative assessments of average thicknesses over different periods and insights into how such averages have been changing over time. The apparent decline in thickness is also consistent with the observed decline in the perennial ice cover.

Satellite observations of ice thickness has also been reported in recent years. It is not possible to measure the thickness from satellite directly but satellite sensors can be used to estimate the height of the freeboard. As discussed in Chapter 4, studies by Comiso et al. (1991) and Wadhams et al. (1992) using near simultaneous aircraft lidar and submarine upward-looking sonar data, demonstrated that the freeboard height can be successfully used to measure the thickness of sea ice. Using satellite altimeter data to measure the freeboard, Laxon et al. (2003) has reported that the average thickness from 1993 to 2001 has been changing at a relatively high variability and the changes are highly correlated to the length of the melt season. A map of the average thickness of Arctic ice from October 1993 to March 2001 as reported by Laxon et al. (2003) is shown n Fig. 7.18. The map shows that the general location of the thicker ice types are the regions north of the Canadian Archipelago and the Greenland Sea. A large part of the Arctic, however is not covered because of limitations in the coverage of the altimeter sensor. Data in the inner regions are currently available and are provided by ICESat laser altimeter data as described in Chapter 4. Recent reports by Kwok et al. (2009) actually show that there has been dramatic changes in the average

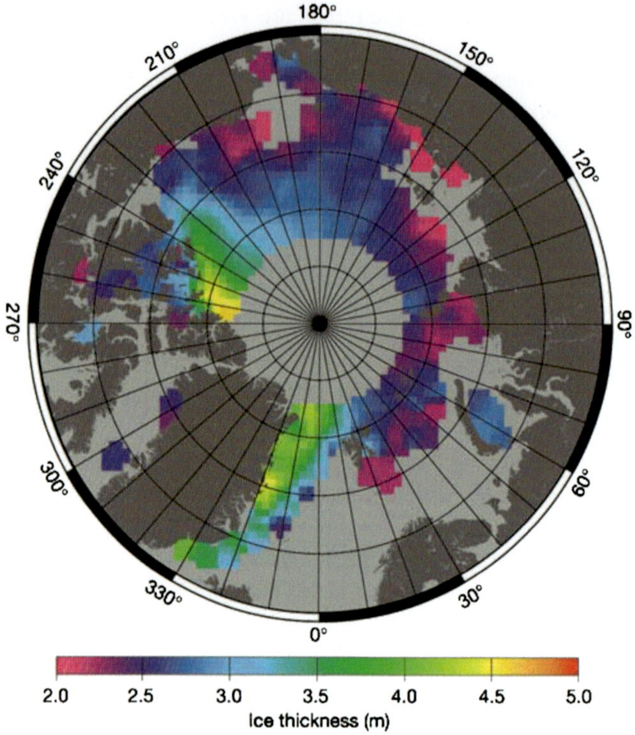

Fig. 7.18 Average thickness of sea ice in the Arctic during the winter period (October to March) from satellite altimeter measurements of freeboard height. The freeboard height was converted to ice thickness using fixed ice, snow and water densities and regional montly snow depth. From Laxon (2003) with permission from Nature

thickness of the Arctic sea ice cover during the period 2003 to 2008. There are uncertainties associated with these measurements due to the unknown density of the sea ice and the thickness of its snow cover. Also the record length of a few years is not suitable for trend analysis. But such observations are important and cannot be ignored in light of the observed changes in the extent of the Arctic sea ice cover.

7.2.5 Drift Characteristics

We have known from buoy studies (Colony and Thorndike 1985) that the Arctic sea ice cover is not static but rather a very dynamic entity with drift velocities that can be as high as several km/day. Knowledge about the drift characteristics of sea ice in the Arctic is important because it provides the means to assess the role of dynamics on the interannual variability of the sea ice cover and, especially, the perennial ice cover. For example, the reductions in the Arctic perennial ice cover in recent years have been attributed in part to the increase in ice flux at the Fram

Strait (Kwok and Rothrock, 1999) and the influx of Pacific Summer Water (PSW) into the Arctic basin (Shimada et al., 2006). The basic drift patterns are sometimes reversed because of changing atmospheric circulation and such reversals can be very favorable to the loss of Arctic sea ice, especially at the Fram Strait. Argos buoy data have provided the most accurate data available for tracking drift velocities. However, the number of buoys in the Arctic has been limited to only a few and is not quite enough to characterize drift patterns for the entire Arctic region. The capabilities for detecting drift patterns, however, have changed with the advent of satellite data. Using time series of such data, several techniques have been developed to track down sea ice features and make it possible to estimate how such features change in location from one time period to another. This was first implemented using SAR data, which provide resolution good enough to track down individual floes but because of the paucity in the coverage on a global scale, AVHRR data were used instead and with greater success. Persistent clouds were the key problems when AVHRR data were used and it was soon realized that passive microwave data provide good enough resolution to enable estimates of drift velocities at reasonable accuracy. Although the resolution of the passive microwave sensors is typically 25 km and the drift of sea ice is about 10 km/day, the velocity of the drift of sea ice has been possible to obtain at reasonable accuracy as explained in Chap. 4, Sec. 4.3.8.

To illustrate the patterns of ice drift and how they change on a month-to-month basis over a winter period, we present, in Fig. 7.19, ice drift data as derived by Liu et al. (1998), using passive microwave brightness temperature data for the period from November 2001 to April 2002. Data during the summer months are not included because of inability of the technique to detect the changes in the drift associated with movements of ice patterns. The technique takes advantage of unique distribution of brightness temperatures associated with ice and surface types, but in the spring and summer brightness temperatures become more uniform because of snow melt and the discrimination, and tracking of patterns becomes more difficult. As indicated in Chap. 2, the direction of the drift of sea ice floes is not identical to that of the wind because of many complications, including the effect of the Coriolis force, water current, water drag, tides, and surface tilts, as discussed in Wadhams (2000). The patterns for the average ice drifts are actually consistent with the drift of surface current as illustrated in Fig. 2.2. The image in November 2001 clearly shows the Beaufort gyre and a relatively strong advection along the Lomonosov/TransEuranic ridge through the Fram Strait and into the Greenland Sea. Similar but weaker patterns are apparent from the December 2001 to April 2002 images. In March 2002, advection of ice from the Lincoln Sea (North of Greenland) is also depicted. Such drift patterns are important to monitor and quantify because the region North of Greenland and the Canadian Archipelago is the area where thick multiyear ice floes are usually located. The loss of these thick floes would make the perennial ice which usually have several years residence time in the Arctic even more vulnerable. Such data can be very useful also for other applications, including the quantification of mass balance and the volume of sea ice that is exported on a monthly basis from the Arctic basin to the Atlantic Ocean.

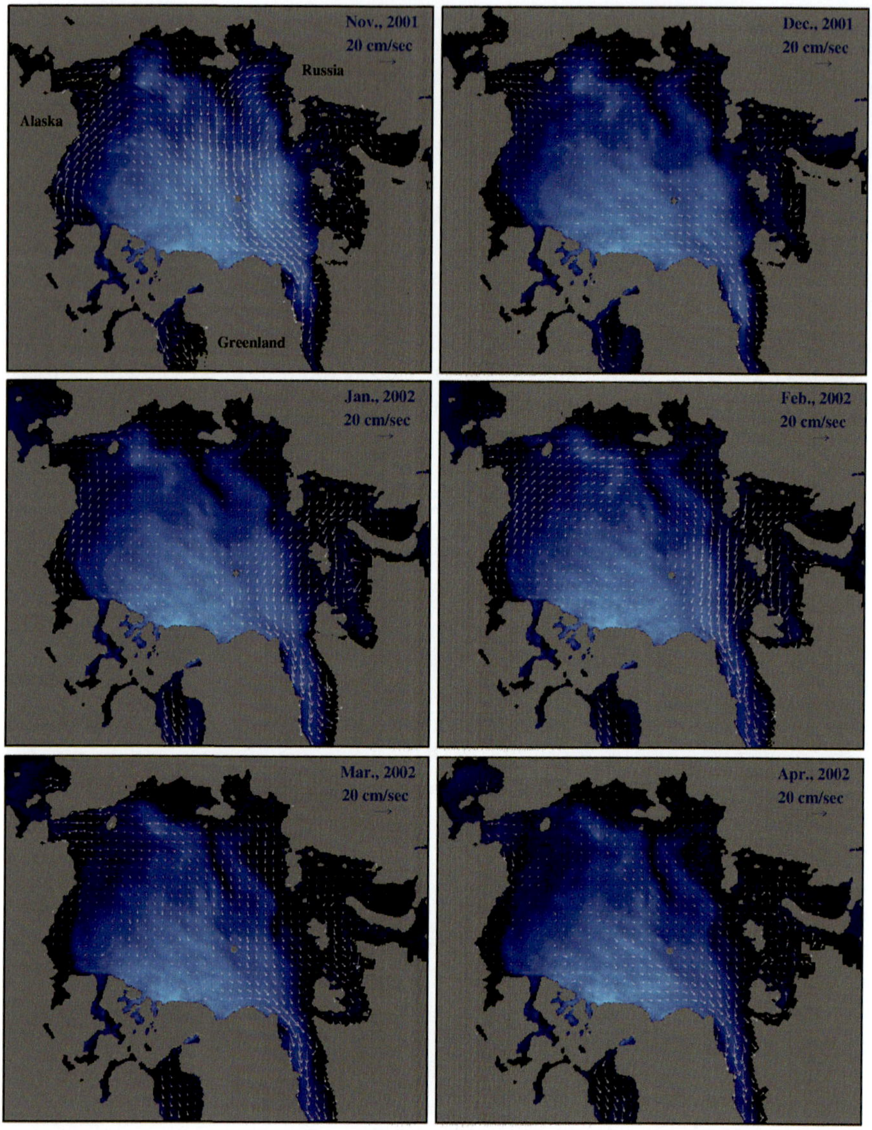

Fig. 7.19 Average monthly-ice drift velocities in the NH from November 2001 to April 2002 as derived from passive microwave data. Courtesy of Anthony Liu and Yunhe Zhao, 2008

7.3 Sea Ice in the Southern Hemisphere

The distribution of the sea ice cover in the Southern Ocean is quite different from those of the NH because of the vastly different configuration of the land masses, as discussed earlier. A set of visible and microwave images, similar to those in Fig. 7.7 but representing sea ice cover in the SH during austral autumn and austral spring is presented in Fig. 7.20. The visible channel data from AVHRR (Fig. 7.20a and b)

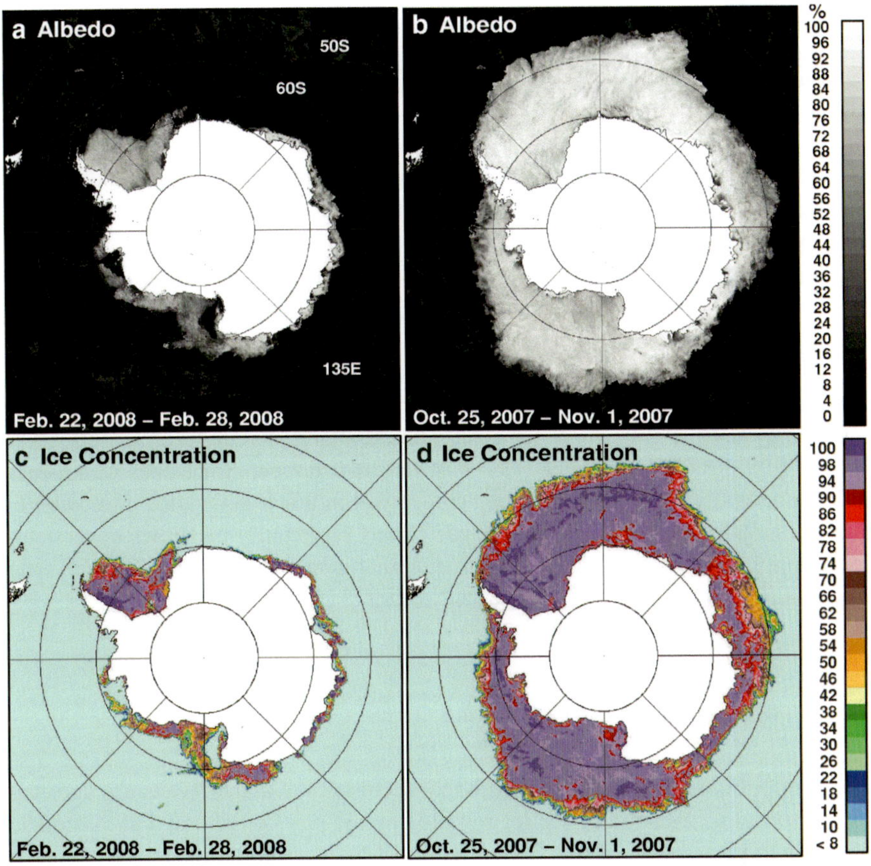

Fig. 7.20 Weekly averages of radiances and sea ice concentration in the Southern Hemisphere using AVHRR visible images on (**a**) 22–28 February 2008 and (**b**) 25 Oct to 1 Nov 2007 and SSM/I ice concentration maps in (**c**) 22–28 April 2008 and (**d**) 25 Oct to 1 Nov 2007

clearly show areas of high consolidation, active ice formation, divergence, and melt during the ice growth and ice decay seasons. Polynya areas, especially near the coastal regions, are shown as relatively darker gray than the thicker ice cover even when they are already covered by new ice, especially if the latter has not acquired a snow cover. At the marginal ice zones, the ice cover is also shown to have intermediate values (grayish), and, at early stages, it may be difficult to discriminate new ice from open water. Infrared data provided by the AVHRR sensor can also be used to obtain complementary information about the ice cover. Both visible and infrared data provide good information about the ice covered regions during clear sky conditions with the visible providing useful data during daytime and the infrared providing good data during both day and night conditions.

The corresponding sea ice cover data, as observed by passive microwave sensors for the same periods as those shown in Fig. 7.20a and b, are presented as color-coded ice concentration maps in Fig. 7.20c and d, respectively. As in the NH, visible and passive microwave data provide very similar information, with the dark and light gray areas in the AVHRR satellite images corresponding to the relatively low and high concentration ice areas, respectively, in the passive microwave data. The spatial changes, as depicted by both visible and passive microwave data for the two periods, reflect basically the expected changes in the physical properties of the ice cover. During the growth period in autumn, large areas covered by new ice look dark gray in the AVHRR image because of the relatively low reflectivity of the latter (Allison 1993; Masson 1991). The same surfaces also have relatively low emissivity (Comiso et al. 1992; Grenfell et al. 1994) and hence low brightness temperatures, providing relatively low ice concentration (Fig. 7.20c). Conversely, during the spring/summer breakup period, the dark gray or low reflectivity areas in Fig. 7.20b correspond to relatively high percentage of open water and therefore low ice concentration areas as depicted by the corresponding passive microwave data (Fig. 7.20d). Low or intermediate ice concentration values in the microwave data in autumn and winter usually reflect the presence of newly formed ice, while similar signatures represent the presence of mixtures of ice and open water in spring and summer. Since open water areas in leads and polynyas are covered by new ice within a few hours after formation in autumn and winter, this is also a more useful way of quantifying the ice cover compared to a strict ice and no ice discrimination that would show near 100% ice cover in practically all regions during ice growth stages.

7.3.1 Mesoscale Characteristics

The complexity of mesoscale processes that occur in the Southern Ocean is depicted by the spatial distribution of sea ice revealed by the pair of Landsat TM (Thematic Mapper) images shown in Fig. 7.21. The set of high resolution images captures in great detail the small-scale and large-scale physical changes in the ice cover from one season to another. When the weather and conditions are right, high resolution visible data, like Landsat, provide excellent information about the spatial redistribution and transformation of the sea ice cover as it gets impacted by atmospheric and oceanic processes. The image in Fig. 7.21a shows typical conditions during early stages of ice growth in the Antarctic, which in this case occurred on 27 March 1989 near the Cosmonaut Sea and Cape Ann (around 50°E). During this time, the only discernible form of ice cover are grease ice, shuga, and small to intermediate pancakes. It is apparent that the ice concentration is generally highest near the continent and lowest to the north, but the growth process is not a simple thermodynamic growth process in which the ice sheet advances continuously to the north. The image in Fig. 7.21a shows a very dynamic process that combines the influence of the atmosphere and the ocean. Eddy formations are apparent in the upper right portion of the image, and the unusual distribution of the ice cover suggests a strong influence of winds, waves, and tides. The image in Fig. 7.21b was taken approximately 8 months later in the same general area and it shows a completely different ice cover that is more typical in spring. During this latter period on November 24, 1989, a much thicker ice cover is apparent and large ice floes are revealed as the ice cover breaks

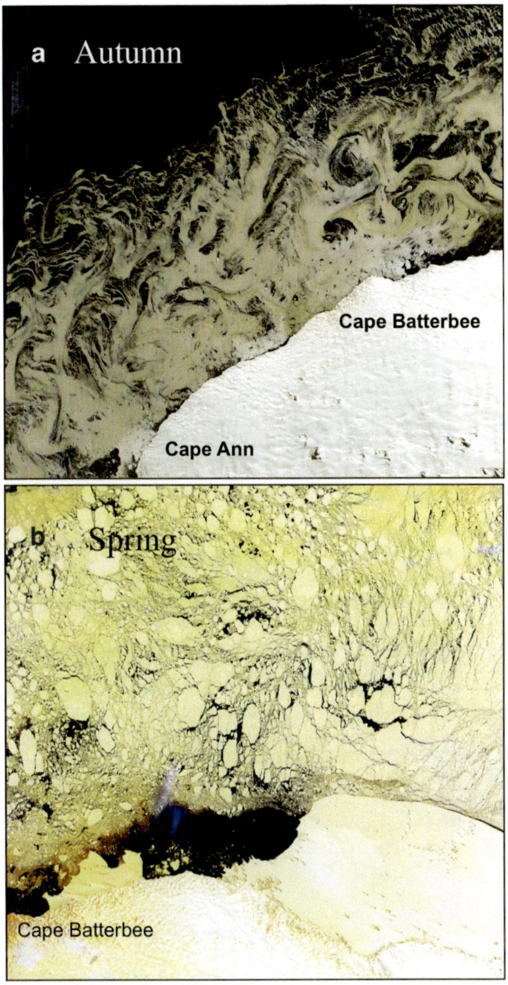

Fig. 7.21 Detailed distribution of the sea ice cover in the Antarctic region as observed by Landsat-7 in (**a**) Autumn on 27 March 1989 and (**b**) Spring on 24 November 1989 over generally the same region. New ice with spatial patterns largely influenced by atmospheric circulation dominates the ice cover in autumn while a much thicker and more defined ice floes dominates the consolidated ice cover in Spring. From Comiso and Steffen (2001) with permission from the American Geophysical Union

up, yielding a complex conglomerate of ice floes of many different sizes and ice types. A spring polynya along the coast is also apparent and was probably initiated mainly by wind forcing (Comiso and Steffen 2001) and in part by ice melt because of the upwelling of warm water in the region. To the south of the polynya, fast ice connected to the continent is still intact with numerous icebergs frozen in the ice. Some of the icebergs were already released to the polynya area which is part of the Magnet Bay. In evaluating seasonal and interannual changes, it is thus important to note that the growth and decay process is not governed by a simple thermodynamic process and as depicted in the images, changes in atmospheric forcing are key factors that in part determines the character and distribution of the ice cover.

Figure 7.22 shows ASAR, MODIS and AMSR-E images at different resolutions during a mid-winter (August 2003) period in the Bellingshausen Sea and it is apparent that the area adjacent to Adelaide Island near the Antarctic Peninsula is a scene of

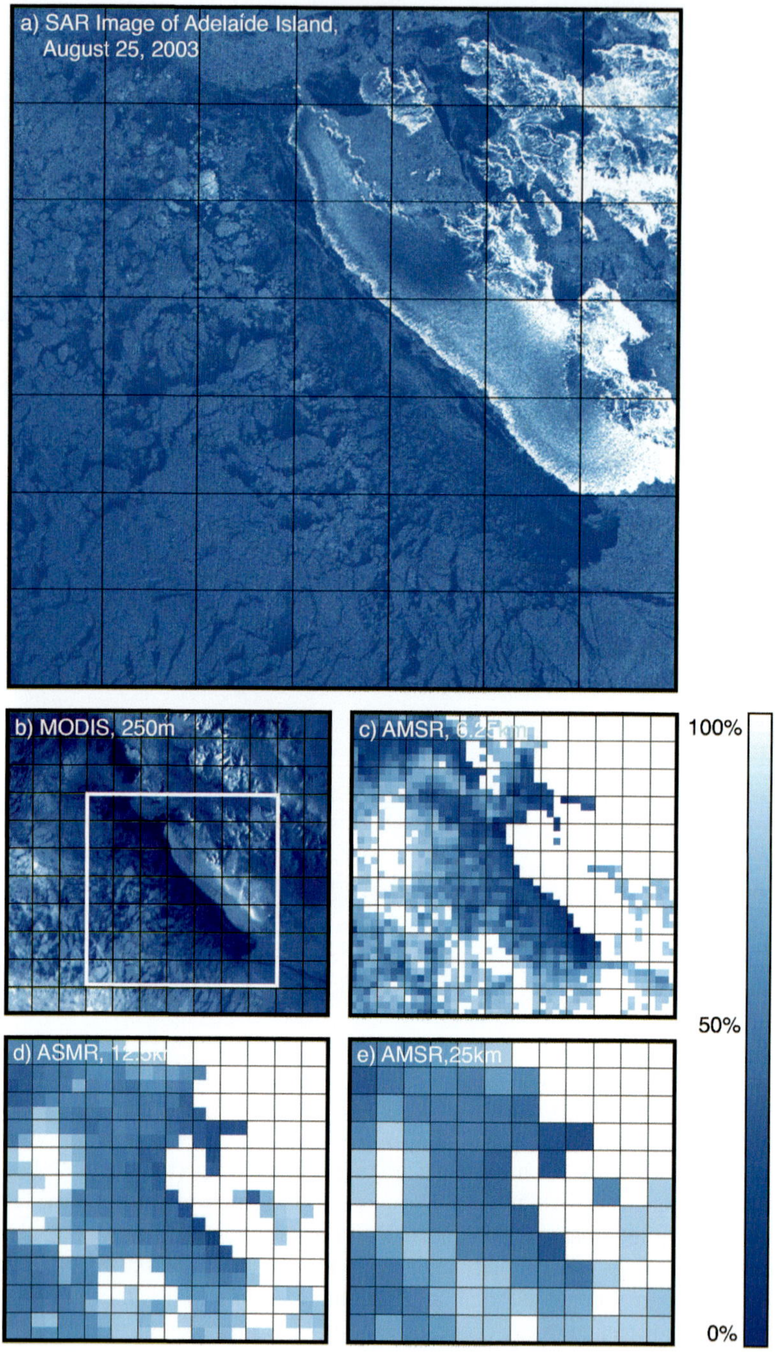

Fig. 7.22 The sea ice cover in the Bellingshausen Sea near Adelaide Island as observed on 25 August 2003 by (**a**) ENVISAT/ASAR, (**b**) EOS-Aqua/ MODIS, and AMSR-E at (**c**) 6.25 km; (**d**) 12.5 km; and (**e**) 25 km resolution. The color scale is only for the AMSR-E data. The ASAR and MODIS data represent the backscatter and reflectivity of the surface and have not been converted to ice concentration. ASAR data, ESA copyright (2003), was provided by Ron Kwok

intense mesoscale process. Strong winds from the continent cause sea ice to be advected to the north and produce coastal polynyas (darker blues). In the polynya region, it is apparent that new ice formation occurs in succession with the thinnest ice (darkest blue) being closest to the island. The ASAR image in Figure 7.22a shows the distribution of the sea ice cover, including the coastal polynya region in great detail. The data in the polynya region provide details about the distribution of open water, grease ice and thin ice in the region but some validation studies are needed to be able to able to accurately sort out the different surface types. A set of images similar to those shown in Fig. 7.6 for the Arctic but using high resolution MODIS data instead of Landsat are also presented. The 6.25 AMSR-E data provided by the 89 GHz channel show good compatibility with the MODIS 250 m data, with much of the detailed features captured by the sensor. This means that the 6.25 km resolution AMSR-E data can be very useful for characterizing the mesoscale features and for estimating associated fluxes in the region. This represents a significant improvement in capability to fully characterize the sea ice cover and quantify its impacts on the ocean and the atmosphere. The 12.5 km AMSR-E data provide a good representation and show the basic features, but many of the small-scale details are lost. As for the 25.0 km image, it is apparent that even if we use the same AMSR-E sensor data and the same set of channels as what is used for the 12.5 km image, the mesoscale features are not captured due to the coarse resolution. In this case, the 25 km data is basically just a degraded resolution version of the 12.5 km data and it is obvious that the former can't provide the right kind of information needed for detailed heat flux studies. Nevertheless, as indicated earlier, the 25 km data provide relatively accurate information about the basic characteristics of the sea ice cover and are useful for large-scale and interannual variability studies.

Sea ice cover images in the SH during minimum and maximum extents, as captured by the AMSR-E high resolution 89 GHz data, are presented in Fig. 7.23. The images show the small-scale features of the MIZ and the inner pack that is made possible through the use of the 6.25 km resolution data. The sea ice cover during the summer minimum as shown in Fig. 7.23a is atypical with a large fraction of the ice located in the Weddell Sea, significant fraction located in the Indian and Western Pacific Oceans and practically no ice in the Bellingshausen Sea. There is also an ice tongue at the Weddell Sea and some coastal polynyas around Antarctica. The ice cover during winter maximum on 3 September 2008, as shown in Fig. 7.23b, shows the semi symmetrical distribution of ice that oftentimes occurs during winter and includes the formation of a corner of a rectangle at the Ross Sea (around 217 E). Also there are some embayments along the MIZ that may represent cyclonic activities. At around 20°E, the ice edge is shown to be discontinuous with the western side a few degrees lower in latitude than the eastern side.

The images presented in Fig. 7.24 are similar to those presented in Fig. 7.8 but located in the Antarctic and taken during a winter period (i.e., 9 September 2002) when the ice surface is still relatively cold and dry. The high resolution MODIS data allow for the identification of large ice floes and icebergs at the MIZ. Cloud cover effect is apparent, but in areas of thin clouds, it is possible to infer the orientation of leads and the general character of the inner pack. The spatial distribution of the ice

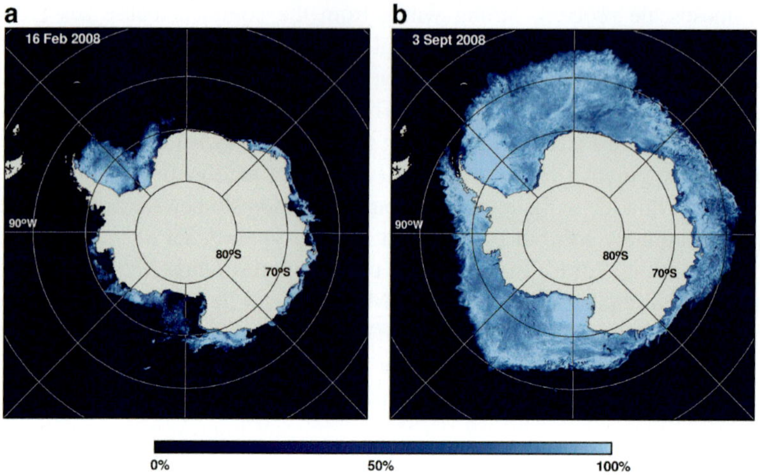

Fig. 7.23 AMSR-E high resolution (89 GHz) images of the sea ice cover in the SH during (**a**) minimum on 16 February 2008 and (**b**) maximum extents on 3 September 2008

cover, as observed in the MODIS image, is better reproduced by the AMSR-E image than the SSM/I image, both at the ice edge and within the pack. Plots along a transect (labeled A to B) across the ice edge are also presented in Fig. 7.24d, and it is apparent that the radiances from MODIS data along this transect show open water signature outside the ice pack, as in Fig. 7.8, but at the ice edge, the signature is not so well defined. The radiances from MODIS increase slightly at the MIZ, suggesting the presence of a wide band of new pancake ice, and then increase with some fluctuations to a maximum value in the inner pack and then vary considerably within the ice pack. The corresponding transects from AMSR-E and SSM/I show more well-defined ice edges with the SSM/I ice concentrations (blue line) showing an ice edge location that is further north than the AMSR-E data (red line) by about 10 km. The ice edge location identified by AMSR-E also appears to be closer to that identified by the MODIS image than that identified by SSM/I. In the MODIS image, there are two strong dips within the ice pack region, which are not identified as reduced concentrations in the AMSR-E and SSM/I transects. The dips likely represent leads that have been refrozen and covered by new ice that have not acquired a snow cover (and hence the relatively low radiances) but thick enough to have an emissivity close to that of thick ice. This result further indicates the need for high resolution data in mesoscale process studies.

7.3.2 Large-Scale Seasonal and Interannual Variability

Multiyear monthly averages of the sea ice cover similar to those shown earlier for the NH are presented in Fig. 7.25 to illustrate the typical spatial distribution of the sea ice cover in the Southern Ocean during the various stages of the ice growth and

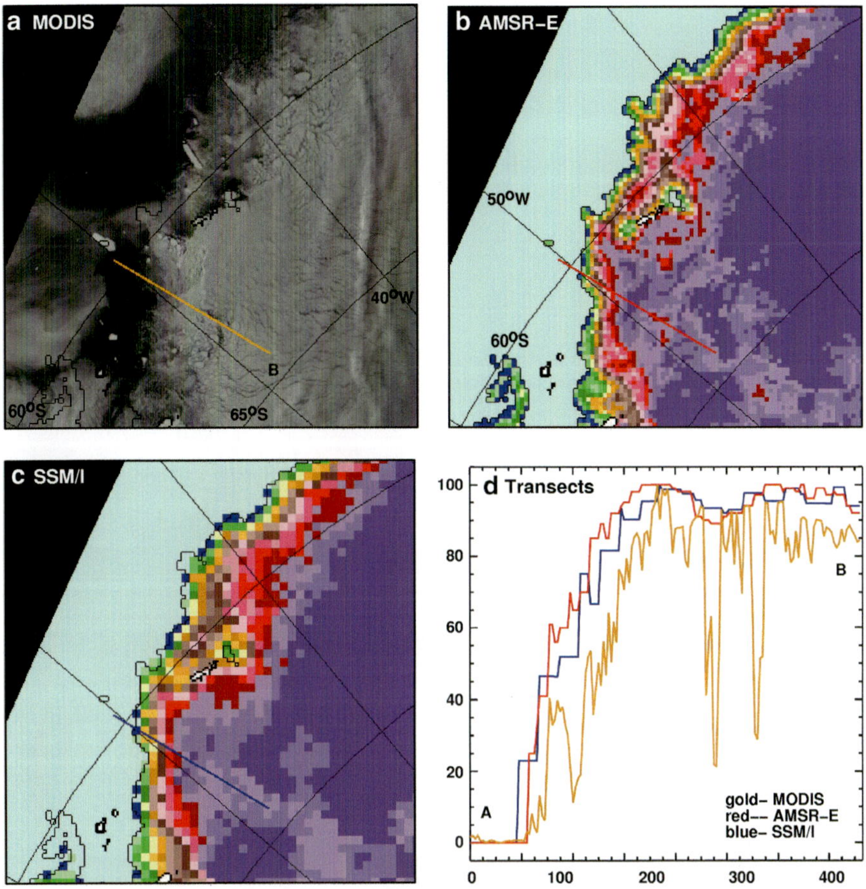

Fig. 7.24 Antarctic sea ice cover on 9 September 2002 as depicted by (**a**) 250 m MODIS visible channel image; (**b**) 12.5 km AMSR-E ice concentration data; and (**c**) 25 km ice concentration SSM/I data. Normalized radiances from MODIS and ice concentrations in % along the line AB (see MODIS image) from ice free ocean to ice covered ocean are plotted in (**d**) From Comiso and Nishio (2008) with permission of the American Geophysical Union

decay seasons. As in the NH images, we use the 12.5 km AMSR-E ice concentration data to provide a representation of the contemporary seasonal sea ice cover at a resolution better than that of the standard historical data. The series of images starts with a mid-summer month (January), which is the warmest month, and the time when the melt rate is relatively high. The minimum extent usually occurs in late February or early March but there are exceptions. The February image shows the general areas where sea ice survives the summer and it is apparent that the two primary locations of the perennial ice, or ice that survives the summer melt, are the Western Weddell Sea adjacent to the Antarctic Peninsula and the coastal regions along the Bellingshausen, Amundsen, and Ross Seas. The areal coverages of the perennial ice in these two regions have been comparable but vary slightly in magnitude and location from 1 year to another (Gloersen et al. 1992). During the

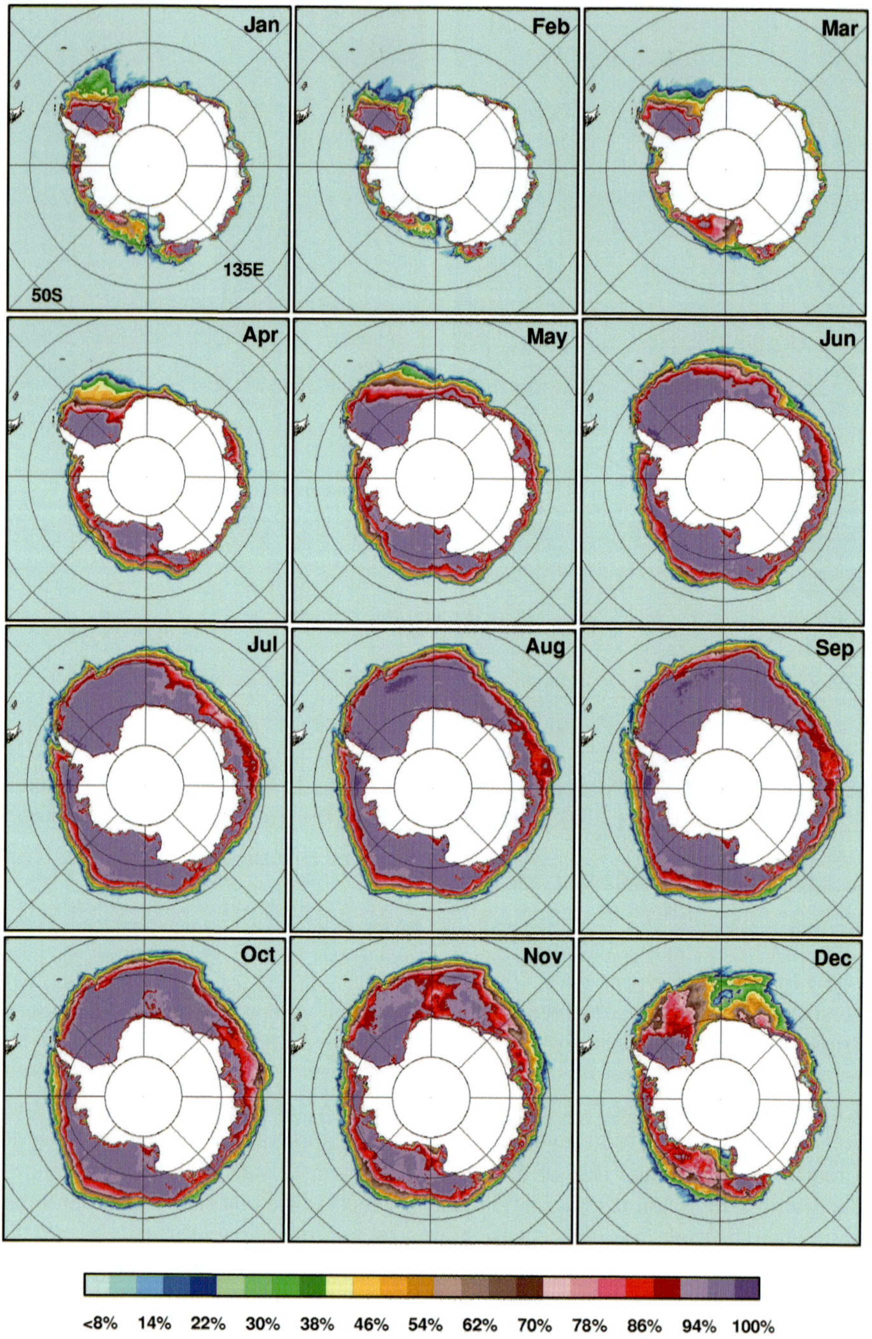

Fig. 7.25 Monthly averages (climatology) of sea ice concentration from January to December in the Southern Hemisphere using EOS-Aqua/AMSR-E data from June 2002 to May 2008

growth period from March through August, the ice cover advances most rapidly to the north from the location of the perennial ice. The ice, however, tend to drift to the northeast, especially in the Weddell Sea, in part, due to the effect of the Antarctic circumpolar current. The rate of advance of the sea ice cover to the north is relatively fast except in some areas where the Antarctic coastline is farthest to the north, as in the Western Pacific (90°E–150°E) and the tip of the Antarctic Peninsula. The shape of the ice cover during ice maxima (i.e., September) is almost symmetrical around the continent with a tendency to form a rectangular corner at about 217°E, which is in part influenced by the bathymetry (Fig. 2.2b) of the region.

Among the most distinctive figures in the inner zone of the ice cover are the reduced or near zero ice concentrations, called polynyas, adjacent to the Ross and Ronne Ice Shelves, at or near the Maud Rise, and near the Cosmonaut Sea, which occur oftentimes during the year. These transient polynyas usually occur anytime from early to latewinter (Zwally et al. 1985; Comiso and Gordon 1998; de Veaux et al. 1993), the importance of which is discussed in Chap. 8. Because they normally appear at different times and in different places during the winter season, they are not depicted very well in the monthly climatological averages. The polynya features that are most evident in the November and December images are the spring polynyas that occur almost consistently in the same regions, suggesting that these regions are covered by relatively thin ice that is vulnerable to melt and ice breakup as the surface temperature goes up in spring. The largest of such polynyas is the Ross Sea polynya which has also been the scene of phytoplankton blooms (Arrigo and Van Dijken 2003; Smith and Comiso 2008) and hence high productivity, reflecting the possible influence of sea ice melt water as discussed in Chapter 9.

To assess quantitatively the seasonal and interannual variability of the sea ice cover in the SH, plots of monthly averages of the ice extent, ice area, and ice concentration from November 1978 to December 2008 are presented in Fig. 7.26. Generally, the seasonal fluctuations are large for the sea ice extent and ice area, as has been mentioned earlier, but it is interesting that the maximum and minimum values for each year are relatively uniform. There are, however, significant fluctuations in the peak and low values mainly due to environmental factors such as surface temperature and winds. The data set also provide the means to examine whether there is a relationship between winter ice and summer ice. A question of interest (as in the Arctic) is whether an extensive winter ice cover would lead to an extensive summer ice cover since there would be more sea ice to melt? Likewise, would an extensive summer ice cover lead to an extensive winter ice cover since there is more ice to start with? Examination of the plots show that during the years when the monthly ice extent and area in the winter were most extensive (i.e., 1980) or least extensive (i.e., 1986), the corresponding values in the subsequent summer were not the most extensive or least extensive, respectively. It is thus apparent that the decay patterns are not significantly influenced by the extent of ice during the preceding winter period. Conversely, the growth patterns are also not significantly influenced by the extent of ice during the preceding summer. This apparently counter-intuitive phenomenon is actually more apparent, especially on a regional basis. For example, in the Weddell Sea, anomalously extensive ice cover in winter is usually followed by anomalously

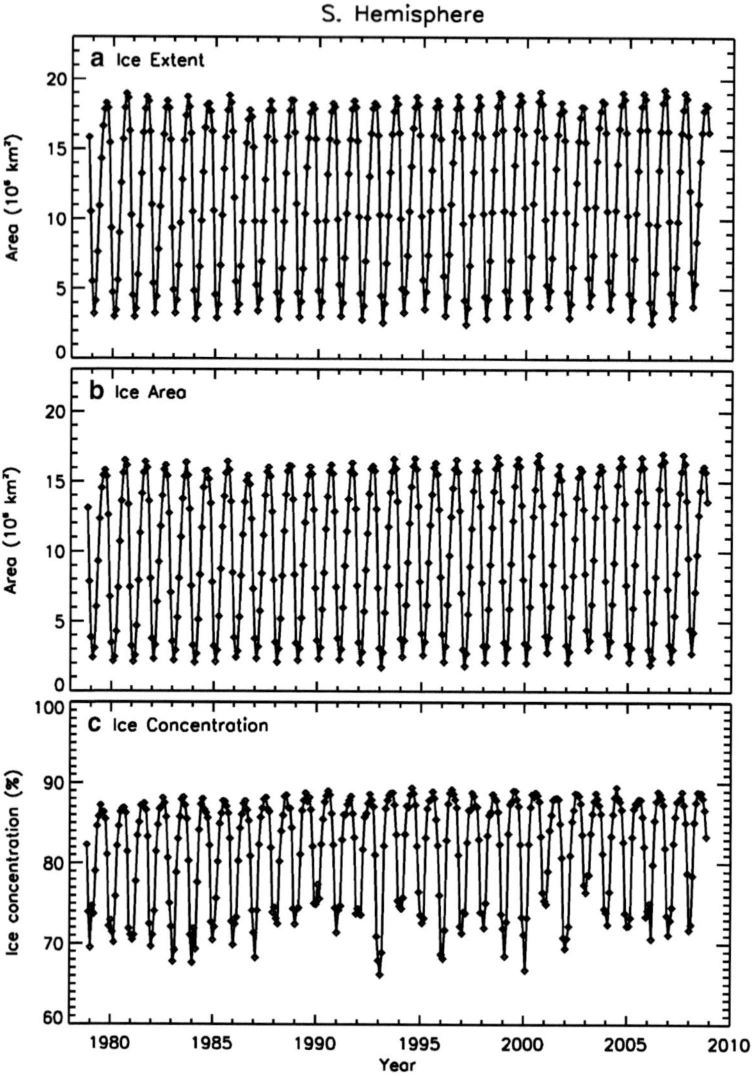

Fig. 7.26 Plots of monthly averages (**a**) sea ice extent; (**b**) sea ice area; and (**c**) sea ice concentration for each month from November 1978 to October 2008 in the Southern Hemisphere using SMMR and SSMI data

low ice cover area in the summer and vice versa, as reported previously by Comiso and Gordon (1998) and Zwally et al. (1983). This means that the interannual variability of the ice cover is likely not influenced by conditioning or its history and more likely by the variability of the atmosphere and the ocean. The minimum values in the ice concentration monthly averages (Fig. 7.26c), which correspond to summer values, show much greater variability than the winter maximum values. During the summer, the sea ice cover is usually not as consolidated as in winter and therefore is more vulnerable to advection and melt. It should be noted that when the monthly ice concentration were unusually low, as in the summer of 1993 and 2000, the monthly ice extent and ice area were also relatively low. Southerly winds that cause divergence and therefore low ice concentration also cause sea ice to be advected to the north where they melt in warmer waters. Low ice extents and ice areas, however, may not be accompanied by low average ice concentrations as in 1997 and 2006. Northerly winds may cause a compaction of the ice cover and hence relatively high ice concentrationas as the ice floes get advected towards the continent and although this may cause ridging and therefore a thicker ice cover, the ice extent and ice area would be lower than normal.

Although the total ice extent and area do not change much in magnitude from one year to another, the shape and location of the ice cover, especially near the ice edge, can have large interannual variations. Such phenomenon is depicted in the set of daily images during winter and summer in Figs. 7.27 and 7.28, respectively. The color-coded images are ice-concentration maps during the days of maximum or minimum extents from 1979 to 2008. As in the Arctic, running 5-day averages of daily extents were used to determine the maximum or minimum extent dates for each year. The first two images (i.e., 1979 and 1980). In Fig. 7.27 provide a good example of how the Antarctic sea ice cover can change so drastically from one winter to another. The ice cover in the Amundsen/Bellingshausen Seas in 1979 goes much further to the north than in 1980. This is also true in the Western Pacific region (90°E–140°E). Despite this, the total extent of sea ice was actually higher in 1980 than in 1979, in part because of more ice in 1980 in the Weddell, Indian, and Ross Seas. Interannual ice cover variability in the Western Pacific region (90°E–160°E) is actually quite substantial with very little ice cover in the region in 1980, 1981, 1989, and 2000. Overall, a quasi-circumpolar symmetry is apparent with the center shifted to the south along the 0°E meridian by about 2°. There are also times as in 2001 when the ice cover shows a semi-triangular shape with corners located at approximately 37°W, 82°E, and 217°E. The shape and size are likely affected by many factors such as winds, tides, and temperature, but the key factor may be the atmospheric pressure modes which may be mode 3 or mode 2 as discussed in Chapter 5. An important factor is also the bathymetry which, as indicated earlier, is likely responsible for the unique protrusion of the ice cover that looks like the corner of a rectangle at 217°E. The set of images shows that the Antarctic sea ice cover during maximum extent is generally confined at latitudes greater than 52°S.

The yearly sea ice cover during minimum extents looks very similar during the period from 1979 to 1988, when the perennial ice west of the Antarctic Peninsula, was primarily located in the Bellingshausen/Amundsen Seas. The sea ice cover as observed by ESMR in the 1970s, was also similar to the 1979 to 1988 data (Zwally et al. 1983). During the period 1989–2003, the perennial ice in the Bellingshausen Sea has diminished substantially as reported by Jacobs and Comiso (1997) and except for a minor recovery in 2001, the ice cover in the region has remained anomalously low. During the period from 1991 to the present, the ice cover in the Amundsen Sea declined significantly as well. During the same time period, the

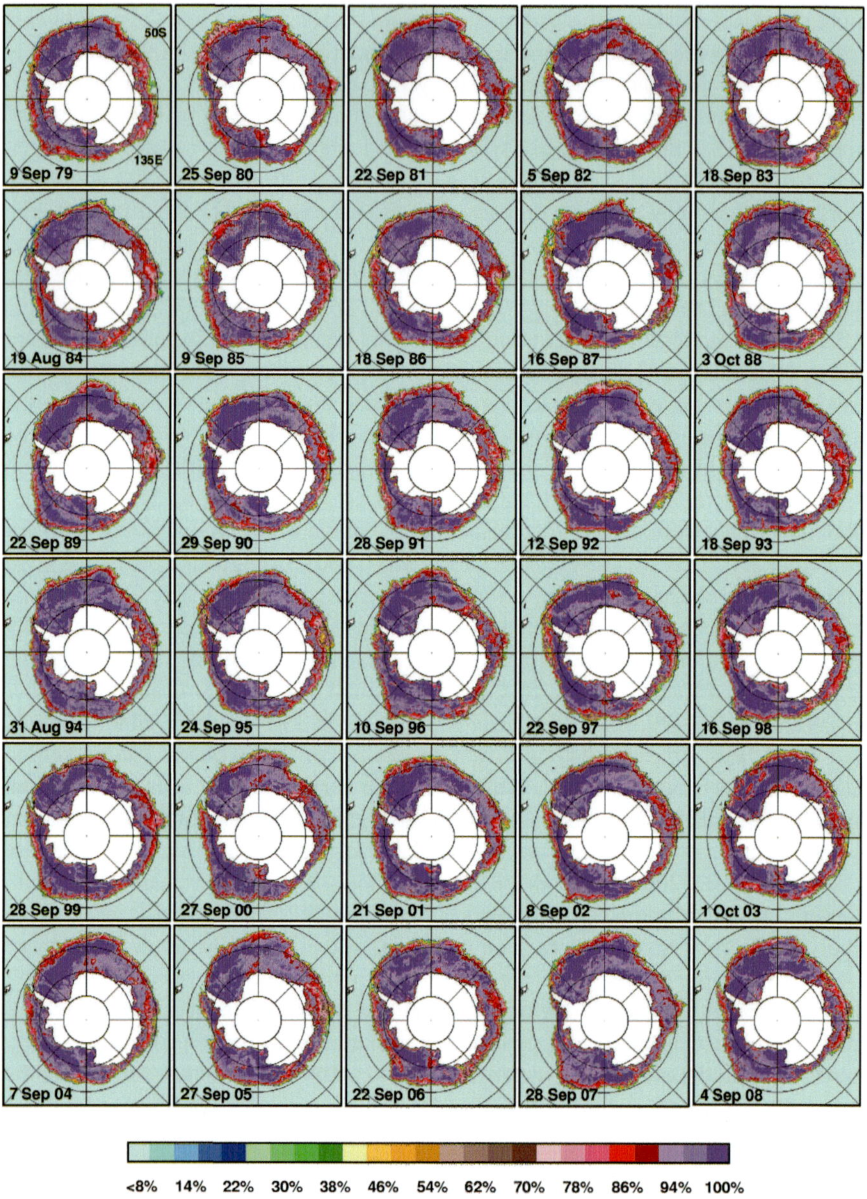

Fig. 7.27 Daily averages of sea ice concentration in the Southern Hemisphere during maximum extent for each year from 1979 to 2008 as derived from SMMR and SSM/I data

perennial ice cover in the Ross Sea region was exceptionally extensive, especially in 2001, 2003 and 2004, suggesting that the summer sea ice cover migrated to the west. The Ross Sea region is usually ice free during the summer, in part because of the formation of a large coastal polynya in early spring along the edge of the Ross

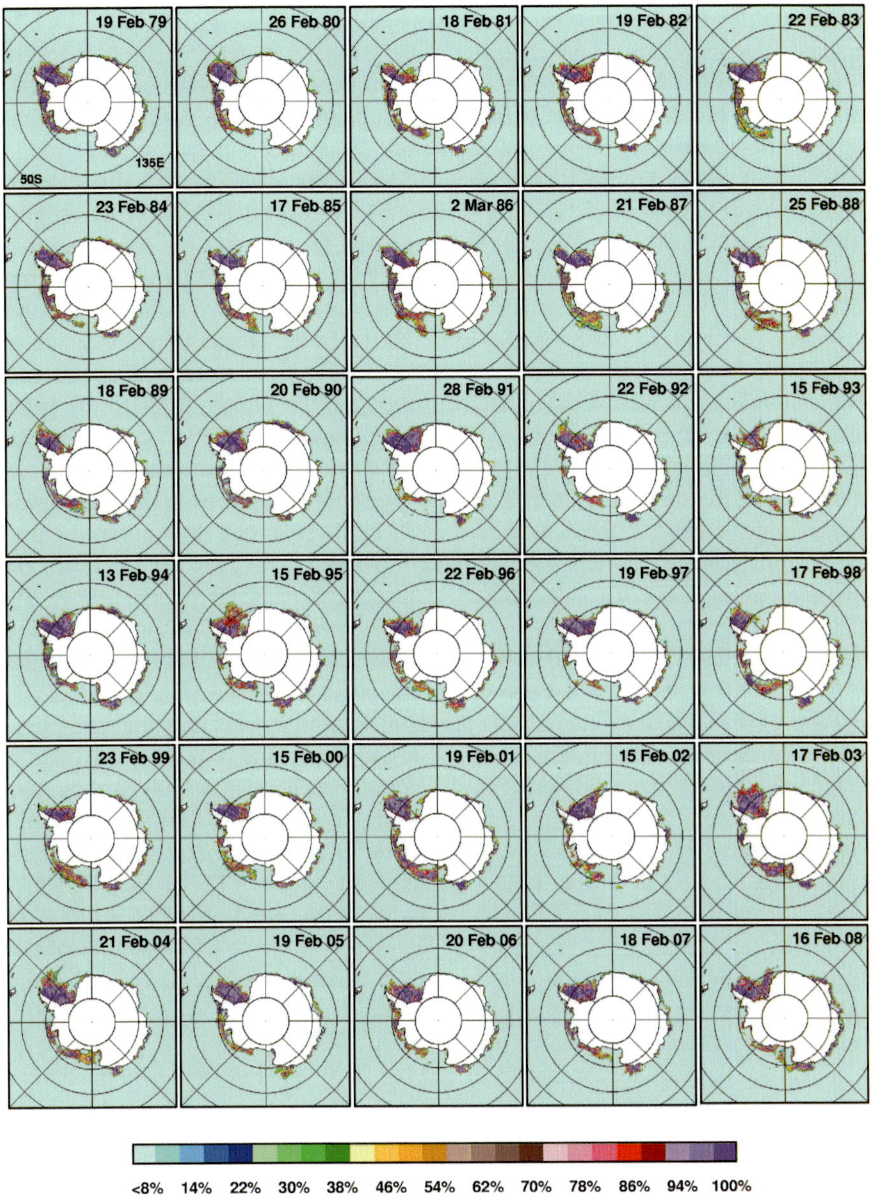

Fig. 7.28 Daily averages of sea ice concentration in the Northern Hemisphere during minimum extent for each year from 1979 to 2008 as derived from SMMR and SSM/I data

Ice Shelf. The large coastal polynya would get larger until sea ice north of it is totally melted and the entire region becomes basically ice free during the summer. The years when this phenomenon did not happen were 1998, 2001, 2003, and 2004.

The calving of very large icebergs in 2000 and 2002 that were grounded for a while in the region also changed the character of the coastal polynya formation and were in part responsible for the presence of sea ice in the coastal regions in 2001, 2003, and 2004. In the Weddell Sea, perennial ice is confined to the ocean shelf region adjacent to the eastern portion of the Antarctic Peninsula. The eastern edge of the ice is in part affected by shelf processes, including the upwelling of warm water in the region that have caused the reductions in the ice cover in the region. The southern portion of the Western Weddell Sea is usually ice covered in the summer, but sometimes the region becomes ice free, as in 1993, 1998, and 2001 because of the influence of the abnormally strong atmospheric forcing in spring that caused the advection of ice to the north (Ackley et al. 2001). In the Western Pacific region at around 145°E, significant sea ice cover at the end of the summer is also apparent. They are mainly located between the continent and the Balleny Islands, the latter serving as a barrier that keeps the ice from being advected farther to the north where they melt. There were some years when there was no ice in the region as in 1987 and 2002, while there were years as in 2008, when the ice cover in the region was a significant fraction of the summer ice cover in the Antarctic. It is intriguing that the ice cover in the eastern side (Western Weddell Sea) and the western side (Bellingshausen Sea) of the Antarctic Peninsula has changed so much from the 1980s to the 2000s. In the 1980s, both sides of the peninsula had substantial sea ice cover but in 2003, 2004, 2007 and 2008, the sea ice cover was mainly in the Western Weddell Sea or in just one side of the peninsula.

Plots of sea ice extent and ice area during maximum extents in winter and minimum extents in the summer are presented in Fig. 7.29a and b, respectively, and it is apparent that the two parameters are coherent to each other in both periods. The maximum extents fluctuate from a minimum value of $18.1 \times 10^6 \text{km}^2$ in 1986 to a maximum value of $19.5 \times 10^6 \text{km}^2$ in 2007, while the ice area changes from a minimum value of $15.8 \times 10^6 \text{km}^2$ in 1986 to a maximum value of $17.1 \times 10^6 \text{km}^2$ in 2007. Linear regression analysis of the data points yielded a trend of 0.7 ± 0.4%/decade for ice extent and 1.2 ± 0.4%/decade for ice area. The range of variability of the minimum extent and area is similar to that of maximum extent and area but the changes are not correlated except that both have relatively low values in 2002. However, increases in maximum values from 2002 to 2005 were followed by decreases in the minimum values from 2003 to 2006 while decreases in maximum values in 2006 and 2007 were followed by increases in the minimum values from 2007 to 2008. High values in the minimum extents and areas occurred in 1983, 1987, 1995, 2001, 2003, and 2008, while low values occurred in 1993, 1997, and 2006. The lack of correlation between the maximum and minimum values is again a manifestation of the important role of atmospheric forcing on the decay and growth processes of sea ice in the Southern Ocean. The trend of the minimum extents shows 2.1 ± 2.6%/decade, while that for area is 3.1 ± 2.9%/decade. The trends for minimum extents and area but are larger than those for maximum extents and areas but the errors are also greater and the trend is likely not as significant. In light of the observed rapid decline of the ice cover in the Arctic, these results are intuitively unexpected.

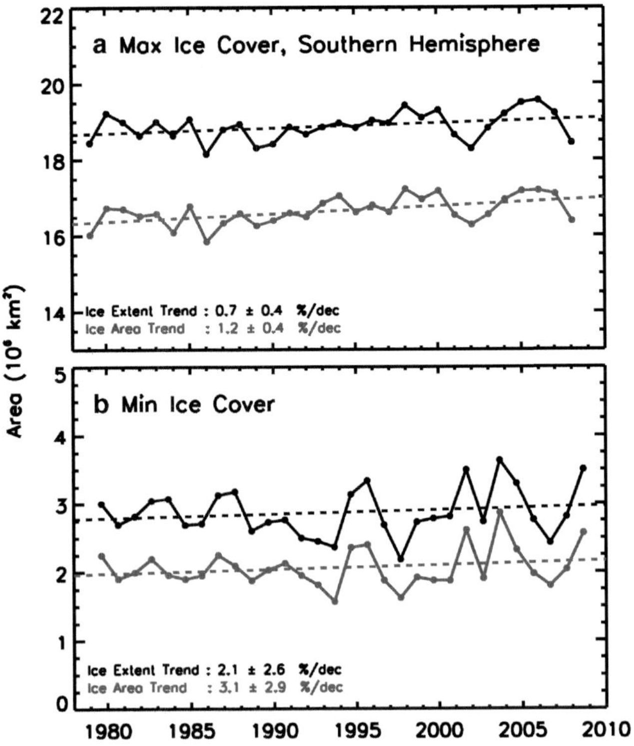

Fig. 7.29 Plots of sea ice extent and ice area during (**a**) maximum and (**b**) minimum ice extents. Dash lines represent the results of linear regression that provide the trend values

7.3.3 Monthly Anomalies and Trends

Results from quantitative analyses of month-to-month averages of extent, area, and ice concentration, as presented in Fig. 7.26 for the period from Nov. 1978 to 2008, indicate that the ice cover in the SH is much more stable than that in the NH. The amplitudes of the extent and area for each ice season are significantly larger that those of the NH and therefore more seasonal but with much less perennial ice cover. Over the 30-year period in which consistently processed passive microwave data are available, the interannual variability of the sea-ice extent and area also appears very slight despite the large interannual variability in the ice concentration, especially during the summer period. In the summer, the ice cover is more vulnerable to winds, waves, and other forcings because of much smaller extents. The average ice concentration is almost constant in the winter at about 83%, compared to about 92% in the Arctic, while the average ice concentration in the summer ranges from 59 to 69%, compared to 72–81% in the Arctic. Note that the distributions for ice extents are again coherent with those of ice area, with the latter being lower in values, as expected.

The plots of monthly averages provide good information about the state of the sea ice cover but they do not reveal the large interannual variability in the spatial distribution of the Antarctic sea ice cover. Color-coded anomaly maps for September and February of each year, from 1979 to 2008, are presented in Figs. 7.30 and 7.31, respectively, to illustrate quantitatively the interannual variability of the SH sea ice cover. As done previously, we use monthly averages of ice concentrations and the

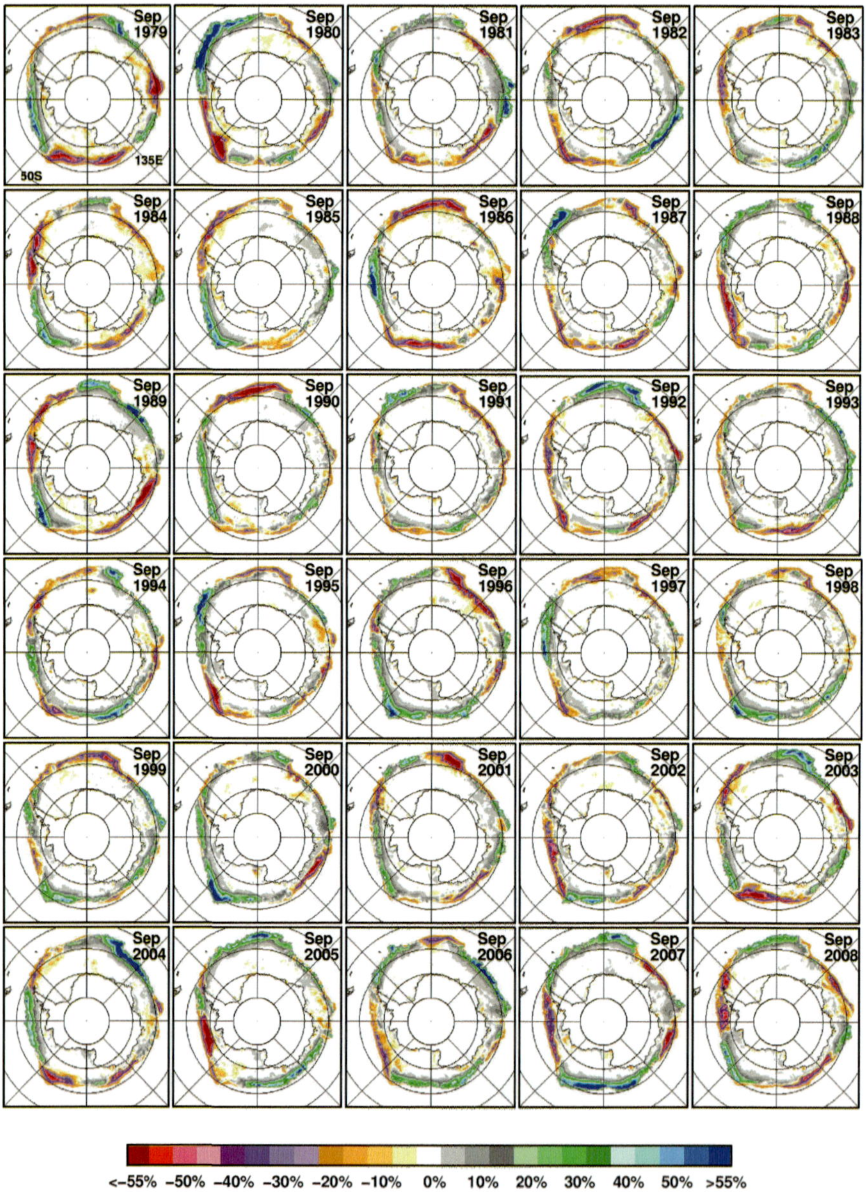

Fig. 7.30 Monthly anomalies of sea ice concentration in the Southern Hemisphere as derived from SMMR and SSM/I data for a winter month (September) from 1979 to 2008

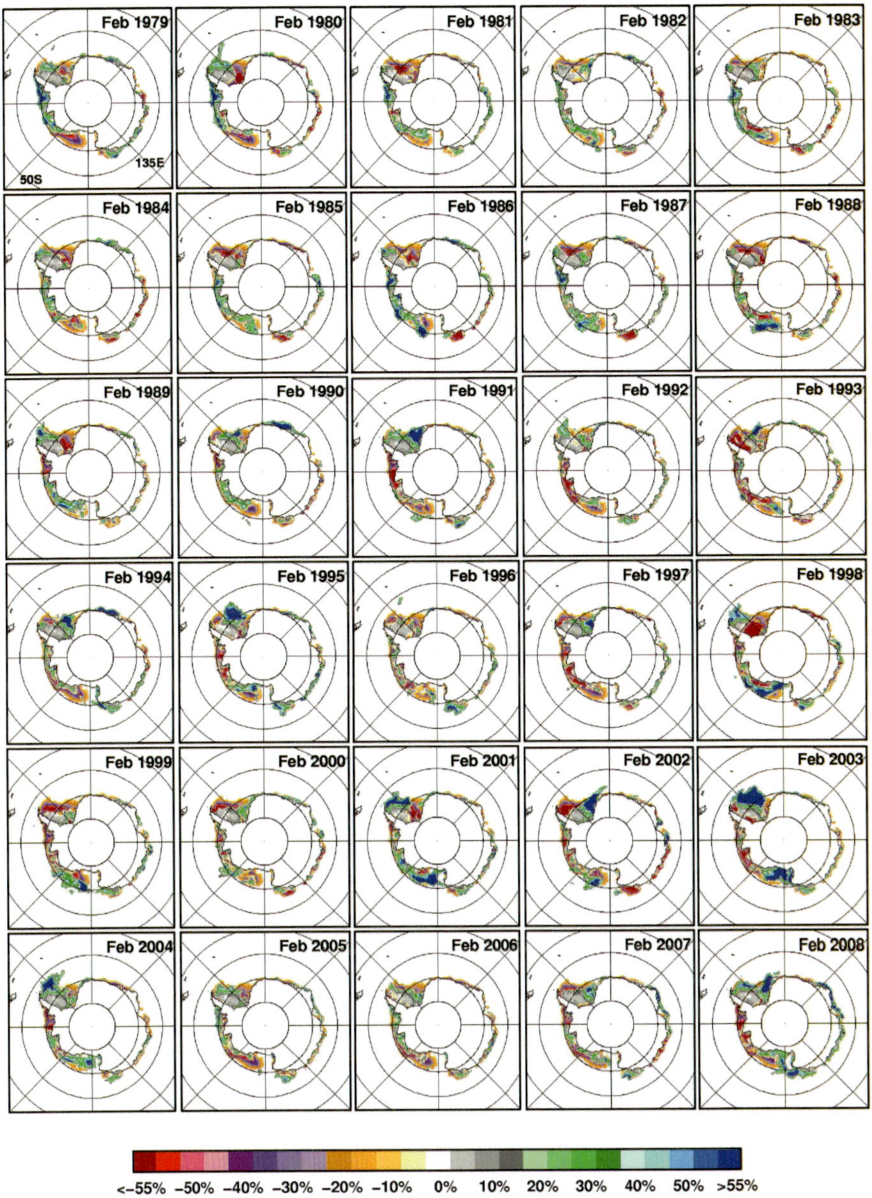

Fig. 7.31 Monthly anomalies of sea ice concentration in the Southern Hemisphere as derived from SMMR and SSM/I data for a summer month (February) from 1979 to 2008

monthly climatology, based on the continuous data from November 1978 to December 2008 to generate the anomalies. The most remarkable phenomenon associated with these images is the patterns of alternating negative and positive anomalies along the edge of the sea ice cover around the continent. From one year to another, these patterns tend to change, with apparent tendency to move around propagating like a wave that has been referred to as the Antarctic Circumpolar Wave (ACW) by White and Peterson (1996). The ACW as mentioned in Chap. 6 in the discussion on monthly anomalies in surface temperature propagates around the continent with a period of about 8 years and a revisit time of 4 years at every point, assuming a mode 2 wave number. It is interesting to note that none of the anomaly maps from 30 years of data in September look identical, although the revisit over a particular area by the wave is supposed to be every 4 years. There are, however, some years when the patterns are similar, such as in 1986 and 1990, and also in 1984 and 2008. The repetition of the patterns in some places suggests that there may be local geographic or oceanographic factors that cause similar features to be reproduced in different years. Although the ACW has been described as an ice edge phenomenon, the anomaly patterns as depicted in Fig. 7.30 and the temperature anomalies in Fig. 6.24 indicate that the effect actually go all the way into the pack. The wave number is also mode 2 only for some years as in 1989 and 2007, and is very often mode 3, as in 1979, 1980, 2003, 2006, and 2008. This phenomenon will be discussed further in Chap. 8 but the set of images shows that although the area and extent of the Antarctic sea ice cover has not been changing much, the spatial distribution of the ice cover, especially at the ice edges, varies substantially each year following various forcings including the Antarctic Circumpolar Wave.

The anomaly maps for a summer month (February) from 1979 to 2006, as presented in Fig. 7.31, provide interesting patterns but not quite the same kind of information as have been apparent from the winter data. Significant ice anomalies are basically confined to regions where the ice cover is located at the end of the melt season. The key regions are the Western Weddell Sea, the Bellingshausen/Amundsen Seas, and the Ross Sea region, but other areas along the coast of East Antarctica occasionally have summer ice as well. In coastal regions where there is usually no ice in the summer, the anomalies may mean late melt period or early freeze-up. The anomalies can be interpreted more accurately using data observed during summer minima (Fig. 7.28). In the Western Weddell Sea, it appears that sea ice is almost always present along the eastern shores of the Antarctic Peninsula in the Western Weddell Sea with 1993, 1998, 2002, and 2003 being among the few exceptions. The high negative anomalies in the southern region in 1998 was caused by a rare spring polynya in the Ronne Ice Shelf coastal region (Ackley et al. 2001), while the one in 2002 was likely caused by the intrusion of warm water that melted the ice at the northern part for this specific year. The polynya events that caused the negative anomalies along the coastal regions in 1993 and 2003 were likely caused by strong easterly winds off the Antarctic Peninsula. The biggest positive anomalies in the region occurred in 1991, 1994, 1995, 2001, and 2003 while the most extensive perennial ice for the entire SH occurred in 2003, followed by 2001 and 1995. To the west of the Antarctic Peninsula, the perennial ice cover was basically confined to the Bellingshausen and Amundsen Seas in the first 2 years but in subsequent years, more ice in the Ross Sea and less ice in the Bellingshausen Sea is observed. The anomalies in the summer sea ice cover were largely positive from 1979 to 1988, and became largely negative from 1989 to 1992 and in 1995, 1997, 1999 and 2004. In February 2003 and 2008, the anomalies were strongly positive at the Weddell and Ross Seas, while it was strongly negative in the Bellingshausen and Amundsen Seas. In other years the anomalies were generally mixtures of positive and negative anomalies.

The interannual variability of the ice extent in the entire hemisphere and in the various sectors, as defined in Zwally et al. (1983) and as described in Chapter 6, can be quantitatively assessed using the plots of the average monthly anomalies presented in Figure 7.32. The peaks and dips in the anomaly plots are again shown

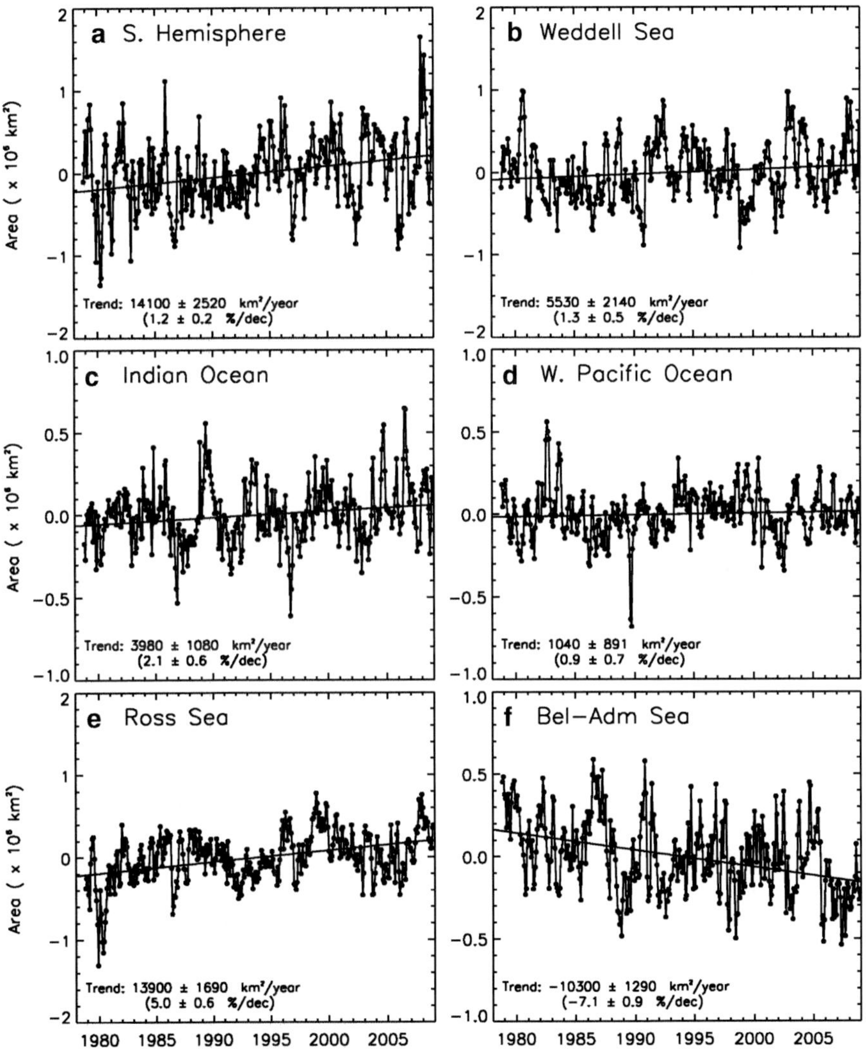

Fig. 7.32 Plots of monthly anomalies in the sea ice extents and trend lines for (**a**) entire Southern Hemisphere and (**b**) Weddell Sea; (**c**) Indian Ocean, (**d**) West Pacific Ocean; (**e**) Ross Sea; and (**f**) Bellingshause/Amundsen Seas Sectors

to occur in different years in the various sectors. This is partly because of the propagating wave (ACW), as described earlier, that causes the anomalies to circle around the continent and visit the different sectors during different time periods. The higher amplitude of the anomalies for the entire hemisphere than those of the individual sectors may be caused by the amplification of the anomalies when contributions from some sectors are in phase.. It is interesting to note that in the Weddell Sea sector, there is a significant negative anomaly in 1999 that appears to be associated with similar anomalies in the other sectors such as the 1998 negative anomaly in the Bellingshausen/Amundsen Seas, the 1997 anomaly in the Ross Seas, and the 1996 anomaly in the Indian Ocean. The lag from one region to another is consistent with the effect of the ACW. In the Indian Ocean sector, there are also negative anomalies in 1986, 1987, and 1988, which were followed by positive anomalies in 1989 and 1990 and again by negative anomalies in 1991 and 1993. In the Western Pacific Ocean sector, the key events are the positive anomalies in 1982 and 1983 and the negative anomaly in 1989. In the Ross Sea, a large negative anomaly in the early 1980s was followed by positive anomalies the following years except for slight drops in 1986, 1991, and 1997. In the Bellingshausen/Amundsen Sea sector, the ENSO years of 1983, 1988, 1992, and 1998 are shown to coincide with negative anomalies confirming observed relationships of ENSO with the ice cover and ACW as described by Peterson and White (1998) and Kwok and Comiso (2002). Note that the dips at the Ross Sea sector (i.e., 1980, 1986, 1992, 1995, and 1997) which is also regarded as a region affected by ENSO (Ledley and Huang 1997), are not coincident with those of the Bellingshausen/Amundsen Sea sector except for 1992. The patterns are relatively irregular at times because the atmospheric circulation in the region is complex and driven by many factors, some of which are regional in origin. For example, in the Weddell Sea, a large positive anomaly in 1981 was followed by a negative anomaly in the subsequent summer, while a negative anomaly in 1991 was followed by 2 years of positive anomalies. Other anomalous patterns have also been cited in the literature (e.g., Hanna 2001).

The anomaly plots allow for quantitative assessments of the trends in the sea ice cover in the Southern Hemisphere. Using linear regression on monthly anomalies from November 1978 to December 2008, the trends in the ice extent are: 1.2±0.2%/decade for the entire hemisphere; 1.3±0.5%/decade for the Weddell Sea Sector; 2.1±0.6 for the Indian Ocean sector; 0.9±0.7 for the West Pacific sector; 5.0±0.6%/decade for the Ross Sea sector and −7.1±0.9%/decade in the Bellingshausen/Amundsen Seas sector. The trends are relatively minor except at the Ross Sea where the trend is positive and the Bellingshausen/Amundsen Seas where the trend is negative. Similar analysis was done for the monthly anomalies for the ice area and the corresponding trends are: 1.2±0.2%/dec for the entire hemisphere; 1.2±0.5%/dec for the Weddell Sea Sector; 2.1±0.6 for the Indian Ocean sector; 0.8±0.7 for the West Pacific sector; 4.9±0.6 for the Ross Sea sector and −7.3±0.9 in the Bellingshausen/Amundsen Seas sector. These results are similar to those of ice extent and overall, they indicate a relatively stable sea ice cover in the Southern Ocean.

The large seasonal variability in the extent (Fig. 7.26) and the ACW make it difficult to interpret interannual changes or trends from the time series of monthly data. Such interannual changes may be evaluated using seasonal averages such as those presented in Fig. 7.2 for the November 1978 to December 2008 period. In the NH, significant interannual variability are apparent during summer (light gray line) and autumn (dash-dot gray line) but during winter and spring, only small changes are apparent, especially during the 1990–2000 period (Fig. 7.2a). The summer averages have higher values than the autumn averages because the latter includes September, which is the month of minimum extent. In the SH, the interannual changes appear minimal for all seasons except for 1980, 1982, and 1997 during summer and autumn. The plots for ice area in both hemispheres are similar in characteristics to those for ice extents, but the average values are significantly lower. In this case, the summer values are lower than the values during autumn because of relatively lower concentrations during the period. The fine gray lines in the plots in Fig. 7.2 represent the trends in the ice cover for the different seasons as derived from linear regression of all the data points. The trends are all negative in the NH and all slightly positive in the SH, as has been reported previously (Parkinson et al. 1999; Zwally et al. 2002), but with slightly different values. This phenomenon suggests that the climates of the two hemispheres are not closely coupled. The results, however, are consistent with predictions from some Global Circulation Models [e.g., Manabe et al. 1992].

To illustrate how the seasonal trends in the SH are distributed spatially, the trend of each pixel for each season was estimated and the results are presented in Fig. 7.33. It is apparent that the Ross Sea area is a region of consistent positive trend for all seasons, while the Bellinshaussen/Amundsens Seas area is the site for consistent negative trend. In the winter and spring it appears that the distribution of the trends depicts a mode 2 system with the positive trends located in the Weddell and Ross Seas while the negative trends are in the Bellingshaussen/Amundsen and Indian/Western Pacific regions. In the summer and autumn, the atmospheric mode is not obvious but the trends are strongly negative in the Bellingshausen/Amundsen Seas while they are strongly positive in the Ross Sea. Overall, however, there are more areas with positive trends than those with negative trends. These results are consistent with the aforementioned results from hemispherical and regional analysis of the trends in sea ice concentration, ice extent and ice area.

7.3.4 Sea Ice Drift Pattern and Thickness

Studies of the drift of sea ice provide additional insights into the role of sea ice in the climate system. For example, the drift of sea ice provides the means to evaluate where they are formed and where they eventually melt and also how rapidly leads open and close. In the Antarctic, previous studies have made use primarily of Argos buoy data (Ackley 1981; Allison 1989; Vihma and Launiainen 1993; Kottmeier et al. 1992). Because of the paucity of buoys available at any one time, most of the studies have been regional, some focusing on the Weddell Sea while some in the Ross Sea. An example of a regional study in the Weddell Sea, is shown on Fig. 7.34 (Kottmeier and Sellman 1996). The data is a composite of results from analysis of

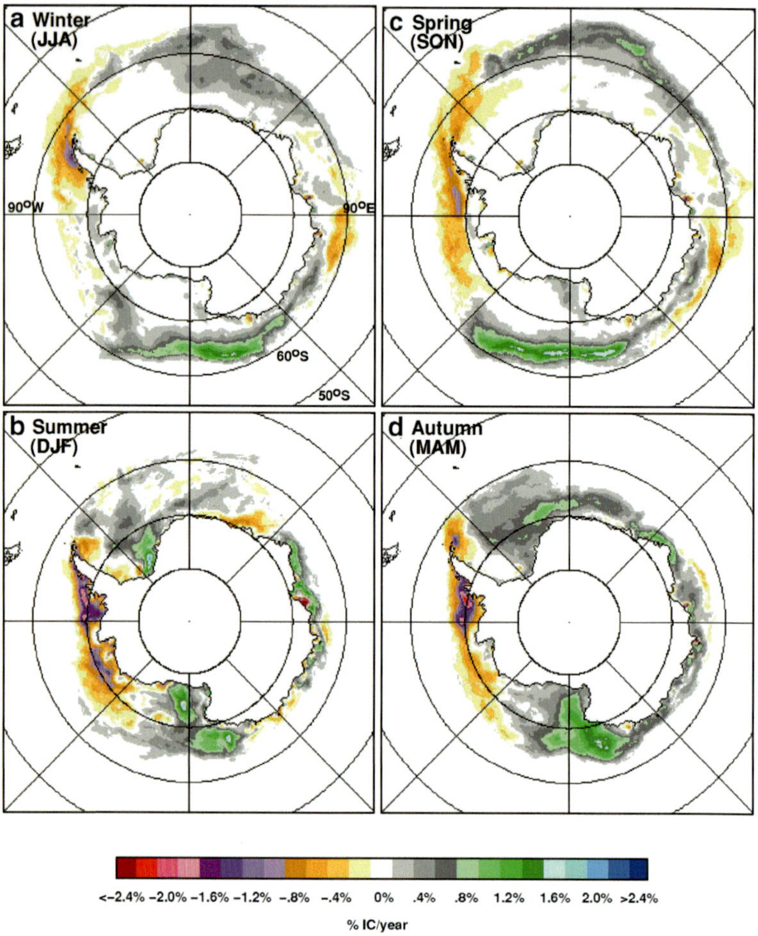

Fig. 7.33 Trends in sea ice concentration in the Southern Hemisphere as derived from SMMR and SSM/I monthly ice concentration data (November 1979 to December 2008) during (**a**) winter; (**b**) spring; (**c**) summer and (**d**) autumn

sea ice buoys deployed during the Winter Weddell Sea Project in 1986, the Winter Weddell Gyre Studies in 1989 and 1992, the Ice Station Weddell in 1992, the Antarctic Zone Flux Experiment in 1994, and several ship cruise in austral summer. The data provide a general direction of ice drift, which in this case defines the circulation of the Weddell Gyre, but because of the lack of a time series of such data, there is no information about interannual variability.

With satellite passive microwave and other data sets, ability to monitor ice drift in the Antarctic on a large scale has become possible. As in the Arctic, data during the summer months are not so reliable and have to be discarded, but in the Antarctic, much of the sea ice cover would disappeared in the summer anyway. The interannual variability of the pattern of the drift of sea ice in the Antarctic is illustrated in

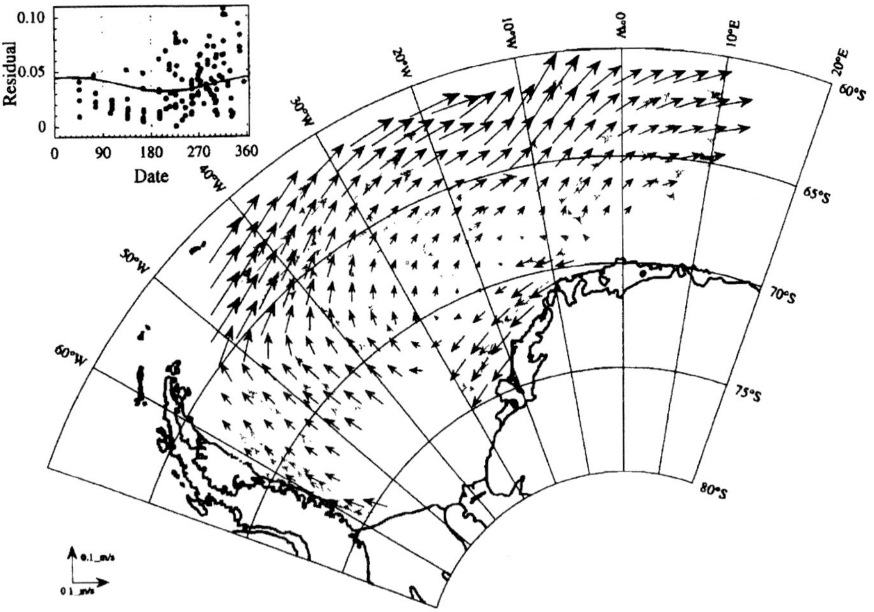

Fig. 7.34 Mean ice drift of sea ice in the Weddell Sea during a thirty day period as derived from buoy data from 1986 to 1994. The residuals (in cm/s) of ice drift speeds from a smooth spatial interpolation is shown in the upper left. From Kottmeier and Sellman (1996) with permission of the American Geophysical Union)

Fig. 7.35, using results from analysis of satellite data. The ice drift data presented are 4-month averages (i.e., from July to October), for each year from 1996 to 2007. The data are derived using a procedure, which makes use of the 85 GHz V-polarization data to track daily ice motion as described in Kwok et al. (1998). Also shown are color-coded contours of sea-level pressure (SLP) winter data (average from July to October) computed from the National Centers for Environmental Prediction (NCEP)-National Center for Atmospheric Research (NCAR) reanalysis output (Kalnay et al. 1996). The drift patterns are shown to be similar for the different years, all showing a cyclonic drift that appears to be persistent during the winter period. The interannual changes in the strength of the drift field also appears to be primarily influenced by changes in the pressure system. Sea ice is shown to drift from the coastal regions of the Ross Ice Shelf, which is a coastal polynya and ice production region, and moves to the north before it gets diverted to the east, following the cyclonic pattern. At the northern reaches near Cape Adare, the sea ice floes that do not follow the gyre may go directly to the north (or northeast) to warm water, where they melt. The strength of drift from near the Ross Sea ice shelf changes from year-to-year and was especially strong in 1998. The drift to the northwest to warmer water was also especially strong in 1999, 2003, 2004, and 2005. Such interannual changes are associated with the location and strength of the sea level pressures as indicated by the isobars. In 2004, when the low-pressure center

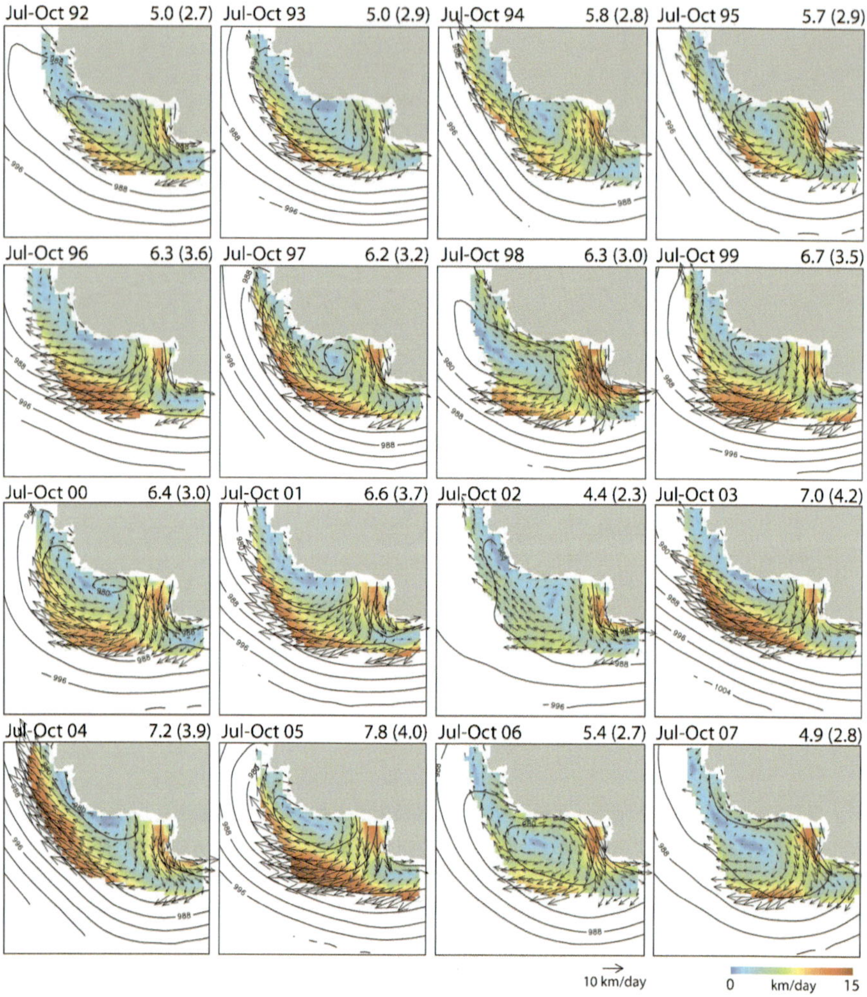

Fig. 7.35 Interannual variability (1992-2007) of mean July-October sea ice motion in the Ross Sea. Isobars of sea-level pressure from NCEP-NCAR products (contour interval: 4 hPa). The quantities on the top right corner are the mean and standard deviation of the drift speed (in km/day) of the vectors within the domain. Courtesy of Ron Kwok of JPL, 2008

was located south of the coastal Antarctica, the drift speed of a band of ice to the south off the Ross Ice Shelf was as high as 7.2 km/day, while that at higher latitudes, going to the northeast to warm water, was about 15 km/day. It is apparent that the drift data provide important information needed for mass balance studies. Through the use of the data, ice production can be assessed through the evaluation of the ice volume coming out of the coastal shelf region, and at the same time, the volume of ice that goes north to the warmer waters can be estimated.

The thickness of sea ice in the Antarctic has not been studied in as much detail as in the Arctic region in part because of the lack of routine measurements such as those provided by upward-looking sonars on board submarines. Estimates of thickness have been based mainly from direct measurements such as those done during field programs (Wadhams et al., 1987). Such measurements have been made initially by drilling hundred of holes in selected ice floes. These measurements have been supplemented by even fewer mooring data in which a moored upward looking sonar is used to estimate the draft of drifting ice floes. More current techniques that had been used to cover much larger areas make use of measurements of the thickness of overturned ice floes as ships are in transit within the ice pack. This utilizes a video camera that is aimed continuously at ice floes and a calibration target located at the side of a ship as described by Worby (1999). Other techniques involves the use instruments mounted on helicopters during field programs. These includes the use of impulse radar as reported by Kovacs and Morey (1986) and the electromagnetic induction technique as described by Holliday et al. (1990) and Hass (2009). Satellite data is obviously needed to obtain to obtain synoptic coverage for all seasons. Although some data exists, their use for this purpose has so far been limited because of the lack of adequate information about the snow cover and overall density of the ice cover. However, significant progress has been made in recent years including the assessments of the freeboard of sea ice has been made using ERS radar altimeter data (Giles et al., 2008) and using ICESat laser altimeter data (Zwally et al., 2008). Getting accurate ice thickness from satellite data is still a challenge and additional research such as a recent aircraft study using concurrent radar and laser data in the Bellingshausen Sea region (Leuschen et al. 2008) could help resolve some of the ambiguities in the interpretation of the data.

References

Ackley, SF (1981) Sea ice atmosphere interactions in the Weddell Sea using drifting buoys. In: Sea Level, Ice and Climatic Change (Proc. Canberra Symposium, December 1979), IAHS Publ. No. 131, pp. 177–191

Ackley SF, Geiger C, King JC, Hunke EC, and Comiso JC (2001) The Ronne Polynya of 1997–1998: Observations of air-ice-ocean interaction, *Ann. Glaciol.* 33:425–429

Alexander V, and Niebauer HJ (1981) Oceanography of the eastern Bering Sea ice edge zone in spring, *Limnol. Oceanogr.* 26:1111–1125

Allison I (1989) The East Antarctic sea ice zone: ice characteristics and drift. *Geo Journal* 18.1:103–115

Allison I, Brandt RE, and Warren SG (1993) East Antarctic sea ice: Albedo, thickness distribution and snow cover, *J. Geophys. Res.* 98(C7):12417–12429

Arrigo KR, Van Dijken GL (2004) Annual changes in sea ice, chlorophyll a, and primary production in the Ross Sea, Antarctica, Deep Sea Res Part II 51:117–138

Bromwich DH and Kurtz DH (1982) Experiences of Scott's Northern Party: Evidence for a relationship between winter katabatic winds and the Terra Nova Bay Polynya, *Polar Rec.* 21:137–146

Bromwich DH and Kurtz DD (1984) Katabatic wind forcing of the Terra Nova Bay polynya, *J. Geophys. Res.* 89:3561–3572

Bromwich DH, Liu Z, Rogers AN, and Va Woert ML (1998) Winter atmospheric forcing of the Ross Sea polynya, in: Jacobs S and Weiss R (Eds.), Oceans, Ice and Atmosphere: Interactions at the Antarctic Continental Margin, *Antarctic Res. Ser.*, vol. 75, AGU, Washington, DC, pp. 101–133

Cavalieri DJ and Martin S (1985) A passive microwave study of polynyas along the Antarctic Wilkes Land Coast, in: Jacobs S (ed) *Oceanology of the Antarctic Continental Shelf*, Antarctic Research Volume 43, pp. 203–226

Colony R, Thorndike AS (1985) Sea ice motion as a drunkard's walk. *J Geophys Res* 90: 965–974

Comiso JC (1986) Characteristics of Arctic winter sea ice from satellite multispectral microwave observations. *J Geophys Res* 91(C1):975–994

Comiso JC (2002) A rapidly declining Arctic perennial ice cover, *Geophys. Res. Lett.* 29(20):1956, doi:10.1029/2002GL015650

Comiso JC (2003) Warming trends in the Arctic, *J. Clim.* 16(21):3498–3510

Comiso JC (2006) Abrupt decline in the Arctic winter sea ice cover, *Geophys. Res. Lett.* 33:L18504, doi:10.1029/2006GL027341

Comiso JC (2009) Enhanced sea ice concentration and ice extent from AMSR-E, *J. Remote Sens. Soc. Jpn.* 29(1):199–215

Comiso JC and Gordon AL (1996) The Cosmonaut Polynya in the Southern Ocean: Structure and variability, *J. Geophys. Res.* 101(C8):18297–19313

Comiso JC and Gordon AL (1998) Interannual variabilities of summer ice minimum, coastal polynyas, and bottom water formation in the Weddell Sea, in: Jeffries M (ed) Antarctic sea ice physical properties and processes, *AGU Antarctic Research Series Volume*, pp. 293–315

Comiso JC and Nishio F (2008) Trends in the sea ice cover using enhanced and compatible AMSR-E, SSM/I, and SMMR data, *J. Geophys. Res.* 113:C02S07, doi:10.1029/2007JC004257

Comiso JC and Steffen K (2001) Studies of Antarctic sea ice concentrations from satellite data and their applications, *J. Geophys. Res.* 106(C12):31361–31385

Comiso JC, Wadhams P, Pedersen LT, Gersten R (2001) The seasonal and interannual variability of the Odden and a study of environmental effects. *J Geophys Res* 106(C5):9093–9116

Comiso JC, Grenfell TC, Lange M, Lohanick A, Moore R, and Wadhams P (1992) Microwave remote sensing of the Southern Ocean ice cover. In: Carsey F (ed) Chapter 12, Microwave Remote Sensing of Sea Ice, pp.243–259, American Geophysical Union, Washington, DC

Comiso JC, Grenfell TC, Lange M, Lohanick A, Moore R and Wadhams P (1992) Microwave remote sensing of the Southern Ocean ice cover. In: Carsey F (ed) Chapter 12, Microwave Remote Sensing of Sea Ice, American Geophysical Union, Washington, D.C., 243–259

Comiso JC, Parkinson CL, Gersten R, and Stock L (2008) Accelerated decline in the Arctic sea ice cover, *Geophy. Res. Lett.* 35:L01703, doi:10.1029/2007GL031972

De Veaux R, Gordon AL, Comiso JC, and Chase NE (1993) Modeling of topographical effects on Antarctic sea ice using multivariate adaptive regression splines, *J. Geophys. Res.* 98(C11):20207–20319

Drinkwater MR and Lytle VI (1997) ERS 1 radar and field-observed characteristics of autumn freeze-up in the Weddell Sea. *J. Geophy. Res.* 102(C6):12593–12603

Eppler D, Anderson MR, Cavalieri DJ, Comiso JC, Farmer LD, Garrity C, Gloersen P, Grenfell T, Hallikainen M, Lohanick AW, Maetzler C, Melloh RA, Rubinstein I, Swift CT, Garrity C (1992) Passive microwave signatures of sea ice, Chapter 4, in: Carsey F (ed) *Microwave Remote Sensing of Sea Ice*, American Geophysical Union, Washington, DC, pp. 47–71

Fetterer FM, Drinkwater MR, Jezek KC, Laxon SWC, Onstott RG, Ulander LMH (1992) Chapter 7: Sea ice altimetry, in: Carsey F (ed) *Microwave Remote Sensing of Sea Ice*, American Geophysical Union, Washington, DC, pp. 111–135

Gloersen P, Campbell W, Cavalieri D, Comiso J, Parkinson C, Zwally HJ (1992) Arctic and Antarctic sea ice, 1978-1987: Satellite passive microwave observations and analysis, *NASA Spec. Publ.* 511

Gordon AL and Comiso JC (1988) Polynyas in the Southern Ocean, *Sci. Am.* 256:90–97

Gordon AL, Visbeck M, and Comiso JC (2007) A link between the Great Weddell Polynya and the Southern Annular Mode, *J. Clim.* 20(11):2558–2571

Grenfell TC, Comiso JC, Lange MA, Eicken H, and Wenshahan MR (1994) Passive microwave observations of the Weddell Sea during austral winter and early spring, *J. Geophys. Res.* 99(C5):9995–10010

Hanna E (2001) Anomalous peak in Antarctic sea-ice area, winter 1998, *Geophys. Res. Lett.* 28:1595–1598

Hass C (2009) Dynamics versus thermodynamics: The sea ice thickness distribution. In: Thomas DN, Dieckmann GS (eds) Sea Ice, pp. 113–152, Wiley-Blackwell, West Sussex, UK

Holiday JS, Rossiter JR, Kovacs A (1990) Airborne measurement of sea ice thickness using electromagnetic induction sounding. in Ayorinde OA, Sinha NK, Sodhi DS (eds) Proceedings of the Ninth International Conference of offshore mechanics and Arctic engineering, pp. 309–315, American Society of Mechanical Engineers, New Hampshire

Jacobs SS and Comiso JC (1989) Satellite passive microwave sea ice observations and oceanic processes in the Ross Sea, Antarctica, *J. Geophys. Res.* 94:18195–18211

Jacobs SS and Comiso JC (1993) A recent sea-ice retreat west of the Antarctic Peninsula, *Geophys. Res. Lett.* 20(12):1171–1174

Jacobs SS and Comiso JC (1997) Climate variability in the Amundsen and Bellingshausen Seas. *J. Clim.* 10(4):697–709

Johannessen OML, Salina EV, Miles MW (1999) Satellite evidence for an arctic sea ice cover in transformation, Science 286:1937–1939

Kalnay E et al. (1996) The NCEP/NCAR 40-year reanalysis project, *Bull. Am. Meteorol. Soc.* 77:437–471

King JC and Harangozo SA (1998) Climate change in the western Antarctic Peninsula since 1945: Observations and possible causes, *Ann. Glaciol.* 27:571–575

Kottmeier C, Sellmann L (1996) Atmospheric and oceanic forcing of Weddell Sea ice motion. *J Geophys Res* 101(C9):20809–20824

Kottmeier C, Olf J, Frieden W, Roth R (1992) Wind forcing and ice motion in the Weddell Sea. *J Geophys Res* 97(D18):20373–20383

Kovacs A, Morey RM (1986) Electromagnetic measurements of multiyear sea ice using impulse radar. Cold Regions Science and Technology 12:67–93

Kwok R (2005) Ross sea ice motion, area flux and deformation, *J. Clim.* 18:3759–3776

Kwok R and Comiso JC (2002) Spatial patterns of variability in Antarctic surface temperature: Connections to the Southern Hemisphere Annular Mode and the Southern Oscillation, *Geophys. Res. Lett.* 29(14), 10.1029/2002GL015415

Kwok R and Rothrock DA (1999) Variability of Fram Strait ice flux and North Atlantic Oscillation, *J. Geophy. Res.* 104(C3):5177–5189

Kwok R, Schweiger A, Rothrock DA, Pang S, Kottmeier C (1998) Sea ice motion from satellite passive microwave imagery assessed with ERS SAR and buoy motions. *J Geophys Res* 103(C4):8191–8214

Kwok R, Comiso JC, Martin S, and Drucker R (2007) Ross Sea polynyas: Response of ice concentration retrievals to large areas of thin ice, *J. Geophys. Res.* 112:C12012, doi:10.1029/2006JC003967

Kwok R, Cunningham GF, Wensnahan M, Rigor I, Zwally HJ, Yi D (2009) Thinning and volume loss of the Arctic Ocean sea ice cover: 2003-2008, *J. Geophys. Res.* 114:C07005

Lange MA, Ackley SF, Wadhams P, Dieckmann GS, Eicken H (1989) Development of sea ice in the Weddell Sea, Ann. Glacio 12:92–96

Laxon S, Peacock N, Smith D (2003) High interannual variability of sea ice thickness in the Arctic region, *Nature* 425(30):947–949

Ledley TS and Huang Z (1997) A possible ENSO signal in the Ross Sea, *Geophys. Res. Lett.* 24:3253–3256

Liu AK, Zhao Y, and Liu WT (1998) Sea ice motion derived from satellite agrees with buoy observations. *EOS Trans.* 79:353, 359

Manabe S, Spelman MJ, and Stouffer RJ (1992) Transient responses of a coupled ocean-atmosphere model to gradual changes of atmospheric CO2, Part II, Seasonal response. *J. Clim.* 5:105–126

Markus T and Cavalieri DJ (2009) The AMSR-E NT2 Sea ice concentration algorithm: Its basis and implementation. *Remote Sens. Soc. Jpn.* 29(1):216–223

Martin S, Drucker R, Kwok R, and Holt B (2004) Estimation of the thin ice thickness and heat flux for the Chukchi Sea Alaskan coast polynya from SSM/I data, 1990-2001, *J. Geophys Res.* 109:C10012, doi:10.1029/2004/JC002428

Martin S, Drucker R, Kwok R, and Holt B (2005) Improvements in the estimates of ice thickness and production in the Chukchi Sea polynyas derived from AMSR-E, *Gephys. Res. Lett.* 31:L005505, doi:10.1029/2004GL0022013

Martin S, Drucker RS, and Kwok R (2007) The areas and ice production of the western and central Ross Sea polynyas, 1992–2002, and their relation to the B-15 and C-19 iceberg events of 2000 and 2002, *J. Mar. Syst.* 68:201–214

Masson R (1991) Satellite remote sensing of polar regions, Belhaven Press, London, UK

Niebauer HJ, Alexander V, and Henrichs SM (1995) A time series study of the spring bloom at the Bering Sea ice edge I. Physical processes, chlorophyll and nutrient chemistry, *Continental Shelf Res* 15(15):1859–1877

Onstott RG (1992) Chapter 5. Sar and scatterometer signatures of sea ice. In: Carsey F (ed) *Microwave Remote Sensing of Sea Ice*, American Geophysical Union, Washington, DC, pp. 73–104

Overland JE and Wang M (2005) The Arctic climate paradox: The recent decrease of the Arctic Oscillation, *Geophys. Res. Lett.* 32:L06701, doi:10.1029/2004GL021752

Parkinson CL and Comiso JC (2008), Antarctic sea ice from AMSR-E from two algorithms and comparisons with sea ice from SSM/I, *J. Geophys. Res.* 113:C02S06, doi:10.1029/2007JC004253

Parkinson CL, Comiso JC, Zwally HJ, Cavalieri DJ, Gloersen P, and Campbell WJ (1987) Arctic sea ice 1973-1976 from satellite passive microwave observations, *NASA Spec. Publ.* 489

Parkinson CL, Cavalieri DJ, Gloersen P, Zwally HJ, and Comiso JC (1999) Arctic sea ice extents, areas, and trends, 1978-1996, *J. Geophys. Res.* 104(C9):20837–20856

Perovich DK, Tucker III WB, and Ligett KA (2002) Aerial Observations of the evolution of ice surface conditions during summer, *J. Geophys. Res.* 107(C10):8048, doi:10.1029/2000JC000449

Peterson RG and White WB (1998) Slow teleconnections linking the Antarctic Circumpolar Wave with the tropica; El Nino-Southern Oscillation. *J. Geophys. Res.* 103(C11):24573–24583

Proshutinsky AY and Johnson MA (1997) Two circulation regimes of the wind-driven Arctic Ocean, *J. Geophys. Res.* 102:12493–12514

Rothrock DA, Yu Y, Maykut G (1999) Thinning of the Arctic sea-ice cover. *Geophys Res Lett* 26:3469–3472

Savage ML and Stearns CR (1985) Climate in the vicinity of Ross Island, Antarctica, *Antarct. J. U.S.* 20:1–9

Smith Jr WO, Nelson DM (1985) Phytoplankton bloom produced by a receding ice edge in the Ross Sea: Spatial coherence with the density field. Science 227:163–166

Smith SD, Muench RD, Pease CH (1990) Polynyas and leads: An overview of physical processes and environment. *J Geophys Res* 95(C6):9461–9479

Stammerjohn SE, Marinson DG, Smith RC, Yuan X, and Rind D (2008) Trends in Antarctic annual sea ice retreat and advance and their relation to El Nino-Southern Oscillation and Southern Annular Mode variability, *J. Geophys. Res.* 113:C03S90, doi:10.1029/2007JC004269

Steig EJ, Schneider DP, Rutherford SD, Mann ME, and Comiso JC (2009) Warming of the Antarctic ice sheet surface since the 1957 International Geophysical Year, *Nature* 457:459–463

Stroeve JC, Holland MM, Meier W, Scambos T, and Serreze M (2007a) Arctic sea ice decline: Faster than forecast. *Geophys. Res. Lett.* 34:L09501, doi:10.1029/2007/GL029703

Stroeve J, Holland MM, Meier W, Scambos T, Serreze MC (2007) Arctic sea ice decline: Faster than forecast. *Geophys Res Lett* 34:L09501, doi:10.1029/2007GL0029703

Sullivan CW, Arrigo KR, McClain CR, Comiso JC, and Firestone J (1993) Distributions of phytoplankton blooms in the Southern Ocean, *Science* 262:1832–1837

Thompson DW and Wallace JM (1998) The Arctic oscillation signature in the wintertime geopotential height and temperature fields, *Geophys. Res. Lett.* 25(9):1297–1300

Turner J, Comiso JC, Marshall GJ, Connolley WM, Lachlan-Cope TA, Bracegirdle T, Wang Z, Meredith M, and Maksym T (2009) Antarctic sea ice extent increases as a result of anthropogenic activity, *Geophys. Res. Lett.* 36:L08502, doi:10.1029/2009GL037524

Tsatsoulis C, Kwok R (Editors) (1998) Analysis of SAR data of the Polar Oceans: Recent Advances. Springer-Verlag, Berlin

Vilma T, Launiainen J (1993) Icer drift in the Weddell Sea in 1990-1991 as tracked by a satellite buoy. *J Geophys Res* 98:14471–14485

Vinje T (2001) Fram Strait ice fluxes and atmospheric circulation 1950-2000. *J. Clim.* 14:3508–3517

Vinje T and Finnekasa O (1986) The ice transport through the Fram Strait, *Norsk Polariinst. Skrifter Rep.* 186:39

Wadhams P (1997) Ice thickness in the Arctic Ocean: The statistical reliability of experimental data. *J Geophys Res* 102(C13):27,951–27,959

Wadhams P (2000) Ice in the Ocean. Gordon and Breach Science Publishers, London, UK

Wadhams P and Comiso JC (1992) The ice thickness distribution inferred using remote sensing techniques. in: Carsey F (ed) Microwave remote sensing of sea ice, AGU Geophuysical Monograph 68, American Geophysical Union, Washington, DC

Wadhams P and Davis NR (2000) Further evidence of ice thinning in the Arctic Ocean, *Geophys. Res. Lett.* 27:3973–3975

Wadhams P, Lange MA, Ackley SF (1987) The ice thickness distribution across the Atlantic sector of the Antarctic Ocean in midwinter. *J Geophys Res* 92(C13):14535–14552

Wadhams P, Tucker W, Krabill W, Swift R, Comiso J, Davis N (1992) The relationship between sea ice freeboard and draft in the Arctic Basin and apllications for ice thickness monitoring. *J Geophys Res* 97(C12):20325–20334

Walsh JE, Chapman WL, and Shy TL (1996) Recent decrease of sea level pressure in the Central Arctic, *J. Clim.* 9:480–486

White WB and Peterson RG (1996) An Antarctic circumpolar wave in surface pressure, wind, temperature and sea ice extent, *Nature* 380:699–702

Worby AP and Comiso JC (2004) Studies of Antarctic sea ice edge and ice extent from satellite and ship observations, *Remote Sens Environ* 92(1):98–111

Yang J and Comiso JC (2007) Unexpected seasonal variability in salinity of the Beaufort Sea upper layer in 1996-1997, *J. Geophys. Res.* 112:C05034, doi:10.1029/2004JC002716

Yang J, Comiso J, Krishfield R, and Honjo S (2001) The role of synoptic storms in the development of the 1997 warming and freshening event in the Beaufort Sea, *Geophys. Res. Lett.* 28(5):799–802

Zwally HJ, Comiso JC, Parkinson CL, Campbell WJ, Carsey FD, and Gloersen P (1983) Antarctic sea ice 1973-1976 from satellite passive microwave observations. *NASA Spec. Publ.* 459, Greenbelt, MD

Zwally HJ, Comiso JC, and Gordon AL (1985) Antarctic offshore leads and polynyas and oceanographic effects, in: Jacobs S (ed) *Oceanology of the Antarctic Continental Shelf*, Antarctic Research Volume 43, pp. 203–226

Zwally HJ, Yi D, Kwok R, Zhao Y (2008) ICESat measurements of sea ice freeboard and estimate of sea ice thickness in the Weddell Sea. *J Geophys Res* 113:doi:10.1029/2007/JC004284

Zwally HJ, Comiso JC, Parkinson CL, Cavalieri DJ, and Gloersen P (2002) Variability of the Antarctic sea ice cover, *J. Geophys. Res.* 107(C5):1029–1047

Chapter 8
Polynyas and Other Polar Phenomena

Abstract The important roles of sensible and latent heat polynyas, the Odden, the Antarctic Circumpolar Wave, polar outbreaks, icebergs, and other features and processes in the polar oceans are discussed in the context of their spatial distributions and frequency of occurrences as inferred from long term and synoptic observations by satellite sensors. Satellite data have provided valuable insights into how these special features could affect the thermohaline circulation and primary productivity of the oceans and cause deep vertical convection, ocean ventilation, storminess, and horizontal transport of salt in the polar oceans. It is apparent that large changes are occurring, including frequency of occurrence of the Odden and the character of the Antarctic Circumpolar Wave. Many questions remain, but some answers are possible through concurrent use of satellite data, in situ data and numerical models.

Keywords Polynyas • Odden • Antarctic circumpolar wave • Polar outbreaks

8.1 Introduction

The sea ice cover is far from being just a continuous sheet of ice that many people would expect in ocean areas, where the surface is below the freezing temperature. As indicated in Chap. 7, the ice cover is actually very dynamic, and it is constantly being altered by winds, waves, tides, temperature, and other external forcing. The ice cover is also relatively thin compared to the depth of the ocean and would break up constantly due to winds, tides, and other factors, to form narrow cracks, called leads. Leads look like rivers in a terrestrial environment, and they are generally linear and very long, extending sometimes into hundreds of kilometers in length. Strong winds also cause the sea ice cover to be advected from coastal regions, islands, or icebergs forming less linear open water areas. These phenomena have been known for a long time (i.e., by Inuits and others) and have been referred to as "polynyas," which is a Russian word for a body of open water surrounded by ice. Satellite data, however, revealed that in addition to these leads and coastal polynyas, there are vast bodies of

J. Comiso, *Polar Oceans from Space*, Atmospheric and Oceanographic Sciences Library 41, DOI 10.1007/978-0-387-68300-3_8,

open water within the ice pack, and in the deep ocean areas. The existence of large open water areas in the middle of the ice packs was first recorded by TIROS satellites in the 1960s, but these satellites carried visible and infrared channels only and full characterization of the features was not possible because of persistent clouds in the polar regions. The polynyas observed during this period were actually those that occurred during the ice breakup in spring. Such spring polynyas are prevalent in the Weddell and Ross Seas in the Southern Ocean and usually occur in areas where the sea ice cover gets melted before that of surrounding areas. The launch of Nimbus-5/EMSR in October 1972 made it possible to monitor the total ice cover in both the Northern and Southern Hemisphere even during periods of darkness in winter and during cloudy conditions. Among the most interesting discoveries made with this sensor is the unusually large winter polynya in the Weddell Sea in the 1970s as has been mentioned earlier and will be discussed in detail in the next section. The data also provided the means to assess with limited accuracy the full extent of large leads and other polynyas including coastal polynyas, along the periphery of the Antarctic continent, the North Sea Polynya near northern Greenland, the St. Lawrence polynya in the Bering Sea, and the New Zemlya polynya near the Eurasian continent.

Satellite data have also led to a greater understanding of an unusual feature in the Greenland Sea called the Odden ice tongue, or simply the Odden, which is usually located near one of the four sites of major deep convection in the world's oceans. Such a feature was known for a long time and was encountered by Nansen, but nobody knew its full extent and temporal variability until the advent of satellite data. Satellite data also enabled the detailed study of the Antarctic Circumpolar Wave (ACW) and other processes that control the circulation and dynamics of the sea ice cover in the Southern Ocean. Although the ACW as originally proposed has been somewhat controversial, it has some scientific merit and is useful in explaining some of the observed variability of the ice cover. Satellite data have also been used to study the marginal ice zone which is a region where air-sea-ice interactions are very intense. It is the region where cold air outbreaks occur in conjunction with polar laws, which are sites where extreme weather events have been known to occur. Big waves and strong winds keep the ice cover from being consolidated in the region. While the ice cover is known to be altered constantly by wind, temperature, tides, currents, and other parameters, it can also be altered by the iceberg calving. Satellite sensors have been able to observe many of the big iceberg events in recent times that led to significant declines in the volume of glaciers and ice shelves. Some of these icebergs were so big that they were stranded in relatively shallow regions thereby restricting the drift of sea ice and altering the direction of surface current. They also serve to create latent heat polynyas, limit the formation of polynyas in other places, and alter the productivity of adjacent regions.

In this chapter, we will provide some insights into the physical characteristics of polynyas, Odden, MIZs, polar lows and the effects of icebergs on the sea ice cover and the ocean. The ability to monitor them in good spatial and temporal detail has made satellite sensors important tools for studying the significance and impacts of these important events and features. In depth understanding of these processes, however, may require analysis of these satellite data in conjunction with field measurements and numerical models.

8.2 Polynyas

Polynyas are regarded as among the most important features of the sea ice cover because of their role in climate change and in many oceanic and atmospheric processes. As discussed previously, they have been postulated to be the main source of the high density and high salinity water (e.g., Zwally et al., 1985) that becomes a part of the World's bottom water that drives the "thermohaline circulation" as depicted approximately in Fig. 2.5. This global circulation system is driven mainly by the differences in water densities attributed to temperature (thermo) and salinity (haline) anomalies. Any change in the formation rate of sea ice that would alter such differences in water densities could thus have a significant impact on the thermohaline circulation. Since the ocean is such a big part of the climate system, a change in ocean circulation would cause a significant alteration of the climate. Polynyas are also regions where vertical convection is initiated, thus enabling the ocean to be ventilated through exchanges of chemicals between the deep ocean and the surface/mixed layer. This is an important process that enables the replenishment of chemicals and/or nutrients in regions where they have been depleted over time. Moreover, polynyas have been observed to be areas of high productivity during the melt of sea ice in spring causing the formation of a stable upper layer of water which is an ideal platform for photosynthesis because of the abundance of sunlight and nutrients in such layers as described by Smith and Comiso (2008). With the plankton being at the bottom of the food web, these are areas where organisms at all tropic levels congregate in search of food. Many of the areas of persistent polynyas in the Arctic have actually been documented as regions where animal and human fossils have been found. They were definitely the hunting sites for indigenous inhabitants of the Arctic and enabled them to survive the harsh and bitter conditions in the region, especially during winter.

Polynyas have been classified as either sensible heat polynyas or latent heat polynyas (Gordon and Comiso 1988). The formation of these two types of polynyas is schematically illustrated in the diagram shown in Fig. 8.1. As the names imply, in sensible heat polynyas, heat is released through a change in water temperature (and therefore can be sensed) while in latent heat polynyas, heat is given up through a change in physical state (from liquid to solid). Sensible heat polynyas are formed through the melting of sea ice in a localized area where the sensible heat of the surface water is relatively high. Such relatively high sensible heat may be due to the upwelling of warm water as could happen in highly convective regions. In winter, the surface air temperature in the polar regions is normally several degrees below melt temperatures and, therefore, some external sources of warm water are needed to keep ice from forming in an ice-free polynya region. In Fig. 8.1, we show that upwelling could be due to the presence of a topographical feature, like the Maud Rise in the Weddell Sea, which serves to deflect relatively warm circumpolar deep water to the surface. Latent heat polynyas are formed near coastal regions, islands, and icebergs and are maintained by strong and persistent winds (e.g., katabatic winds), that cause the displacement of ice away from the shoreline and the formation of open water areas, as illustrated in the schematics. The surface of the open water area gets frozen almost immediately in such environment, and it is called latent heat polynya because

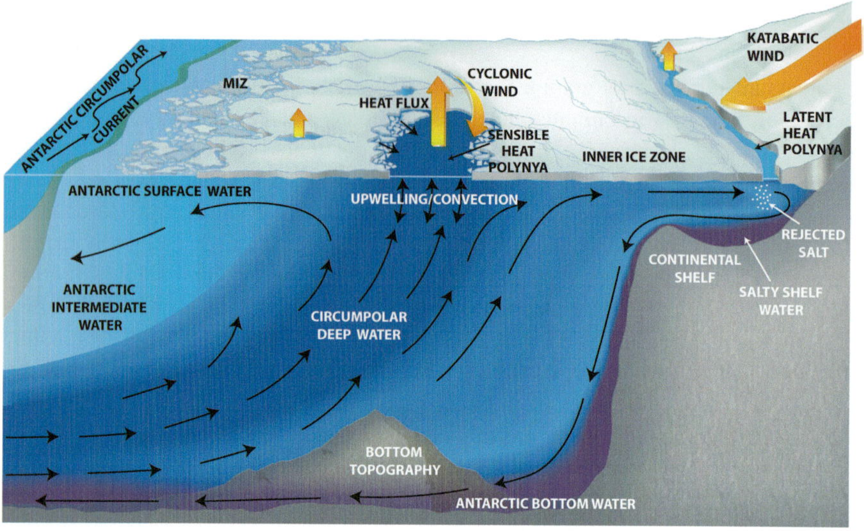

Fig. 8.1 Schematic diagram illustrating the formation of latent and sensible heat polynyas (adapted from Gordon and Comiso, 1988). The black arrows show ocean circulation including the upwelling of warm water that leads to the formation of sensible heat polynyas. From Comiso and Drinkwater (2007) with permission of Cambridge University Press

"latent heat" is released during the transformation of liquid water to ice. This may actually be the more important type of polynya since it causes the formation of cold high density water through the ejection of brine during ice formation as illustrated in Fig. 8.1. The process leads to the formation of Antarctic bottom water that becomes part of the thermohaline circulation as mentioned earlier. Also, because of the persistence of wind and the almost continuous process of polynya formation, these regions are usually regarded as ice factories and are also the source of a large fraction of the ice cover (Zwally et al. 1985; Markus et al. 1998; Comiso and Gordon 1998).

8.2.1 Sensible Heat Polynyas

The most spectacular sensible heat polynya ever observed in the two hemispheres has been the large Weddell Polynya in the Southern Ocean (Zwally and Gloersen 1977; Carsey 1980; Gordon and Comiso 1988). It was spectacular because it was unusually large, averaging about 250,000 km^2 in winter and that it stayed unfrozen for the entire sea ice season in an area where the surface temperature usually gets as low as −35°C (see Fig. 6.23). Furthermore, it occurred in approximately the same area continuously during three consecutive ice seasons (i.e., 1974, 1975, and 1976). Fortuitously, it occurred after the launch of Nimbus-5/ESMR, and satellite data were available to enable continuous and consistent observations of the variability and persistence of this phenomena (Zwally et al. 1983). It was initially assumed that the Weddell Polynya was a frequent feature of the Southern Ocean since throughout the

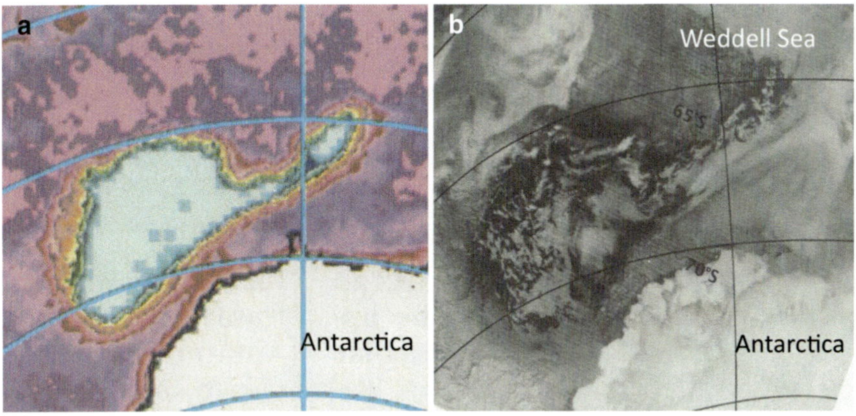

Fig. 8.2 The Weddell Polynya in September 1975 as observed by (**a**) ESMR and (**b**) VHRR sensors. In (**a**), Purples light blues represent open water. while in (**b**), black represents open wahter while white represents ice covered areas or clouds. These oases of open water have been well known to Inuits and other natives in the Arctic region for generations. Archaeologic studies in the north have revealed the presence of human and animal remains in the vicinity of polynyas. From Gordon and Comiso (1988) reproduced with the permission of Scientific American)

period that ESMR was in operation it was almost always there, first as a strong spring polynya in 1973 and then as a persistent winter polynya in the following three years. Since then, however, the polynya has not recurred. This led to speculations that there was a misinterpretation of the brightness temperature data from the satellite passive microwave sensor, which at that time was a relatively new instrument. Figure 8.2 shows near simultaneous images of the polynya from two independent observations providing an indisputable evidence that it was a real phenomenon with an average size comparable to that of Italy. The images show an unobscured view of the polynya as provided by passive microwave data (Fig. 8.2a) and the visible channel of the NOAA Very High Resolution Radiometer (VHRR) data in the third week of August 1975. The VHRR sensor is the predecessor of the AVHRR sensor and was one of the few imaging sensors then available. The polynya as depicted by the VHRR data was partly covered by clouds which were likely formed in part by unusually high heat, and humidity fluxes from the exposed surface of the ocean in the polynya region. The two images complement each other with VHRR data providing high resolution ice and surface temperature information while microwave data providing continuous and full coverage of the sensible heat polynya.

The impact of such a large polynya event has not been fully assessed because of the lack of in situ measurements during the time period when it occurred. Preliminary assessments, however, suggest that the consequence of such a feature can be profound. Hydrographic measurements in 1977 in this region of the waters of the Weddell Sea revealed that the ocean temperature down to 3,000 m depth had cooled by half a degree centigrade after the polynya event (Gordon and Comiso 1988). Figure 8.3 shows the profiles for potential temperature, salinity and density, before 1973 and after 1977. The difference in deep ocean temperature between the two periods is a manifestation that significant convection or "overturning" took place,

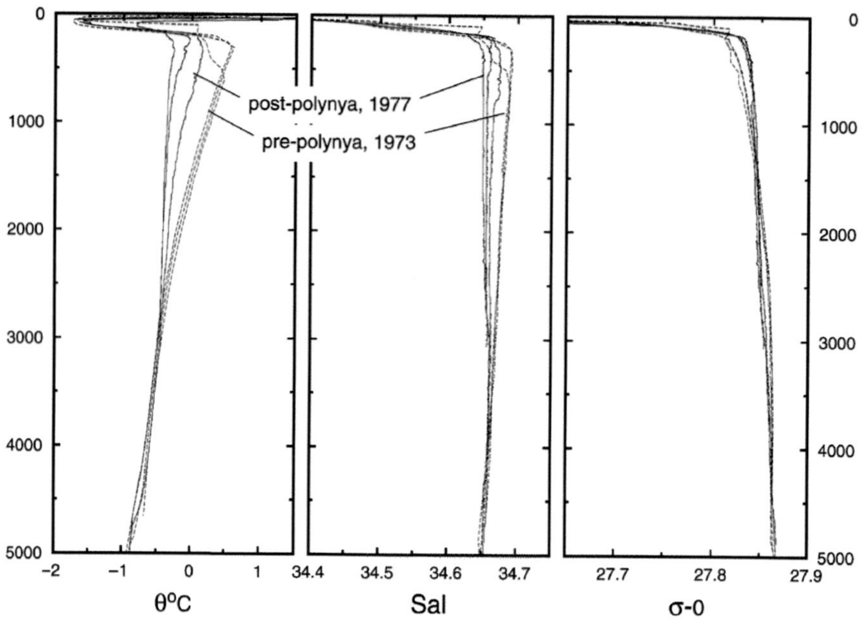

Fig. 8.3 Profiles of potential temperature, salinity (psu) and density before 1973 and after 1977 in the Weddell Sea. The large Weddell Polynyas occurred during the years in between. From Gordon et al. (2007) with permission of the American Geophysical Union

allowing heat to be drawn from the water column by the cold atmosphere. Through the convective process, the deep ocean gets "ventilated," thereby restoring oxygen levels that are depleted by the oxidation of organic material, as well as introducing elements of anthropogenic origin when the water is exposed to the atmosphere. Cyclonic winds and the upwelling of relatively warm water may be needed to initiate its formation, but it takes ocean convection to keep the surface water warm in the large ice-free region for a relatively long period (Martinson et al. 1981).

Simulation studies using coupled sea ice – ocean physical models have indicated the importance of cyclonic winds in at least the initial stages of the formation process, but wind by itself is not able to explain the sustained duration or reappearance of the phenomenon (Lemke et al. 1990; Hibler and Ackley 1983; Parkinson 1983). Oceanographic conditions are therefore believed to play a significant role, through the upwelling of warm water over bottom topographic features, like the Maud Rise, maintained in this case by the Weddell Gyre. The presence of a large ocean eddy at the Maud Rise during the occurrence of the polynya was postulated by some (Holland 2001) to explain the persistence. Since this type of sustained polynya has not reappeared since the 1970s, the debate over the precise formation mechanism and frequency occurrence continues.

A recent report by Gordon et al. (2007) provides new insights into the possible cause of the Weddell Polynya. The atmospheric circulation in the Antarctic is governed by the Southern Annular Mode (SAM), which is also called Antarctic

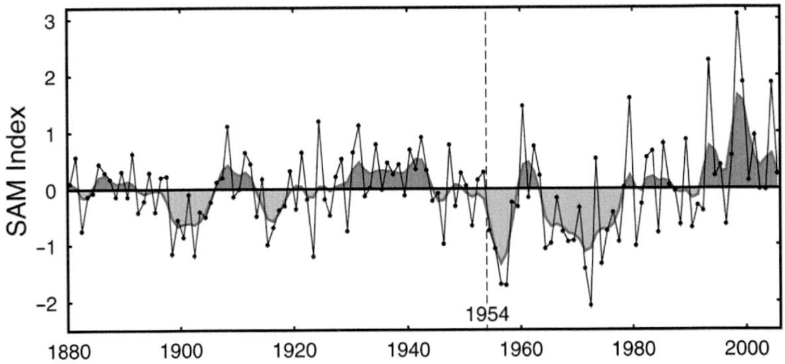

Fig. 8.4 Southern Annular Mode (SAM) indices based on sea level air pressure (SLP) for various sectors of the Southern Ocean. Yearly values are shown in the *black line* connecting annual values. The thicker *grey line* and *gray pattern* denotes the 5 year low pass filtered data. The index prior to 1954 (*dash line*) is computed without Antarctic SLP stations and thus more uncertain. From Gordon et al. (2007) with permission of the American Geophysical Union

Oscillation (AAO), and the atmospheric variability south of 30°S can be studied using SAM indices which is the difference in pressure of an annular area at 40°S and that at 65°S. The variability of the SAM index from 1880 to 2002 is shown in Fig. 8.4 to fluctuate from positive to negative phases and vice versa, and in the 1970s, it is shown that the SAM index was consistently negative. In the 1980s and 1990s, it was mainly positive. The SAM indices have been shown by Visbeck and Hall (2004) to be related to the circulation pattern of the ocean and sea ice in the region. It was postulated by Gordon et al. (2007) that the Weddell polynya may have been induced by the observed prolonged negative phase of SAM during the 1970s. Such prolonged negative phase of SAM may have led to colder and drier atmospheric conditions that could cause the formation of a salty surface layer that reduces the stability of the pycnocline. This in turn would enable the upwelling of warm water at the Maud Rise which is near the center of the observed polynya and cause the initiation of the Weddell Polynya that is sustained through deep-reaching convection during the winter period (see Fig. 8.1). According to the Gordon et al. (2007) study, the positive phase since 1980 that led to warmer, wetter conditions over the region may have kept the polynya from forming. It would be interesting to know if a large Weddell Polynya also formed in 1900, 1915 and 1955 when SAM as also in its negative phase.

Although the large Weddell polynya has not recurred since 1976, the Weddell Sea has been the site of transient and smaller polynyas, usually centered about the Maud Rise. Such polynya, which has been referred to as the Maud Rise polynya, sometimes stays on for about a week or two, disappears for several weeks and may reappear again during the same winter season. However, a spring polynya almost always occurs in the same location. The frequency statistics of the occurrence of the polynya has been studied in conjunction with the topography of the region by De Veaux et al. (1993) and the results show good correlations of the location of polynya occurrence with a sloping side of the Maud Rise. Figure 8.5 shows ice concentration maps from AMSR-E using the 89 GHz and the standard products as well as MODIS visible channel data

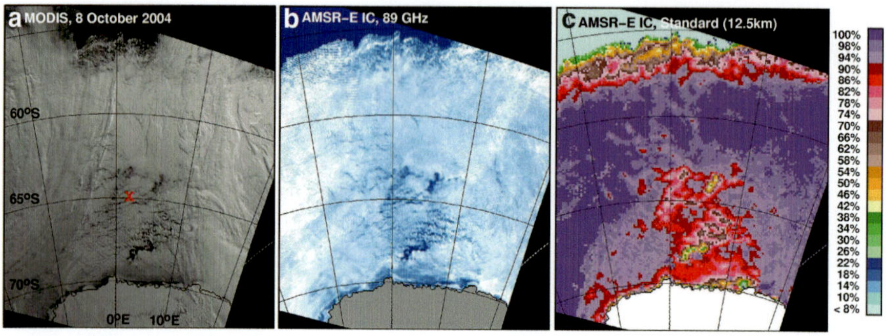

Fig. 8.5 A Maud Rise transient polynya at the Maud Rise as observed on 8 October by (**a**) EOS-Aqua/MODIS, (**b**) AMSR-E high resolution (89 GHz) channels; and (**c**) AMSR-E as derived by the standard algorithm at 12.5 km resolution

on 10 October 2004, and it is apparent that the Maud Rise (5°E, 66°S) is surrounded by a halo of reduced ice concentration. It has been observed by Fahrback et al. (2004) who made use of mooring data at around 64°S that warm events (greater than 1°C) occasionally occur in the region. Such warm events would cause a thinning of sea ice and the vulnerability of the region to the formation of spring polynyas.

The length of time and frequency of the occurrence of the polynya feature in the region for each year from 1979 to 2007 is presented in Fig. 8.6a, and it appears that the feature is active every year but with different lengths of persistence. Among the most persistent Maud Rise polynyas were those that occurred in 1992, 1994, and 2004 followed by those in 1980, 1982, 1992, 2000, 2001, and 2002. Sometimes they occur at the beginning of winter and sometimes in the middle of winter. Since they almost always occur in spring, the ice cover in the region is likely relatively thin indicating relatively warm water in the region and suggesting the important role of topography in the upwelling of warm water and in the formation of the Maud Rise polynya. Also shown in Fig. 8.6b, c are the bathymetry map of the Weddell Sea region and the winter polynya in 1994 which occurred during a winter field program showing relatively warm water in the region (Muench et al. 2001).

A similar transient polynya has been occurring east of the Maud Rise in the Cosmonaut Sea, and it has been the subject of investigation for sometime (Comiso and Gordon 1987; Takizawa et al. 1994; Arbetter 2004; Bailey et al. 2004; Prasad et al. 2005). It has been considered as primarily a sensible heat polynya, because it usually forms far enough from the coastal regions and in the deeper part of the ocean. It was postulated that the primary cause of the polynya is a stretching of the water column in the area as a result of a vorticity conserving interaction of the Antarctic Circumpolar Current with the coastal current (Comiso and Gordon 1996). It was estimated that a water column stretching can cause an upwelling of warm water at a rate of 2.6×10^{-4} cm/s, which is much higher than the regional Ekman upwelling rate of 5.7×10^{-5} cm/s (Gordon and Huber 1990). Subsequent studies indicate the important role of the atmosphere in the polynya formation. It was pointed out by Arbetter (2004) that the formation of the polynya is usually accompanied by a drop in sea level pressure

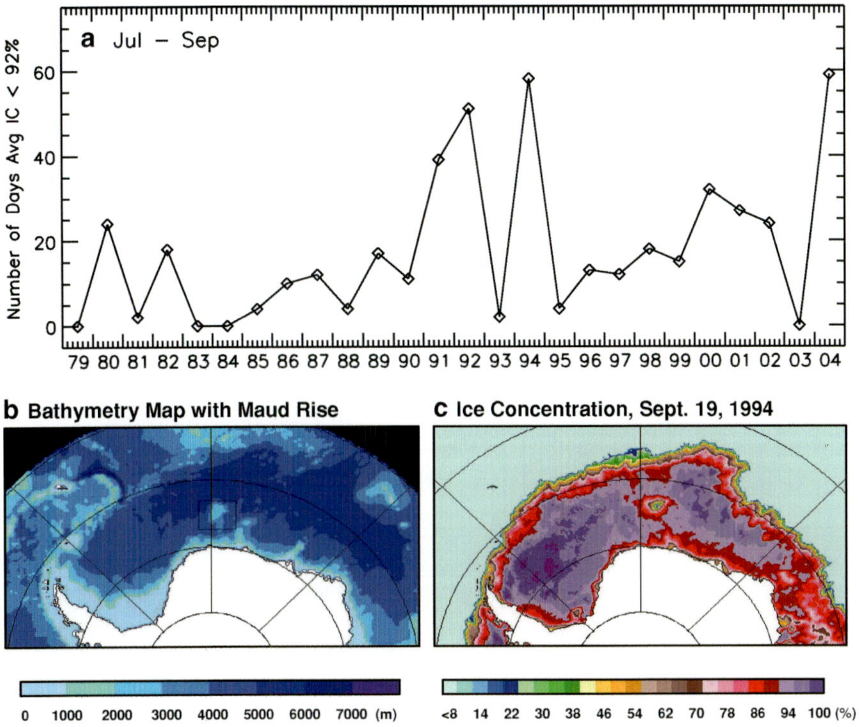

Fig. 8.6 (**a**) Plots of number of days when the ice concentration averages within the Maud Rise region (*smaller box* in map **b**) is less than 92%; (**b**) bathymetry of the Weddell Sea region; and (**c**) ice concentration map on 19 September 1994 with a transient polynya at the Maud Rise. From Gordon et al. (2007) with permission of the American Geophysical Union

and an increase in atmospheric wind divergence. A modeling study by the same group (Bailey et al. 2004) was able to simulate the presence of both sensible heat and latent heat polynyas (also mention by Comiso and Gordon 1996). The model results indicated that the polynya was initiated by winds and maintained by oceanic heat. More detailed modeling was also done by Prasad et al. (2005) who pointed out that the presence of a storm could be the key since such a storm could raise the −1.6°C isotherm depth by 30 m through wind-driven mixing, making sufficient oceanic heat to enter the mixed layer from below, thereby preventing or delaying ice formation.

One of the most persistent version of the Cosmonaut Polynya occurred in 2008 and is depicted by the set of images presented in Fig. 8.7. The set illustrates the evolution of the polynya (located at around 64°S and 45°E) from date of occurrence on 16 July to termination date on 9 August. The images made use of AMSR-E data, which provide better resolution than the SSM/I data used in previous studies and an improved representation of spatial distribution of the polynya. The higher resolution also enables a more accurate quantification of the actual open water area and the fluxes between the surface and the atmosphere. The area of ice-free water in this polynya is estimated to be about 30,000 km^2 and the length of occurrence

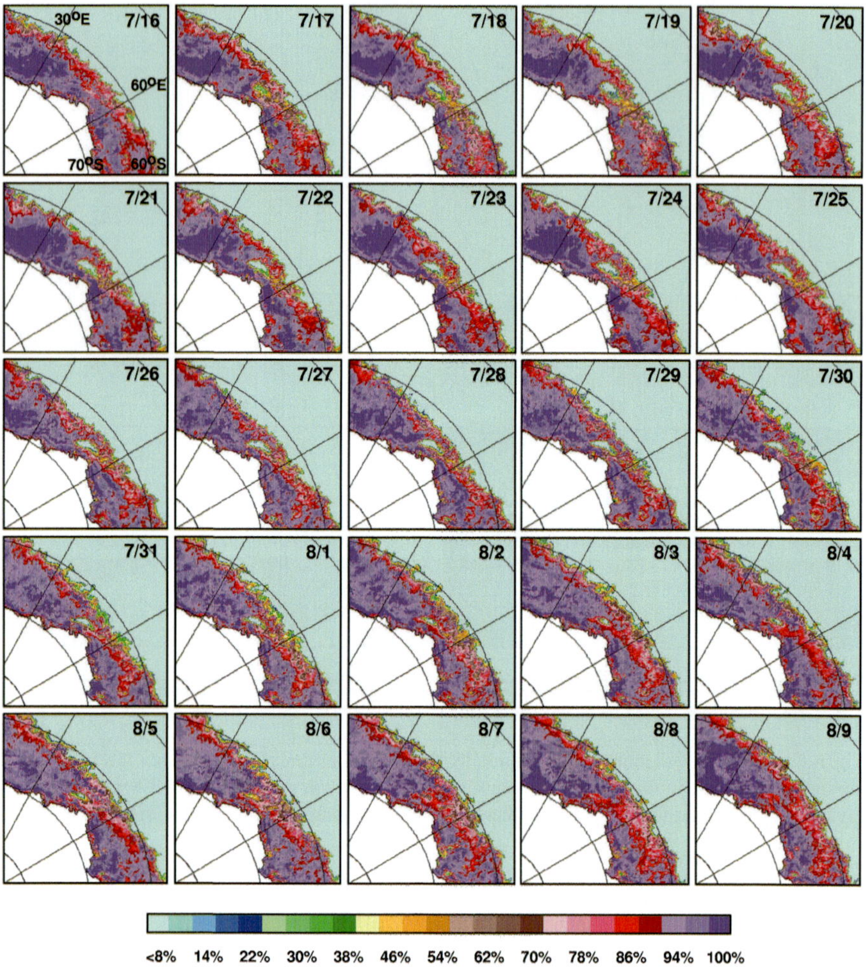

Fig. 8.7 Daily ice concentration maps in the Indian Ocean illustrating the development of the Cosmonaut Polynya from 16 July to 9 August in 2008. The polynya is the area of zero or low concentration ice in the middle of the pack at around 50°E

is about 23 days, which is a period longer than previously reported (e.g., Comiso and Gordon 1996). Also, the polynya is narrow and more defined than those reported in previous studies. Previous studies indicate the presence of an embayment in the area before polynya formation and the role of atmospheric forcing in the initiation and maintainance of the polynya. It is interesting to note that this polynya formed after the region was already consolidated as depicted in the 16 July image. The polynya grew in size during the next three days, maintained almost the same size for about two weeks and then decreased in size from 4 August to 7 August. During the entire period, no obvious occurrence of a cyclone or storm was detected. The usual embayment that precedes the formation was also not present. The relative stability of the ice cover during the period suggests that an oceanographic source of heat as discussed by Comiso and Gordon (1996) was likely a key to the occurrence of the polynya. Unfortunately, in situ measurements are not available during the period to verify this hypothesis. It appears that this polynya is so far the closest in terms of formation characteristics and persistence to the large Weddell polynya in the 1970s. The much smaller size, compared to that of large Weddell polynya may be the reason why it did not survive the winter as explained by Comiso and Gordon (1987). Full understanding of this sensible heat polynya which has been occuring more frequently than similar features in other places could lead to new insights into the mystery of the large Winter Weddell Polynya.

8.2.2 *Latent Heat Polynyas*

Many latent heat polynyas are formed every year along the coastal regions of Antarctica, caused by strong winds off the ice shelves and glaciers. The lack of a northern boundary for the sea ice cover in the Antarctic makes the advection of sea ice to the north basically unrestricted. On the other hand, the ice cover in the Arctic basin is surrounded almost entirely by land which keeps the ice from drifting to the south, and the ice circulation basically confined to the Arctic basin. Most of the latent-heat polynyas in the Arctic are found adjacent to islands like the St. Laurence Island and Novaya Zemlya and it has been observed that they contribute significantly to bottom and intermediate water formation (e.g., Martin and Cavalieri, 1989).

In the Southern Hemisphere, due to extremely cold air temperatures, and strong winds blowing off the Antarctic continent, exposed open water areas do not survive as ice-free surfaces for a long time. Depending on the wind, temperature and wave conditions frazil ice crystals accumulate rapidly at the ocean surface and the surface water quickly freezes as discussed in Chapter 2 to form a sheet of thin ice called *nilas*. Such new ice is formed almost continuously in some regions, where katabatic winds are common as in the Terra Nova Polynya (Kurtz and Bromwich 1983; Bromwich and Kurtz 1984). By katabatic winds (which come from a Greek word katabatikos meaning going downhill), we refer to those that usually originate atop a plateau or a glacier where the density of air increases with lower temperatures. As the air gets accelerated due to gravity from a high elevation, their velocity could exceed hurricane force winds when they reach the bottom. Katabatic winds are commonly found over the glaciers and ice sheets of Greenland and Antarctica. The overall impact of Antarctic latent heat polynyas on ice production has been examined by Wu et al. (2003) using a numerical general circulation model and the results show consistency with estimates using satellite observations.

As indicated earlier, an important process occurs beneath the surface in active coastal polynya areas. Formation of sea ice is accompanied by the rejection of brine (salty liquid) that in turn increases the saltiness (or *salinity*) of the underlying water (see Fig. 8.1). When polynyas are formed in shelf regions where the water is relatively shallow, as in the Western Weddell and Ross Seas, the saltier waters collect on the shelf. Eventually, the density of the shelf water matches or exceeds that of the surrounding waters, and the water sinks off the shelf slope to become part of the Antarctic bottom water. There are, however, many complications to this concept, including the exchange with fresh water from the melt of ice shelves and the residence time of the water in the shelf region. Attempts to better understand the process have led to many field programs including the Antarctic Slope (ANSLOP) project (Gordon et al. 2009).

Among the most persistent regions of coastal polynya formation are those in the Ross Sea adjacent to the Ross Sea Ice Shelf and in the Western Weddell Sea near the Ronne Ice Shelf. These two regions are also located in relatively shallow shelf regions where brines ejected during ice formation can accumulate and the salinity of the water mass increases until it reaches that of the Antarctic bottom water. An example of a coastal polynya is illustrated in the high resolution Landsat image of the Western Weddell Sea near the Ronne Ice Shelf presented in Fig. 8.8. The image shows detailed distribution of the ice cover including thick ice floes in the northern (top left) regions and different stages of new ice near the coastline represented by different shades of gray (or different albedo). The initial sea ice cover gains thickness through thermodynamic processes and the thicker the ice is the higher the albedo, and hence the different gray levels in the image. Because the ice gets constantly pushed by winds, ice types in lighter gray are farther to the north while those in darker gray are closer to the coastline. Some rafting in the thin ice region is also evident in the image represented by a narrow

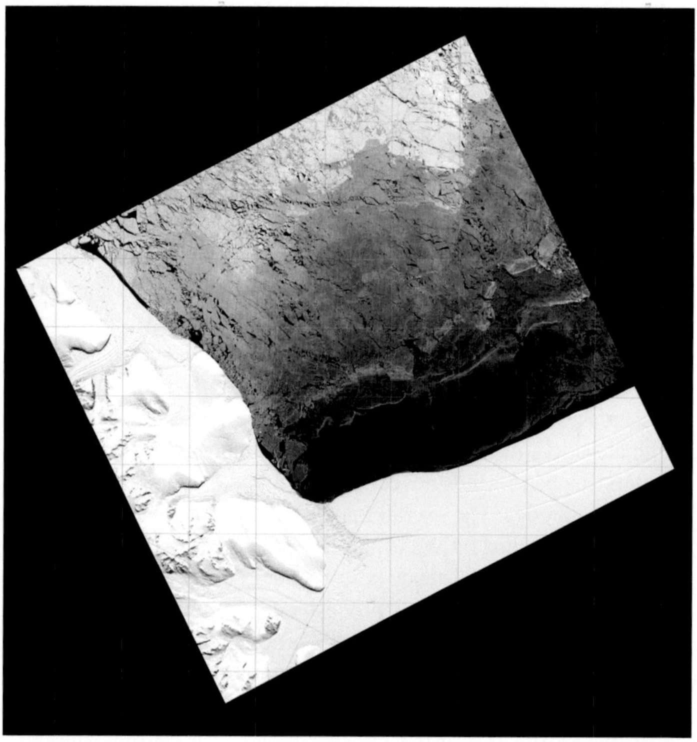

Fig. 8.8 A Landsat-7 image illustrating the formation of a coastal polynya along the Ronne Ice Shelf in the Western Weddell Sea. Strong winds pushes the sea ice cover to the north leaving behind ice free water that becomes the site of new ice formation (dark gray in the image)

linear pattern of lighter gray. The process gets repeated during the autumn and winter making it an almost continuous ice production area during the period.

Among the most useful tool that is currently available to study the characteristics and long term variability of coastal polynyas and their impact on ice production and bottom water formation is satellite passive microwave data. Because of its persistence, the Ross Sea polynya has been the subject of several investigations in recent years (e.g., Bromwich et al. 1998; Martin et al. 2007; Kwok et al. 2007). A set of images of sea ice concentrations from passive microwave data in the Ross Sea during maximum extent for each year of satellite observation in the winter period is presented in Fig. 8.9. It is apparent that the sensor, which has a 25-km

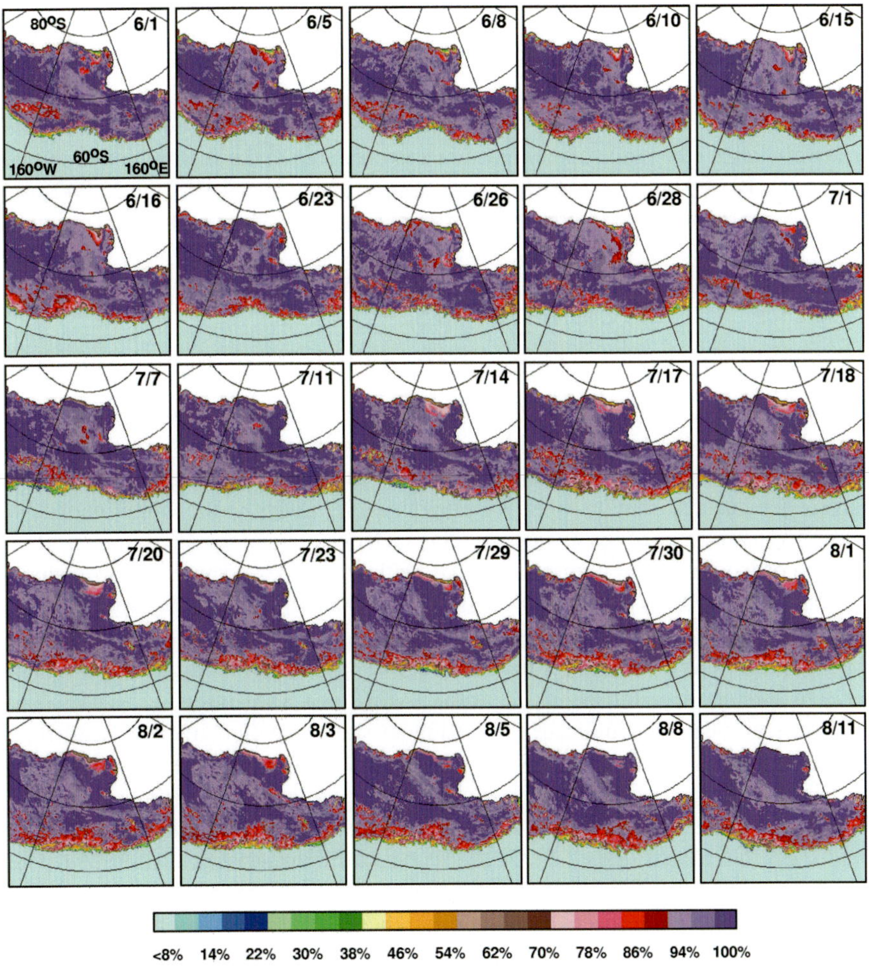

Fig. 8.9 Daily sea ice concentration maps in the Ross Sea from 18 August 1979 to 6 September 2007 illustrating Ross Sea coastal polynya formation events. The polynya is represented by reduced ice concentrations along the edge of the Ross Ice Shelf and is sometimes affected by the presence of icebergs grounded in the vicinity as in 2000 and 2002

resolution (the size of the little squares in Fig. 8.8), is not able to provide the spatial details of the coastal polynya as effectively as the Landsat data. Also, the polynya is relatively small during this time period, because the ice cover is at its maximum extent, and the total mass of the ice cover offers the highest resistance to wind-forced advection. There are times when the polynya was relatively large as in 1989, 1998, and 2005, and times when it was relatively small as in 1990, 1991, and 2003. Since these are daily images and the strength of the wind varies from one day to the next, the extent of the coastal polynya for each year is not good representation of polynya activities during the year. Instead, the data illustrates how the polynya can vary from one day to another, taking into account interannual changes in wind and ice drift circulation and the impact of iceberg calving.

The polynya variability for each year is better demonstrated by the plots of open water in the shelf area region (where bathymetry shows <1,000 m) for each year as presented in Fig. 8.10. As indicated, the open water area in the region fluctuates on a daily basis, and sometimes there are periods of enhanced open water area. This would be the periods when winds are unusually strong, and the polynya areas are relatively large. The range of estimated open water area in the shelf region is shown to be from 10,000 km^2 to 50,000 km^2 with the average being close to the 25,000 km^2 which is similar to those inferred by Martin et al. (2007). The yearly salinization in the Ross Sea shelf region can be estimated using the daily open water areas in the region as estimated from daily ice concentration data. Such estimates can be made using a formula used in Zwally et al. (1985):

$$\Delta S = sTAR_i / 0.1\ h$$

where ΔS is the salinization of the total volume of shelf water within the study region scaled by 0.1 to units of psu, s is the salt rejection in gcm^{-2} for each meter of sea ice formation (a value of 2.5 gm cm^{-2} is used) assuming that the newly formed ice has a salinity of 10.0, *T* is the duration in days of the winter period, A is the observed ratio of open water to total surface area of each study region, R_i is the rate of ice formation in meters per day within the open water area (two values, 0.10 and 0.17 m/d are used for low and high estimates), and h is the average water column thickness which in our case is estimated to be about 518 m. The results for the salinization as derived from the shelf region (<1,000 m) during each year from 1979 to 2008 are listed in Fig. 8.10 for the SMMR and SSM/I time series data. The two vertical dash lines indicate the estimated start time (left line) and end time (right line) of ice production in the region. The low estimates for yearly salinization are shown to range in value from 0.06 to 0.14 psu, while the high estimates go from 0.12 to 0.24 psu. The values are about half the estimates by Zwally et al. (1985), but for a different period (i.e., 1973–1976) and a different sensor called ESMR. The estimates by Zwally et al. (1985) for the Ross Sea were 0.258 and 0.439 psu for the low and high values, respectively. The difference is caused mainly by much lower estimate of open water area in the polynya region as inferred from the multichannel ice concentration algorithm, compared with that estimated by the one-channel ESMR algorithm, used in the Zwally study. The ESMR algorithm is especially sensitive to the thin ice and low temperatures common in the Ross Sea coastal area.

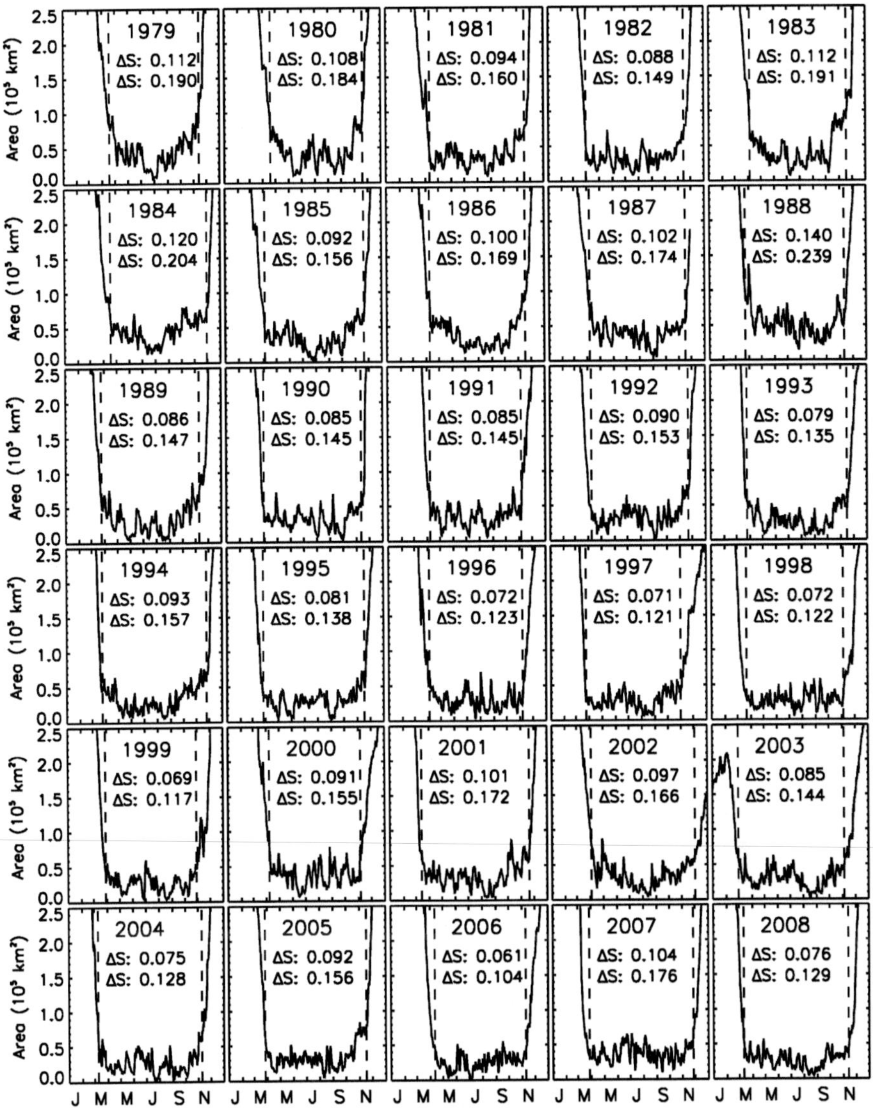

Fig. 8.10 Estimates of daily polynya areas for each year from 1979 to 2008 and associated salinization of water masses in the region calculated using a simple model. Upper and lower estimates of salinization are provided for each year

The results show high interannual fluctuations in salinization with relatively high values in the 1980s, low values in the 1990s, and more moderate values in the 2000s. Regression analysis show trends of approximately −0.14 psu per decade from 1979 to 2008. An alternative technique recently developed by Martin et al. (2007), makes use of thin ice thickness estimated from SSM/I 37 GHz channels and conjunction with ECMWF data. Preliminary results show more than a doubling in salinization (0.075 to 0.156 psu) from 1992 to 2007. During the same period, the estimates shown in Fig. 8.9 show comparable results but the trend is not as high. The estimates of salinization in the 1980s made use of data primarily from the SMMR sensor, which provides data that is not as accurate as those from SSM/I and AMSR-E data. Slight bias or error in ice concentration in the SMMR data could cause significant errors in the estimate of salinization. The approximate agreement of two techniques in the estimate of salinization is, however, encouraging and although the trend result from 1992 to 2008 do not represent a long-term trend, it provides useful information about what is currently going on in the region that may affect the ice production and bottom water formation.

Monthly averages of the narrow-band albedo in the Ross Sea region from 1979 to 2008 are presented in Fig. 8.11 to illustrate the interannual variability of spring

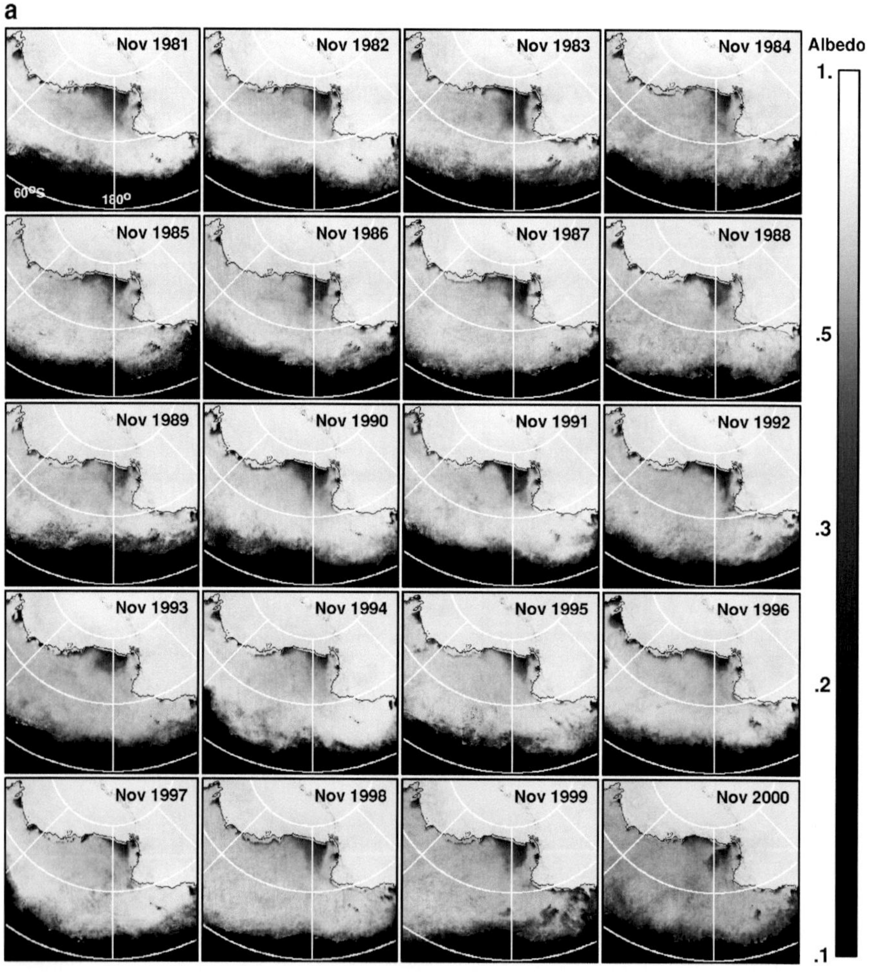

Fig. 8.11 (**a**) Monthly averages of AVHRR albedo data illustrating spring polynyas in the Ross Sea from 1979 to 2000. Changes in polynya formation is apparent after iceberg calvings in 2000 and 2002

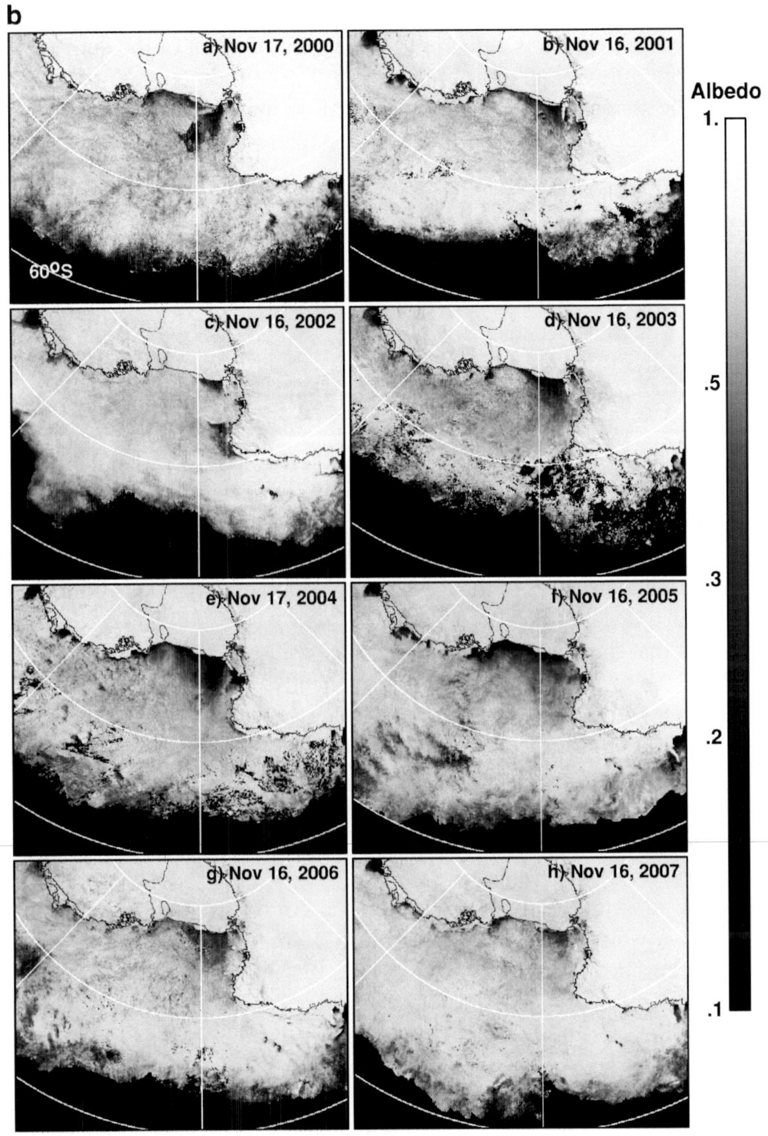

Fig. 8.11 (continued) **(b)** Weekly averages of albedo data in the Ross Sea during spring from 2000 to 2007)

polynya in the Ross Sea. The set of images for 1979–1999 are for time periods when there were no unusual external forcing like massive iceberg calvings. Such events happened in 2000 and 2002 which as discussed later in this chapter caused significant alteration in the coastal polynya formation. In Fig. 8.11, it is quite clear that even during the years from 1979 to 1999, large interannual variability of the Ross Sea spring polynya is apparent. Sometimes, the polynya is expansive as in 1981, 1983, 1986, 1990, 1991, and 1999, and sometimes it is be relatively smaller as in 1989, 1994, 1996, and 1997. Also, sometimes, the width of coastal polynya activity is equal to the width of the front of the entire ice shelf as in 1981, sometimes about two-thirds as in 1982, and sometimes only about half as in 1984. The shape of the coastal polynya also changes every year. Interannual variations in wind intensity and direction are likely the major factors but since these are monthly averages, the features are integrated effects of the daily polynya activities during the month. Also, the upwelling of warm water in the region could be a factor, but the interannual variability and scale of this effect is basically unknown.

The year 2000 was a very special year for the Ross Sea and especially the spring polynya region, because it was the year when a huge iceberg called B15 broke up from the Ross Ice Shelf. The impact of the presence of this iceberg in 2000 on the spring polynya is quite large as shown in Fig. 6.11. In 2000, a big polynya to the south of the iceberg was formed and the usual spring coastal polynya is relatively small mainly because sea ice in the region could not moved freely since it was being blocked by the grounded iceberg. In 2001, the spring coastal polynya was also relatively small in part because of the presence of the iceberg. In 2002, another huge polynya, called C19, broke up from the Ice Shelf and the phenomenon similar to that of 2000 happened, but only this time, the open water polynya is located farther north near Cape Adaire behind the big iceberg B15, which at that time has moved hundreds of km to the north of the shelf. From 2003 to 2007, the spring polynya became practically normal again, with the extent in 2007 quite small compared with those in other years. While the impacts of icebergs come and go, it is good to see that some systems recover and go back to their original characteristics after the perturbation (which in this case are the icebergs) has disappeared. Although salinization and ice production in the region have been changing, it is also apparent that the iceberg events did not make a long term impact on the system.

Among the most interesting polynyas in the Arctic region is the North Water (NOW) polynya which has been regarded as a latent heat polynya. The polynya is among the earliest discovered polynya in the Arctic and is likely the most persistent and the largest in the region. Direct observations by Muench and Sadler (1973) have indicated the advection of ice in the region has been caused by strong northerly winds confirming its classification as a latent heat polynya. However, measurements by Steffen and Ohmura (1985) have revealed an upwelling of relatively warm water that may originate from the northward-flowing West Greenland Current suggesting that there is also a sensible heat polynya component. Modeling studies by Mysak and Huang (1992) indicate that there are actually two time scales in the polynya formation. One is the latent heat polynya component, which is characterized as a relatively fast one, of the order of days, as described by Smith et al. (1990). The other one is the sensible heat component which is a relatively slower one, on the order of weeks, that is associated with the upwelling of heat off Greenland. A recent version of the polynya with possibly the two components is shown in Fig. 8.12 using data from the AMSR-E 89 GHz Channel (see arrow). The polynya changes in size almost every day (not shown) as observed regularly at a good temporal resolution

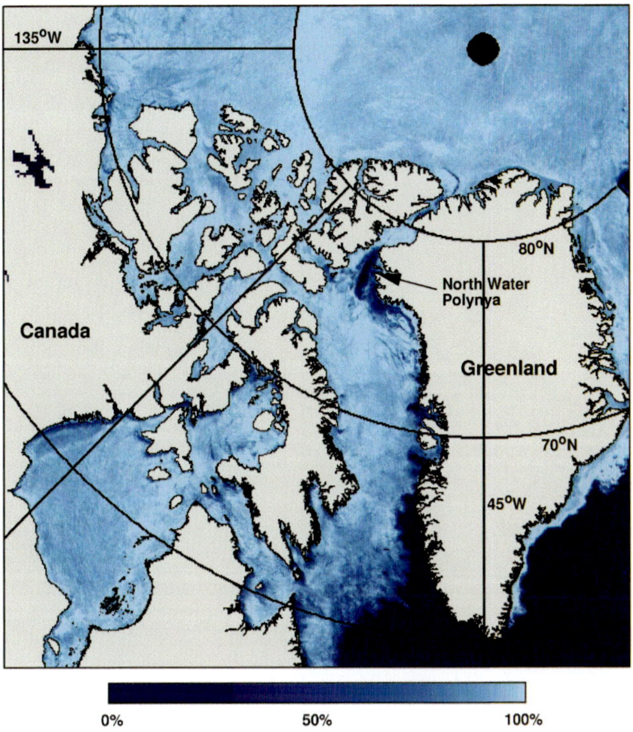

Fig. 8.12 The North Water Polynya as observed by AMSR-E 89 GHz radiometers on 23 January 2008. The polynya is one of the largest and most persistent in the Northern Hemisphere

by the Aqua/AMSR-E sensor. The part of the polynya that changes the most on a daily basis is likely the fast version of the polynya, while the one that does not change much with time is the slower polynya. Time series studies using AMSR-E data indicates that it is the middle/offshore polynya that changes much from one day to the next confirming the study of Mysak and Huang (1992). This may be the only polynya of its kind in the Arctic region, and like the Cosmonaut polynya in the Antarctic it has both sensible heat and latent heat polynya components. Other polynyas in the Arctic have also been studied, such as the Siberian Shelf Polynya (Martin and Cavalieri 1989) and the Chukchi Sea Polynya (Martin et al. 2004). These polynyas are equally interesting, because they are expected to generate high salinity and high density water that could alter the characteristics of the Arctic Intermediate Water.

8.3 The Odden in the Greenland Sea

An ice feature that looks like a tongue extending beyond the expected limits of the Greenland Sea ice cover has been an object of curiosity and interest for a long time (Nansen 1902; Nansen 1924; Koch 1945). It has been called the "Odden" and consists of an ice cover that is usually about 1,300 km long and covers as big an

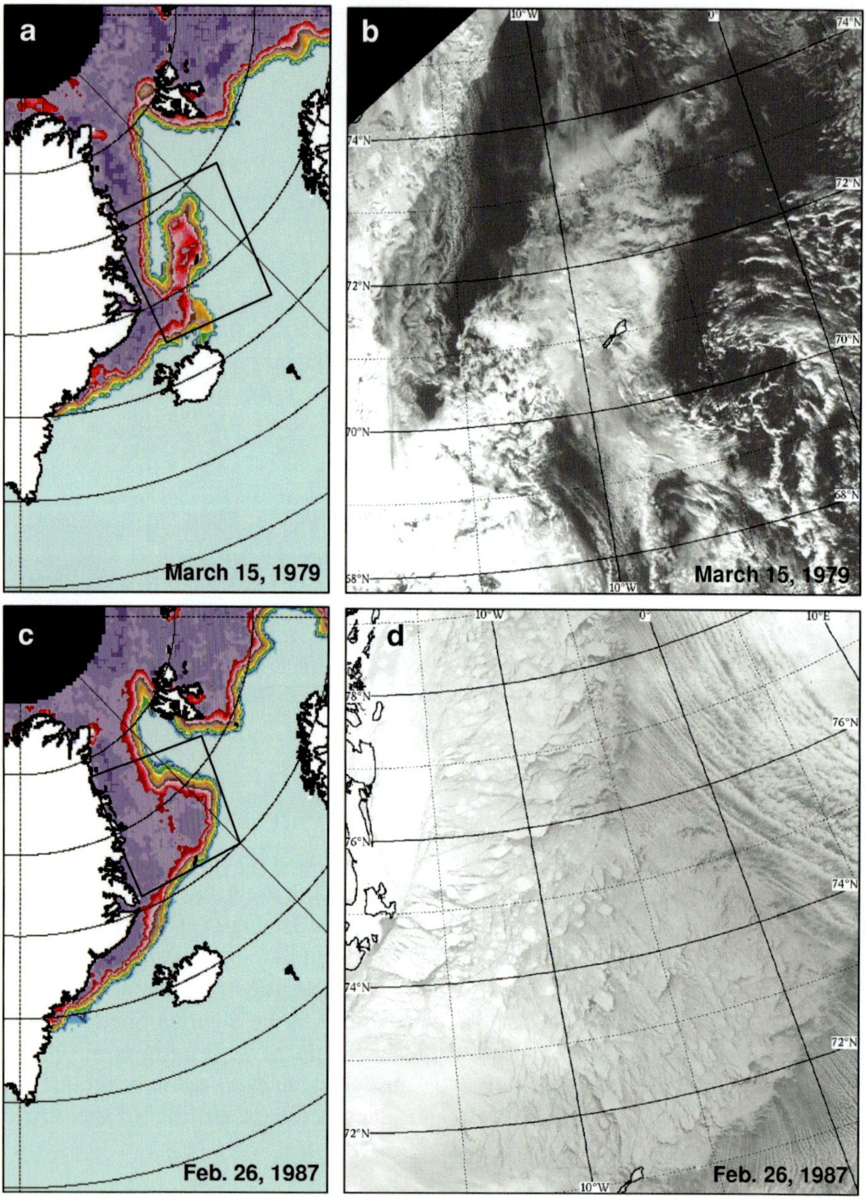

Fig. 8.13 Spatial characteristics of the Odden as observed on 15 March 1979 by (**a**) SSM/I and (**b**) AVHRR and on 26 February 1987 by (**c**) SSM/I and (**d**) AVHRR sensors. The shape, extent and ice cover of the Odden depends on many factors including wind, temperature and ocean current. From Comiso et al (2001) with permission of the American Geophysical Union

area as 330,000 km^2. It turned out that the feature is of great importance since the site of formation is adjacent to one of only four known deep ocean convection regions. The specific location of the convection is the embayment formed by the

Odden and called Nordbukta. It has been postulated that Nordbukta serves the same role as a latent heat polynya in that it is an area where new ice is formed constantly causing the formation of dense water masses that sinks and, therefore, could initiate deep ocean convection (Toudal and Coon 2004). To illustrate how the Odden is represented in both microwave and visible data, Fig. 8.13 shows the feature during two different stages: one when it appears like a tongue as on 15 March 1979 and Nordbukta was ice free and the other on 26 February 1987 when Nordbukta is completely covered by ice. The visible image from AVHRR in Fig. 8.13 provides enough resolution to make it apparent that the ice cover in the Odden area consists primarily of loose pancakes and new ice, with the ice concentration much higher in February than in March. A different but a rare type of Odden has also been identified by Wadhams and Comiso (1999). This type occurs late in the ice season and consists mainly of old ice advected through the East Greenland current and by the Jan Mayen current. The net effect is also different in that it produces meltwater that stabilizes the water column and enhances productivity in the region.

The oceanographic and atmospheric processes associated with the Odden phenomenon have been investigated in detail through field work (Wadhams and Wilkenson 1999; Gascard 2002; Johannessen and Lygre 1996), through modeling (Toudal and Coon 2001, 2004; Roach et al. 1993), and through analysis of satellite data (Carsey and Roach 1994; Toudal et al. 1999; Comiso et al. 2001). The occurrence of the Odden has been shown to be quite unpredictable. Sometimes it occurs early in the winter, disappears in mid-winter, and reappear later in the season, and sometimes it does not occur until late in the winter or early in the spring periods (Wadhams and Comiso 1999). The development of the ice feature during a good Odden season is illustrated in Fig. 8.14. Each image represent daily ice concentrations but only the days when significant changes in the Odden occurs are presented. The early stages show a distinct "tongue-like feature" until 17 December when there was so much ice that the embayment (i.e., Nordbukta) became covered with ice as well. Nordbukta was either partially or completely covered by sea ice until around 20 February when the original shape of the feature started to re-emerge. The feature persisted until it started to separate from the rest of the Greenland Sea ice cover on 17 April 1997 and completely disappeared on 2 May 2007. It is apparent that during the Odden ice season, daily changes in the size and shape of the Odden are observed indicating its vulnerability to changes in environmental conditions including wind, temperature, and ocean current. Such sensitivity to the various environmental parameters also makes the Odden looks very different from one year to another.

The frequency of occurrence and also the character and persistence of the Odden have been documented accurately by satellite data. From 1978 through 1998, the feature appeared every year except in 1983, 1994, and 1995. From 1999 to 2008, it occurred only once as a fully developed feature in 2004 (Fig. 8.15). It is apparent from Fig. 6.12 and 6.28 in Chapter 6 that the sea surface temperature in the Greenland Sea was anomalously warm in recent years starting in 2001. Some connections of warmer temperature with the frequency of the Odden is likely but needs to be studied. Because the Odden has been linked with the occurrence

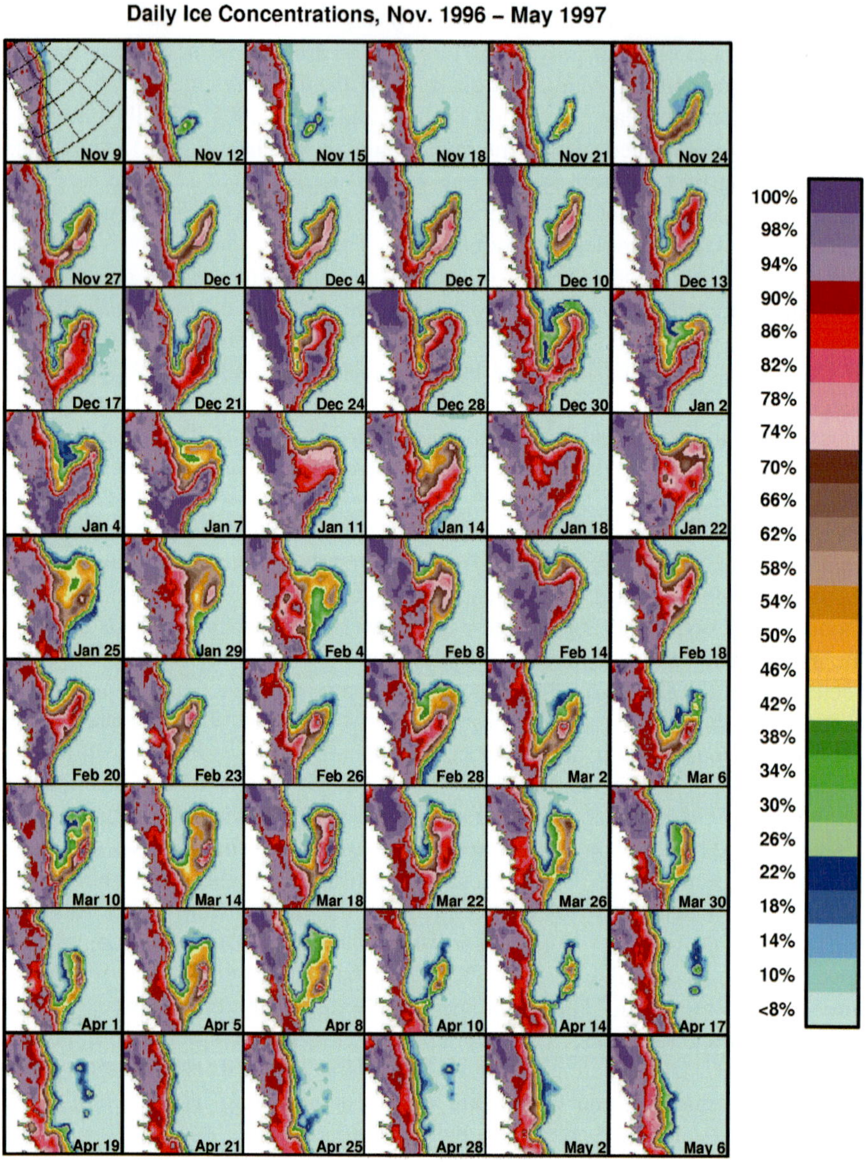

Fig. 8.14 Development of the Odden ice tongue during one sea ice season (i.e., from November 1996 to May 1997). From Comiso et al (2001) with permission from the American Geophysical Union

of deep ocean convection in the region, it is important to know as well how strong the link is and to what extent, the gradual decline in occurrence is going to affect deep ocean convection in the region. Previous studies showed that deep ocean

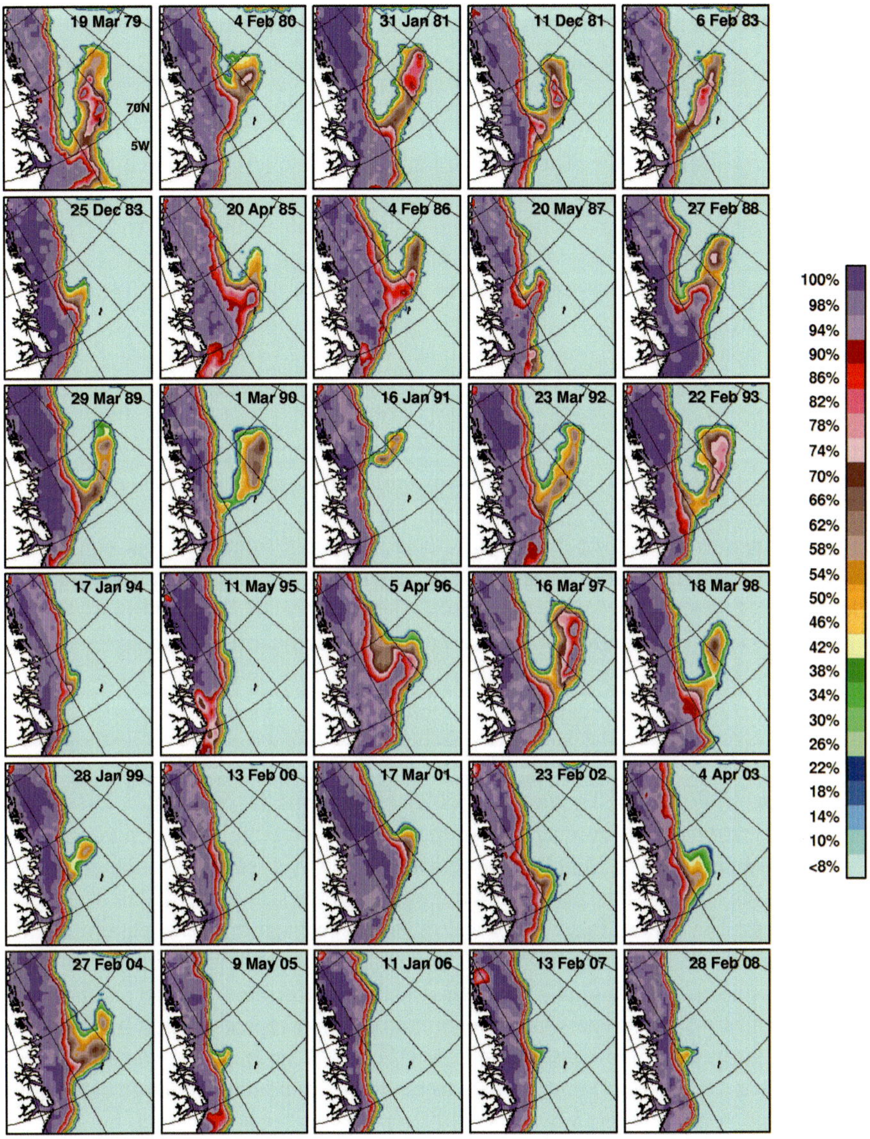

Fig. 8.15 Yearly maximum extent at the Greenland Sea from 1979 to the 2008. There have been some years during the satellite era when the Odden did not occur as in 1995 and 2000. From 2005 to 2008, the Odden had not been active

convection had occurred in Nordbukta even when the Odden was not present (Johannessen and Lygre 1996). More field studies in the region are needed to gain a full understanding of the phenomenon.

8.4 Antarctic Circumpolar Wave and Other Polar Processes

The Southern Ocean is the unifying link of the three major oceans of the world and allows for exchanges of water masses at all depths. It is also the site as already mentioned several times of the Antarctic Circumpolar Wave (ACW) which as initially characterized by White and Peterson (1996) is an eastward propagating anomaly with a period of 4–5 years and takes about 8–10 years to encircle the Antarctic continent. The phenomenon was identified using sea level pressure, wind stress and sea surface temperature for the period 1985 to 1994, and sea ice edge data from 1979 to 1991. The presence of such a wave like phenomenon is also apparent in the monthly anomaly maps of surface temperatures shown in Fig. 6.24 and the monthly anomaly maps of ice concentration shown in Fig. 7.30. The ACW has been linked with ENSO by some investigators (Peterson and White 1998; Ledley and Huang 1997; Kwok and Comiso 2002) with the correlation varying depending on the location of the study area. Peterson and White (1998) postulate that the anomalies are directly linked with the ENSO cycle through an anomalous vertical convection and a regional overturning cell in the trophosphere. It is this same cell that causes the formation of fast planetary waves which carries ENSO signals around the Southern Hemisphere.

The eastward propagation of the ice cover has been confirmed in some studies using buoy data (e.g., Wadhams 2000; Massom 1992). A well-behaved ACW that follows the postulated characteristics would be useful since it would enable one to make projections about the short-term behavior of the sea ice cover. An updated analysis of the ACW as manifested in the sea ice cover for the period 1978 to 2008 is presented in Fig. 8.16. The image is color coded with the different colors representing deviations from an average location of the ice edge. In this analysis, we deleted the February and March data, because the ice has retreated to the continent in most places during this period and sea ice anomalies in these places would be meaningless. The data in Fig. 8.16 reveal the same propagating wave pattern identified by White and Peterson (1996) for the period 1979–1989 and with the positive anomalies (in green) changing longitudinal locations from one year to another. For the period 1984–1993 (4 years later), a similar pattern is repeated (dash line) but not as consistent as the 1979–1989 period. A 1987–1997 propagating wave is not so apparent, but a 1991–2000 one appears to be a good candidate although somewhat inconsistent. A propagating wave from 1995 to 2004 is not so apparent, but may be because the wave became stationary at around 210E during the 1998–2001 period. Beyond this time period, a propagating wave as described by White and Peterson (1996) is not revealed by the diagram. The diagram shows instead that the ACW propagates a lot slower or that it became mainly a stationary wave. The assumption of a mode-2 phenomenon made by White and Peterson (1996) is not without any problem. A mode-2 phenomenon would show the positive and negative anomalies along the ice edge repeated twice around the continent. The monthly ice concentration anomaly map in Fig. 7.30 shows that the wave is not a strictly mode-2 phenomenon. Sometimes, it is repeated three times suggesting a mode-3 system instead and for

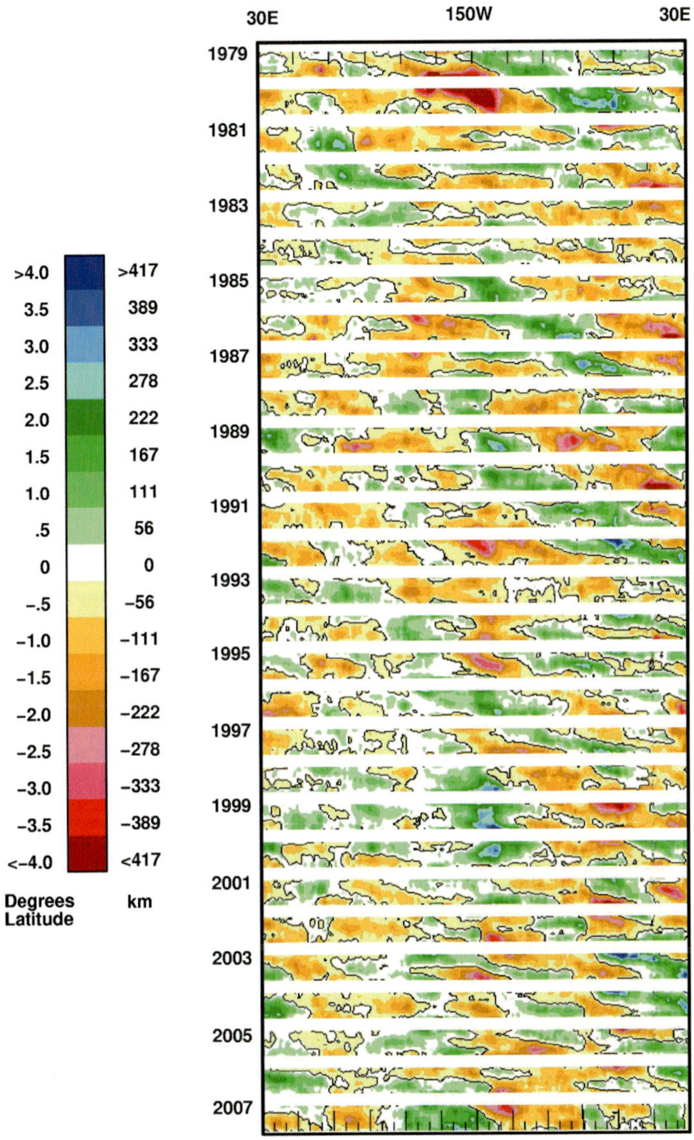

Fig. 8.16 Anomalies of monthly sea ice concentration at the ice edge depicting the Antarctic Circumpolar Wave using passive microwave data from 1979 to 2007

some years, it is not even apparent what mode the system is in. Also, there was a big ENSO event in 1998 when the wave was stationary, and the system was more like a mode-3 system. The irregular patterns after the 1979–1988 phenomenon reflect the unpredictable nature of the atmospheric pressure system and the wind circulation in the region. The circumpolar nature of the wave has actually been challenged by Yuan and Martinson (2000). Using a complex EOS analysis, they indicated that the sea ice anomaly propagates eastward in the Pacific and the Weddell Sea region though not continuously through the Drake passage. The signal then becomes insignificant and less consistent in the Indian and Western Pacific Ocean. Such an observation is also reflected to some degree in Fig. 8.16. A later study by Yuan and Li (2008) also stressed the importance of the mode-3 pattern and the influence of the Pacific South America (PSA) pattern, SAM, ENSO, and the Antarctic Dipole (ADP) on the sea ice variability. Other studies, however, have shown evidence that the Antarctic Circumpolar Wave has been around during the past 50 years (Melice et al. 2005). The definition of the term may be a problem since it is difficult to find a regular pattern in the region. The problem is further complicated by the expected influence of the Ozone Hole, which has apparently altered the pressure patterns and the vorticity of the winds in the Western Antarctic region (Turner et al. 2009). The ACW and the alternating negative and positive anomaly pattern is a useful concept and an important tool for analyzing the variability of the sea ice cover in the Antarctic. More research on the subject is obviously necessary, and it would be interesting if the wave starts to behave as it did from 1979 to 1988.

8.5 MIZ Processes and Polar Lows

The sea ice cover limits heat and humidity fluxes between the ocean and the atmosphere while the ocean and the atmosphere limit the extent and coverage of the ice cover. The heat exchanges are very strong in thin or open water areas within the pack (Maykut 1978), and it is also very strong at the Marginal Ice Zone (MIZ). The MIZ is the region that includes the ice edge and the adjacent the ice cover that is not consolidated because of the influence of waves and wind. The MIZ is usually covered by individual pancakes of various sizes that are packed together almost continuously or grouped in bands a few km wide and several hundred kms long, usually parallel to each other. The presence of sea ice serves to dampen the waves while the drift of sea ice, about 90% of which is submerged, serves to stir the upper layer of the ocean and cause the mixed layer to be more homogeneous. Winds cause either convergence or divergence of the ice cover thereby decreasing or increasing the width of the MIZ, respectively. Surface temperature is on the average right at or near the melting (or freezing) point at the MIZ region, and slight changes in temperature can alter the location of the ice edge and the distribution and character of the ice cover. The region is also very rich in biology in part, because of the presence of sea ice and changes in the location of the marginal ice zone could greatly alter the productivity of a region.

The interannual variability of the location and character of the marginal ice zone are depicted in the monthly anomaly maps in both hemisphere and in winter and summer (Chap. 7). In Fig. 7.14 and 7.30, it is apparent that the location of the MIZs can change by several hundred km from one year to another. Even in the summer, the location of the marginal ice zone can vary significantly interannually especially during the last few years. In the Southern Ocean, the MIZ is influenced mainly by ice edge processes like the ACW as described earlier. At a certain longitude, the MIZ could advance or retreat substantially on account of the ACW. The ACW is associated with pressure patterns that changes not only the dynamics of the sea ice cover, but also the dynamics of the ocean waves that penetrate through the MIZ. Ocean waves affect not just the width of the MIZ, but also the floe size distributions of the ice cover. The mechanics of ocean-ice interaction have been the subject of many investigations, and it has become apparent that it is relatively complex and requires in situ measurements, numerical modelling and satellite observations as discussed in detail by Wadhams (2000).

The MIZ is also the site of many interesting polar processes, such as cold air outbreaks and polar lows (Carleton and Carpenter 1990; Mitnik 1992; Hartmann et al., 1997; Kolstad and Bracegirdle 2008). There is usually a large gradient of air temperatures from sea ice-covered oceans to open oceans. As the wind blows from ice covered to open oceans the cold air sucks large amount of energy out of the open ocean. This results in what has been called marine "cold-air outbreaks." This phenomenon is easily identified in satellite images through the presence of "cloud streets" or well defined bands of clouds adjacent to the ice edge as illustrated in the Terra/MODIS image shown in Fig. 8.17a. The two-dimensional cloud streets are present in the passive microwave TB data (Fig. 8.17b), which is sensitive to water vapor differences between the streets. In an extended area, and using wind data from QuikScat/SeaWinds for the same period, the cloud streets are organized to form "roll vortices," as described by Hartmann et al. (1997) and Renbrew and Moor (1999). A fraction of the upward heat flux in the subcloud layers is carried out by organized roll motions as depicted in Fig. 8.17c, where the enhanced winds are shown in red cloud streets are organized by "roll vortices," aligned approximately with low-level winds as indicated in Fig. 8.17c.

The term "cold-air outbreak" has been generally defined as the occurrence of two or more days during which the local mean daily surface air temperature is at least two standard deviations below the local wintertime mean temperature (Vavrus et al. 2006). In the polar regions, it has also been referred to as marine cold-air outbreak, because of the distinctive cloud characteristics. Figure 8.18 shows another example of cold-air outbreak but at the Sea of Okhotsk on 26 January 2008 as captured by Terra/MODIS and ALOS/PALSAR data. The PALSAR data provide a better distinction of the difference sea ice types and provide good supplementary information about the cloud patterns. To quantify this phenomenon, a cold-air outbreak index has been utilized (e.g., Kolstad and Bracegirdle 2008), which is defined as the difference of surface physical temperature and the temperature at 700 hPa (about 2,000 m altitude). The 700 hPa level is used because it is above the boundary layer and thus less likely modified by surface fluxes. Through this technique, it is possible to investigate statistically cold-air outbreaks for different years and assess how the intensity varies interannually and determine if there is a trend.

Polar lows are small-scale cyclones that form over the polar regions during the cold season, and they usually follow the occurrence of cold-air outbreaks. In the Arctic they have been called Arctic hurricanes because they form spiral patterns often with an eye that are similar to those of tropical hurricanes when observed

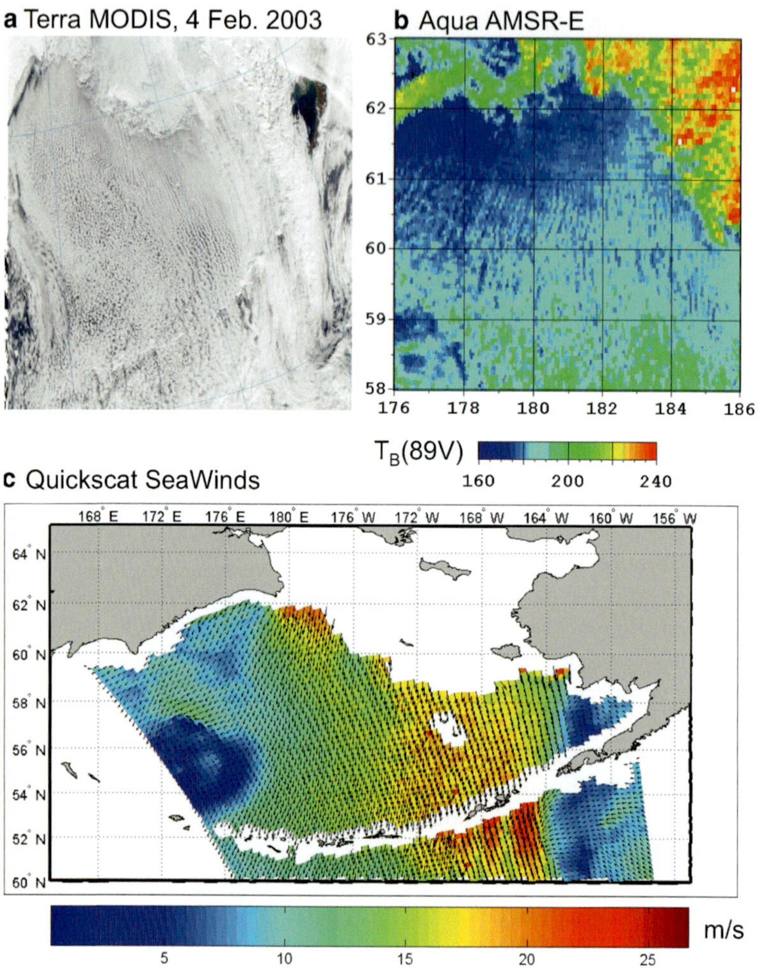

Fig. 8.17 Cold air outbreak in the Bering Sea on 4 February 2003 as depicted by (**a**) EOS-Terra/MODIS; (**b**) EOS-Aqua/AMSR-E; and (**c**) QuikSCAT SeaWinds. Courtesy of Leonid Mitnik, 2008

from space. They are relatively small depressions about a few hundred kilometers in diameter that usually form in the cold water in winter in both polar regions. However, they are known to cause adverse weather conditions in the polar regions with winds at near hurricane intensity of about 90 km/h and can have significant impact on ocean circulation (Condron et al. 2008). Examples of such polar lows in two regions of the Arctic are shown in Fig. 8.19 using NOAA-9/AVHRR and Terra/MODIS images. It is intriguing that the spiral features in the two images go in opposite directions with the Barents Sea version being cyclonic which the Bering Sea version is anti-cyclonic. This has been observed to be a regional phenomenon with those in the Atlantic Ocean side being generally cyclonic and those in the Pacific Ocean side being anti-cyclonic. It was not until the 1980s

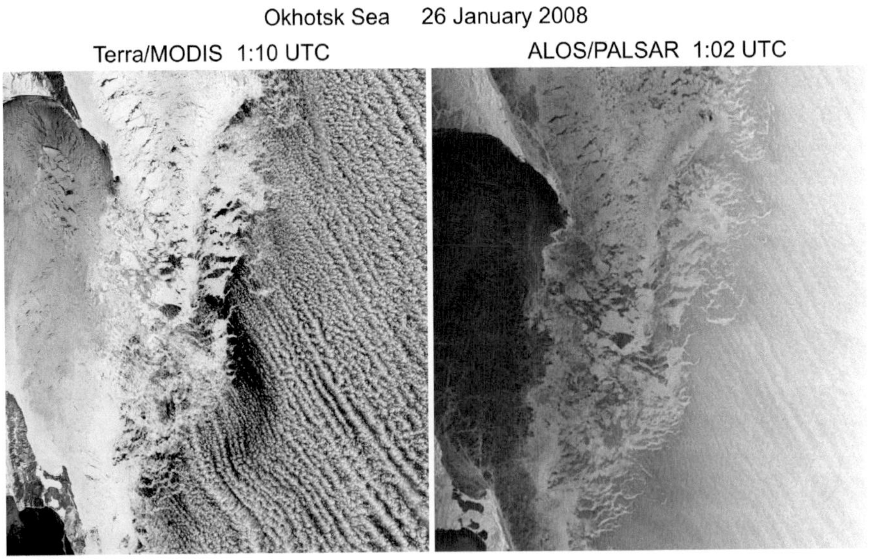

Fig. 8.18 Cold-air outbreak at the Okhotsk Sea on 26 January 2008 as observed by Terra/MODIS and ALOS/PALSAR. Courtesy of Leonid Mitnik of Vladivostok University, 2008 and NASA

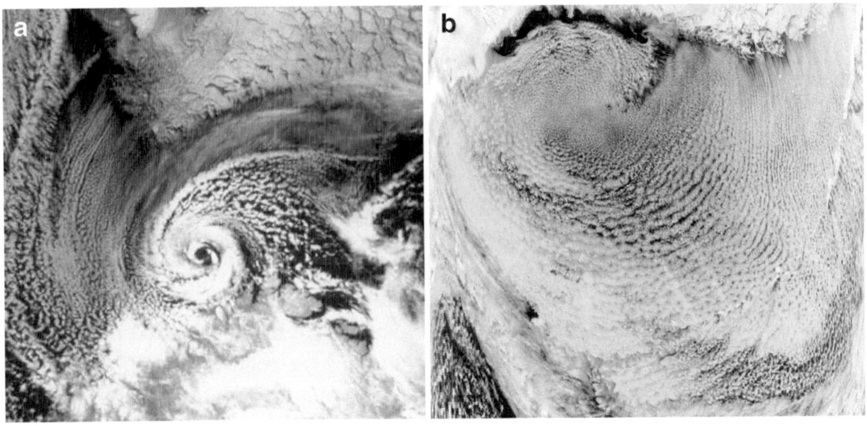

Fig. 8.19 Polar low and cold air outbreak as depicted by (**a**) NOAA-9 AVHRR data on in the Barents Sea on 27 February 1987 and (**b**) by Terra/MODIS data on 21 January 2006 in the Bering Sea. Courtesy of NASA

when scientists started using satellite remote sensing images to monitor these phenomenon that it was recognized that polar lows occur frequently in the polar regions. Polar lows have been the subject of many investigations (e.g., Sardie and Warner 1983; Carleton and Carpenter 1990). A comprehensive review of the history and current understanding of the phenomenon is provided by Rasmussen and Turner 2003).

The impact of polar lows on ocean circulation could be significant as pointed out by Condron et al. (2008). The lack of representation of these mesoscyclones in models apparently led to underestimation of atmospheric forcing at the air-sea boundary. The incorporation of these features in a numerical model indicates enhanced surface latent and sensible heat fluces and a dramatic increase in cyclonic rotation of the Nordic Sea gyre (Condron et al. 2008). The difference from original is about 4 times and caused the Greenland Sea deep water formation to increase by about 20% in 1 month making open ocean convection in the region much more active. It also led to an increase in the volume transport of intermediate and deep water overflowing the Denmark Strait.

8.6 Distribution of Icebergs and Impacts

Icebergs are not sea ice since they originate from glaciers, ice shelves, or ice sheets but they are important components of the ocean ice cover. They affect the sea ice cover and the ocean in many ways depending on size, distribution, and location. For example, they can be grounded and if large enough, can restrict the drift of sea ice and cause open water to form as the sea ice shifts its direction and speed. The open water can in turn serve as an ice factory and takes the role of a latent heat polynya. As the wind changes direction, the location of this polynya would change as well. They can also be massive, tens to hundreds of km in size, and when melted would alter the salinity of the mixed layer of the surrounding water. In some cases, icebergs can affect the productivity of a region (e.g., Ross Sea). Ability to monitor icebergs globally would be very useful, especially for mass balance, sea ice, productivity, and other studies. This has not been done in the past because of the lack of high resolution satellite data that provides global coverage. High resolution data are usually the desired, if not the required data set to ensure that most if not all icebergs are taken into account. SAR, ASTER, Landsat, or similar types of data would be ideal, but currently they lack global spatial coverage at a good enough temporal resolution. The closest to global coverage are those provided by MODIS and similar systems which provide observations at a relatively high resolution (250 m). Because of the persistence of clouds in the polar regions, monitoring icebergs using visible data requires analysis of extensive amount of data and dedicated efforts.

The availability of time series of satellite images provide the means to monitor not just the icebergs, but also the development in the glacier or ice sheet that leads to the iceberg calving. One example is the set of images shown in Fig. 8.20 starting with the 16 September 2000 image in Fig. 8.18a indicating the evidence of a fracture in the Pine Island glacier. The Pine Island Glacier is one of the most active glaciers in the continent and one of the largest dischargers of ice in recent years. A widening of the fracture was captured more than a year later on 4 November followed by an almost separation on 12 November 2001. This is actually the best time to assess the amount of mass released by the continent to the ocean since the actual area can be established quantitatively and if the average thickness is known, the actual mass

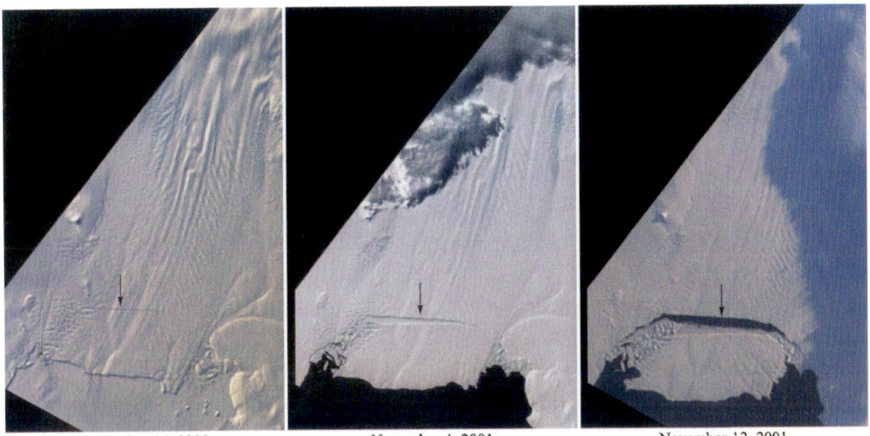

Fig. 8.20 Sequence of EOS-Terra MISR images illustrating the birth of a large tabular Iceberg from Pine Island Glacier. A large crack was formed on 16 September 2000, widened on 4 November and separated as an iceberg on 12 November. Courtesy of NASA

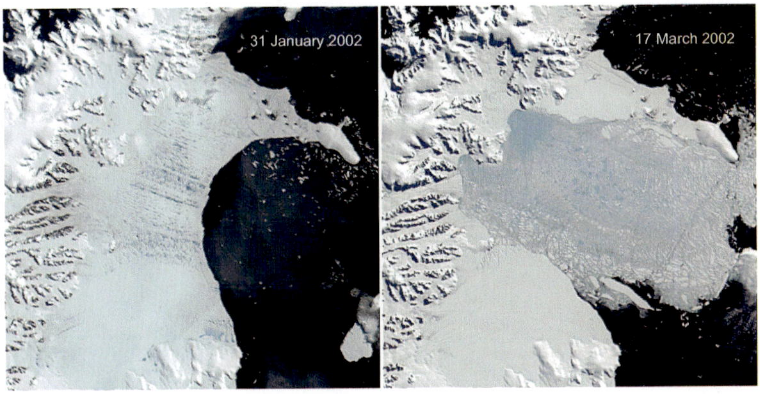

Fig. 8.21 Larsen ice shelf before its collapse in summer on 31 January 2002 and after its collapse in early autumn on 17 March 2002. On 17 March 2002, the sea ice has grown and icebergs are embedded within the pack. Courtesy of NASA

can be estimated. After the calving event, the iceberg can break into pieces and accurate accounting of the contributions of various pieces may not be trivial. Through detailed monitoring of the continental boundary, it is thus possible to assess potential mass of icebergs that are expected to calve in a near distant future as well as their locations.

Another example is the breakup of the Larson Ice Shelf as illustrated by the sequence of images shown in Fig. 8.21. The breakup of the ice shelf caused a significant increase in the area covered by sea ice in the region, and at the same time caused the introduction of enormous number of icebergs. The icebergs were eventually advected to the east along the Weddell gyre and because of their number, altered the flow of the sea ice cover. The Larson Ice Shelf is located in the east side of the Antarctic Peninsula which is regarded as an area of climate anomaly in the Antarctic region (King et al.; Jacobs and Comiso 1997) and has been the site of unusual number of crevasses. This contributes to the breakup of the Ice Shelf into many small icebergs.

The most spectacular iceberg recorded occurred at the Ross Ice Shelf on 17 March 2000. Called B15, the iceberg that came off the Ross Ice shelf had a total area of 11,000 km^2 and is regarded as the largest iceberg ever recorded. The iceberg, shown in Fig. 8.22, was 295 km long, 37 km wide and approximately 200 m thick yielding a total volume of 2,200 km^3. As indicated, part of the iceberg broke up, and this occurred again in 2002 and 2003. The biggest part called B-15A drifted away slowly from the Ross Island, then past the Drygalski ice tongue in April 2005, and ran aground off Cape Adare in Victoria Land, where it broke into several smaller pieces on 27–28 October 2005. The availability of almost continuous coverage by satellite sensors made it possible to monitor the iceberg from time of formation to the time it gets to the warmer seas where it melts.

An iceberg almost as spectacular as the B-15 and called C-19 came off the Ross Ice Shelf in May 2002 and was nearly as big (i.e., 200 km by 32 km by 200 m).

Fig. 8.22 The Ross Sea ice cover and the B15 iceberg which calved from the Ross Ice Shelf on 17 March 2000. The iceberg is regarded as the biggest free floating iceberg ever observed. Courtesy of NASA

Considered the second biggest iceberg ever, this iceberg drifted off the shelf and become grounded in a position almost perpendicular to B-15 for many months. During this time, the circulation of sea ice was inhibited in the region. The iceberg drifted several hundred kms to the north between September and October and became stationary up to the first half of January 2003 Fig. 8.23. The unusual size and orientation caused large amount of sea ice to be trapped in spring making the Ross Sea coastal polynya very different from those of other years as discussed previously. It also disrupted normal ocean circulation. In January, the iceberg moved northward rapidly, passed Cape Adana, and then broke into two pieces called C-19A and C-19B with the bigger portion (C-19A) staying at the coast of Victoria Land near Dumont d'Urville for about two years. By September 2005, C-19A started drifting northward passing by an area adjacent to the Balleney Island in March 2006. The impact of these icebergs on the latent heat polynya formation is illustrated in Fig. 8.11 (see November 2000 and 2002 images). The impact on the primary productivity can be profound as discussed by Arrigo et al. (2002) and in the next chapter.

Icebergs in the Northern Hemisphere are also abundant, but are normally found near glaciers and some ice shelf regions. The decline in volume of most of the glaciers in North American and other regions has been reported and is likely to continue (Abdalati 2006). There are thousands of glaciers in North America alone, and a significant fraction are located at the Ellsmere Island at the Canadian

Fig. 8.23 The Ross Sea cover with the C19 iceberg (top right) on 1 January 2003 as observed by the DMSP/OLS sensor. Also shown is B15 near the bottom. C19 is regarded as the second biggest iceberg. Courtesy of DMSP

Fig. 8.24 Icebergs at Dobbin Bay off Ellsmere Island on 31 July 2000 as observed by EOS Terra/ASTER. From Comiso and Parkinson (2004). Courtesy of APS, NASA and JAXA

Archipelago. Among the most notable iceberg that was found in the Arctic Ocean was the ice island, called T3, which was about 30 km in diameter and 50 m thick and broke off around 1935. It calved off the Ward Hunt Ice Shelf in Northern Ellesmere Island. After it was discovered in 1947, a research station was established on it in 1952 and was manned almost continuously until the ice island exited from the Arctic through Fram Strait in 1984. Other such ice islands have also been detected in the Arctic region.

In addition to ice islands, there are numerous icebergs coming of various glaciers in the region. Figure 8.24 illustrates the type of glaciers and icebergs that are normally found in the Elsmere Island. The image shows icebergs coming out of the Eugenie glacial tongue with icebergs being dumped into Dobbin Bay as captured by the Terra/Advanced Spaceborne Thermal Emission and Reflection Radiometer (ASTER). Also shown are the intricate inter-connectivity of various glaciers in the region and how some of them are linked to or possibly maintained by a common source. It is apparent that in this particular area, the rate of iceberg production is very high, and the icebergs comes in different sizes and can be relatively small.

The image in Fig. 8.22 makes it apparent that the inventory and monitoring of every iceberg globally even with satellite data is not trivial, but not impossible with current technology. Such a venture would require the use of one-meter resolution systems to cover areas like Dobbin Bay and the Larson Shelf region where icebergs can be too small to be identified by conventional systems. The lifetime of most icebergs is relatively long and the temporal resolution must be high enough to cover the occurrence of new calving and the melt of older ones. For mass balance studies, however, the ones that count the most are the relatively big ones and most of the icebergs of importance are likely possible to detect using the MODIS 250-m resolution data. Persistent cloud cover in some regions, however, will limit the temporal resolution. In the Arctic region, increases in the calving of icebergs are expected because of observed warming in the region (Comiso 2006). In the Antarctic, more calving is expected in the vicinity of the Antarctic Peninsula which is known to have significant warming, but it is not clear what will happen along the coastline of East Antarctica where some cooling has been observed (Comiso 2000; Doran et al. 2002). The calving of big icebergs from the Ross Ice Shelf in recent years, indicates that it is difficult to assess what to expect.

References

Abdalatti W (2006) Recent changes in high-latitude glaciers, ice caps and ice sheets, Weather *61*(4):95–101.

Arbetter TE, Jynch AH, Bailey DA (2004) Relationship between synoptic forcing and polynya formation in the Cosmonaut Sea: 1. Polynya climatology. *J Geophys Res 109*, C-4022, doi:10.1029/2003JC001837.

Arrigo KR, van Dijken GL, Ainley DG, Fahnestock MA and Markus T (2002) Ecological impact of a large Antarctic iceberg. *Geophys Res Lett 29*(7):10.1029/2001GL014160.

Bailey DA, Lynch AH, Arbetter TE (2004) Relationship between synoptic forcing and polynya formation in the Cosmonaut Sea. *J Geophys Res 109*, C04023, doi:10.1029/20003JC001838.

Bromwich DH, and Kurtz DD (1984) Katabatic wind forcing of the Terra Nova Bay polynya, *J. Geophys. Res 89*, 3561–3572.

Bromwich, D.H., Z. Liu, A.N. Rogers, M.L. Va Woert (1998) Winter atmospheric forcing of the Ross Sea polynya, In: Jacobs, S., Weiss, R. (Eds.), Oceans, Ice and Atmosphere: Interactions at the Antarctic Continental Margin, *Antarctic Res. Ser 75*, AGU, Washington, DC., pp 101–133.

Carleton, A.M. and D.A. Carpenter (1990) Satellite climatology of 'Polar Lows' and bradscal climatic associations for the Southern Hemisphere. *Int. J. Clim. 10*, 219–246.

Carsey F (1980) Microwave observation of the Weddell Polynya. *Mon Wea Rev 108*(2):2032–2044.

Carsey F. and A.T. Roach (1994) Oceanic convection in the Greenland Sea Odden region as interpreted in satellite data, In The Polar Oceans and their role in shaping the global environment, ed by: O.M. Johannessen, R.D. Muench and J.E. Overland, *AGU Geophysical Monograph 85*, Washington, DC., pp. 211–222.

Comiso, J.C. (2000) Variability and trends in Antarctic surface temperatures from in situ and satellite infrared measurements, *J. Climate*, *13*(10), 1674–1696.

Comiso J.C., Parkinson C.L. (2004) Satellite observed changes in the Arctic. Physics Today 57(8):38–44.

Comiso J.C. (2006) Arctic warming signals from satellite observations, Weather, 61(3):70–76.

Comiso, J.C., and A.L. Gordon (1987) Recurring polynyas over the Cosmonaut Sea and the Maud Rise. *J. Geophys. Res 92*(C3), 2819–2834, 1987.

Comiso, J.C., and A.L. Gordon (1996) The Cosmonaut Polynya in the Southern Ocean: Structure and variability, *J. Geophys. Res., 101*(C8), 18297–19313, 1996.

Comiso, J.C., and A.L. Gordon (1998) Interannual variabilities of summer ice minimum, coastal polynyas, and bottom water formation in the Weddell Sea, in Antarctic sea ice physical properties and processes, *AGU Antarctic Research Series Volume,* edited by M. Jeffries, pp 293–315, 1998.

Comiso JC, Parkinson CL (2004) Satellite observed changes in the Arctic. *Physics Today 57*(8) 38–44.

Comiso, J.C., P. Wadhams, L.T. Pedersen, and R. Gersten (2001) The seasonal and interannual variability of the Odden and a study of environmental effects. *J. Geophys. Res., 106*(C5), 9093–9116.

Condron, A., G.R. Bigg, and I.A. Renfrew (2008) Modeling the impact of polar mesocyclones on ocean circulation. *J. Geophys. Res 113*, C10003, doi:10.1029/2007JC004599.

De Veaux, R., A.L. Gordon, J.C. Comiso, and N.E. Chase, (1993) Modeling of topographical effects on Antarctic sea ice using multivariate adaptive regression splines, *J. Geophys. Res., 98*(C11), 20207–20319.

Doran PT, Priscu JC, Lyons WB, Walsh JE, Fountain AG, McKnight DM, Moorhead DL, Virginia RA, Wall DH, Clow GD, Fritsen CH, McKay CP, Parsons AN (2002) Antarctic climate cooling and terrestrial ecosystem response, *Nature* doi:10.1038/nature710.

Fahrbach E, Hoppema M, Rohardt G, Schroder M, Wisotzki A (2004) Decadal scale variations of water mass properties in the deep Weddell Sea. *Ocean Dynb 54*:77–91.

Gascard JC, Watson AJ, Messias MJ, Olsson KA, Johannessen T, Simonsen K (2002) Long lived vortices as a mode of deep ventilation in the Greenland Sea. *Nature 416*:525–527.

Gordon, A.L., and J.C. Comiso, (1988) Polynyas in the Southern Ocean, *Scientific American, 256,* 90–97.

Gordon AL, Huber B (1990) Southern Ocean winter mixed layer. *J Geophys Res 95*(C7): 11655–11672.

Gordon, A.L., M. Visbeck, and J.C. Comiso (2007) A link between the Great Weddell Polynya and the Southern Annular Mode, *J. Climate., 20*(11), 2558–2571.

Gordon AL, Orsi AH, Muench R, Huber BA, Zambianchi E, Visbeck M (2009) Western Ross Sea continental slope gravity currents. *Deep-Sea Res II 56*:796–817.

Hartmann, J., C. Kottmeir, and S. Raasch (1997) Roll vortices and boundary-layer development during a cold air outbreak. *Boundary-Layer Met., 84*, 45–65.

Hall A, Visbeck M (2002) Synchronous variability in the Southern Hemisphere atmosphere, sea ice and ocean resulting from Annular Mode. *J Climate 15*:3043–3057.

Hibler III WD, Ackley SF (1983) Numerical simulation of the Weddell sea pack ice. *J Geophys Res 88*(C5):2873–2887.

Holland DM (2001) Explaining the Weddell Polynya – a large ocean eddy shed at Maud Rise. *Science 292*:1697–1700.

Jacobs, S. and J.C. Comiso (1989) Satellite passive microwave sea ice observations and oceanic processes in the Ross Sea, Antarctica, *J. Geophys. Res., 94*, 18195–18211.

Jacobs, S.S., and J.C. Comiso (1997) Climate variability in the Amundsen and Bellingshausen Seas, *J. Climate, 10*(4), 697–709.

Johannessen OM, Lygre K (1996) Observations of convective chimneys, in European Subpolar Ocean Programme. In Wadhams P, Wilkinson JP, Wells SCS (Eds) Sea Ice - Ocean Interactions, pp. 262–294, Commission of the European Communities, Cambridge, UK.

Koch L (1945) The East Greenland Ice, C.A. Reitzels Forlag, Copenhagen.

Kolstad EW, Bracegirdle TJ (2008) Marine cold-air outbreaks in the future: an assessment of IPCC AR4 model results for the Northern Hemisphere. *Clim Dyn 30*:871-885.

Kurtz, D.D. and D.H. Bromwich, 1983. Satellite observed behavior of the Terra Nova Bay Polynya. *J. Geophy. Res., 88*(C14), 9717–9722.

Kwok, R., and J.C. Comiso (2002) Spatial patterns of variability in Antarctic surface temperature: Connections to the Southern Hemisphere Annular Mode and the Southern Oscillation, *Geophys. Res. Lett., 29*(14), 10.1029/2002GL015415.

Kwok, R., J.C. Comiso, S. Martin, and R. Drucker (2007) Ross Sea Polynyas: Response of ice concentration retrievals to large areas of thin ice, *J. Geophys. Res., 112*, C12012, doi:10.1029/2006JC003967.

Ledley TS, Huang Z (1997) A possible ENSO signal in the Ross Sea. *Geophys Res Lett 24*:3253–3256.

Lemke P, Owens WB, Hibler III WD (1990) A coupled sea ice-mixed layer-pycnocline model for the Weddell Sea. *J Geophys Res 95*:9513–9525.

Markus T, Kottmeier C, Fahrbach E (1998) Ice formation in coastal polynyas in the Weddell Sea and their impact on oceanic salinity, in Antarctic Sea Ice: Physical processes, interactions and variability. *AGU Antarctic Research Series., 74*:273–292.

Martin S, Cavalieri DJ (1989) Contributions of the Siberian shelof polynyas to the Arctic Ocean intermediate and deep water. *J Geophy Res 94*(C9):12,725–12,738.

Martin S, Drucker R, Kwok R, Holt B (2004) Estimation of the thin ice thickness and heat flux for the Chukchi Sea Alaskan coast polynya from SSM/I data, 1990–2001. *J Geophys Res 109*:C10012, doi:10.1029/2004/JC002428.

Martin, S., R. Drucker, R. Kwok and B. Holt (2005) Improvements in the estimates of ice thickness and production in the Chukchi Sea polynyas derived from AMSR-E, *Gephys. Res. Lett., 31*, L005505, doi:10.1029/2004GL0022013.

Martin, S., R.S. Drucker and R. Kwok (2007) The areas and ice production of the western and central Ross Sea polynyas, 1992–2002, and their relation to the B-15 and C-19 iceberg events of 2000 and 2002, *J. Marine Systems, 68*, 201–214.

Martinson DG, Killworth PD, Gordon AL (1981) A convective model for the Weddell Polynya. *J Phys Oceanogr 11*:466–488.

Massom RA (1992) Observing the advection of sea ice in the Weddell Sea using buoy and passive microwave data. *J Geophys Res 97*(C10):15559–15572.

Maykut GA (1978) Energy exchange over young ice in the Central Arctic. *J Geophys Res 83*:3646–3658.

Mitnik LM (1992) Mesoscale coherent structures in the surface wind field during cold air outbreaks over the Far Eastern seas from the satellite side looking radar. *La Mer 30*:287–296.

Melice JL, Lutjeharms, JRE, Goose H, Fichfet T, Reason CJC (2005) Evidence of the Antarctic circumpolar wave in the sub-Antarctic during the past 50 years. *Geophy Res Lett 32*:L14614, doi:10.1029/2005GL023361.

Muench RD, Sadler HE (1973) Physical oceanographic observations in Baffin Bay and Davis Strait. *Arctic 26*: 73–76.

Muench RD, Morison JG, Padman L, Martison DG, Schlosser P, Huber B, Hohmann R (2001) Maud Rise revisited. *J Geophys Res 106*:2423–2440.

Mysak LA, and Huang F (1992) A latent- and sensible-heat polynya model for the North Water, Northern Baffin Bay, *J Phys Ocean 22*:596–608.

Nansen F (1902) The oceanography of the North Polar Basin. Norwegian North Polar Expeditgion 1893-1896. *Scientific Results III*:1–427.

Nansen F (1924) With seals and Bears: My first journey to the Arctic Seas. Jacob Dybwads Forlag, Oslo.

Parkinson CL (1983) On the development of the Weddell Polynya in a sea ice simulation, *J. Phys. Ocean 13*(3):501–511.

Peterson RG, White WB (1998) Slow teleconnections linking the Antarctic Circumpolar Wave with the tropica; El Nino-Southern Oscillation. *J Geophys Res 103*(C11):24573–24583.

Prasad TG, McClean JL, Hunke EC, Semtner AJ, Ivanova D (2005) A numerical study of the western Cosmonaut polynya in a coupled ocean-sea ice model. *J Geophys Res 110*, C10008, doi:1029/2004JC002858.

Renbrew, I.A., and G.W.K. Moor (1999) An extreme cold-air outbreak over the Labrador Sea: Roll vortices and air-sea interaction, *Monthly Weather Rev, 127*, 2379–2394.

Roach ATK, Aagaard K, Carsey F (1993) Coupled ice-ocean variability in the Greenland Sea, *Atmosphere-Ocean 31*:319–337.

Sardie JM, Warner TT (1983) On the mechanism for the development of polar lows. *J Atmospheric Science 40*(4):869–881.

Smith SD, Muench RD, Pease CH (1990) Polynyas and leads: An overview of physical processes and environment. *J Geophys Res 95*(C6):9461–9479.

Steffen K, Ohmura A (1985) Heat exchange and surface conditions in North Water, northern Baffin Bay. *Ann Glaciol 6*:178–181.

Toudal L, Coon MD (2001) Interannual variability of the sea ice induced salt flux in the Greenland Sea. *Ann Glaciol 33*:385–390.

Toudal LP, Coon MD (2004) A sea ice model for the marginal ice zone with an application to the Greenland Sea. *J. Geophys. Res 109*: C03008, doi:10.1029/2003JC001827.

Toudal L, Hansen KQ, Valeur H, Wadhams P, Aldworth E, Comiso JC (1999) Mapping of ice in the Odden by satellite and airborne remote sensing. *Deep Sea Research 46*(6–7), 1255–1274.

Turner J, Comiso JC, Marshall GJ, Connolley WM, Lachlan-Cope TA, BracegirdleT, Wang Z, Meredith ZM, and T. Maksym T (2009) Antarctic sea ice extent increases as a result of anthropogenic activity. *Geophy Res Lett 36*, L08502, doi:10.1029/2009GL037524.

Vavrus S, Walsh JE, Chapman WL, Portis D (2006) The behavior of extreme cold air outbreaks under greenhouse warming. *Int. J. Climatol 26*:1133–1147.

Visbeck M, Hall A (2004) Reply. *J Climate 17*:2255–2258.

Wadhams P (2000) Ice in the Ocean. Gordon and Breach Science Publishers, London, 251p.

Wadhams P, Wilkinson JP (1999) The physical properties of sea ice in the Odden ice tongue. *Deep-Sea Res II 46*(6-7):1275–1300.

Wadhams P, Comiso JC (1999) Two modes of the Odden ice tongue, *Geophys. Res. Letters, 26*(16), 1999.

White WB, Peterson RG (1996) An Antarctic circumpolar wave in surface pressure, wind, temperature and sea ice extent. *Nature 380*:699–702.

Wu X, Budd WF, Allison I (2003) Modelling the impacts of persistent Antarctic polynyas with an atmosphere-sea-ice general circulation model. *Deep-Sea Research II 50*:1357–1372.

Yuan X, Li C (2008) Climate modes in southern high latitudes and their impacts on Antarctic sea ice. *J Geophys Res 113*:C06591, doi:10:1029/2006JC004067.

Yuan X, Martinson DG (2000) Antarctic sea ice extent variability and tis Global connectivity. *J Clim., 13*:1697–1717.

Zwally J, Gloersen P (1977) Passive microwave images of the polar regions and research applications. *Polar Record 18*(116):431–450.

Zwally, H. J., J. C. Comiso, C. L. Parkinson, W. J. Campbell, F. D. Carsey, and P. Gloersen (1983) Antarctic Sea Ice 1973–1976 from Satellite Passive Microwave Observations, *NASA Spec. Publ., 459.*

Zwally, H. J., J. C. Comiso, and A. L. Gordon (1985). Antarctic offshore leads and polynyas and oceanographic effects, In: *Oceanology of the Antarctic Continental Shelf*, ed. by S. Jacobs, *Antarctic Research 43*:203–226.

Chapter 9
Variability of Phytoplankton Pigment Concentrations and Primary Productivity

Abstract The polar oceans have been regarded as regions of high biological activity but moderate primary productivity. During spring and summer, they are sites of extremely high phytoplankton concentrations in part because of the abundance of nutrients and the presence of stable layers of meltwater that provide an ideal platform for photosynthesis. The productivity is more limited in autumn and winter when extended darkness and relatively cold surface temperature prevail and much of the region is covered by sea ice. Analyses of more than a decade of ocean color SeaWiFS data have yielded many interesting results about the spatial distribution and interannual variability of phytoplankton pigment concentration and primary productivity.

Keywords Phytoplankton concentration • Primary productivity

9.1 Introduction

The color of the ocean as revealed by satellite sensors has been a subject of great interest because it is associated with a tiny single-celled organism called phytoplankton. These microscopic floras that include diatoms, other algae, and bacteria are at the bottom of the food-web, and the ability to detect large scale plankton concentrations has made it possible to assess regionally and globally their characteristics, spatial distribution, and concentration. Being a photosynthetic plant, the overall impacts of these organisms are also more profound than previously envisioned, since they have the ability to ingest large amounts of carbon dioxide from the atmosphere which eventually gets stored in the deep sea. They can therefore act as a "carbon sink" or a "biological pump" that could serve to suppress the current positive trend of carbon dioxide in the atmosphere. Their potential role in the carbon cycle and climate change may thus be quite significant as revealed by recent studies (Behrenfeld et al., 2006).

The open ocean regions adjacent to the sea ice edge have been the focus of many studies because of the relatively high phytoplankton concentration and high productivity observed in these regions. Earlier ship studies have indicated that the seasonal

J. Comiso, *Polar Oceans from Space*, Atmospheric and Oceanographic Sciences Library 41, DOI 10.1007/978-0-387-68300-3_9,

fluctuation of phytoplankton distribution is high in these regions, where the plankton appear to be well adapted to their environment (Sakshaug and Skjodal 1989). The seasonality is in part associated with the seasonality in the availability of sunlight and nutrients which regulate phytoplankton photosynthesis and biomass accumulation. Solar heating and associated changes in sea surface temperatures follow a gradual seasonal pattern in the ice-free ocean, and phytoplankton growth is normally acclimated to ambient temperatures. However, as with other organisms, there are physiological limits to tolerance and with the range of values observed in polar regions, sea surface temperature could be an important factor.

An intriguing concept that has been postulated to affect the phytoplankton growth and distribution is the seasonal retreat of sea ice in spring and summer. The melt of sea ice during this time period has been considered an important event in phytoplankton dynamics since the low density melt water forms a relatively stable upper layer, where the amount of light available for phytoplankton photosynthesis is abundant (Alexander and Niebauer 1981; Sakshaug and Skjodal 1989). Assuming ample nutrients and iron, such a stable layer can be a platform of high growth rate for phytoplankton. The actual impact of the melt of sea ice on phytoplankton abundance has been studied but results are usually inconclusive (Wang et al. 2005; Arrigo et al. 2002). It appears that there are other factors that affect the growth rate of planktons since correlation studies of ice melt with plankton concentration indicated relatively weak correlations or inconsistent results. Among such factors is the typically persistent cloud cover that reduces the amount of incident solar radiation available for photosynthesis. Also, the stratification of the water, whether from freshening or solar heating, impedes the vertical transport of nutrients and limits the resupply of nutrients from below the surface layer. In this regard, knowledge about the regional patterns of wind stress is important, since it is the latter that can affect the redistribution of nutrients through upwelling or horizontal advection. Equally important is the availability of iron, which has been observed to be limiting phytoplankton growth in the subarctic North Pacific (Martin and Fitzwater 1988). In the relatively shallow shelves with high river runoff and abundant crustal sources, availability of iron is likely not a problem. But in deeper waters, well removed from continental shelves, iron limitation may be a significant factor.

A good understanding of how the variability of plankton concentration is related to the variability of the sea ice cover is important because the sea ice cover has been changing considerably in recent years as discussed in Chap. 7. There are regions where the sea ice cover has been declining rapidly, such as the Arctic and the Bellingshausen Seas (Comiso et al. 2008; Zwally et al. 2002; Jacobs and Comiso 1997), but there are also regions, such as the Bering Sea and the Ross Sea, where the sea ice extent has been increasing. Some of the most dramatic changes in the sea ice cover that have been directly observed occurred during the last decade in the Arctic (Comiso et al. 2008; Stroeve et al. 2007; Comiso 2002), and it is fortuitous that during the same period, continuous ocean color data are available. This allows for the study of observed changes in plankton concentration that may be directly correlated with changes in the sea ice cover and other environmental parameters.

The satellite ocean color sensor that currently provides the longest and the most comprehensive data on plankton distribution is the SeaWiFS sensor which was launched on 1 August 1997. The data set from this sensor is the set that is at present most suited for ocean color time series studies. The historical pigment data set provided by SeaWiFS has also undergone the most extensive validation. The retrieved plankton concentrations from SeaWiFS data have been compared with those from overlapping Terra/MODIS and Aqua/MODIS data, and all these sensors provide generally similar spatial and temporal distributions. The length of the data set can be extended through the use of OCTS (1996–1997) data and the discontinuous and spatially sparse ocean color data set from the Nimbus-7 CZCS sensor (1978–1986). However, there are many unresolved issues associated with accuracy, bias and data gaps as discussed in Comiso et al. (1993). For uniformity and ease in interpretation, we will use the SeaWiFS data solely for the time series and variability studies. There was an instrument problem that made SeaWiFS data unavailable in January, February, March, June, July, and August 2008. For completeness, we used MODIS data during this period to fill in the gaps and avoid discontinuity in the time series after noting that the two sensors provided consistent pigment concentrations during periods of overlap.

9.2 Ocean Color in the Northern Hemisphere

9.2.1 Seasonal and Interannual Variations in Pigment Concentration

The accumulation and availability of a continuous time series of SeaWiFS ocean color data from 1997 to the present allow for detailed examination of the spatial distributions as well as seasonal and interannual variability of the plankton concentrations. Although the record of about 12 years is relatively short, it is the set with the longest record length that is available for creating a climatology. Color-coded maps of such monthly climatologies of chlorophyll concentrations from January to December in the Northern Hemipshere are presented in Fig. 9.1. The set of images provides a general overview of the plankton distribution in the region as well as the month-to-month variability and the seasonality of the distribution. The color-coding depicts low pigment concentration regions as blues and greens, while high concentration regions are represented by yellows, oranges, and reds. The averaging technique is similar to those used in previous studies for visible and infrared data presented in previous chapters. It is also the same technique utilized in previous studies (e.g., Comiso et al. 1993; Arrigo et al. 2002; Smith and Comiso 2008). In particular, the monthly averages are not true monthly averages but averages of measurements under cloud free conditions only. Thus, different data elements (pixels) are averages over slightly different time period. Some biases when compared with true monthly averages are expected especially in areas where clouds are persistent. However, they are in part minimized because of the relatively high sampling rate for each day, especially

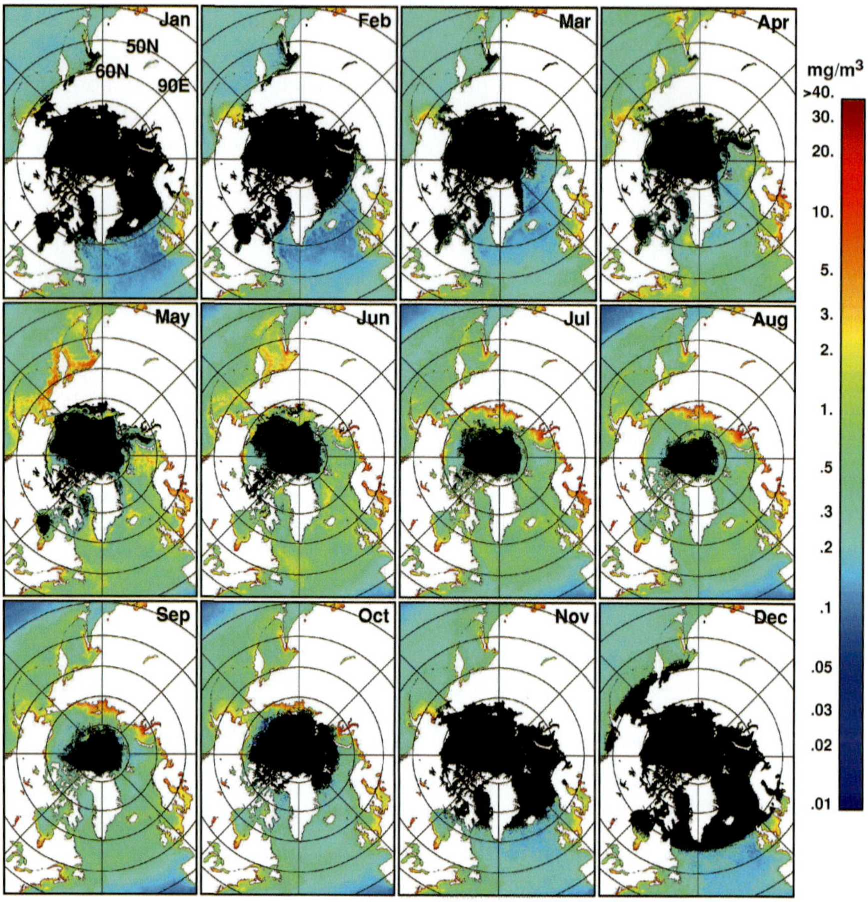

Fig. 9.1 Monthly averages (climatologies) of pigment concentrations in the Northern Hemisphere from January to December using SeaWiFS data from August 1997 to July 2008. Black areas represent no data due to darkness, clouds or sea ice

in the polar regions, as discussed in Chap. 4. Averages for multiple years are also done in a similar way, and there is an additional issue that is associated with the ever changing shape and location of the sea ice cover. The climatological pigment concentration maps show less sea ice cover than those depicted by the monthly data for each year because as the ice cover moves around from one year to another, there are more areas exposed for pigment concentration observations and hence more extensive spatial coverage. This means that for some pixels, the climatology (or multiyear averages) for some summer months represents averages only for the years when the pixel was ice free during the month. Another issue is an unknown bias associated with the masking of sea ice. Ocean color data are retrieved only in areas where sea ice concentration is not significant, and this usually means the ice free open ocean areas beyond the ice edge. Thus, the contribution of planktons in ice free waters within the ice pack (e.g., the marginal ice zone) are not included in the analysis.

The set of climatological monthly images shown in Fig. 9.1 allows for a quick visual assessment of how the spatial distribution of the pigment concentration typically changes from one month to another and from one season to another. The evolution of the patterns of blooms from one month to the next as the sea ice melts in the peripheral seas and in the Arctic basin is clearly depicted in the images. Bloom patterns behind retreating ice edges are apparent in the western side in the Bering Sea (east of 180°E) and the Okhotsk Sea, where the pigment concentrations and their spatial coverage develop into their peak values as the sea ice cover retreats to the Arctic basin. The patterns of bloom formation in both sides of the Pacific and Atlantic Oceans are shown to be very seasonal. In the Pacific Ocean region, pigment concentration intensifies and expands in late winter and spring, and reaches maximum values in April in the Bering Sea and in May at the Okhotsk Sea and other locations. The bloom period in the Atlantic Ocean side also starts in April and the peak period occurs approximately in May. In this region, significant blooms start to occur near the southern tip of Greenland in Baffin Bay and at the Barents Sea. In May, extensive bloom patterns around Greenland and Iceland as well as the rest of Barents Sea are observed. In June, the blooms are still near peak values but the distribution is more uniform. In July, the bloom areas shift to the Arctic basin and the values are especially high near the coastal regions. The coastal blooms extend further to the north in August, but it is apparent that the bloom patterns do not follow the retreat of the ice cover (black area) in the basin, which reaches minimum values about the middle of September. This is not consistent with the expected enhancements in plankton concentration associated with the introduction of meltwater from the retreat of the sea ice cover. The lack of plankton growth further north in the Arctic basin illustrates the complexity of the plankton dynamics and the need to consider other parameters in the interpretation of the ocean color data. The intense bloom patterns occur generally in the shallow shelf regions, where the supply of nutrients and iron are likely not limited because of wind-induced upwelling and river run-off. In the deeper waters to the north, the supply of nutrients and iron are probably very limited not only because of the greater distance from the source but also because much of the nutrients and iron are entrained with dense water created during the ice formation and end up near the bottom of the ocean, where there is little chance for them to get back to the surface.

During the relatively short period when a significant part of the Central Arctic is ice free, the peak of the bloom period appears to occur in August although the maximum open water area is not reached until the middle of September. The biological activity in the entire Central Arctic basin had been difficult to evaluate because of the presence of the perennial ice cover, but the latter has been declining rapidly, as indicated in Chap. 7, Sec. 7.2.2, and in recent years, a significant part of the region has been exposed to give us an opportunity to study the chlorophyll distribution in the Arctic basin. The exposed areas in August and September include those in the Chukchi and Beaufort Seas, where the plankton concentrations in both shallow and deep ocean regions can be evaluated. The distribution of phytoplankton blooms in the Arctic basin is unexpectedly asymmetric with the plankton concentrations in the eastern region a few times higher than those in the western regions. This phenomenon is persistent throughout the summer period and early autumn. It also appears that the deep ocean regions of the Arctic basin do not show the same high chlorophyll values that are apparent in the Siberian, Laptev, and Kara Sea regions. In October, the Arctic Basin is almost completely covered by sea ice. By November, the only regions with relatively high plankton concentrations are the Bering Sea and the southern part of Hudson Bay.

The climatological monthly data also indicate significantly different patterns in the development of blooms between the North Pacific and the North Atlantic Oceans at intermediate latitudes. In the North Pacific, there are intense blooms in spring, but they are generally well defined and confined mainly to the relatively shallow Bering and Okhotsk Seas. On the other hand, the spring blooms in the Atlantic are much more widespread and extend over basically a large fraction of the region north of 45°N. In the summer, both regions show relatively lower chlorophyll concentrations than in spring, while in autumn, the concentrations are slightly more elevated than in summer. In summer and autumn, a halo of relatively high pigment concentrations is apparent at high latitudes (i.e., 50°N in the Atlantic Ocean and 40°N in the Pacific Ocean) showing a strong contrast in the plankton distribution between intermediate and polar latitudes. In December, a general decline in the pigment concentrations is apparent caused in part by cold surface temperatures and short daylight hours.

Although the link of the melt of sea ice to plankton concentrations has not been well established, it is apparent that the concentrations are relatively high in areas that were ice covered in the previous winter in regions such as the Sea of Okhotsk, the Bering Sea, the Labrador Sea, and the Barents Sea. The high plankton concentration values are generally located in or adjacent to the seasonal sea ice region. The connection is thus likely but other conditions must be satisfied to ensure that the effect of meltwater on plankton growth can be assessed. The seasonal sea ice region in the Arctic basin is generally located in shelf regions especially in the Russian side, where the water is relatively shallow. Wind forced Ekman upwelling enables nutrients near the bottom of shallow waters to recirculate back to the surface, where they are available for consumption. Also, iron is likely more available near coastal areas. Thus, in the Kara, Laptev, and East Siberian Seas, it is likely that there are no nutrient or iron limitations. On the other hand, despite the abundance of meltwater in the deep ocean away from coastlines, the plankton concentrations are relatively low likely because of limited supply of nutrients and iron.

Adjacent to the coastal areas and near the mouth of big rivers, the interpretation of the derived pigment concentrations becomes difficult because of possible contamination of the water by particulates other than planktons. Water that contains a dominance of these other particulates have been referred to as Case-2 water as discussed in Chap. 4. Such particulates may consist of CDOM, suspended matter, mud, or other particles that may come from river run-offs. Reflected sunlight signals from the bottom of very shallow parts of the ocean can also be a factor. Ability to discriminate Case-1 from Case-2 water (or to estimate the error caused by the presence of Case-2 water on the estimated plankton concentrations) would be desirable, but such techniques have not matured. Furthermore, there is a general paucity of in situ case-2 data in the Arctic, and there is no means to effectively validate results. The distributions presented are the results from analysis of data as derived from the use of the global case-1, SeaWiFS algorithm (i.e., OC5).

The climatology as presented in Fig. 9.1 shows that the peak bloom period occurs in May. To illustrate how the phytoplankton pigment distribution changes during this peak bloom period, monthly averages for May from 1998 to 2008 are presented in Fig. 9.2. As can be inferred from the climatological averages (i.e., Fig. 9.1), the Bering and Okhotsk Seas are among the regions with consistently high pigment concentrations during spring. These are seas that are at least partially covered by sea ice in winter. Considerable year-to-year variations are, however, apparent, especially around the Kamchatka Peninsula, where pigment concentrations are usually elevated. During some years (e.g., 2001, 2002, 2004, and 2006) bloom features have consistently high values around the peninsula while in other years (2003, 2005, 2007, and 2008),

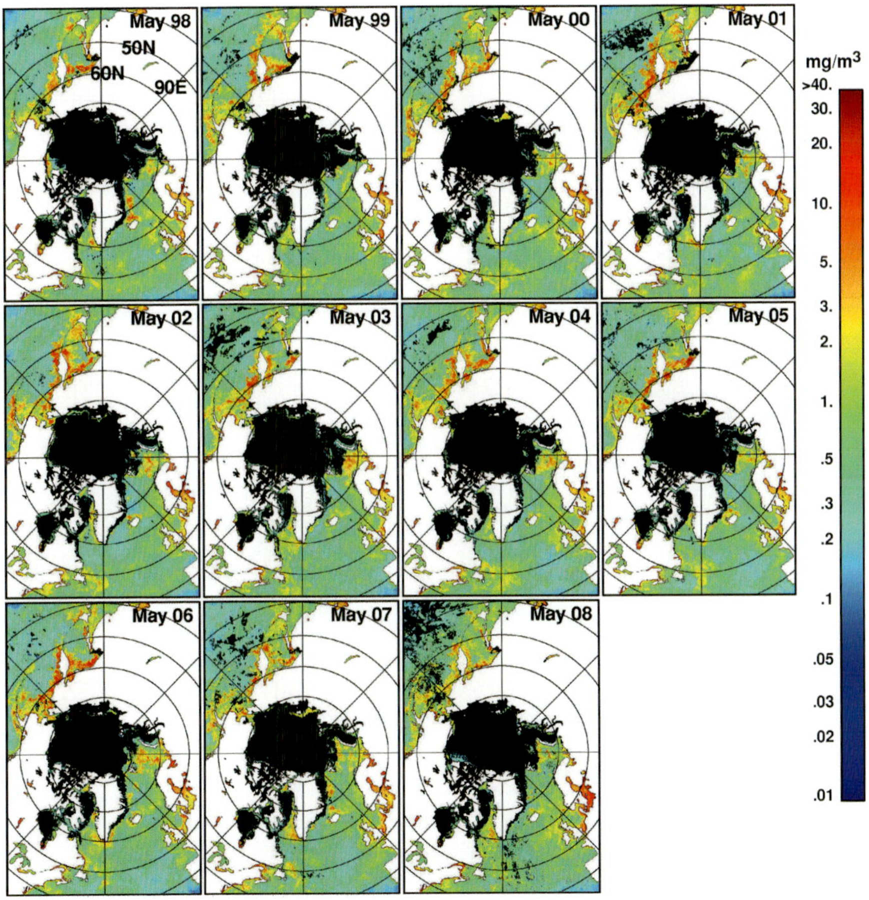

Fig. 9.2 Monthly pigment concentrations in the Northern Hemisphere during a spring month (May) for each year from 1998 to 2008 using SeaWiFS data

the plankton concentrations in the same regions are not so high. In other regions, the location and intensity of the blooms are even less predictable. For example, elevated pigment concentrations are apparent in the Greenland Sea and around Iceland in 1998 and 2007, but in other years, the pigment concentrations in these regions were more modest. It is, however, apparent that the pigments tend to be elevated along shallow coastal areas. In the deep oceans, some bloom features usually appear in the Northern Pacific Ocean to the west of Hokkaido (Japan) or in the Northern Atlantic Ocean to the north of Norway.

Quantitative comparison of the bloom patterns for the different years can be made through the use of the color-coded monthly anomaly maps presented in Fig. 9.3. Abnormally high pigments are represented in purples or reds, while abnormally low pigments are represented in blues or greens. Areas of missing data (in black) are areas covered by sea ice or areas with persistent cloud cover as in the

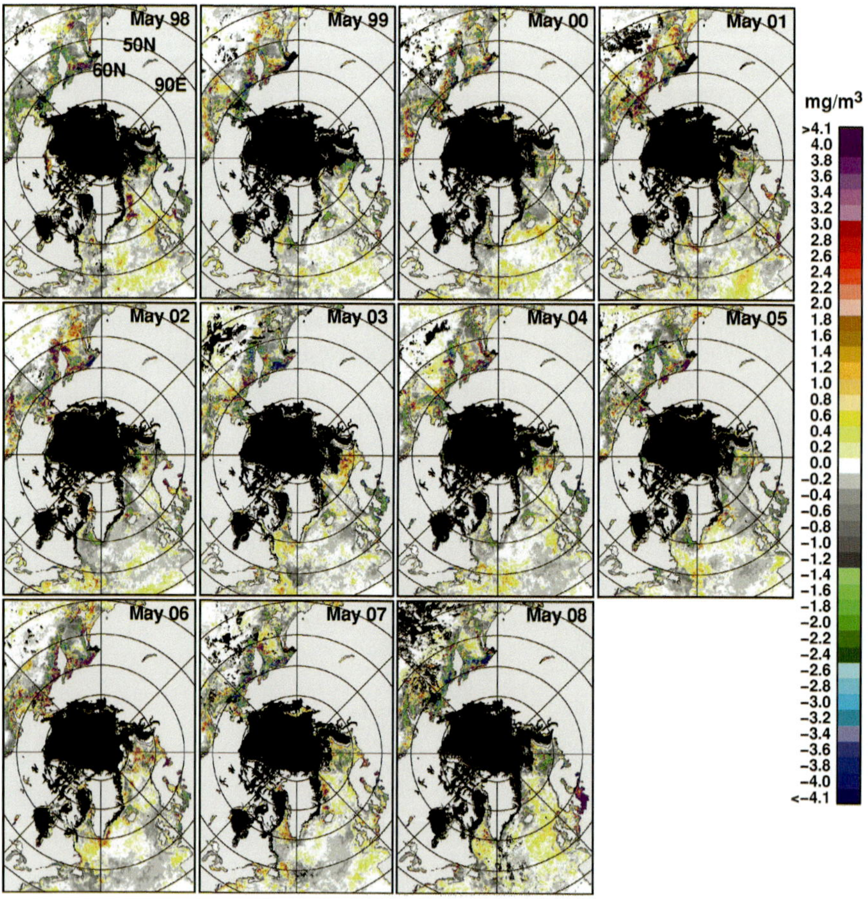

Fig. 9.3 Monthly Anomalies of pigment concentrations in the Northern Hemisphere during a spring month (May) for each year from 1998 to 2008 using SeaWiFS data

open oceans at relatively low latitudes. The images generally show mixtures of positive and negative anomalies in most regions. This is caused by interannual shifts in the location of blooms on account of changes in wind directions and other environmental factors. However, there are regions where the location and magnitude of bloom patterns are relatively consistent. For example, in the Pacific Ocean region, it is apparent that the pigment concentration anomalies in the Bering and Okhotsk Seas are mainly positive in 2001 and 2002 while they are mainly negative in 1998, 2007, and 2008. In the Atlantic Ocean side, the pigment anomalies are predominantly positive in the Barents Sea in 2003 and 2006 and mainly negative in 1999, 2005, and 2008. Overall, although the plankton concentrations are generally elevated at high latitudes, especially near the ice edges, the monthly anomalies need to be interpreted carefully and in the context of the dynamics of the plankton and the location of the bloom patterns as revealed by the images in Fig. 9.2.

Monthly averages of plankton concentrations in September for each year from 1997 to 2007 are presented in Fig. 9.4. September is not only the time when the sea ice melt period ends, but also when ice freeze-up starts. During the month, the plankton concentrations at high latitudes (>40°N) remain relatively high but not as high, as in spring. The highest values as derived from satellite data are located in the Arctic basin especially along the coastal regions of the East Siberian, Laptev, and Kara Seas. It is apparent that in the Arctic Ocean, the asymmetry in the distribution of planktons between the eastern and western regions, as cited previously in Fig. 9.1, is a persistent phenomenon despite large interannual variations in the spatial distribution and extent of the sea ice cover. The distribution of plankton concentrations along the edges of the sea ice cover (in black) is not consistently high. Although the melt of sea ice may be a factor in the early summer period when the sea ice continues to retreat to the north, the plankton concentrations near the ice edge becomes relatively low farther away from the coastline. Comparing with the bathymetry map of the Arctic (Fig. 2.1a), it appears that the pigment concentration starts to go down in deeper parts of the ocean. As indicated earlier, limitations in the access to nutrients and iron are likely the cause for the reduction of plankton concentration in the deeper ocean regions. Nutrient limitation in the region is in part caused by a process during ice formation, in which relatively dense water masses are formed through brines rejection and as these water masses sink, nutrients near the surface get entrained and end up near the bottom of the ocean. Once the nutrients get in deep ocean regions, there is a little chance that they would surface again and be available for consumption by plankton near the surface. The Eastern region is on the average much more shallow (because of a very wide shelf area) than the Western region, and the observed asymmetry in the pigment distribution is likely associated with bathymetry. A halo defining the location of enhanced pigment concentration at high latitudes compared to those at low latitudes, as described earlier, is evident and appears to be generally located at around 45°N in the Atlantic Ocean and 40°N in the Pacific Ocean. The location of the halo does not coincide with bathymetric features but in the Pacific Ocean, the location is approximately near the oceanic fronts (Lynn, 1986) where nutrients are made available through the upwelling of intermediate or deep water. In the Atlantic Ocean, surface water from the south cools down at

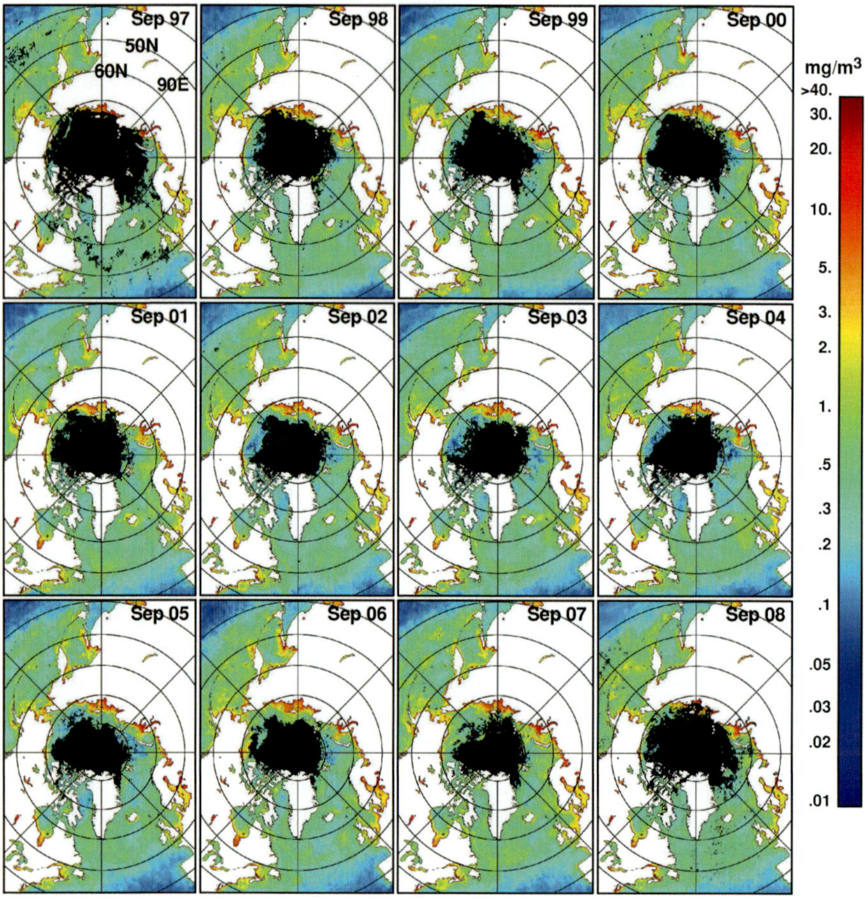

Fig. 9.4 Monthly averages of pigment concentrations in the Northern Hemisphere during an end of summer month (September) for each year 1998 to 2008 using SeaWiFS data

approximately the same latitude and downwells to deeper areas likely entraining a significant fraction of the nutrients near the surface.

Monthly anomaly maps for September are presented in Fig. 9.5 and provide the means to better quantify the interannual variations of the spatial distribution of plankton concentrations during the period. It is during this time period when the sea ice cover in the Northern Hemisphere has declined to its minimum extent and meltwater is abundant. It is apparent that the interannual variations in plankton concentrations are relatively moderate with the anomalies being predominantly positive in 1998, 2002, 2006, and 2007 while the anomalies were predominantly negative or neutral during other years. In September 1997, the sea ice cover was relatively low but the image for this year has the most missing data likely because of persistent cloud cover or improper sea ice masking. The year with the most positive anomalies appears to be 2006 followed closely by 2007. These are years when the extents of the sea ice

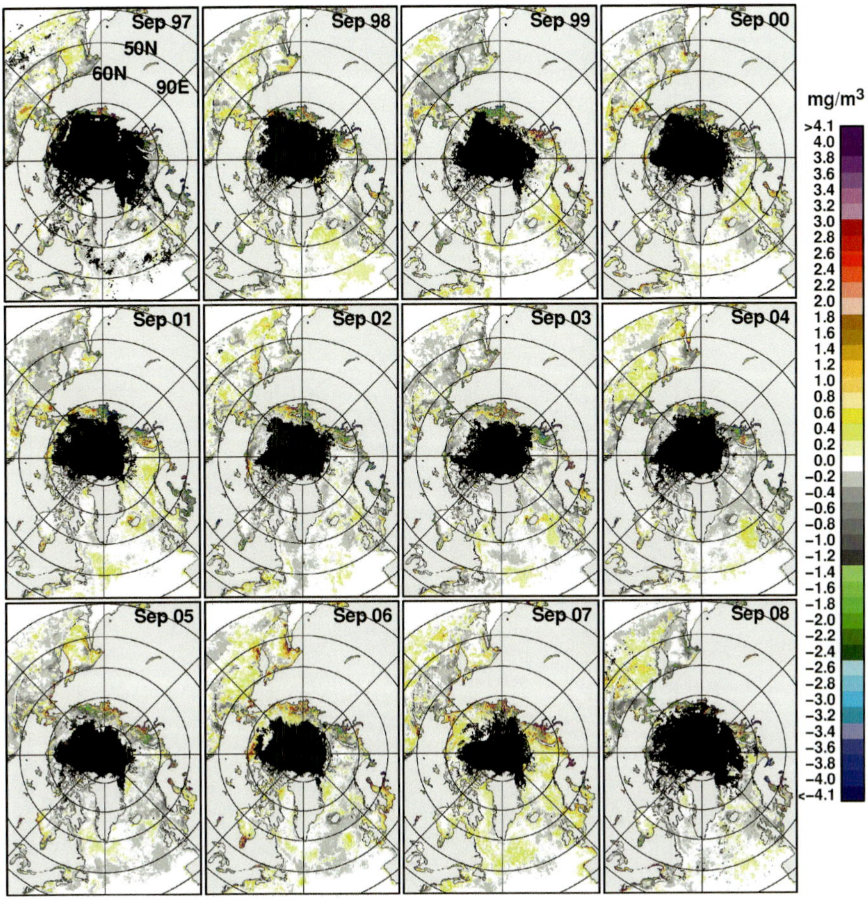

Fig. 9.5 Monthly Anomalies of pigment concentrations in the Northern Hemisphere during the end of summer month (September) for each year fom 1998 to 2008 using SeaWiFS

cover were relatively low and the SSTs were relatively high. Note that the open water data in the Arctic Basin in 2008 are much less than in 2007, although the extents of the ice cover for the two years were very similar as indicated in Chap. 7. Again, the lack of data in 2008 could be the result of more persistent clouds, but other factors like improper masking of the sea ice cover could contribute.

Yearly averages of the phytoplankton distributions in the Northern Hemisphere from 1998 to 2007 are presented in Fig. 9.6. For the Northern Hemisphere, our yearly averaging goes from January to December, while in the Southern Hemisphere, the averaging goes from September of one year to August the following year. In the estimates of yearly productivity, the same definition applies. The strategy is to obtain yearly averages that cover the entire bloom period from spring to summer. In the Southern Hemisphere, the peak bloom period in the spring and summer go from November of one year to March the following year. The yearly images

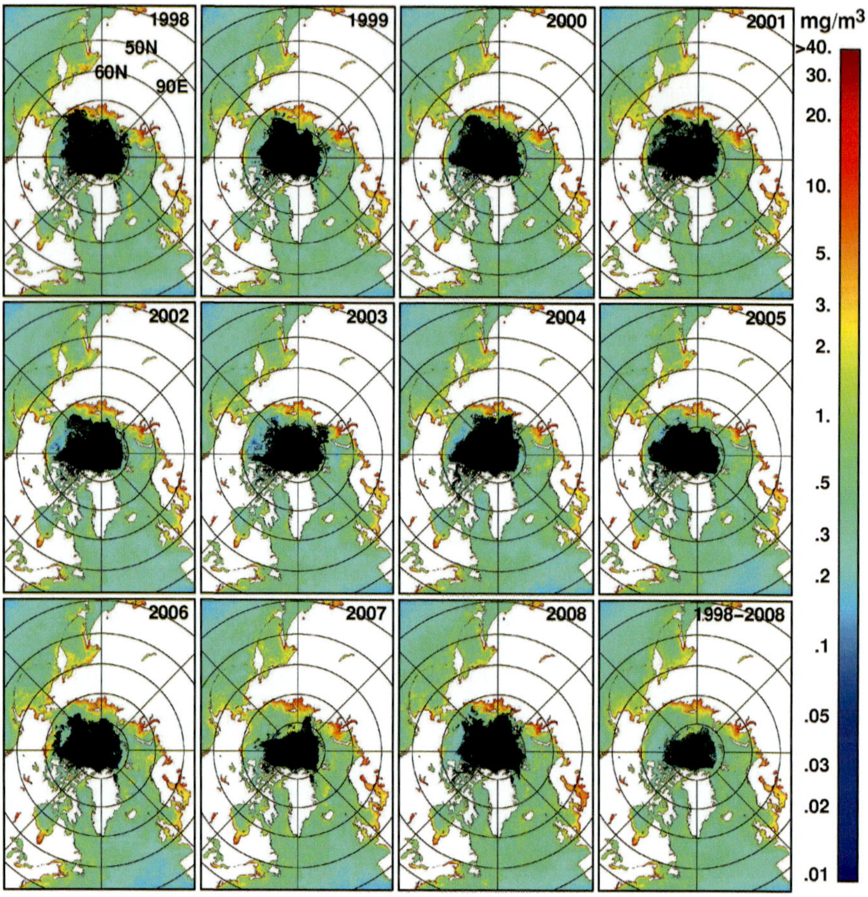

Fig. 9.6 Yearly averages (January to December of each year) and multiyear average of pigment concentrations from SeaWiFS from 1998 to 2008

presented in Fig. 9.6 indicate that overall, the spatial distributions of plankton concentrations do not change much from one year to another in the Northern Hemisphere. Subtle changes are apparent in the North Atlantic Ocean and North Pacific Ocean, but they are mainly changes in location and the magnitude of the blooms and the patterns appear basically the same. Inside the Arctic basin, the data represent primarily summer month data since the region is covered by sea ice for much of the year. The interannual variability in the region is thus very similar to those of September as shown in Fig. 9.4, but there are differences because of significant contributions from the months of June, July and August. Comparing the plankton distribution with the bathymetry map in Fig. 2.1a, it is apparent that enhanced distributions inside the Arctic Basin usually occur in relatively shallow waters. As indicated earlier, shallow waters enable the resupply of nutrients from the bottom making it possible for the blooms to be maintained. Blooms can be advected by surface currents or strong winds and brought to some deep water areas but the images in Fig. 9.6 do not show a good manifestation of such a process. The last image in Fig. 9.6 is the multiyear average of all the available SeaWiFS data up to 2008 and can be regarded as the yearly average representing typical spatial distribution of planktons in the Northern Hemisphere at high latitudes during the 1998 to 2008 period. This image is basically the average of all the yearly images and was generated by taking the average of pigment concentrations in each pixel using all available SeaWiFS data from 1998 to 2008.

The yearly anomaly maps of chlorophyll *a* concentrations as presented in Fig. 9.7 enable a more quantitative evaluation of interannual variability of the bloom patterns. In the Arctic basin, mixtures of positive and negative anomalies are apparent in the Eastern Region and primarily in the Laptev and Kara Sea regions, where pigment concentrations are usually highly elevated. This is an indication that the enhanced bloom areas move around from one year to another. However, in 2006, 2007, and 2008, it appears that positive anomalies are much more dominant in the Arctic Basin than in other years, indicating enhanced and more expansive distributions during

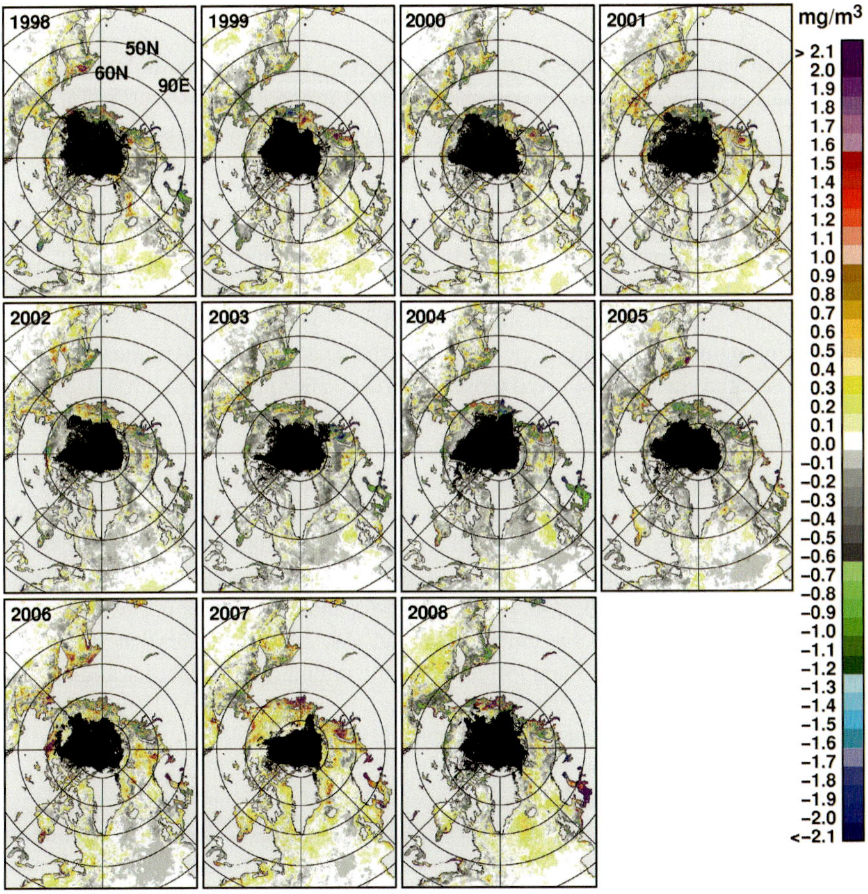

Fig. 9.7 Yearly anomalies of pigment concentrations from SeaWiFS from 1998 to 2008. The anomalies were estimated using the multiyear average data shown in Fig. 9.6

these years. Positive anomalies are also generally dominant in other regions beyond the Arctic basin during these three years except for the Bering Sea in 2007 and 2008 and the Sea of Okhotsk in 2008. Positive anomalies were also dominant overall in 2001 but mainly in the peripheral seas. It is interesting to note that positive anomalies were unusually high near the mouth of the MacKenzie River in 2006, while in 2007 and 2008, they were unusually high near the Lena, Yenisey, and Ob rivers. It was also abnormally low near the mouth of the MacKenzie River in 2002 and also low near the Lena, Yenisey and, Ob rivers in 2003 and 2005. This phenomenon suggests that river runoff and therefore case 2 water may be important issues in these regions for some years but not every year. Additional research is needed to evaluate these effects, but it should be noted that the overall impact appears localized and that the strength of river runoff impact varies significantly from one year to another.

9.2.2 Temporal and Regional Changes

Interannual variability of the spatial distribution of biomass accumulation in the Arctic and peripheral seas has been characterized in the previous section for each year from 1998 to 2007. To quantify seasonal and interannual fluctuations of chlorophyll concentrations, monthly averages of plankton pigment concentrations in the Northern Hemisphere are presented in Fig. 9.8a. As expected, the peak value for each year occurs in May, while the low values occur in winter. The increase in average value during spring is quite rapid, reaching maximum values in about two months. The values stay at a relatively high level until September, at which time the ice free surfaces in the Arctic basin starts to freeze. It is apparent that there is significant interannual variability especially with the peak values. The highest monthly average occurred in 2001 followed closely by 2006, 2008, 2007, and 2002. It should be noted that although the peak values usually occur in May, there are some exceptions as in June 2007. It is also interesting that when the monthly peak values were relatively low, as in 2003–2005, the monthly minimum were also relatively low. Monthly anomalies for the same data are presented in Fig. 9.8b together with a 5-month running average (in gray), and it is apparent that there was a cyclical variation of the anomalies with time during the eleven year period with peaks in 1998, 2001, 2006 and 2007. Also, note that the highest anomaly values occurred in July 2007 followed by June 2006, December 1998, and May 2001. It appears that any month of the year can have a significant influence on the interannual variability and trends. The black line is a linear fit to the data indicating a positive trend of 0.072 ± 0.015 mg/m^3/decade during the 11-year period. This modest trend should be considered in the context of the relatively short record length and the cyclical pattern.

To gain insights into the temporal variability of the plankton distributions, weekly maps of ocean color were generated. The improved temporal resolution derived from weekly composites provide a means to better assess the length of a bloom period and the spatial distribution when the average pigment concentration reaches its peak. Although daily averages can also be generated from the satellite

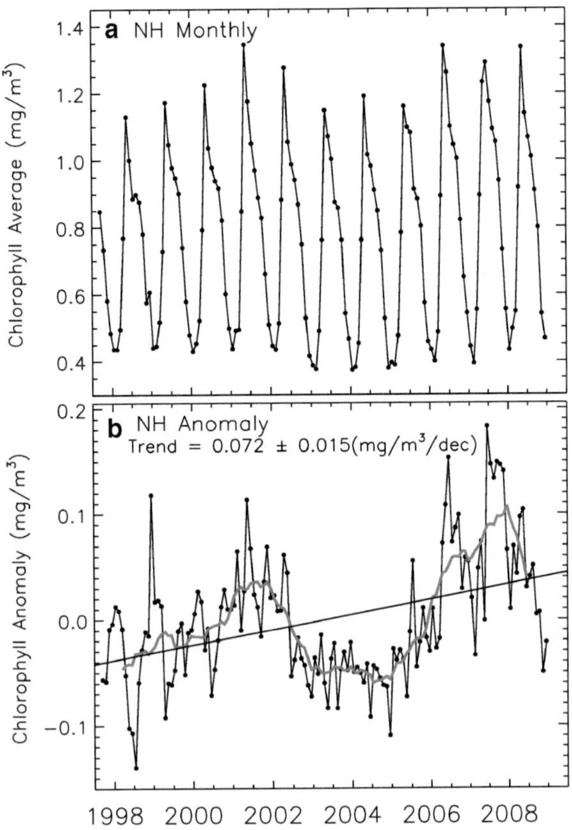

Fig. 9.8 Plots of (**a**) monthly averages and (**b**) monthly anomalies of pigment concentration in the Northern Hemisphere from August 1997 to December 2008. A yearly running average (in gray) and a trend line from results of a linear regression on the anomaly data are also shown

data, they are not useful for time series studies because of large data gaps caused by extensive and persistent cloud cover in the region. Also, to get an idea how the plankton concentrations change in different parts of the Arctic, nine relatively productive but highly variable areas were identified as study areas, with boundaries as indicated by different colors as shown in Fig. 9.9. The study areas include peripheral seas such as the Bering, Okhotsk, Barents, Greenland, and Labrador Seas, and bays such as Hudson Bay and Baffin Bay, all with some sea ice cover during the winter. Two sub-areas in the Arctic basin are also included, namely: the Eastern region that includes the Siberian Sea, Laptev Sea, and Kara Sea, and the Western region that includes the Chukchi and Beaufort Seas and the Canadian Archipelago. These study areas are between 65 to 80°N and from 70 to 180°E for the Eastern Arctic Region and 180 to 280°E for the Western Arctic region. For comparison with results from other seas, a North Atlantic study area, as indicated in the map, is also chosen as a part of this study. In the time series, there are gaps in the data in some regions on account of darkness and lack of data. Thus, although there are regions, like the Sea of Okhotsk, the Labrador Sea, and the North Atlantic, where ocean color data are available nearly all year, other regions, like the Eastern and Western

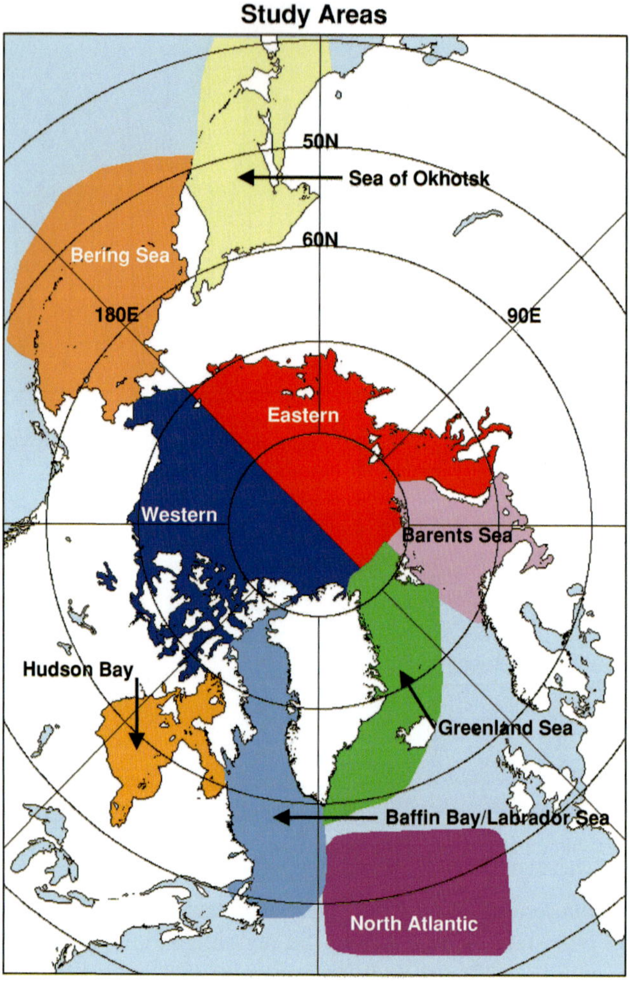

Fig. 9.9 Map of the various sectors in the Northern Hemisphere that are used as study areas. For sea ice studies, the Barents Sea study area also includes the Kara sea while the Western and Eastern sectors are combined to form the Arctic sectors

Arctic, and Hudson Bay, have substantial gaps in winter when they are effectively in darkness or covered with sea ice.

Averages of chlorophyll concentrations within the boundaries of the aforementioned nine study areas were calculated for each week from January 1998 through December 2008, and the results are presented in Fig. 9.10. For improved ability to assess interannual variations by season, the plots are color-coded such that weekly data for each season are represented by different colors. In particular, winter, spring, summer, and autumn are represented in the color scheme by different shades of blue, green, yellow, and red, respectively. The plots presented in Figs. 9.10a to 9.10i show clearly that the timing of the occurrence of peak of the bloom period is different

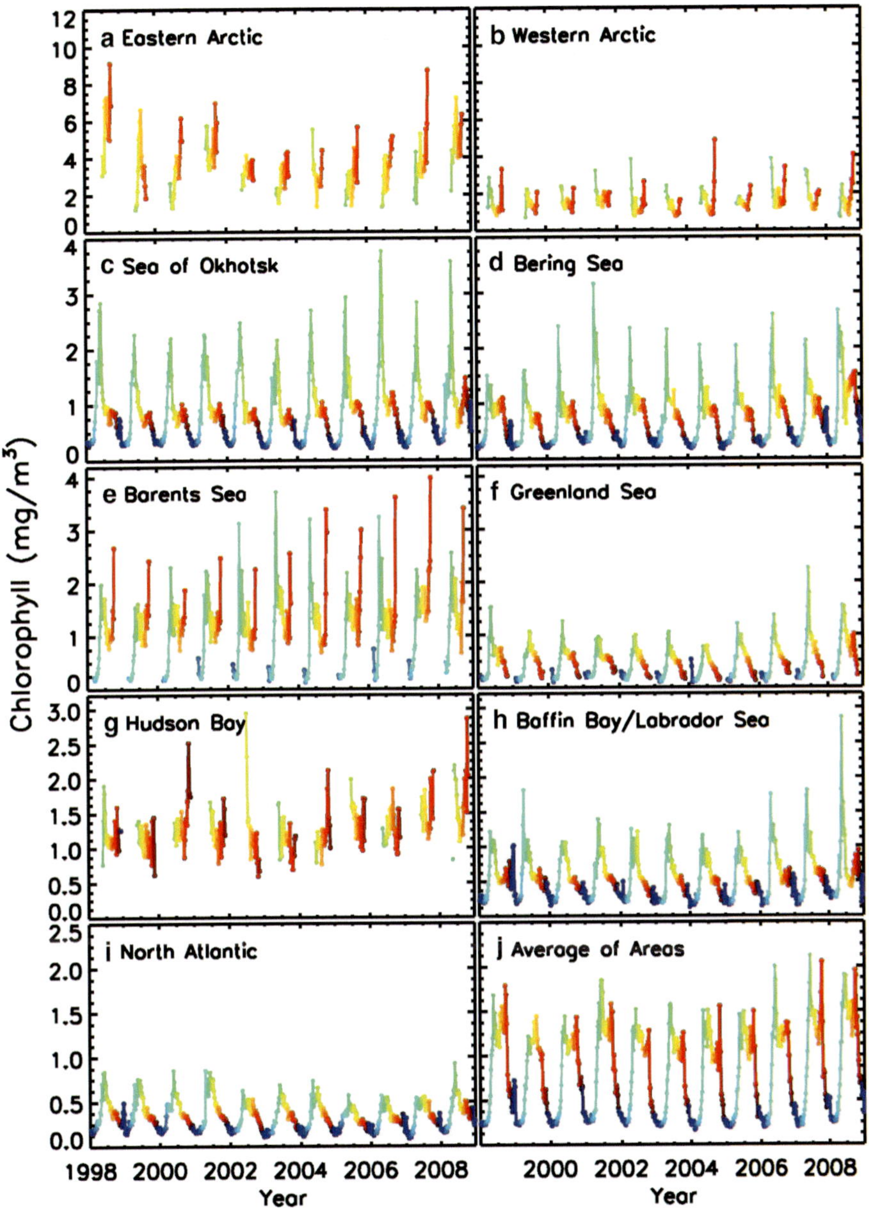

Fig. 9.10 Plots of monthly average pigment concentration in the various sectors (see Fig. 9.9) in the Northern Hemisphere with the various seasons represented in different colors

for the different sectors. Before discussing the data from the individual regions, it is useful to examine first the averages of the distributions of all the study areas as presented in Fig. 9.10j. The plot provides the means to get a general overview of the seasonal variability of plankton concentrations in the entire Northern Hemisphere. It is apparent that there is one dominant peak in spring and another one in the late summer/ early autumn period. Large interannual variability is apparent with the spring blooms (in green) varying from about 1.2 to 2.3 mg/m^3 with the relatively high values during the 11-year period occurring in 2001, 2006, 2007 and 2008. Large interannual variability is also observed during the late summer and early autumn periods with the relatively high values occurring in 1998, 2007 and 2008. The plankton concentrations in the Eastern and Western Arctic are presented in Fig. 9.10a, b, show quantitatively the large contrast in the plankton distributions of the Eastern and Western regions as described earlier. The pigment concentrations in the Eastern region are on the average about 3 times higher than those in the Western region during the summer and autumn when the regions have the least ice cover. Relatively high values appears in the spring season but these are averages over much smaller open water areas that are generated as the sea ice cover begins to retreat to the north. The seasonal development for each year looks different partly because of interannual differences in the onset of ice melt, location of ice retreat, and weather patterns. The main phytoplankton blooms in the high Arctic areas occur when the ice cover starts to break up and the availability of both sun light and nutrients are near seasonal maxima. Increases in pigment concentrations during late autumn (in red) in parts of the Arctic Ocean, when the ice cover starts to advance and nutrients are low, have not been validated and may be artifacts caused by frazil ice scavenging of cells with accumulation of grease ice.

The corresponding monthly plots for the other seven study areas are presented in Fig. 9.10c–i. The averages for spring (in green) in the Sea of Okhotsk, the Bering Sea, and the Barents Sea are shown to be highest in 2006, 2001, and 2003 for the three regions, respectively, but the distributions are generally similar in these three regions. The averages are also similar in summer and autumn (orange and red) in the Okhotsk and Bering Seas, but in the Barents Sea, the averages in autumn are comparable to or sometimes higher than those of spring. The high values in this region during the autumn are located in the shallow waters north of Russia and again, there are likely some contaminations in the signal by case 2 waters. The variability of the plankton distributions in the Greenland Sea, Baffin Bay/Labrador Sea, and the North Atlantic study areas as depicted in Figs. 9.10f, 9.10h, and 9.10i, respectively, are similar to those of the Bering Sea with the peak values occurring in the spring but the magnitude of the averages is significantly less than those of the other regions. In the Hudson Bay study area, the averages are not so consistent with the average spring values varying by as much as a factor of two. This suggests that some factors like the supply of nutrients may change from one year to another, as may be caused by upwelling events which are usually influenced by wind patterns.

Although we have only about 11 years of data, it is useful to identify patterns of change in plankton concentration since other environmental parameters like the sea ice cover and surface temperatures have been changing significantly in recent years. Monthly anomalies in the same study regions were estimated and presented in Fig. 9.11. Large interannual variability in the pigment anomalies is apparent in the Eastern Arctic in summer but not in the Western Arctic. Interannual variability in spring is also large in the Sea of Okhotsk, Bering Sea and Barents Sea but not in

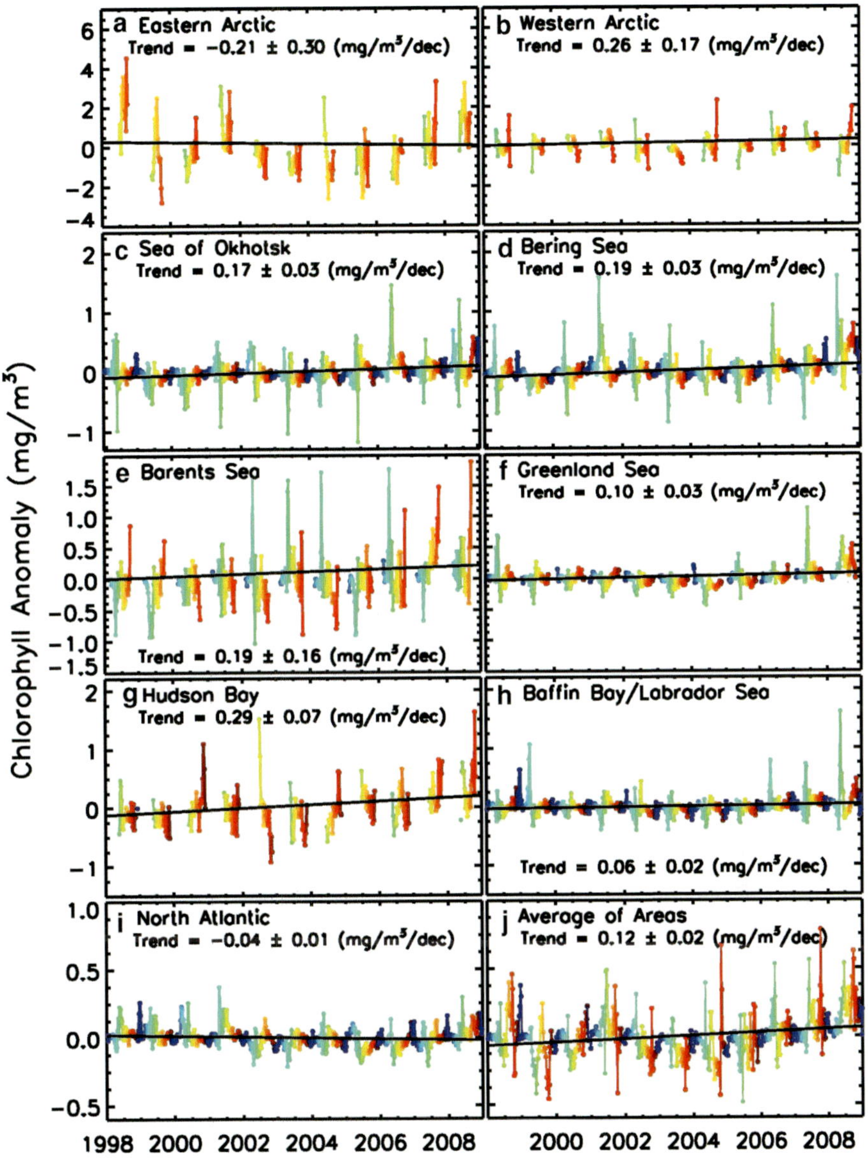

Fig. 9.11 Monthly anomalies of average pigment concentration in the various sectors in the Northern Hemisphere with the various seasons represented in different colors

the Greenland Sea, Baffin Bay/Labrador Sea and Baffin Bay. The specific reasons for the different variability for the different regions is not known but they are likely associated with regional environmental factors. It is interesting to obtain contrasting trends in the Eastern and Western Arctic regions with the former showing negative trend of -0.2 ± 0.3 mg/m^3/decade and the latter showing positive trend of

0.3 ± 0.2 mg/m^3/decade. The negative trend in the Eastern region is mainly due to an anomalously high value in 1998. It is thus important to interpret the trends in the context of a relatively short data set and the large interannual variability in some regions. The Western region has been a region where the summer ice extent has declined rapidly in recent years and also where sea surface temperatures have increased as well. The trends in the plankton concentrations in the Okhotsk Sea, Bering Sea, and Hudson Bay are all positive and significant at 0.17 ± 0.03, 0.19 ± 0.03, and 0.30 ± 0.08 mg/m^3/decade, respectively. The trend in the sea ice cover has also been declining over these same regions in recent years suggesting a possible connection. The trends at the Barents Sea, Greenland Sea, and Baffin Bay are also positive, but relatively minor and insignificant at 0.19 ± 0.16, 0.10 ± 0.03, and 0.06 ± 0.02 mg/m^3/decade, respectively. For comparison, the trend at the North Atlantic is negative at -0.04 ± 0.01 mg/m^3/dec. When data from all these study areas are combined together, the overall trend is 0.12 ± 0.02 mg/m^3/dec.

9.2.3 Interannual Changes in Primary Productivity

The maps plankton concentrations observed from space by satellite ocean color sensors provide an assessment of biological activities going on in the upper layer of the ocean. As described in Chap. 4, the parameter that is needed to understand the aquatic biological system is what is referred to as primary productivity, which is defined as the amount of bacterial, algal, or plant biomass built up over time through the process of photosynthesis (Thomas 2004). This parameter which provides an assessment of how much organic matter is produced by photosynthetic material is usually expressed in units of carbon fixed by photosynthesis, per unit area of space or volume, per unit of time. Primary productivity maps in the Northern Hemisphere for each year from 1998 to 2008 have been generated through the use of a model as described in Chap. 4 and presented in Fig. 9.12. The technique utilizes the pigment concentrations derived from SeaWiFS data, Photosynthetically Active Radiation (PAR), surface temperature, and other parameters as discussed by Behrenfeld and Falko-wski (1997). Overall, it appears that the Oceans at high latitudes are not very productive especially in the Arctic basin. The main reasons for this apparently low productivity are: (a) the presence of an ice cover and the long periods of darkness in much of the winter months; (b) relatively low pigment concentrations in the deep ocean part of the Arctic; and (c) persistence of cloud cover in the region. The results are generally consistent with yearly primary productivity estimates in the Arctic Ocean basin by Pabl et al. (2008) but there are some differences because they used a different algorithm for primary productivity. Because of the ice cover, a large fraction of the Arctic Ocean is exposed to the sun only for a few weeks during the summer months. Depending on latitude location, the contribution to the primary productivity during the winter months can be very limited because of darkness. Since as defined, the primary productivity is the integrated contribution during the

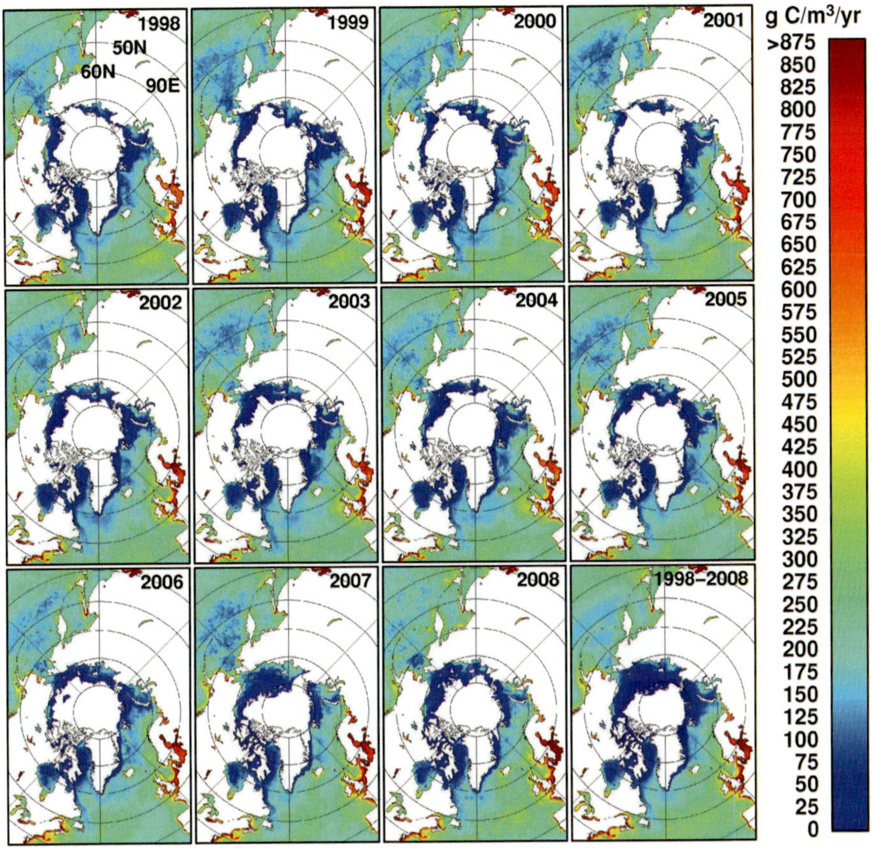

Fig. 9.12 Yearly primary productivities in the Northern Hemisphere as estimated from SeaWiFS, AVHRR data and a model from 1998 to 2008 and the average of all years

entire annually cycle, the overall yearly contribution is small for some regions because of relatively short periods of primary production. Thus, the regions in the Arctic basin close to the North Pole have unusually low productivity. They also correspond to the deeper parts of the Arctic where limitations in nutrient and iron limitations are likely the norm. As indicated previously, the perennial ice extents in the 1998–2008 period were much lower than those in the 1980s and early 1990s. This was especially true in 2007 when the area of the perennial ice was almost 40% less than average extent. Yearly variations in productivity are more apparent in the peripheral seas than in the Arctic basin. For example, the productivity at the Sea of Okhotsk is shown to be relatively higher in 2005, 2006, and 2007 than earlier years. The productivity at the Bering Sea is also shown to be enhanced in 2003 and 2004 compared to other years. In the Arctic basin, the productivity is relatively low and affected by the extent and location of the sea ice cover. It appears that the basin was most productive in 2007 when the sea ice cover retreated the farthest north.

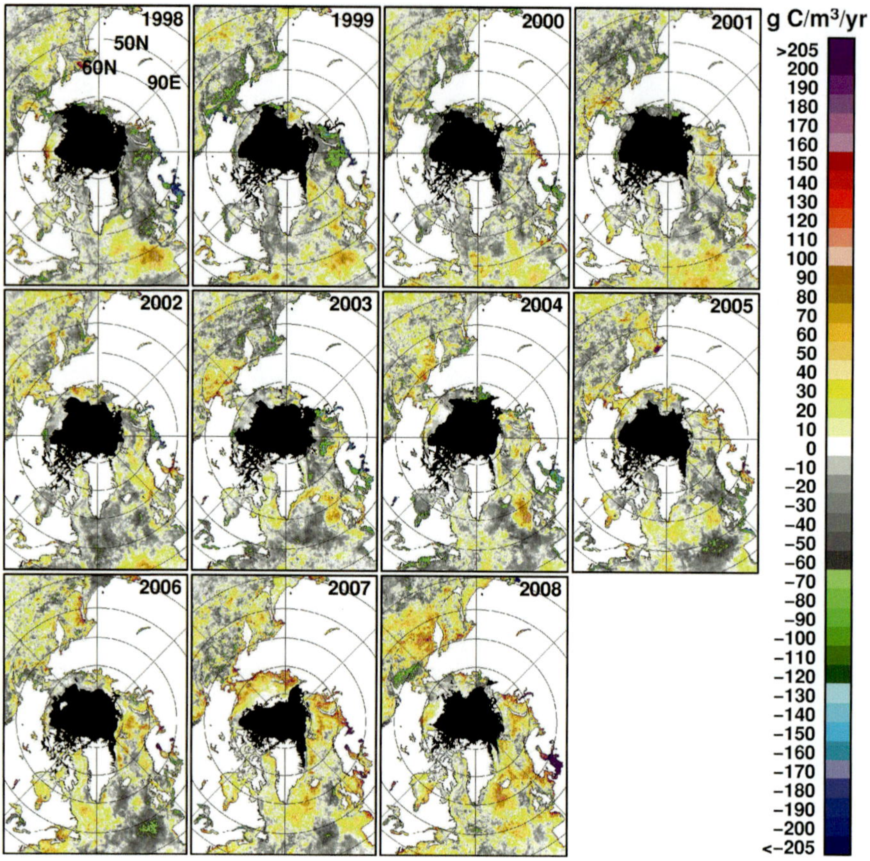

Fig. 9.13 Yearly anomalies of primary productivities in the Northern Hemisphere as estimated from SeaWiFS, AVHRR data and a model from 1998 to 2009

The maps of yearly anomalies as presented in Fig. 9.13 provide a means to better quantify the interannual changes in the productivity. It is apparent that the anomalies are predominantly negative in 1998, 1999, and 2000 at latitudes greater than 50°N, whereas it was predominantly positive in 2007 and 2008 for the same region. In the Atlantic Ocean and the regions south of 50°N, positive anomalies are dominant in 1998, 1999, 2000, and 2001 whereas in the same region, negative anomalies are dominant for the other years. The predominance of positive anomalies in North Pacific is apparent in 2006, 2007, and 2008, whereas in other years it was basically neutral or primarily negative. In 1998 the anomalies in the Arctic basin were generally negative except in the Beaufort Sea when the ice cover started to retreat substantially in the region. The perennial ice cover was a record low in 2005 and during this time the anomalies in primary productivity were generally positive. A dramatic decline in the perennial ice cover occurred in 2007, and it is apparent that the anomalies in primary productivity were strongly positive in the Arctic Basin during this period. The anomalies were also strongly positive in the Arctic Basin in 2008

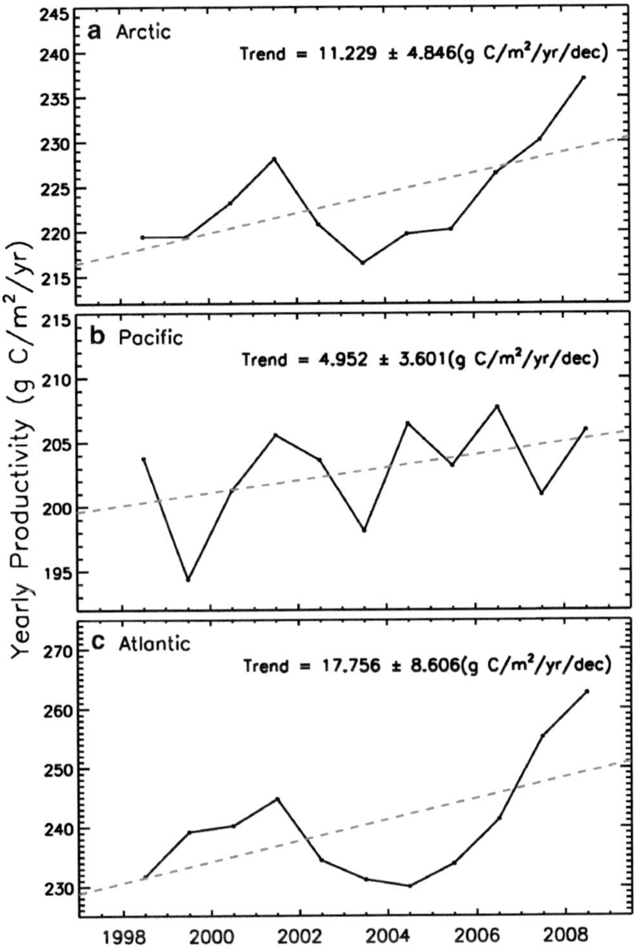

Fig. 9.14 Plots of yearly primary productivity in the (**a**) Pan-Arctic, (**b**) Pacific Ocean, and (**c**) Atlantic Ocean

although to a lesser degree, and during this year, they were strongly positive in the Northern Pacific and Atlantic Oceans as well. It is apparent that during the recent years when the sea ice cover was going through considerable changes, the primary productivity in the region was also changing considerably.

Plots of the average yearly productivity in the Pan-Arctic region and also separately in the Pacific and Atlantic Oceans are presented in Fig. 9.14a–c, respectively. The productivity in the Pan-Arctic region is shown to increase from 219 to 228 gC/m^2/year in 1998–2001, and then it decreased to 217 gC/m^2/year in 2003 and this was followed by increases up to 237 gC/m^2/year in 2008. The yearly productivity in the Pacific Ocean is a lot more variable with significant declines occurring in 1999, 2003, 2005, and 2007. The yearly productivity in the Atlantic Ocean is very similar to that for the Pan-Arctic region except that the decline from 2001 continued to 2005 followed by a more steep increase up to 2008. In the Pacific Ocean region, the yearly productivity was between 194 to 208 gC/m^2/year from 1998 to 2008, while that for the Atlantic Ocean was from 230 to 245 gC/m^2/year until it went up significantly to 255 gC/m^2/year in 2007 and

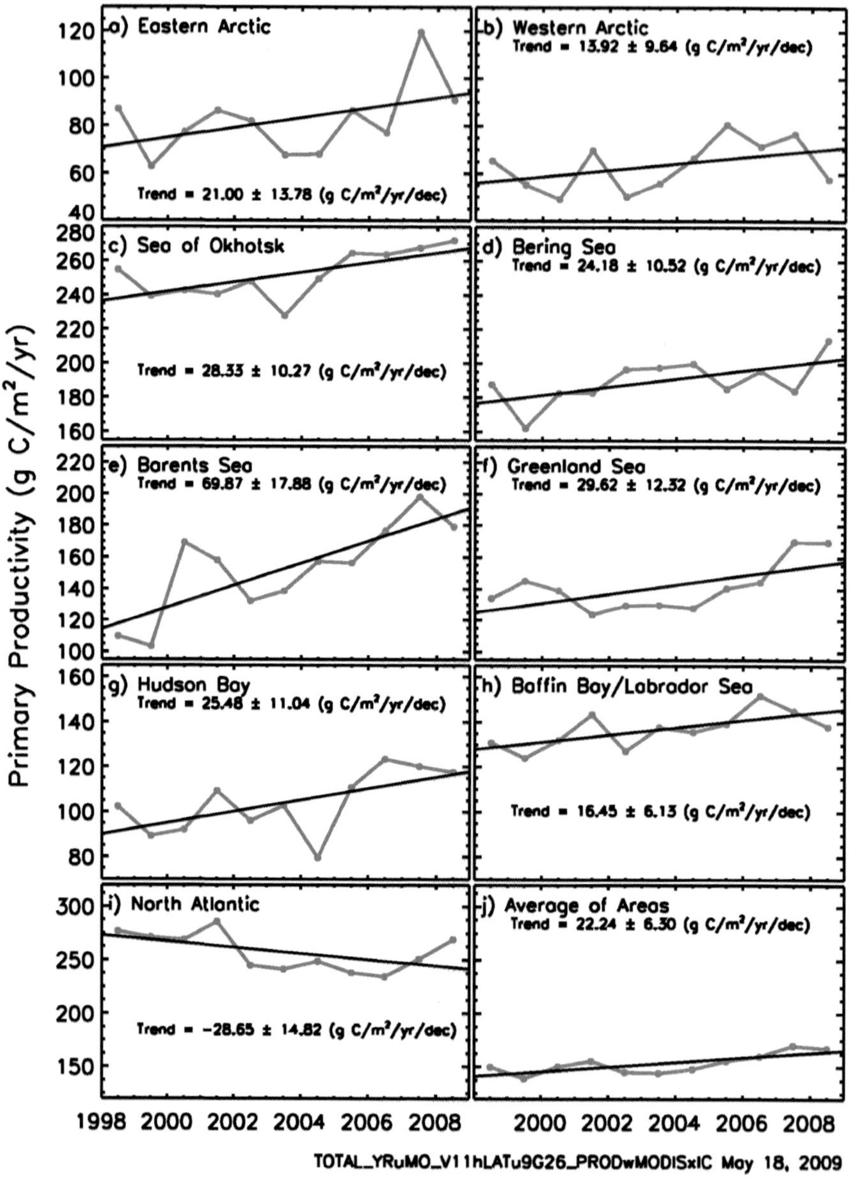

Fig. 9.15 Plots of yearly primary productivity in (**a**) Eastern Arctic; (**b**) Western Arctic; (**c**) Sea of Okhotsk; (**d**) Bering Sea; (**e**) Barents Sea; (**f**) Greenland Sea; (**g**) Hudson Bay; (**h**) Baffin Bay/ Labrador Sea; (**i**) North Atlantic; and (**j**) all sectors

262 g C/m^2/year in 2008. The trends in productivity in the three regions are all positive and are 11.2±4.8, 5.0±3.6, and 17.8±8.6 g C/m^2/year/decade for the Pan-Arctic, Pacific Ocean, and Atlantic Ocean, respectively.

Averages of yearly productivity in the different study areas of the Arctic and in the combined areas are presented in Fig. 9.15 as well to provide a quantitative estimate of interannual variability and trends. Significant interannual fluctuations are apparent, especially in the Eastern Arctic, Sea of Okhotsk, Barents Sea, and Hudson Bay. Results from regression analysis of the data (black line) indicate a positive trend for all sectors except the North Atlantic region which has a trend of – 28.6±14.8 g C/m^2/year/decade. The trend is most positive at the Barents Sea at 69.9±17.9 g C/m^2/year/decade followed by more moderate trends of 29.6±12.3, 28.3±10.3, 25.5±11.0, 24.2±10.5, and 21.0±13.8 g C/m^2/year/decade in Greenland Sea, Sea of Okhotsk, Hudson Bay, Bering Sea, and Eastern Arctic, respectively. The trends in the Baffin Bay/Labrador Sea and the Western Arctic sectors are even more moderate at 16.4±6.1 and 13.9±9.6, respectively. The overall trends when all the areas are combined is 22.2±6.3 g C/m^2/year/decade. It is remarkable that the trends in regions that are partly covered by sea ice during the year are all positive, while the region that is away from the ice edge is negative. Although the record length is relatively short, we may already be observing the effect of the changing sea ice cover and SST on the productivity in the region.

9.3 Ocean Color in the Southern Hemisphere

The productivity of the Southern Ocean was basically unknown until the advent of satellite ocean color data. Early studies using Nimbus-7/CZCS pigment data indicate that the yearly averages of plankton concentrations of the ocean region near 40°S are greatly enhanced compared to other regions (Comiso et al. 1993; Sullivan et al. 1993; Moore and Abbott, 2000; Arrigo et al. 2008). Such results were especially significant in light of the possible role of the Southern Ocean as a carbon sink (Sarmiento and leQuere 1996; Anderson et al. 1998; Takahashi et al. 2002). The Southern Ocean has also been observed as the site of a large diversity of birds, marine mammals, fishes, and microorganism, which are all dependent on the productivity of the region. Furthermore, it is the site of extensive sedimentary deposits of biogenic material and of an external forcing usually referred to as the "Ozone Hole," that may significantly impact the productivity of the region (Neale et al. 1998, 2009).

The phytoplankton productivity, growth, and biomass are at the same time expected to be affected by the sea ice cover as discussed earlier. Unlike the Arctic, the sea ice cover in the Antarctic is like a ring of contiguous sea ice distribution surrounding the continent in winter. As the ice cover retreats to the south during spring and summer, stratified melt water areas are introduced that may cause enhanced phytoplankton growth and accumulation as described earlier (Smith and Nelson 1985). During maximum extent, the sea ice cover in the Antarctic is significantly greater than that in the Arctic, while during minimum extent, the sea ice cover in the Antarctic is much less than in the Arctic. This means that much more melt water is released during the rapid decay of sea ice in the Antarctic than in the Arctic. This in turn provides a good opportunity to assess the strength of the ice-phytoplankton bloom relationship.

The study of ocean color in the Southern Ocean requires a grid that covers a larger part of the Southern Hemisphere than the typical SSM/I grid that we used for other parameters. The gridding technique is still the same but the nominal resolution is 6.25 by 6.25 km as was done with the Arctic ocean color data and is extended to cover regions south of 30°S. The extended grid enables the study of the distribution of blooms over a region that includes the northern boundary of the Southern Ocean. Such grid was also used in Chapter 6 for SST and in previous studies of chlorophyll *a* distribution in the region by Comiso et al. (1993) and Sullivan et al. (1993) using CZCS data. The region includes the subtropical frontal zone that is bordered by the north and south subtropical fronts and the polar fronts (Belkin and Gordon (1996); Moore et al. 1999; Sikes et al. 2009) which are highly dynamic regions where frontal and vertical exchanges occurs. Such exchanges allow for a resupply of the surface water with nutrients and other chemicals.

9.3.1 Seasonal and Interannual Patterns of Plankton Concentration in the Southern Ocean

The seasonality of the plankton concentration in the Southern Ocean is depicted by the multiyear averages for each month from January to December (Fig. 9.16) using SeaWiFS data from August 1997 to July 2008. The areas in white in the images correspond to land or missing data on account of darkness, sea ice, or clouds during the period. A pulsating characteristics of the plankton distribution in the monthly images during a seasonal cycle is evident. The pigment concentrations are shown to have their highest values in the summer and their lowest values in the winter. A halo of relatively high pigment concentrations is apparent in the images with an edge boundary at around 35°S. The most intense bloom regions are apparently the coastal regions of South America and Antarctica. Relatively high concentrations are also apparent near South Africa, New Zealand, and the Kuerguélen Island (50°S, 70°E). In January and February, the bloom patterns are quite intense in some continental boundaries and especially in the Ross and Weddell Seas where coastal polynyas occur in spring. During these months, the patterns are also more expansive in part because of less sea ice cover. In March, the plankton blooms are still quite intense with patterns in some areas similar to those in February. In April and May, significant reductions in plankton concentration are apparent especially around 50°S. It is also apparent, however, that during the winter months of June, July, and August, the pigment concentrations in the Southern Ocean are relatively low except in the northern regions (e.g., near the South Georgia and the South Orkney Islands and the areas adjacent to the sea ice edges). Longer periods of darkness, colder temperatures, and adverse weather conditions are likely contributing factors. In September and October, blooms develop in the shallow regions near the Faukland, the South

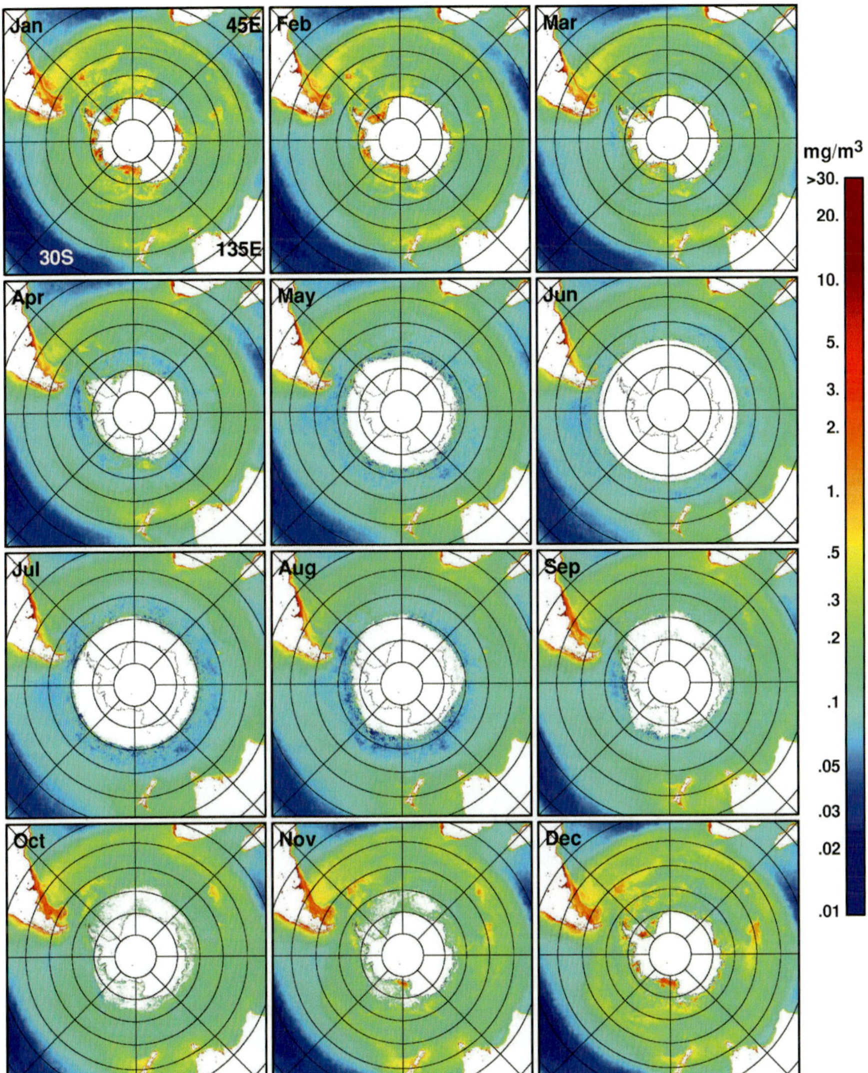

Fig. 9.16 Monthly averages (climatologies) of pigment concentrations in the Southern Hemisphere from January to December using SeaWiFS data from August 1997 to July 2008

Georgia, and the South Orkney Islands and in the relatively shallow areas in the East Atlantic side. In November, the blooms spread out and become more expansive, and at this time bloom patterns starts to appear behind the melting sea ice cover especially in the Ross Sea coastal regions. In December, the pigment concentrations reached or are close to their peak values with the highest pigment concentrations located in the Western Weddell and Ross Seas, near the coastal areas. The patterns in December are usually similar to those of January except for the effect of more sea ice in the former. Overall, it appears that the months from October through the Austral summer to April are the months of most intense phytoplankton activity and, therefore, of highest productivity in the region.

To gain insights into the interannual variability of the bloom patterns, monthly averages in December, when the blooms attain peak or near peak values, are presented in Fig. 9.17 for the years 1997 to 2008. Meltwater from the retreat of the ice cover is usually most abundant during this period, and it appears that the location

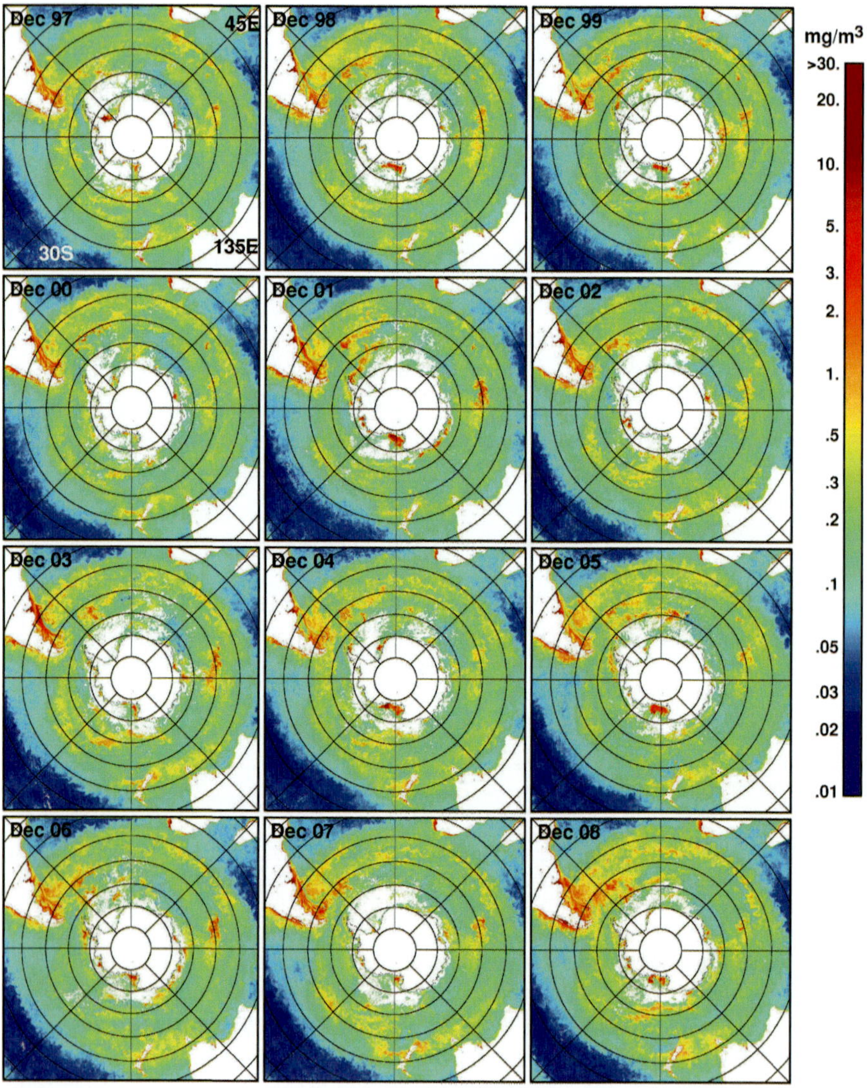

Fig. 9.17 Monthly pigment concentrations in the Northern Hemisphere during an early summer month (December) for each year from 1998 to 2008 using SeaWiFS data

of blooms within the ice covered regions are very much dependent on where the coastal polynyas are usually located. In December, the location of such polynyas is shown to vary considerably from one year to another. When it opens early in spring, the Ross Sea coastal polynya is shown to be covered mainly by high concentrations of phytoplankton. The exceptions were in 1997, when the Ross Sea polynya was relatively small on account of high sea ice concentration in the region during this period, and in 2000 and 2002, when big icebergs broke out of the Ross Ice Shelf. The presence of these icebergs kept the conventional type of Ross Sea coastal polynya as described previously (Zwally et al. 1985; Jacobs and Comiso 1989) from forming. These icebergs had a big impact on the productivity of the region as described by Arrigo et al. (2002) and Smith and Comiso (2008). The spatial extent and distribution of the bloom patterns also depend on the size and the date of onset of melt of the polynya. It is apparent that the plankton concentration in the Ross Sea region was unusually high in 1998, 1999, 2001, 2004, and 2005. In 1997, 2003, 2006, and 2007, the spatial coverage of relatively high pigment concentrations was not as high, in part because of more sea ice coverage in the region during the period. Another polynya region of interest is the Western Weddell Sea coastal region, where an intense bloom is shown in 1997, when a rare coastal polynya in front of the Ronne Ice Shelf was formed early in the spring season (Ackley et al. 2001). Other bloom patterns near the coast in the region occurred in 2000 and 2004 but further to the east. In 2001, the edge of the sea ice cover retreated to the south earlier than other years in the Western Weddell Sea, and a relatively intense bloom appeared behind it suggesting a relationship of blooms with ice melt. During the same year, intense blooms also appeared at the Falkland Islands and the Kerguélen Island. Similar patterns at the ice edges also occurred in 2005 and 2006. Unusually intense blooms at the Pine Island Bay (around 110°W) are also apparent in 2001, 2002, and 2006 likely associated with an early polynya formation.

The monthly maps in Fig. 9.17 provide the means to assess qualitatively the interannual variability of the spatial distribution of the pigments. The December anomaly maps presented in Fig. 9.18, more quantitative assessments of the changes. High positive anomalies are apparent in much of the Ross Sea coastal regions in 1998, 2001, and 2005. On the other hand, high negative anomalies are prevalent in the same region in 2000 and 2002 indicating the strong influence of iceberg calving on the productivity of the region. Other years show a mixture of positive and negative anomalies in the region with 1999 and 2004 showing more positive than negative anomalies while 2003, 2006, and 2007 show more negative than positive anomalies. The enhanced plankton concentration to the north of the Western Weddell Sea ice edge in 2001 is apparent with the positive anomalies averaging about 0.8 mg/m^3. The anomalies to the north of the Eastern Weddell Sea ice edge in 2005 are even more positive averaging about 1.8 mg/m^3. The region is in the vicinity of a topographical feature called the Maud Rise, which as discussed in Chapter 8 is an area where upwelling of warm water is possible. Such upwelling would also enable the resupply of the surface water with nutrients and other chemicals. It is also apparent that there are years, as in 1999, 2001, 2006, and 2007, when some coastal areas of East Antarctica have enhanced plankton concentrations.

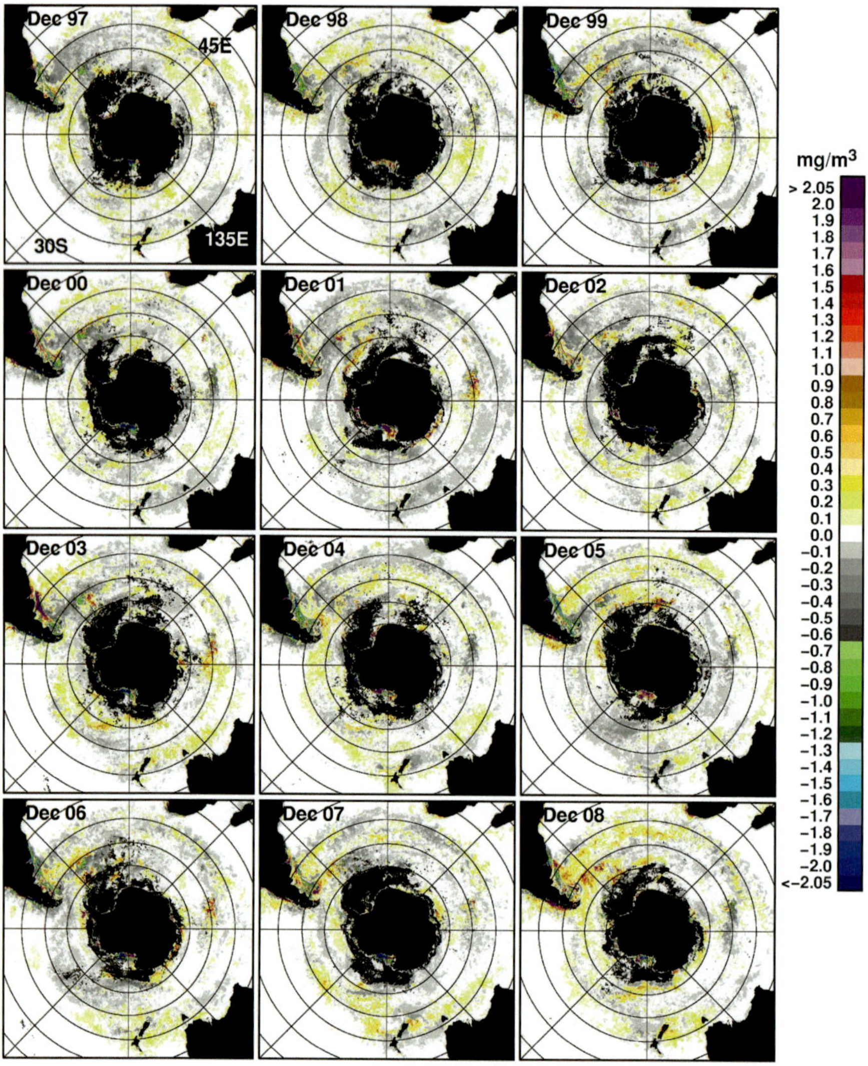

Fig. 9.18 Monthly Anomalies of pigment concentrations in the Southern Hemisphere during an early summer month (December) for each year from 1998 to 2008 using SeaWiFS data

Monthly averages of pigment concentration in March, representing distributions at the end of summer when the surface waters are relatively colder, are presented in Fig. 9.19 for the years 1998–2008. The climatological monthly averages shown in Fig. 9.16 indicate that the plankton concentrations in March are significantly reduced compared with January or February. The patterns, however, are similar during these months suggesting a relatively long residence time for the blooms. The March images show the same ring of blooms at around 35°S as the December images, but the concentrations are relatively lower. Also, at 60 to 70°S, there are

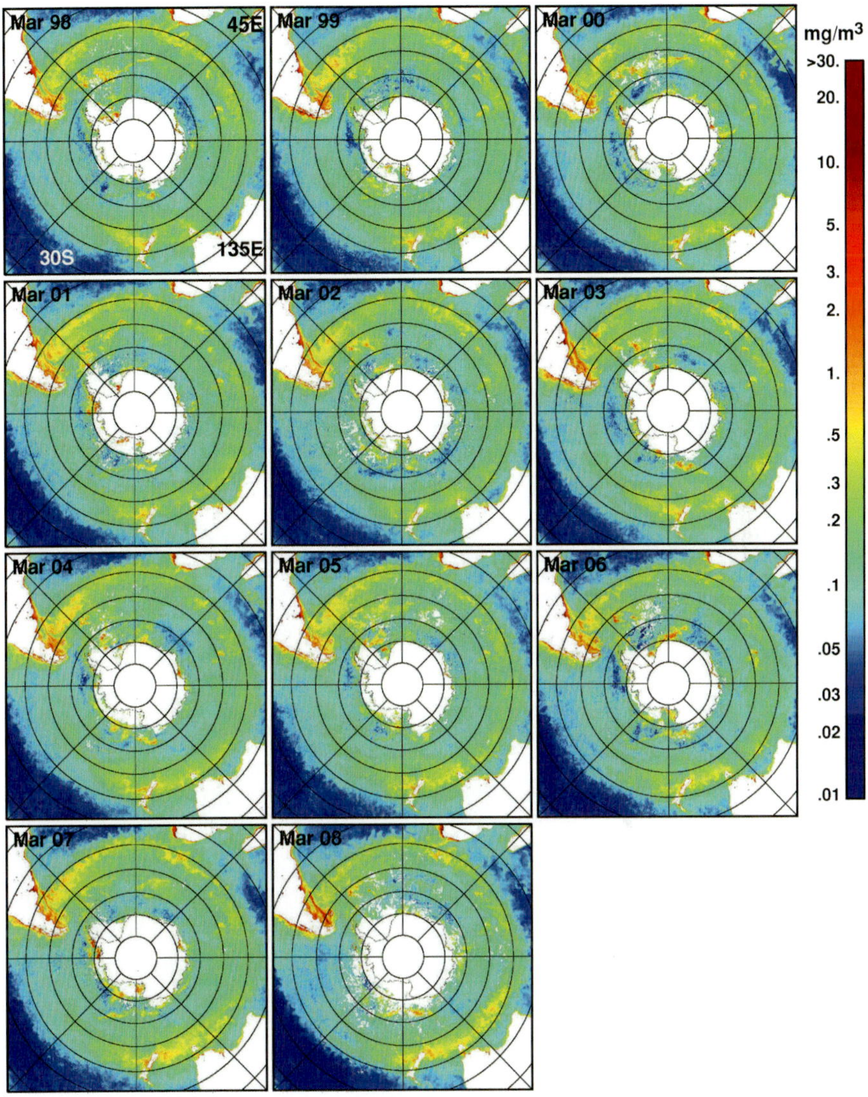

Fig. 9.19 Monthly averages of pigment concentrations in the Southern Hemisphere during an end of summer month (March) for each year 1998 to 2008 using SeaWiFS data

areas where the pigment concentrations are relatively low, suggesting that there might be iron limitations in these regions. The blooms around the continent observed in the December images have also been considerably diminished with a few years like 2001, 2003, 2004, 2005, 2006, and 2007 showing some remnants of the summer blooms. The most intense bloom feature is again that which surrounds South America and especially the eastern side. The pattern goes through significant interannual variability, and sometimes the blooms gets extended toward the

Faukland Islands as in 1999, 2001, 2002, 2004, 2005, and 2007. A bloom pattern to the east of New Zealand is apparent for most years but disappears some years as in 2002 and becomes part of a more extended bloom towards Australia in other years as in 2007 and 2008. It is apparent that in March, most of the biological activities in the region occur north or 50°S. The monthly anomaly maps for March, presented in Fig. 9.20, show that the interannual variabilities are relatively minor. The anomalies are strongly positive near the eastern coast of South America as in 2003 and 2008, but are mainly negative during other years. Positive anomalies

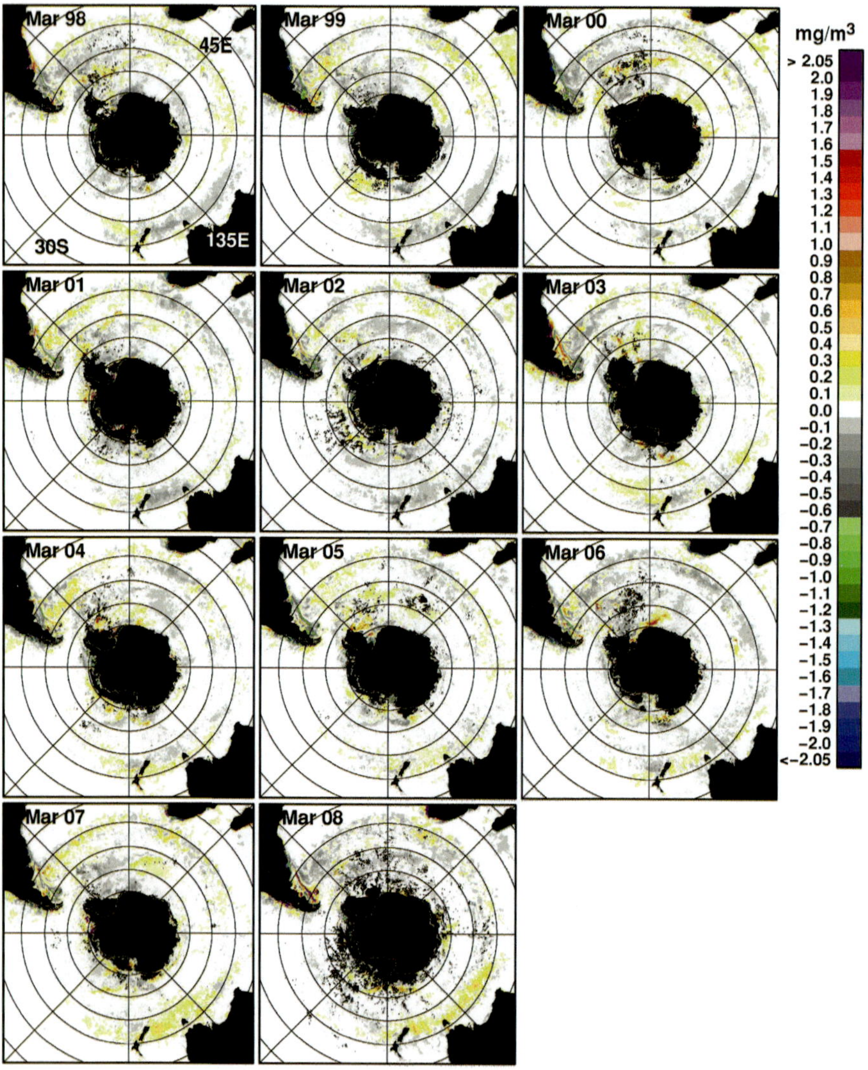

Fig. 9.20 Monthly Anomalies of pigment concentrations in the Southern Hemisphere during the end of summer month (March) for each year fom 1998 to 2008 using SeaWiFS

are also apparent in the New Zealand and South Australia area. But overall, no pattern of change is apparent during the period.

To gain additional insight into the interannual variability yearly averages of pigment concentrations from 1997 to 2008 are presented in Fig. 9.21. As mentioned earlier, the yearly average is still a 12-month average but starts in August of one year and ends in July of the following year. The halo of enhanced phytoplankton blooms at around 35°S is apparent in each of the yearly maps, but to the south of

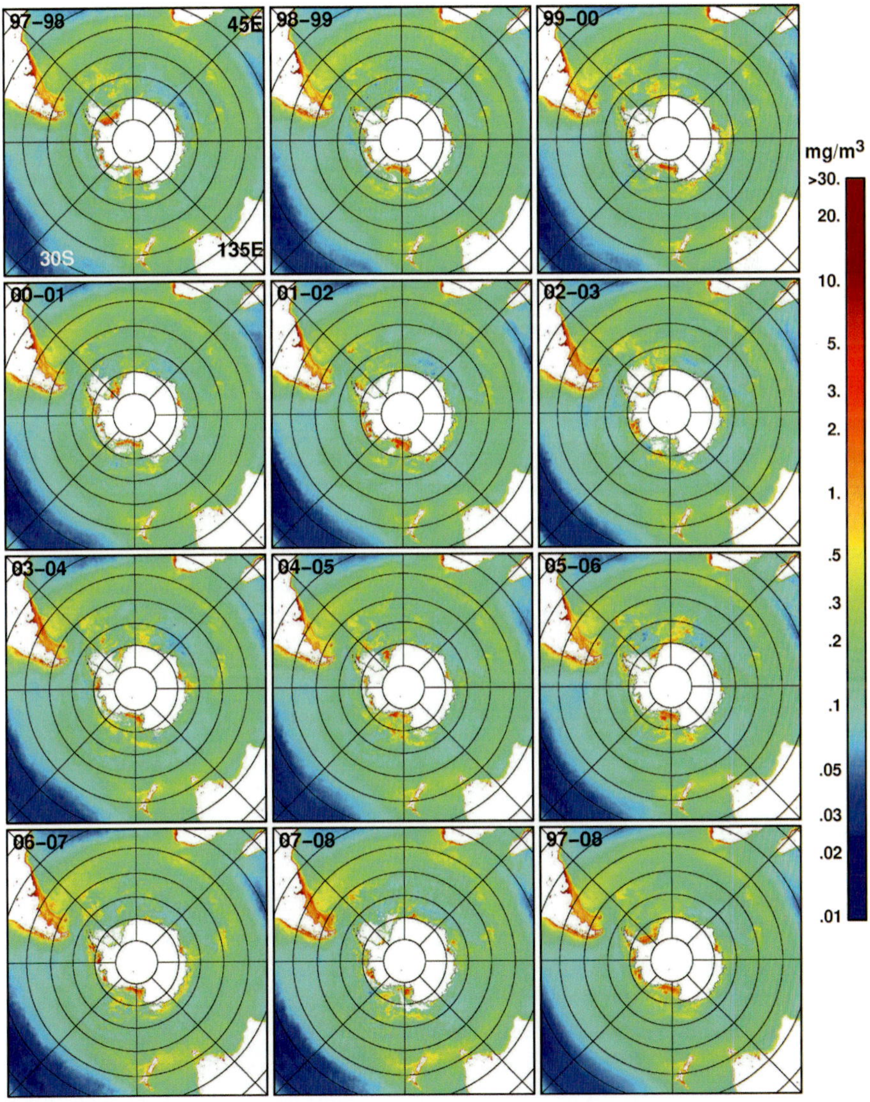

Fig. 9.21 Yearly Averages (August of one year to July the following year) and multiyear average of pigment concentrations from SeaWiFS from 1998 to 2008

this halo, the patterns of blooms and the spatial features are different for different years. High plankton concentrations are apparent in various locations around the continent, but the spatial coverage and locations of these features change each year. The pigment concentrations between 70°S and 55°S are surprisingly lower than in other areas inside the halo. This is an area where the pigment concentrations are expected to be relatively high since it is a general area where the seasonal ice cover is located and therefore where abundant supply of meltwater should be available during the spring and summer. Again, as in the Arctic, this illustrates that the abundance of one parameter needed for phytoplankton growth does not necessarily guarantee an enhanced growth. In particular, the availability of iron and nutrients has been an issue for this region. Iron is usually transported by winds as dust particles from continental land areas, and iron may be limiting in these regions because of their lack of proximity to land areas. Also, during the formation of seasonal ice, brines are released causing the formation of dense water that may sink and entrain some of the nutrients to the deeper parts of the oceans where they are no longer available for consumption. The presence of a stable layer of melt water may also serve to restrict wind induced upwelling of nutrients to the surface. But it is interesting that for some years, there are bloom patterns in the region as in 2003–2004 and 2005–2006. There are still many unknowns and the lack of ample in situ data makes it difficult to come up with accurate interpretation.

The yearly anomaly maps presented in Fig. 9.22 provide the means to assess the magnitude of the inter-annual variability of the pigment concentrations. The yearly anomalies are shown to be most significant around the continent and where the seasonal ice cover usually resides. Significant areas of positive anomalies are apparent in the 1999–2000 image and followed by the 2002–2003 image. Negative anomalies are relatively prominent in the 1998–1999 and the 2000–2001 images. Large positive anomalies are also apparent in the 2007–2008 image but mainly to the east of Argentina. The Ross Sea is also a region with the unpredictable anomalies with the coastal areas being significantly negative in 2000–2001 and 2002–2003, when the calving of B15 and C19 occurred. Overall, it appears from the yearly maps that the interannual variability is relatively modest.

9.3.2 *Temporal Variability and Trends*

Plots of monthly averages of Chlorophyll plankton distributions in the Southern Ocean for the regions >60°S are presented in Fig. 9.23a. Winter data is not included because of long darkness in the region during the period. It is apparent that there is a significant interannual variability with peak values fluctuating from 0.35 to 0.49 mg/m^3. The highest average occurred in January 2006 followed by January 2000 and December 2001. The plots also indicate that the season for plankton blooms, as measured by the width at half height, are longer for some years than other years. Monthly anomalies for the same period are presented in Fig. 9.23b, and it is apparent that the interannual variability is significant and that the average pigment

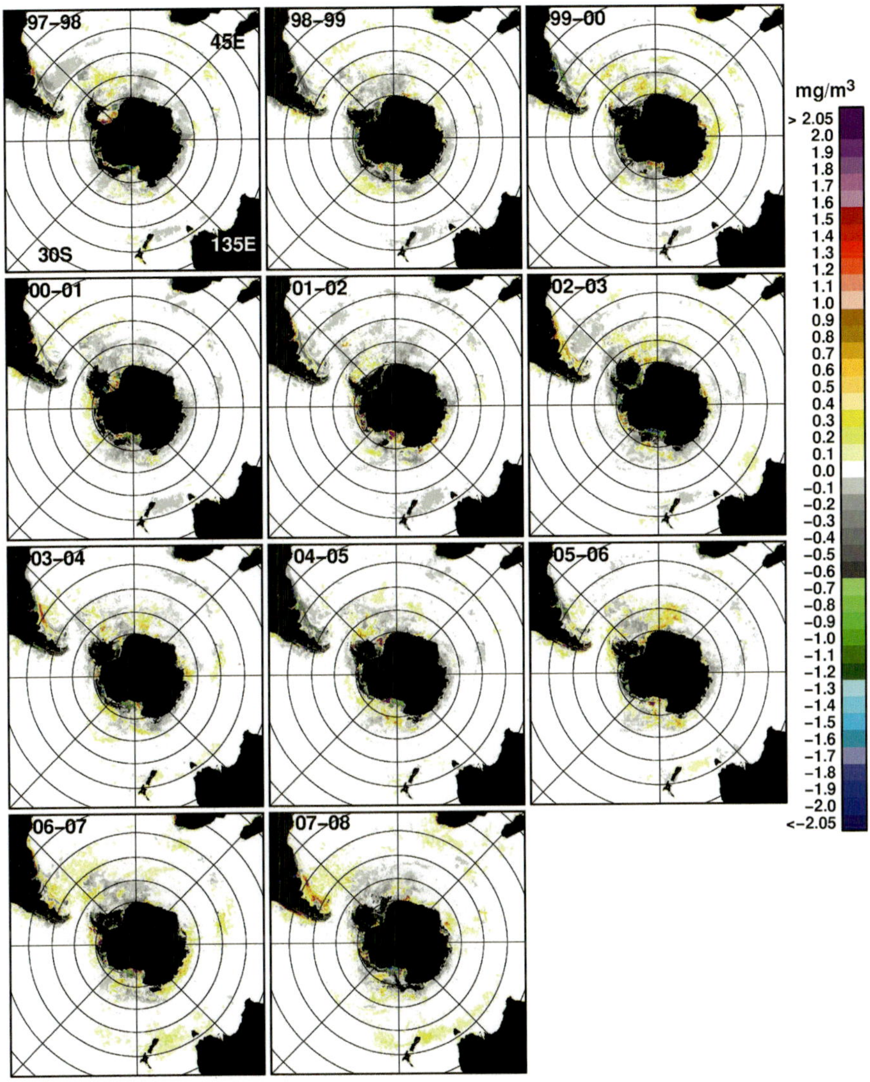

Fig. 9.22 Yearly anomalies of pigment concentrations from SeaWiFS from 1998 to 2008. The anomalies were estimated using the multiyear average data shown in Fig. 9.6

concentration has been increasing at the rate of 0.022 ± 0.009 mg/m^3/decade during the 1997 to 2008 period.

For comparison, a plot of monthly averages of chlorophyll concentrations over the region 35–50°S is presented in Fig. 9.24a. The distribution also shows considerable interannual variability but with the highest values for the peaks occurring in different years. The study area is at low enough latitude to make it possible to include the winter months. The range of variability is similar but lower than that of

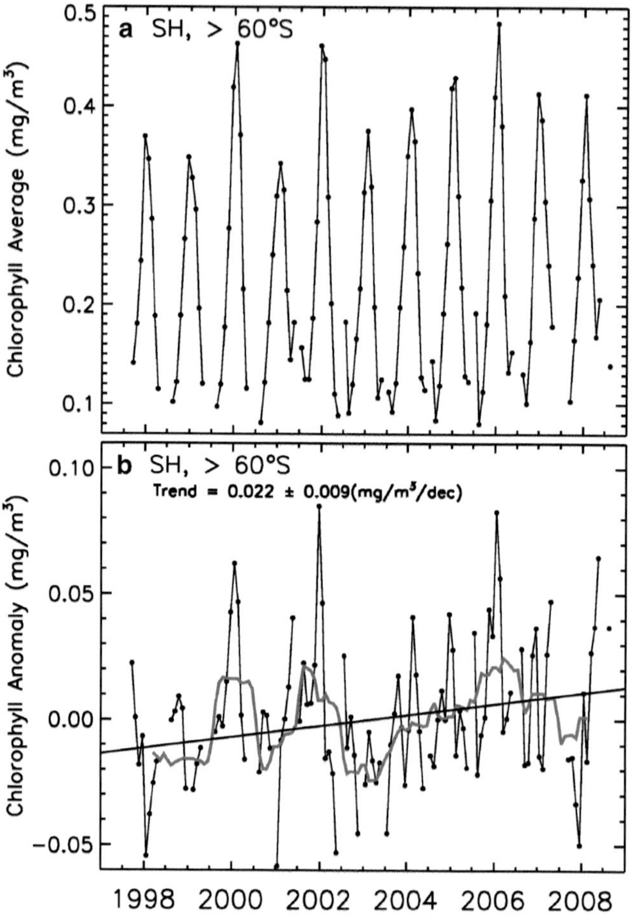

Fig. 9.23 Plots of (**a**) monthly averages and (**b**) monthly anomalies of pigment concentration in the Southern Hemisphere for the ocean region >60°S

the study region at higher latitude shown in Fig. 9.23a. The monthly anomalies at the lower latitudes as shown in Fig. 9.24b also show significant interannual fluctuations with a suggestion of a cyclical variability and a trend of 0.02 ± 0.004 mg/m^3/decade.

Monthly anomalies of Chlorophyll concentration for the different sectors in the Sourthern Ocean as defined by Zwally et al. (1983) and discussed in Chap. 7 are shown in Fig. 9.25. The study areas are confined only to >50°S, and it is apparent that the trends are positive and significant in three sectors while negative and insignificant in the other two sectors. The trends are 0.010 ± 0.011, 0.011 ± 0.007, and 0.014 ± 0.008 mg/m^3/decade for the Weddell Sea, Ross Sea, and Bellingshausen/Amundsen Seas sectors and −0.005 ± 0.005 and −0.006 ± 0.006 mg/m^3/decade for the Indian Ocean and West Pacific Ocean sectors. It is interesting to note that the trends in pigment concentration in the various sectors are less positive (if not negative) than those in Figs. 9.22 and 9.23.

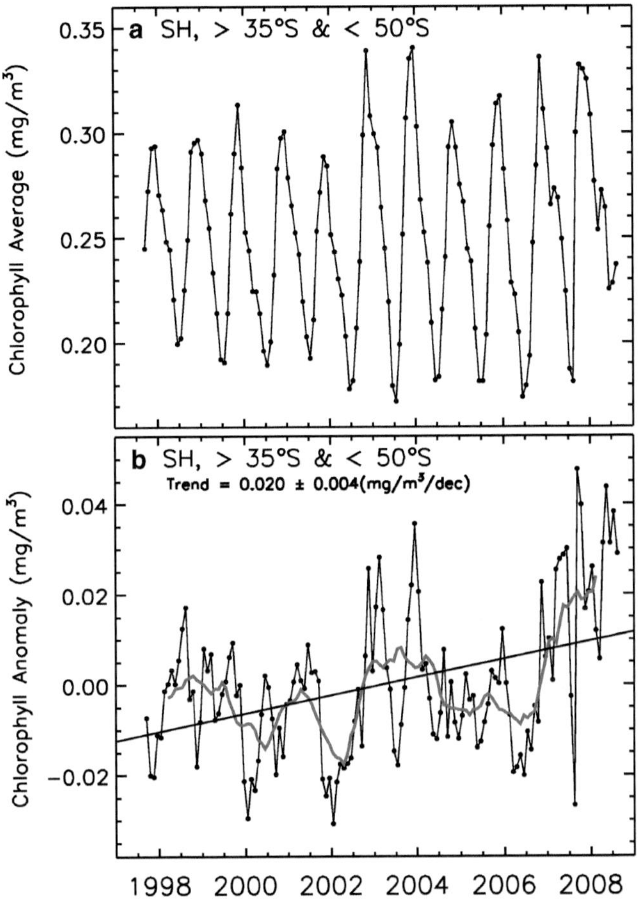

Fig. 9.24 Plots of monthly anomalies of pigment concentration in the Southern Hemisphere with 5 month running average and trend

9.3.3 *Yearly Patterns of and Changes in Primary Productivity*

Modeling studies have stressed the importance of the Southern Ocean as a carbon sink (Takahashi et al. 2002). The primary productivity of the Southern Ocean south of 50°S has been studied previously using SeaWiFS data from 1997 to 2006 (Smith and Comiso 2008; Arrigo et al. 2008). The results show relatively low productivity in much of the region with the high productivity areas basically confined to the coastal regions of the Antarctic continent, but only for a relatively short period of time. The yearly productivities of the extended region, as described earlier, using the same algorithm used by Smith and Comiso (2008) which utilized the technique developed by Behrensfeld and Falkowski (1997) as discussed in Chap. 4, are presented in Fig. 9.26. It is apparent that the primary productivities for each year are greatly

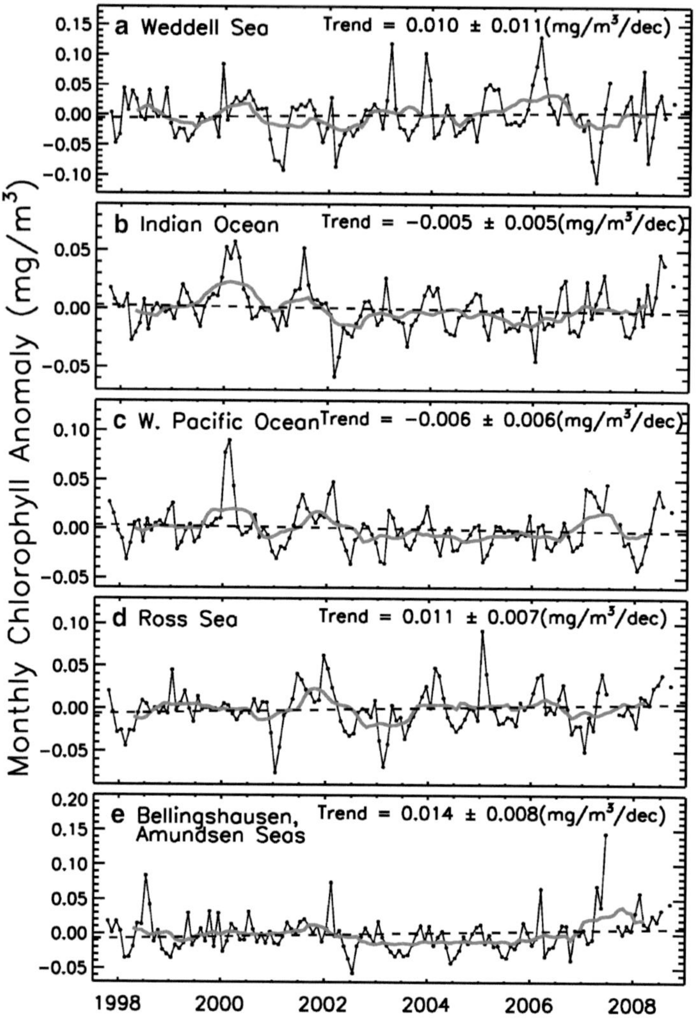

Fig. 9.25 Plots of monthly anomalies of pigment concentration in various sectors of the Southern Hemisphere with 5 month running average and trend

enhanced just barely north of 50°S with a ring of high productivity along the 35°S latitude. South of 50°S is an area of much less productivity in part because the area is covered by sea ice or is in darkness for a large fraction of the year. It is apparent that the enhanced productivity starts at South America, goes around a circle but only up to a little beyond New Zealand since after that, the productivity is significantly reduced between this area and South America. This region is part of the Bellingshausen/Amundsen Sea sector as defined in Chap. 7, which has been identified as an area of climate anomaly in the Southern Ocean by Jacobs and Comiso (1997). The other high pigment concentration areas identified in the previous section are shown in the yearly maps as relatively low productivity areas mainly because they

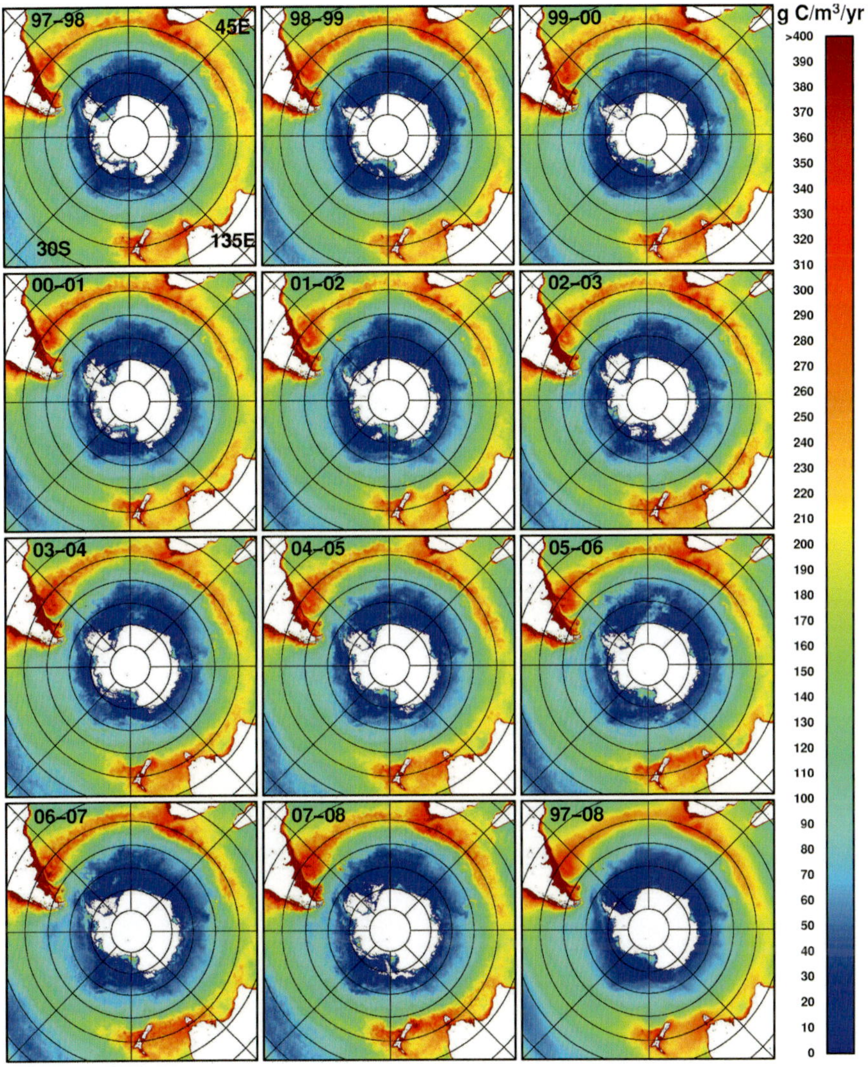

Fig. 9.26 Yearly primary productivity (August of one year to July the following year) in the Southern Hemisphere from 1998 to 2008 as derived using SeaWiFS data, AVHRR data and a model

are at higher latitudes and subject to longer darkness. Despite the yearly averaging, the yearly productivity maps show some of the distinct mesoscale characteristics along the ring indicating that the region is a highly active region with eddies and frontal activities. The productivity is shown to be specially high near land areas (e.g., South America, Africa, Australia and New Zealand), which may in part be due to the abundance of iron in these regions and in part due to the presence of case-2 water. The inability to effectively separate effects of case-2 water in these regions is unfortunate and is considered as a part of the uncertainty in the analysis.

Yearly variabilities are apparent in the maps shown in Fig. 9.26, but they are normally subtle except in the seasonal sea ice region. The yearly changes are more apparent in the yearly anomaly images presented in Fig. 9.27. In the 1997–1998 average, the anomalies are mainly positive with some negative anomalies located in the Falkland Island and New Zealand area. The anomalies are also primarily positive anomalies for the years 1998–1999, 1999–2000, 2000–2001, 2002–2003, and 2003–2004. Conversely, the anomalies are primarily negative for the other years and most especially in 2006–2007. It is interesting that the map for 2007–2008 shows strong positive

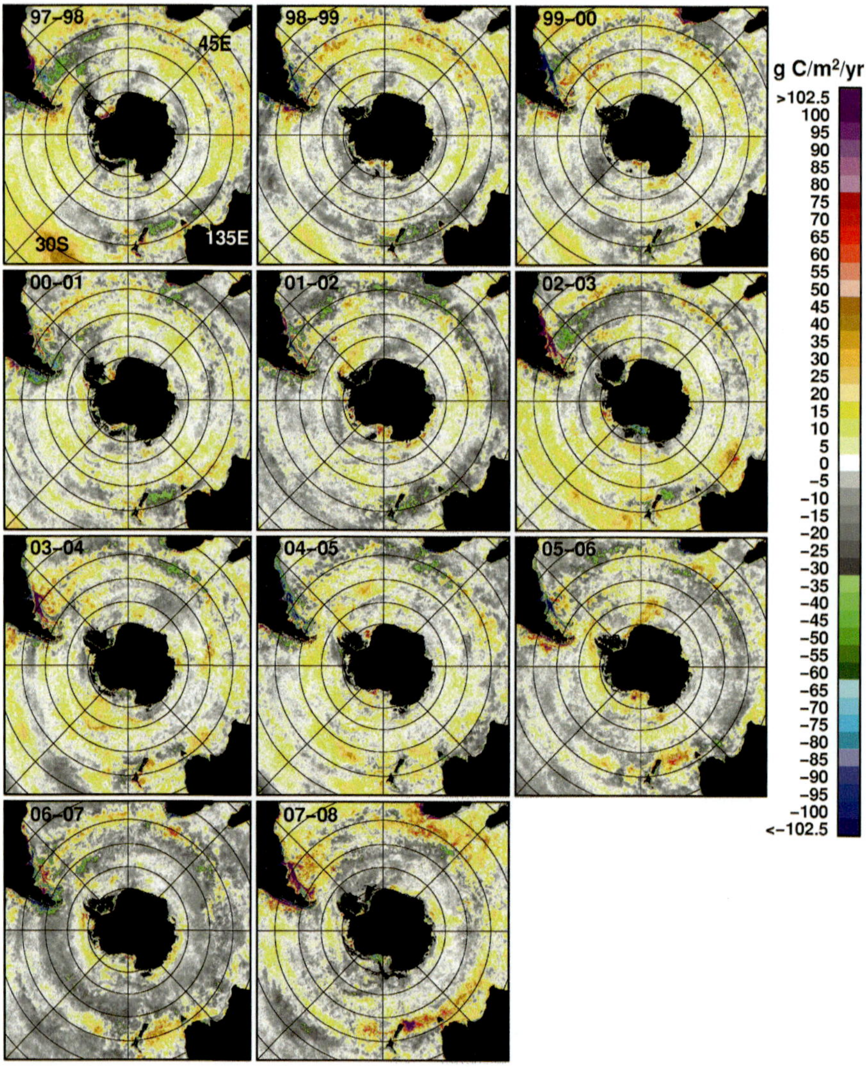

Fig. 9.27 Yearly anomalies in primary productivity in the Southern Hemisphere as derived from SeaWiFS data, AVHRR data and a model

anomalies near land areas and negative anomalies in most other areas. Equally interesting is the anomaly map in 1997–1998 which shows almost the reverse effect.

Plots of averages over the five sectors, as defined earlier, and for the region >60°S are presented in Fig. 9.28. In the Weddell Sea sector, the yearly production shows a sequence of increases and decreases every year until it started to increase to maximum value from 2003 to 2006 and then decrease after that to a relatively low value in 2008.

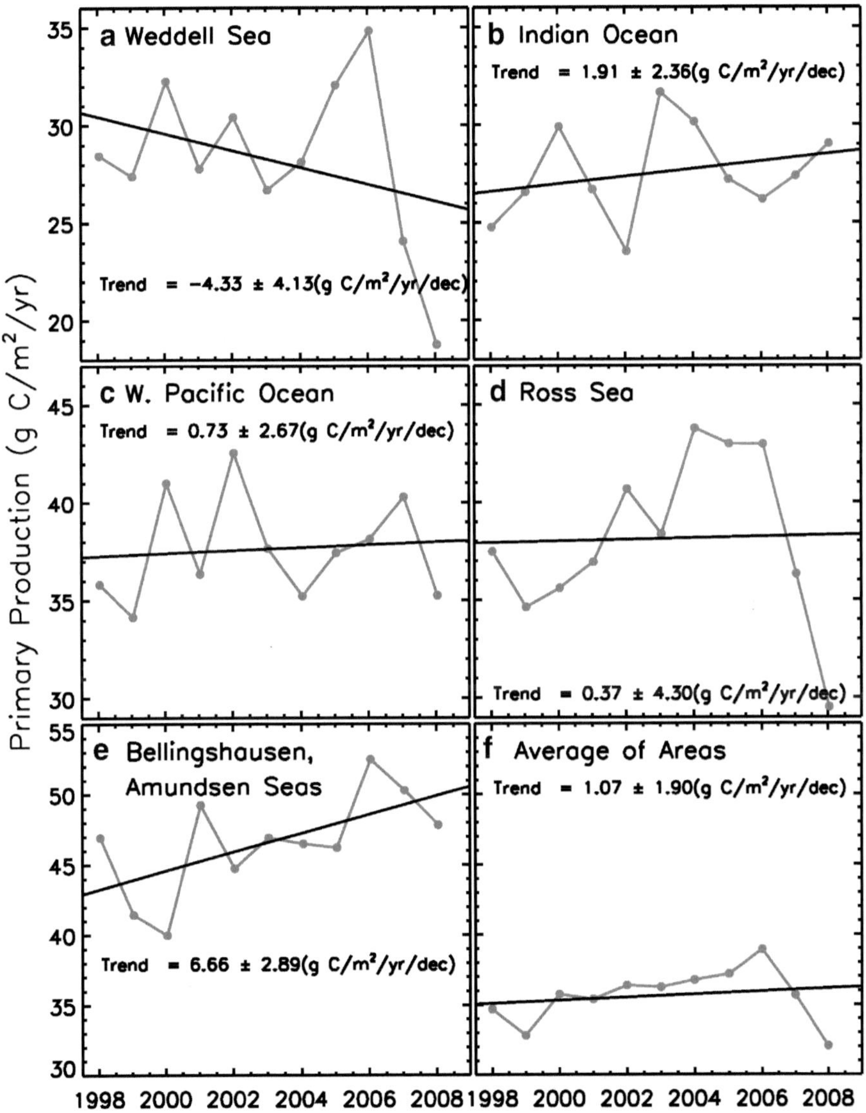

Fig. 9.28 Plots of primary production for the region >60°S in the (**a**) Weddell Sea; (**b**) Indian Ocean; (**c**) West Pacific Ocean; (**d**) Ross Sea; (**e**) Bellingshausen/Amundsen Seas; and (**f**) combined sectors

In the Indian Ocean, a sequence of increases and decreases are also apparent, but this time at an interval of every two years. In the Western Pacific Ocean, the pattern is like that of the Weddell Sea except that the low value was in 2004 and going to a peak value in 2007 before the relatively low value in 2008. Except for the decline in primary productivity from 1997 to 1998 and from 2001 to 2002, the yearly averages in the Ross Sea sector went steadily up until 2004 and was almost constant up to 2006 before it dropped in 2007 and to its lowest value in 2008. The plot for the Bellingshausen/Amundsen Seas sector shows significant variability as well with maximum value occurring in 2006 as in the Weddell Sector but declining more moderately after that. The averages of the yearly primary productivities in all the study areas are presented in Fig. 9.28f and the plot show almost uniform values until the decline in 2008. The trends in productivity for the different sectors are all positive except for the Weddell Sea sector, which has a negative trend of -4.3 ± 4.1 g C/m^2/year/dec. The others have varying positive trends of 6.7 ± 2.9, 1.9 ± 2.4, 0.7 ± 2.7, and 0.4 ± 4.3 g C/m^2/year/dec. The average of the different sectors has a trend of 1.1 ± 1.9 g C/m^2/year/dec. The trends are mainly insignificant considering the large interannual variability, short record length, and relatively large errors. The cause of the declines in 2007 and 2008 in the Ross and Weddell Seas is not known but it may be associated with the global cooling and increases in the sea ice cover in these regions during these years.

For comparison, plots of the average yearly productivity in the same sectors but for the area >35°S are presented in Fig. 9.29. The plots show similar interannual variability but very different patterns with the 2008 data point showing consistent increase from the 2007 value in all sectors. Also, this time we get positive, but insignificant trends of $2.3\ \pm3.2$, 1.4 ± 3.9, and 0.6 ± 3.3 g C/m^2/year/decade in the Weddell Sea, Ross Sea, and West Pacific Ocean sectors, respectively while the trends are -7.2 ± 3.0 and -1.4 ± 3.9 g C/m^2/year/decade for the Indian Ocean and Bellingshausen/Amundsen Seas Sectors, respectively. The trend for the combined areas is negative but insignificant at -0.9 ± 2.75 g C/m^2/year/decade.

The SeaWiFS data set provided detailed information about the spatial and temporal variability of the pigment concentration and primary productivity of the polar oceans. Much still need to be learned about the significance of the pigment distributions, including the relative contribution of case 1 and case 2 water in coastal areas. We also need to know more about the bias associated with the lack of data during cloudy conditions and inability to retrieve pigment concentration and productivity in partly ice covered areas. Our assessment of interannual variability and trends is just very preliminary since the record length of the data set is still very short. However, the data have generally provided insights into the spatial and temporal variability of the pigment concentrations and net primary productivity. We also observed that there are links of the pigment concentration to sea ice, or the availability of melt water during ice retreat, but other conditions must be satisfied including the availability of nutrients and iron, irradiance, and possibly the right surface temperature ranges before the links can materialize. Our estimates of yearly primary productivity may need to be improved knowing that the optical properties of polar oceans are different from those of extra polar regions and that there are known shortcomings of current productivity models. However, we hope that our results can serve as the baseline for future retrievals.

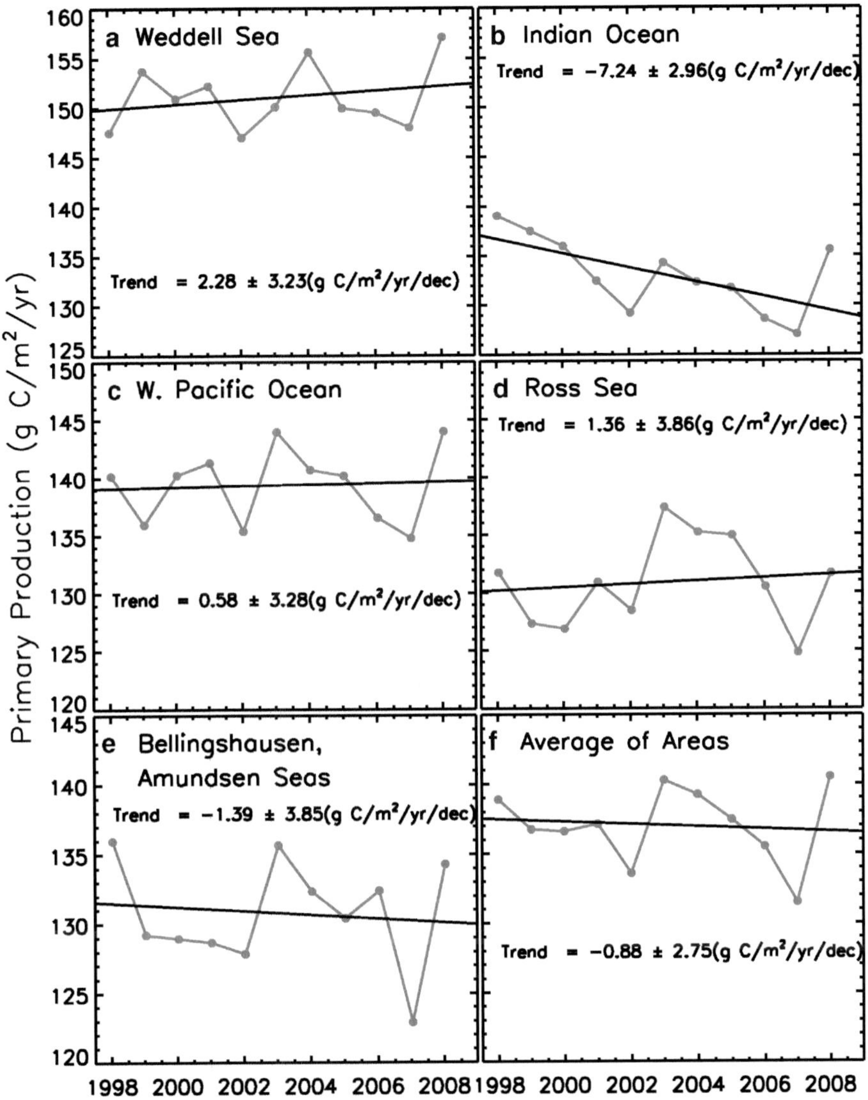

Fig. 9.29 Plots of primary production for the region > 35°S in the (**a**) Weddell Sea; (**b**) Indian Ocean; (**c**) West Pacific Ocean; (**d**) Ross Sea; (**e**) Bellingshausen/Amundsen Seas; and (**f**) combined sectors

References

Ackley SF, Geiger CA, King JC, Hunke EC, Comiso JC (2001) The Ronne polynya of 1997/98: observations of air-ice-ocean interaction. Ann Glaciol 33:425–429.

Alexander V, Niebauer HJ (1981) Oceanography of the eastern Bering Sea ice-edge zone in the spring. Limnol Oceanogr 26:1111–1125.

Anderson L, Jones EP (1992) Tracing upper waters of the Nansen Basin in the Arctic Ocean. Deep Sea Res 39:S425–S434.
Anderson LG, Olsson K, Jones EP, Chierici M, Fransson A (1998) Anthropogenic carbon dioxide in the Arctic Ocean: Inventory and sinks. J Gophys Res 103(12):27707–27716.
Arrigo KR, van Dijken GL, Ainley DG, Fahnestock MA (2002) Ecological impact of a large Antarctic iceberg. Geophys Res Lett 29(7), doi:10.1029/2001GL014160.
Arrigo KR, Van Dijken GL (2004) Annual changes in sea ice, chlorophyll a, and primary production in the Ross Sea, Antarctica, Deep Sea Res. Part II 51:117–138.
Arrigo KR, van Dijken GL, Bushingsky S (2008) Primary production in the Southern Ocean, 1997–2006. J Geophys Res 113:C08004, doi:10.1029/2007/JC004551.
Belkin IM, Gordon AL (1996) Southern Ocean fronts from the Greenwich meridian to Tasmania. J Geophys Res 101(C2):3675–3696.
Behrensfeld MJ, Falkowski P (1997) Photosynthetic rates derived from satellite-based chlorophyll concentrations. Limnol Oceanogr 42:10–20.
Comiso JC (2002) A rapidly declining Arctic perennial ice cover. Geophys Res Lett 29(20):1956, doi:10.1029/2002GL015650.
Comiso JC (2003) Warming trends in the Arctic. J Climate 16(21):3498–3510.
Comiso JC, Parkinson CL, Gersten R, Stock LV (2008) Accelerated decline in the Arctic sea ice cover. Geophy Res Lett 35:L01703, doi:10.1029/2007GL031972.
Comiso JC, Kwok R, Martin S (2010) Variability and trends in ice extent and ice production in the Ross Sea. Deep Sea Res (submitted).
Comiso JC, Maynard NG, Smith WO, Jr., Sullivan CW (1990) Satellite ocean color studies of Antarctic ice edges in summer/autumn. J Geophys Res 95(C6):9481–9496.
Comiso JC, Cavalieri D, Parkinson C, Gloersen P (1997) Passive microwave algorithms for sea ice concentrations. Rem Sens Environ 60(3):357–384.
Comiso JC, McClain C, Sullivan C, Ryan J, Leonard CL (1993) CZCS pigment concentrations in the Southern Ocean and their relationships to some geophysical parameters. J Geophys Res 98(C2):2419–2451.
Cota G, Wang G, Comiso JC (2004) Transformation of global satellite chlorophyll retrievals with a regionally tuned algorithm. Rem Sens Environ 90:373–377.
Jacobs S, Comiso JC (1989) Satellite passive microwave sea ice observations and oceanic processes in the Ross Sea, Antarctica. J Geophys Res 94:18195–18211.
Jacobs SS, Comiso JC (1997) Climate variability in the Amundsen and Bellingshausen Seas. J Climate 10(4):697–709.
Martin JH, Fitzwater SE (1988) Iron deficiency limits phytoplankton growth in the north-east Pacific subarctic. Nature 331:341–43.
Martin JH, Fitzwater SE, Gordon RM (1990) Iron Deficiency limits phytoplankton growth in Antarctic waters. Global Biogeochemical Cycles 4:5–12.
Mitchell BG, Brody E, Yeh EN, McClain C, Comiso JC, Maynard NC (1991) Meridional zonation of the Barents Sea ecosystem inferred from satellite remote sensing and in-situ Bio- optical observations, Pro Mare Symposium. Polar Res 10(1):147–162.
Moore JK, Abbott MR (2000) Phytoplankton chlorophyll distributions and primary production in the Southern Ocean. J Geophys Res 105(C12):28709–28722.
Moore JK, Abbott MR, Richman JG (1999) Location and dynamics of the Antarctic Polar Front from satellite sea surface temperature data. J Geophys Res 104(C2):3059–3073.
Neale PJ, Davis RF, Cullen JJ (1998) Interactive effects of ozone depletion of vertical mixing on photosynthesis of Antarctic phytoplankton. Nature 392:585–589.
Neale PJ, Jeffrey WH, Sobrino C, Pakulski, J, Phillips-Kres J, Baldwin AJ, Franklin LA, Kim HC (2009) Inhibition of phytoplankton and bacterial productivity by solar radiation in the Ross Sea Polynya. In: Smithsonian at the Poles: Contribution to International Polar Year, Science, (eds) Krupnik I, Lang MA, Miller SE, Smithsonian Institution Scholarly Press, Washington, DC., pp. 299–308.
Pabl S, van Dijken GL, Arrigo KR (2008) Primary production in the Arctic Ocean, 1998–2006. J Geophys Res 113:C08005, doi:10.1029/2007/JC004578.

Sakshaug E, Skjodal HR (1989) Life at the ice edge. Ambio 18(1):60–67.

Sarmiento JL, leQuere C (1996) Oceanic carbon dioxide uptake in a model of century scale global warming. Science 274:1346–1350.

Sikes EL, Howard WR, Samson CR, Mahan TS, Robertson LG, Volkman JK (2009) Southern Ocean seasonal temperature and subtropical front movement on the South Tasman Rise in the late Quaternary, Paleoceanography 24, PA2201, doi:10.1029/2008PA001659.

Smith W Jr., Comiso JC (2008) The influence of sea ice primary production in the Southern Ocean: A satellite perspective. J Geophys Res 113:C05S93, doi:10.1029/2007JC004251.

Smith W Jr., Comiso JC (2009) Southern ocean primary productivity: Variability and a view to the future. In: *Smithsonian* at the Poles: Contributions to the International Polar Year Science (eds) Krupnik I, Lang MA Miller SE, Smithsonia Institution Scholarly Press, Washington, DC., pp. 309–318.

Smith WO Jr., Nelson DM (1985) Phytoplankton bloom produced by a receding ice edge in the Ross Sea: Spatial coherence with the density field. Science 227:163–166.

Stroeve J, Holland MM, Meier W, Scambos T, Serreze MC (2007) Arctic sea ice decline: Faster than forecast. Geophys Res Lett 34:L09501, doi:10.1029/2007GL0029703.

Sullivan CW, McClain C, Comiso JC, Smith WO Jr. (1988) Phytoplankton standing crops within an Antarctic ice edge. J Geophys Res 93:12487–12498.

Sullivan CW, Arrigo KR, McClain CR, Comiso JC, Firestone J (1993) Distributions of phytoplankton blooms in the Southern Ocean. Science 262:1832–1837.

Takahashi T et al (2002) Global sea-air CO_2 flux based on climatological surface ocean pCO_2 and seasonal biological and temperature effects. Deep Sea Res. Part II 49(9–10):9–10.

Thomas DN (2004) Frozen Oceans: The floating world of pack ice. Natural History Museum, London.

Wang J, Cota GF, Comiso JC (2005) Phytoplankton in the Beaufort and Chukchi Seas; Distributions, dynamics, and environmental forcing. Deep Sea Res II 52:3355–3368.

Wilson PR, Ainley DG, Nur N, Jacobs SS, Barton K, Ballard G, Comiso JC (2001) Adelie penguin population growth in the South Pacific Sector of Antarctica: Relation to sea-ice extent and the Southern Oscillation. Mar Ecol Prog Ser 213:301–309.

Zwally HJ, Comiso JC, Gordon AL (1985) Antarctic offshore leads and polynyas and oceanographic effects. In: Oceanology of the Antarctic Continental Shelf (ed) Jacobs S, Antarctic Research Volume 43, American Geophysical Union, Washington, DC., pp. 203–226.

Zwally HJ, Comiso JC, Parkinson CL, Campbell WJ, Carsey FD, Gloersen P (1983) Antarctic Sea Ice 1973-1976 from Satellite Passive Microwave Observations. NASA Spec. Publ. 459, Washington, DC.

Zwally HJ, Comiso JC, Parkinson CL, Cavalieri DJ, Gloersen P (2002) Variability of the Antarctic sea ice cover. J Geophys Res 107(C5):1029–1047.

Chapter 10
Decadal Changes, Correlations, and Trends

Abstract While we are still grappling with fundamental questions about the root causes of observed changes in the polar regions, satellite sensors have provided us with extraordinary tools for evaluating the scope of these changes and for understanding how changes in the different observables may be related to each other. We analyze satellite or reanalysis data over a 26-year period in which data on surface temperature, sea ice cover, albedo, and sea level pressure are concurrently available. The difference of 13-year segments of the different variables are shown to be coherent and it is apparent that changes in one parameter are also observed in the other parameters. Correlations of surface temperature with sea ice concentration and albedo are very strong and possible connections between chlorophyll concentration with the ice cover and surface temperature have been identified in the early spring period. Drastic declines in the Arctic perennial and multiyear ice cover are also observed suggesting that profound changes are going on in the polar oceans.

Keywords Decadal changes • Correlations • Trends

10.1 Introduction

A global warming of about 0.07 K per decade has been observed during the last century (Hansen et al. 2001; Jones et al. 1999; Karl and Trenberth 2003), and since 1978, the warming rate has increased to about 0.2 K per decade. If this trend continues, the Earth will warm by about 2 K within this century, and the impacts are feared to be profound. The primary cause of warming is still a subject of contentious debate (e.g., Broecker 1992; Spencer 2008) but the IPCC 2007 report has indicated that there is now a 90% confidence level that the culprit is the increasing the level of atmospheric greenhouse gases. This confidence level is considerably higher than the 50% indicated in the previous report released in 2003. The warming observed since 1978 is of special interest because it coincides with the highest rate of increase in greenhouse gases observed. It is also fortuitous that since the 1970s

J. Comiso, *Polar Oceans from Space*, Atmospheric and Oceanographic Sciences Library 41, DOI 10.1007/978-0-387-68300-3_10,

we have had continuous observations of the Earth and the Sun by satellite sensors. Analysis of solar data has indicated that the solar irradiance goes through cyclic fluctuations with a period of about 11 years and an amplitude of about 0.2 W/m^2 and since 2001 the irradiances have been going downwards (Frohlich 2000) while the Earth's surface temperature continued to increase.

Modeling studies have also indicated that global warming would be amplified in the Arctic by as much as 3–5 times (Holland and Bitz 2003), primarily due to the ice-albedo feedback effects. In the previous chapters, we have shown that the warming rate as derived from 26 years of satellite data is about 0.7 K per decade at latitudes inside the Arctic circle and about 0.6 K per decade at >60°N latitudes. This rate is about 3 times the global rate and is consistent with modeling projections. We have also shown that the sea ice cover has been declining at a negative rate of about −3 to −4% per decade. Other areas in the Arctic are showing the effect of such warming as well, with the snow declining at 2% per decade (Armstrong and Brodzik 2001), the surface melt in Greenland increasing at 11% per decade (Abdalati 2006), the glacier volume in Alaska alone is estimated to be about 96 km^3/year (Arendt et al. 2002) and the permafrost thawing (Romanovsky and Osterkamp 2006). The variability of the Arctic ice cover has been tied up with the variability of the Arctic Oscillation index, but in recent years, the AO has gone into neutral (Overland and Wang 2005) while the Arctic ice cover, especially the perennial ice, has been declining rapidly.

In the Southern Hemisphere and inside the Antarctic circle, the warming rate is much more modest at about 0.1 K per decade. This is consistent with reports of cooling in parts of the Antarctic continent (Comiso 2000; Doran et al. 2002). The amplification factor due to ice albedo feedback is likely not as strong in this region as in the Arctic in part because of a very different environmental condition. In the Arctic, the high albedo regions include snow covered and sea ice areas that are quite extensive during the winter and spring period. After the snow has melted in the summer, the high albedo areas are primarily the sea ice cover in the Arctic Ocean, the glaciers, and the Greenland ice sheet. As the sea ice cover declines in the summer, more liquid water gets exposed to solar heating, causing a warmer Arctic Ocean that inhibits the formation of ice in the subsequent winter, causing further retreat in the sea ice in the following summer. In contrast, the high-albedo regions in the Southern Hemisphere during the summer are mainly the snow covered continent of Antarctica and sea ice, with the sea ice contributing only about 13% of the total ice surface area covered by either ice or snow. Thus, even a 10% decline in the summer sea ice cover (as in the Arctic) in the Southern Hemisphere would contribute to only about 1.3% decline in the albedo of the ice covered region. Such change would have a very little impact on the SST of the Southern Ocean. Moreover, the albedo of the continent does not change much because only a small fraction around the periphery is subject to surface melt. Meanwhile, the Antarctic sea ice cover has not been changing much, and if at all, it has been increasing slightly, as indicated in Chap. 7. Thus, an ice-albedo feedback would go in the opposite direction, and could lead to some cooling. The observed small increase in surface temperature and the increase in ice extent have been associated with changes in the Southern Annular Mode (SAM) attributed to the combined effects of increasing greenhouse gases and the development of the Antarctic ozone hole (Arblaster and Meehl 2006; Miller et al. 2006; Turner et al. 2009). Although the ozone hole is an austral spring phenomenon, the impact on the tropospheric flow is greatest during the summer and autumn (Thomson and Solomon 2002). The relative contributions of greenhouse gases and the ozone hole are still under study but the trends in temperature and sea ice in the region illustrate the complexity of the polar climate system.

In this chapter, we wish to gain additional insights into the satellite-observed changes in surface temperature, sea ice, albedo, and ocean color in the polar regions. We will show how some of the parameters are related to each other explicitly and how they contribute in tandem to an improved knowledge about our changing polar oceans. The mapping to a common grid also facilitates the process of comparing the temporal and spatial changes detected in each climate variable. It also enables the quantitative comparison of the changes of these variables in different areas and the assessment of the consistency and coherence of such changes as well as the strength of the coupling of these variables. We will discuss possible connections of the observed changes with the Northern Hemisphere Annular Mode (NAM), better known as the Arctic Oscillation, and also the Southern Hemisphere Annular Mode (SAM) and how the latter might have contributed to the observed patterns of variability and trend in the satellite data.

10.2 Decadal Changes in Surface Temperature, Sea Ice, Albedo, and Sea Level Pressure

In the previous chapters, we showed remarkable interannual changes in three key polar parameters, namely, surface temperature, sea ice, and albedo. It is physically intuitive to expect that these parameters are strongly related to each other since we know that an increase in surface temperature is expected to cause a decline in the sea ice cover, which in turn would cause a decline in albedo. However, the relationships are not so predictable because the system is dynamic, and the impact of one variable on the other is sometimes difficult to track down, especially if there is a time lag. A study of the three parameters concurrently in conjunction with sea level pressure is thus desirable to be able to assess how strongly they are coupled together and how the observed changes are impacted by sea level pressure. It is also interesting to note that the impact of warming has not been uniform throughout the polar regions. In particular, while considerable warming has been occurring in some regions, significant cooling has been going on in other (sometimes adjacent) regions. Similar phenomena have also been occurring with the sea ice cover and albedo. While realizing that atmospheric circulation and oceanic processes are integrated in the system in a complex fashion, it is useful to be able to observe cuncurrent changes in the geophysical parameters and understand how such changes can be used to gain insights into the phenomenon and identify the likely drivers of the change.

Detailed analysis of the variability of each parameter has been discussed separately in previous chapters using as long a time series as is available. Because of different record lengths, it is sometimes difficult to assess whether the changes observed in one parameter are consistent or related to those observed in another parameter. To minimize ambiguity in interpretation, we will show the results of analysis of all four parameters over an identical time period, which, in this case, is the overlapping data from August 1981 to July 2007 for the Arctic and from January 1982 to December 2007 for the Antarctic. Although data after December 2007 are available, we need an even number of years to fit the specific type of analysis that will be presented. We will also demonstrate that the few years of difference do not matter much and that the results provide basically the same conclusions. We recognize that the historical satellite data represent a relatively short period and may not reflect long-term changes in the climate system. These data, however, provide a unique spatial and global coverage, and could be considered as the starting core of a climate record that will accumulate in the years to come and will eventually provide a clearer picture of the true direction of changes in the polar regions.

10.2.1 Northern Hemisphere

The overall changes and trends in surface temperature, sea ice cover, and albedo in the Arctic from 1981 to 2007 are summarized and depicted in the set of images presented in Figs. 10.1–10.3. The sets of images represent the same time periods when satellite data were available for all three variables (i.e., August 1981 to July 2007), and therefore provide concurrent information from different variables over the same period. Figure 10.1a represents the average of the first half of the surface temperature data record (1981–1994), while Fig. 10.1b represents the average for the second half (1994–2007). The multiyear averaging is similar to the yearly averaging described in Chap. 6 and is such that it starts from mid-summer of the first year and ends in the summer of the last year (i.e., from August of 1 year to July the last year). In particular, Fig. 10.1a corresponds to the average of all data from August 1981 to July 1994, while Fig. 10.1b shows the average of all data from August 1994 to July 2007. It is remarkable and at the same time reassuring that these two images have such a strong resemblance to each other. Many distinct surface features are reproduced with high fidelity from one time period to another indicating that despite the aforementioned difficulty in masking persistent clouds in the ice covered regions, the retrieved data clearly represent surface parameters. The strong similarity of the two images suggests that the spatial distribution of surface temperature in the entire region has not changed much since 1981. The images show regions where the temperatures are consistently warm and regions where they are consistently cold. A close examination of the two images reveals subtle but significant shifts in the location and distribution of the isotherms. The difference map shown in Fig. 10.1c (i.e., data in Fig. 10.1b minus data in 10.1a) allows for a quantitative evaluation of how surface temperature had changed in the Arctic from one 13-year period to the next. The change data have been normalized and the color scale has been adjusted accordingly to show decadal (i.e., 10-year) changes. Positive changes are represented by

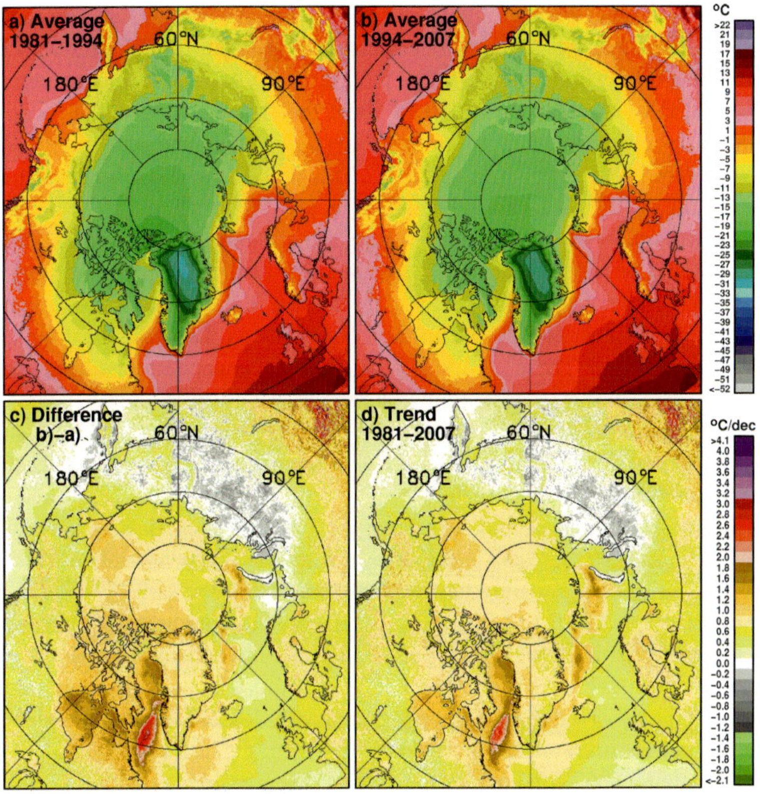

Fig. 10.1 (**a**) Average of surface temperature from August 1981 to July 1993; (**b**) average of surface temperatures from August 1993 to July 2007; (**c**) (**b**) minus (**a**); and (**d**) trend in temperatures for each data element from August 1981 to July 2007 using monthly data. Surface temperatures were derived from AVHRR data

warm (orange and red) colors, while negative changes are represented by cool (gray and green) colors. The difference map reveals that, generally, the changes are overwhelmingly positive, indicating a warming Arctic, with the most positive changes located in the western Arctic (e.g., Beaufort Sea), North America, Baffin Bay, Greenland, Greenland Sea, and Barents Sea. An unexpected feature in the difference map is the large areas of modest cooling in parts of Russia. This is caused in part by unusually warm conditions in the 1980s as revealed by the anomaly maps shown in Fig. 6.12. The difference map is a simple way to evaluate the overall changes from one time period to another during the satellite era. For comparison, the trends in each pixel element, inferred using a linear regression analysis on each pixel in the monthly anomaly data, are presented in Fig. 10.1d. It is apparent that the trend map looks almost identical to the difference map and that either technique provides good information about decadal changes in various parts of the Arctic region. Although the trends vary significantly from one region to another, the dominance of positive trends is apparent especially in North America, Greenland, and sea ice covered areas. We have shown in Chap. 6 that the average temperature north of 60°N and north of the Arctic Circle increased by about 0.60°C and 0.66°C per decade, respectively, during the satellite recording period. The difference and trend maps provided by Fig. 10.1c and 10.1d show the specific geographic location of areas that have changed the most (or the least) during the time period.

The magnitude of the trends has large spatial variability and is most similar in the continental areas (e.g., Alaska, Canada, Europe, and Western Russia). The exception is Eastern Russia (or Northern Asia), where some cooling is evident, as indicated earlier. It should be noted that a shift in surface type can lead to large changes in the trends. For example, high positive trends are located in the seasonal ice regions and especially in the Chukchi Sea, Baffin Bay, Hudson Bay, Labrador Sea, Greenland Sea, and Barents Sea. These are areas where sea ice concentrations have changed from near 100% to near 0% during the period, as discussed in Chap. 7. The trends in surface temperature are also shown to be relatively minor in some areas such as the Bering Sea. Not surprisingly, these are also areas where the sea ice cover did not change much during the study period.

A similar set of images but using sea ice concentration data is presented in Fig. 10.2. The images represent averages over the sea ice growth and decay seasons (i.e., August of one year to July the following year) and the same 13-year periods. Because of large seasonal and interannual variations, the gradients of concentration across the marginal ice zones, as reflected in these multiyear averages, are not so strong. The ice edges are also not as well defined as those in the daily images because of the averaging of ice edges that are not consistently located in the same place. The two images in Fig. 10.2a and 10.2b basically represent the history of the sea ice cover during the satellite era with each of the 13-year averages depicting a typical yearly ice cover during each period. It is again remarkable that the two images are so similar. The edges of high concentration areas in the Central Arctic have moved slightly to the north from the first 13-year period to the second period, and some of the peripheral seas show slightly lower concentrations during the second period than the first period. But overall, the general locations of consolidated and unconsolidated ice cover are basically the same for the two periods. The difference map presented in Fig. 10.2c reflects decadal changes quantitatively, and it shows mainly negative values, indicating an almost consistent decline in ice concentration over the entire Arctic region. Some positive but relatively subtle anomalies are apparent and occurred in the Bering Sea and the Sea of Okhotsk. The difference map has also been normalized to show decadal change instead of 13-year change. The negative changes are regionally dependent with the biggest change occurring at the Baffin Bay/Labrador Sea area, followed by significant changes in the Eastern and Western parts of the Arctic Basin and also at Hudson Bay and Greenland Sea. These are the same areas where the trends in surface temperature shown in Fig. 10.1 are highly positive.

The trend in the ice concentration for each data element was also estimated for the same period for the entire region and the results, as presented in Fig. 10.2d, show patterns that are very similar to those shown by the difference maps. The good agreement is notable and indicates that as with the temperature maps, the decadal changes can be quantified by either method. It is interesting to note that the most negative

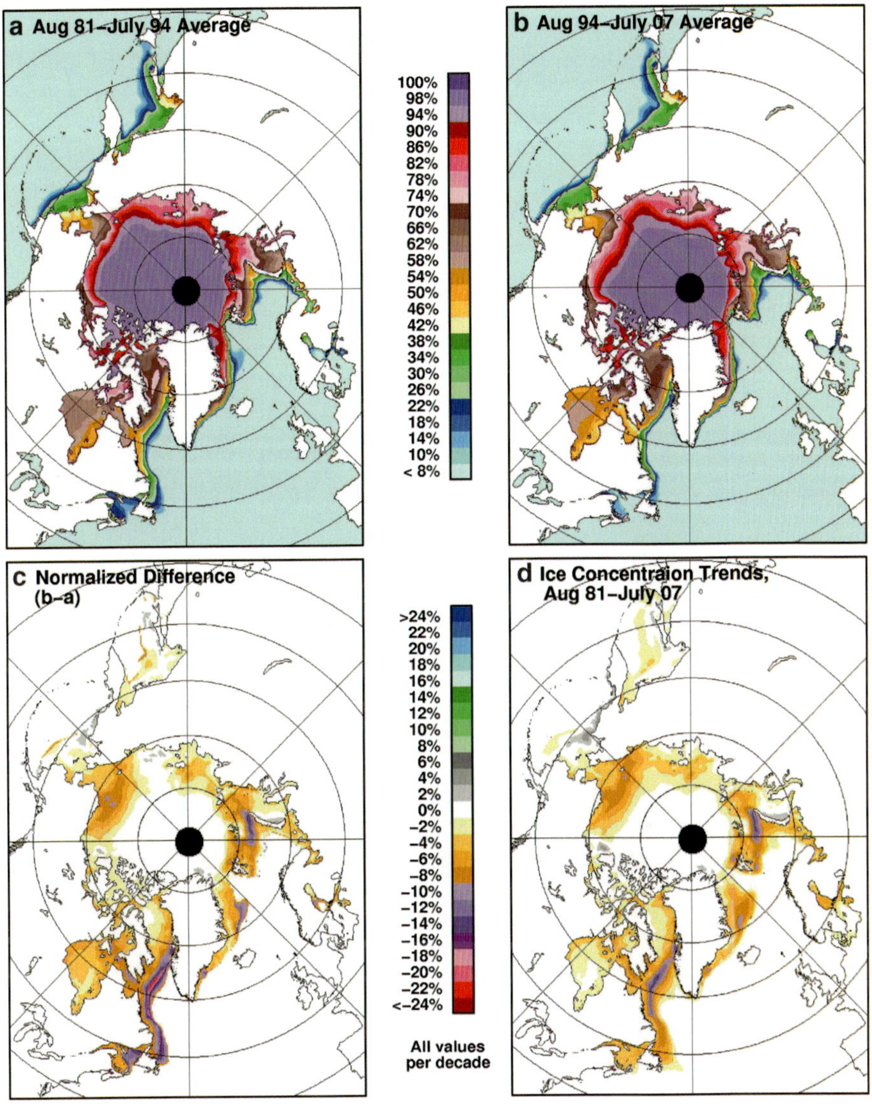

Fig. 10.2 (**a**) Average of ice concentrations August 1981 to July 1993; (**b**) average of sea ice conctrations from August 1993 to July 2007; (**c**) (**b**) minus (**a**); and (**d**) trends in sea ice concentrations for each data element from August 1981 to July 2007 using monthly data. Ice concentrations were derived from SMMR and SSM/I data

trends occur at the Barents Sea, Labrador Sea, Greenland Sea and Baffin Bay. The changes in these regions are associated more with winter and spring ice cover, and the declining ice extents in these seasonal ice regions during the period. While the biggest decline in the Arctic sea ice cover, has been in the perennial ice cover as exhibited during the summer ice minima, the negative anomalies do not appear as strongly in the trend because the decline of ice to summer minimum values occurs at a relatively short period. It is evident, however, that the increased open water areas in the Chukchi Sea and Beaufort Seas during summer in recent years are reflected in the difference and trend data. The patterns are consistent with the regional trend results presented in Chap. 7. Some of the discrepancies between the difference and trend maps maybe associated with the long term averaging that tend to smear the data in some regions. A minor factor could be the change in sensor from SMMR to SSM/I in 1987, which were intercalibrated during overlap; however, the two sensors do not have exactly the same attributes, including resolution.

Changes in the narrow-band albedo at 0.6 μm in the region are presented in the same format as the previous two figures in Fig. 10.3. Because of missing data due to darkness during autumn and winter months, the averages were done only for the

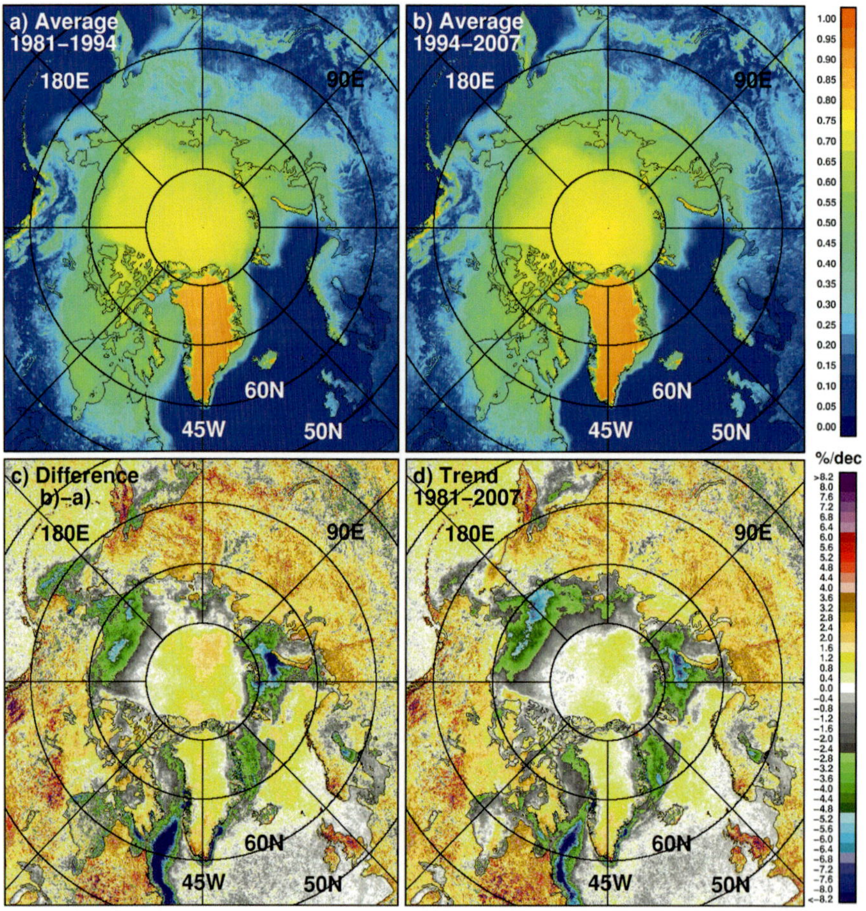

Fig. 10.3 (**a**) Average of albedos from August 1981 to July 1993; (**b**) average of albedos from August 1993 to July 2007; (**c**) (**b**) minus (**a**); and (**d**) trends in albedos for each data element from August 1981 to July 2007 using monthly data. Narrow-band albedos were derived from AVHRR data

months from March through September and for the same set of years. The multi-year averages presented in Fig. 10.3a and 10.3b again indicate a generally good consistency between these two maps of albedo. The good agreement of many distinct spatial features, including those over land, is again apparent. It is interesting to note that the patterns of high albedo over the ice covered ocean are similar to those of sea ice concentration (see Fig. 10.2a and 10.2b), while patterns over land are similar to those of surface temperature (Fig. 10.1a and 10.1b). This reflects the high albedo of sea ice over ocean regions and that of snow over land. It also indicates that snow is most persistent where the average surface temperature is relatively low. The difference map (Fig. 10.3c) shows changes that are very similar to those of the sea ice cover with significant changes evident in the Chuckchi/Beaufort Seas, Barents Sea, Greenland Sea and Baffin Bay. Using the same set of data (March to September only), the trends for each pixel were also calculated, and the resulting trend map (Fig. 10.3d) shows basically the same features as the difference map but of slightly different magnitude in some areas (e.g., near ice edges). The high contrast in albedo between sea ice and open water makes these results very similar to those of the ice concentration data. It is apparent that the Odden in the Greenland Sea shows up in both difference and trend maps, although it is weakly represented in the multiyear average maps. The slight difference in the representation of the trends in the ice concentration and albedo maps over the oceans are associated with differences in sensitivities of the infrared and microwave radiation to different surfaces, such as those over new ice, snow covered ice, barren ice, and ice with melt ponds.

Overall, the observed changes and trends in surface temperature, sea ice cover and albedo in the Arctic region are consistent with the previously noted warming trends and decline in the sea ice cover. The analysis of all three variables over the same period provides the means to better understand how changes in one variable are reflected in the changes in the other variables. Just a qualitative comparative analysis of the images shows many connections on a regional basis and also over the entire PanArctic region. For example, a warming in the Chukchi Sea of about 0.8°C per decade occurred concurrently with a decline in ice concentration in the same region of about −6% per decade and in the decline of the albedo of the surface by about −3% per decade. The strength of the relationship is not always high because the sea ice cover is very dynamic and the albedo is sensitive to surface properties. In particular, changes in surface temperature because of ice dynamics may not be reflected in the ice concentration data while changes in albedo on account of changes in surface characteristics my not be observable in the ice concentration and surface temperature data. It is interesting that the changes in surface temperature and albedo in the land areas are generally consistent with the changes in surface temperature and surface albedo in the adjacent sea ice cover. Since surface temperature is primarily influenced by surface air temperature, which is an atmospheric parameter, changes in surface temperature are not expected to be affected by land/ocean or other boundaries. However, as the snow gets melted in the land areas, the temperature of adjacent sea ice cover may be distinctly different. The albedo of the land surfaces compared to those of sea ice would be expected to be different as well.

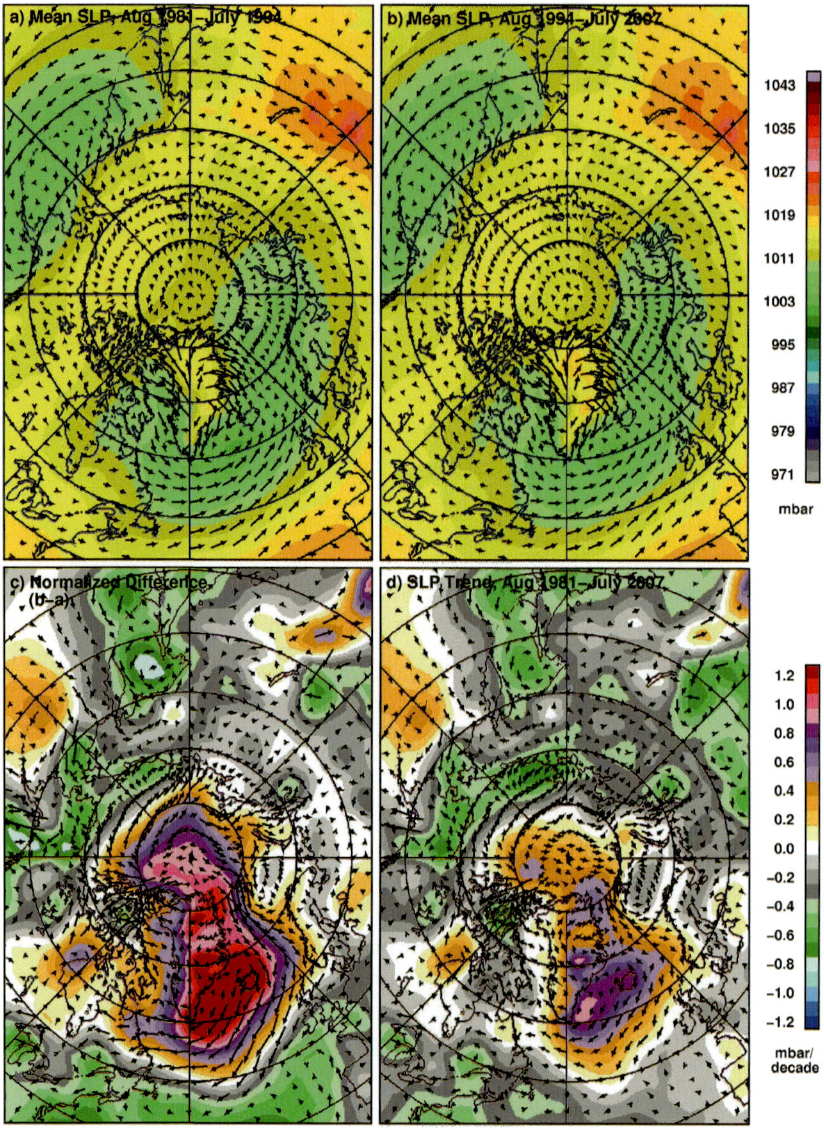

Fig. 10.4 (**a**) Average of sea level pressure and winds from August 1981 to July 1993; (**b**) average of sea level pressure and winds from August 1993 to July 2007; (**c**) (**b**) minus (**a**); and (**d**) trend in sea level pressure and winds for each data element from August 1981 to July 2007 using monthly data. Sea level pressure and winds were derived from NCEP reanalysis data (Kalnay et al., 1996)

The sea level pressure and wind data for the same periods shown in Figs. 10.1 and 10.2 are presented in Fig. 10.4. The 13-year averages for the pressure are remarkably similar despite large interannual variations as indicated in Chap. 5. The pressure system at high latitudes is defined by two low pressure areas, one in the North Atlantic and the other in the North Pacific. The pressure maps also show two relatively high pressure areas, one in mid-Asia and the other one in the Atlantic Ocean at mid-latitudes. The sizes and shapes of these features as depicted in the first 13-year period are mainly reproduced in the second period. The difference maps, however, show significant changes from one period to another. In the North Atlantic/Greenland region, the change in pressure is mainly positive, indicating that the lows in this region are not as deep in the more recent decade than in the previous one. In the North Pacific region, the changes are also positive but the magnitude of change is much smaller. Positive changes are also apparent in parts of North America (i.e., near Hudson Bay), the Central Arctic and parts of Asia. The rest of the changes are negative. The trend maps shown in Fig. 10.4d shows similar results but generally smaller decadal changes. Overall, the results show that on average, the pressure and wind patterns did not change much from one period to another. A deepening of the lows in the Chukchi/Beaufort Sea area might, in part, have caused more stormy weather and less ice cover in the region (Yang et al. 2004; Comiso et al. 2008).

For more quantitative comparisons, plots of SST for each month of the year using the thirteen year averages from 1982 to 1994 and from 1995 to 2007 are presented in Fig. 10.5. Averages of SSTs north of the Arctic Circle (>66.5°N) and at latitudes >50°N are shown separately in Fig. 10.5a and b, respectively. The differences in the average SST north of the Arctic Circle during the first and second periods are relatively small. The first period, represented by the red line, is usually colder than the second period, represented by the blue line, except in August, September, and November, when the first period was slightly warmer than the second period. The first period was colder than the second period in December and January by 0.2° or greater. For comparison, we also plotted the data for the years 2007 (gray) and 2008 (black). In 2007, the monthly SSTs were warmer, especially in May to August, than the historical averages for each month except October and November, while in 2008, the SSTs were significantly higher than historical averages in January, May, and June but lower in March and April. The distinctly higher temperatures in July and August in 2007 may have contributed to the abnormally low perennial ice in 2007. For the expanded study area (i.e., >50°N), the average SSTs in the first period is clearly consistently lower that those of the second period (Fig. 10.5b). The difference is on the average about 0.2 K going up to about 0.5 K in August. The SSTs for 2007 and 2008 are comparable with those of the second period except in March and April when the 2008 SSTs were almost the same as those of the first period. The colder SST in the spring of 2008 was likely a factor in the slower decline in the ice cover during the spring period of this year than in 2007.

The corresponding monthly temperatures for the surface ice temperatures (SIT) during the same periods are shown in Fig. 10.5c and d. The SIT in this case represents surface temperatures only of sea ice. Again, two sets of plots are presented, one for the region north of the Arctic Circle and the other for the entire sea ice covered region. The SITs for both periods have much larger seasonal fluctuation and are relatively more similar (than the corresponding SSTs) with the earlier period consistently colder than the latter period. The biggest discrepancies occur in November, December, January and March, when the second period was warmer than the first by about 1.0°C. The temperatures in 2007 were almost consistently warmer than those of the two historical periods, with the exceptions being in January and May, when the SIT was slightly lower than that of the second period. The SIT in 2008 was

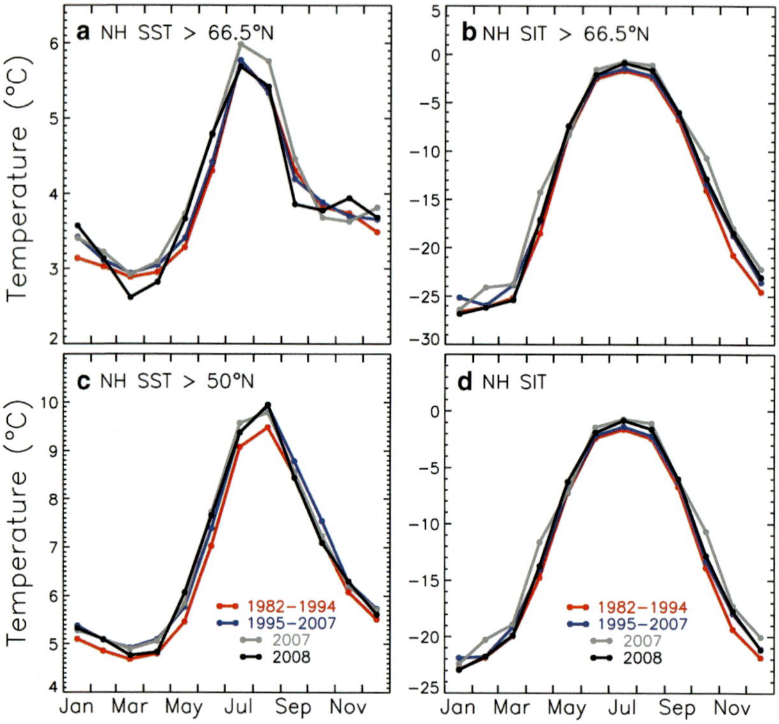

Fig. 10.5 Monthly averages of SST over 13-year periods in (**a**) inside the Arctic Circle (>66.5°N) and (**b**) >50°N. SIT monthly averages over 13-years in sea ice covered regions (**c**) inside the Arctic Circle (>66.5°N) and (**d**) the entire Northern Hemisphere

lower in January, became similar to the first period in February and March, and then was higher than those of the two periods from April to August. The plots of SIT for the entire Northern Hemisphere, as presented in Fig. 10.5d, show slightly higher values, but the relative distributions are almost identical to the corresponding plots for the region inside the Arctic Circle. A slight difference occurred in January when the SIT in 2007 was almost the same as that of the second period. In both study regions, the SIT was unusually high in February, April and October 2007.

The 13-year averages of the sea ice extent and ice area for each day of the year are presented in Fig. 10.6a and 10.6b, respectively. It is apparent that the monthly averages in extent and area are consistently higher in the first period than in the second period by an average of about $0.8 \times 10^6\,km^2$ with the biggest difference occurring during the end of the summer period. It is also apparent that the monthly values in 2007 were consistently lower than those of the two periods by about $1.5 \times 10^6\,km^2$. On the other hand, the monthly values in 2008 were a lot more consistent with those of the second period starting in February, but started to deviate in June and almost caught up with 2007 in September. The ice extent and ice areas in September are shown to have much lower values than those of the 13-year averages.

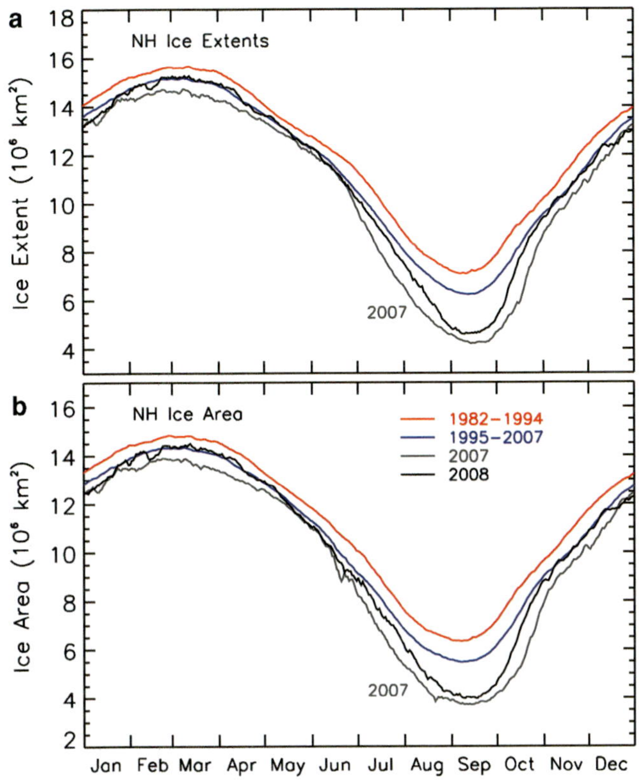

Fig. 10.6 Thirteen year averages of daily (**a**) ice extents and (**b**) ice area in the Northern Hemisphere from January to December, depicting the seasonal variation in the sea ice cover. The 13-year periods are 1982 to 1994 and 1995 to 2007. For comparison, daily values for single years in 2007 and 2008 are presented showing much lower ice cover in the summer during these years than the 13-year averages

Comparing the plots presented in Figs. 10.5 and 10.6, it is apparent that generally, the changes in surface temperature is consistent with expected changes in the sea ice extent and ice area. The plots for SSTs and SITs show that the first period was generally colder than the second period and that those for 2007 and 2008 were also unusually high. The plots for sea ice also show that the ice extents and ice areas are consistently lower in the second than those in the first period. It appears that the negative correlation of SST with the sea ice cover is stronger than between SIT and the sea ice cover. This is especially the case in the summer, and the relatively weak correlation with SIT is mainly because SIT during the summer is close to melt temperature and does not change much with time. The unusually high SST inside the Arctic Circle in 2007 has been confirmed by other studies (Shibata et al. 2009) and is likely a factor in the dramatic decline of the ice cover during the summer of 2007. The low ice cover in the summer of 2008 may, in part, be caused by conditioning of the mixed layer of the ocean associated with enhanced solar flux in 2007 and, in other recent years, in the Arctic Basin.

10.2.2 Southern Hemisphere

The decadal changes observed in the Antarctic are less dramatic than those observed in the Arctic. Environmental factors make the climate in the region very different and changes in the system are likely revealed in a different way. Some issues associated with the observation of climate change in the region have been examined by Jacobs (2006) and Hanna (1996). New insights about the state of climate in the region have recently emerged (e.g., Steig et al. 2009; Monaghan et al. 2008; Turner et al. 2009). In the Antarctic region, the corresponding sets of images of 13-year averages and trends for surface temperature, sea ice and albedo, as recorded by satellite data, are presented in Figs. 10.7–10.9. The sets of images depict the overall climate conditions

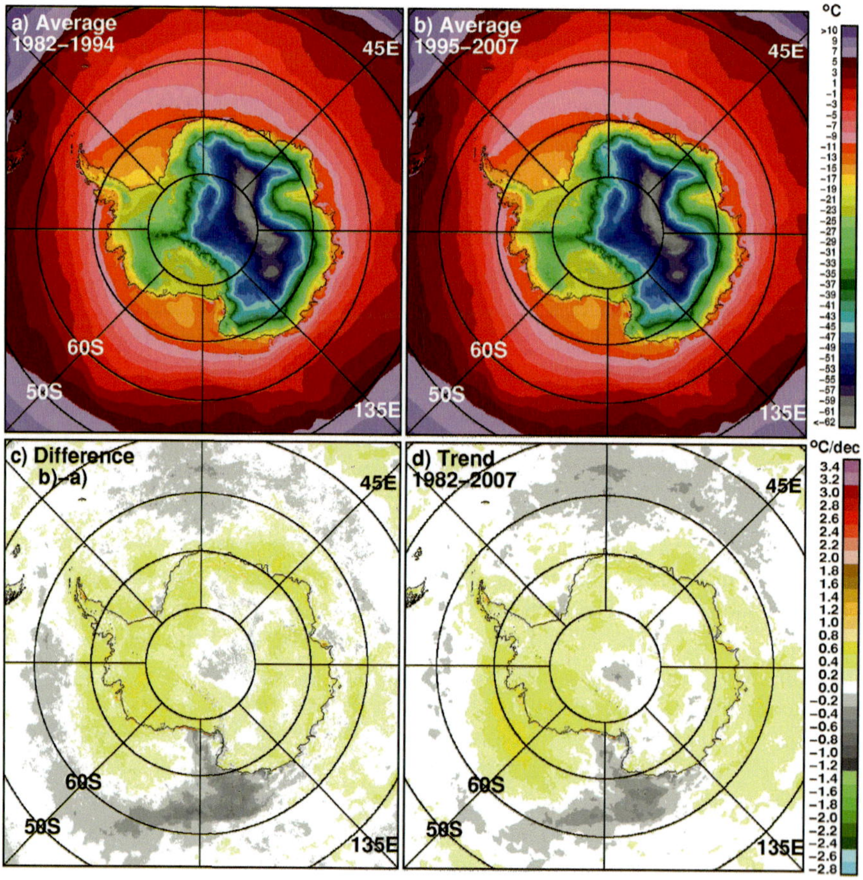

Fig. 10.7 (**a**) Average of surface temperature in the Southern Hemisphere from January 1982 to December 1993; (**b**) average of surface temperatures from January 1994 to December 2007; (**c**) difference of averages shown in (**b**) and those in (**a**); and (**d**) trend in temperatures for each data element from January 1982 to /December 2007 using anomaly monthly data

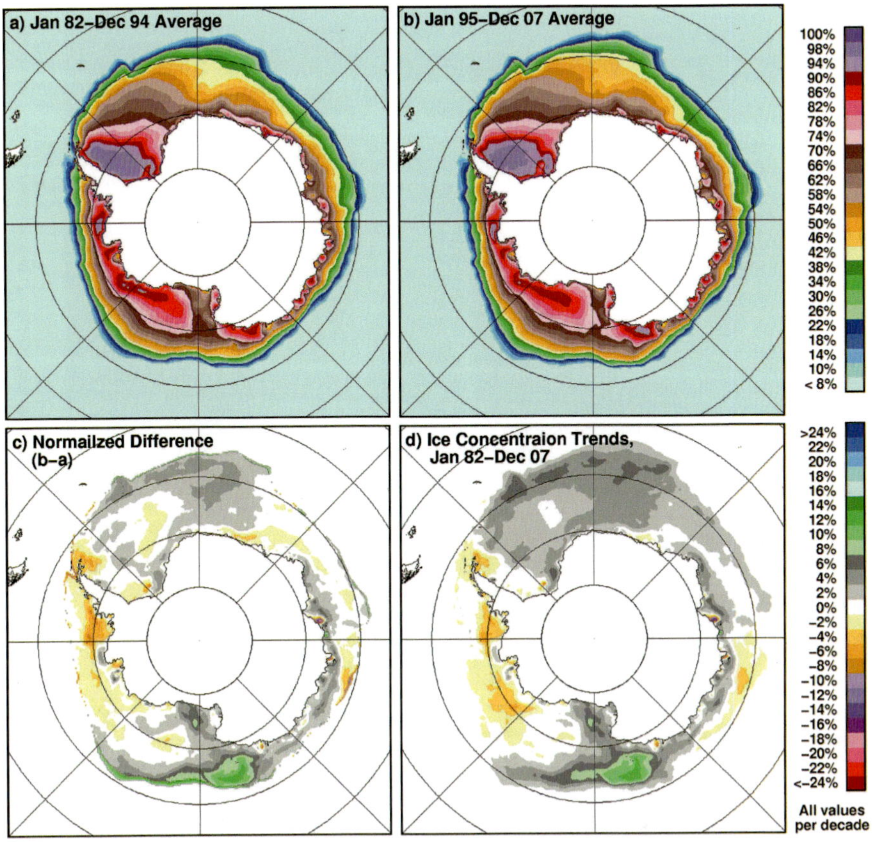

Fig. 10.8 (**a**) Average of ice concentration in the Southern Hemisphere from January 1982 to December 1993; (**b**) average of ice concentration from January 1994 to December 2007; (**c**) the difference of averages shown in (b) and those shown in (**a**); and (**d**) trend in ice concentration for each data element from January 1982 to /December 2007 using monthly anomaly data

in the Antarctic from 1982 to 2007. Since the ice season in the two hemispheres occurs 6 months apart, the averaging starts in January 1982 and ends in December 1994 for the first set and starts in January 1995 and ends in December 2007 in the second set. As indicated earlier, the distributions of the contours of the different parameters are quite different from those of the Northern Hemisphere because of environmental and other factors. The temperature contours over land follow approximately the elevation contours of Antarctica with lowest temperatures usually located at the high altitude regions. The −2°C contour also follows approximately the general location of the edges of the sea ice cover. As in the Northern Hemisphere, the average of surface temperature for the Southern Hemisphere as derived from the first 13 years of satellite data (Fig. 10.7a) looks strikingly similar to the average from the next 13 years of satellite data (Fig. 10.7b). This is despite the more dynamic nature of the surface in the region compared to that of the Arctic and large interannual changes in

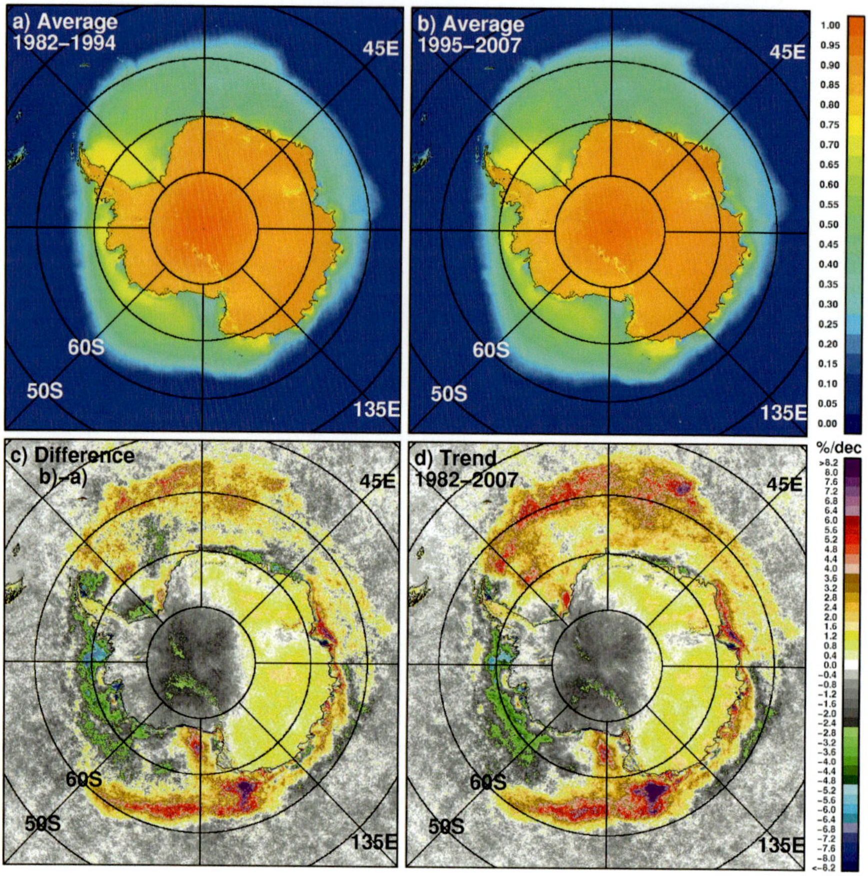

Fig. 10.9 (**a**) Averages of narrow-band albedo in the Southern Hemisphere from 1981 to 1994; (**b**) averages of albedo from 1994 to 2007 and (**c**) difference of averages in (**b**) and those in (**a**); and (**d**) trends in albedo as inferred from monthly anomaly data from 1981 to 2007

the shape and location of the sea ice cover. Many surface features in the continent depicted in the first period are reproduced in detail in the second period, again indicating that the retrieved data are representative of radiative emissions from the surface.

Some subtle differences in surface temperature exist between the two periods and these are better quantified in the difference map shown in Fig. 10.7c. The difference map shows slightly warmer temperatures in the second set of data than the first in much of the continent, including the Antarctic Peninsula, West Antarctica, and the periphery of East Antarctica. Some cooling is apparent at the East Antarctic plateau, in the Ross Sea and in other parts of the Southern Ocean. A feature of interest is an area of warming in the Amundsen Sea, which is surrounded by an area of cooling north of this region and in the Ross Sea. This is a very active region in which ice is declining in the eastern part while ice production is increasing in the western region. The distributions of the sea ice cover is likely influenced by the location and characteristics of this warming area. As in the Arctic, the trends in each pixel were generated and the corresponding map, as presented in Fig. 10.7d, show patterns that are very similar to those of the difference map. The trend map shows slightly more warming in the Amundsen Sea and less cooling in the Ross Sea and the Antarctic Plateau than the difference map, but overall, information contents from both maps are basically the same.

The 13-year averages of ice concentrations for the early and later part of the satellite record are presented in Fig. 10.8a and 10.8b, respectively. Despite large interannual variations in the location of the circumpolar ice edges in the winter and also in the summer, as discussed in Chap. 7, the two 13-year averages look surprisingly similar. Considering large interannual changes caused by interannual shifts in circulation patterns and other environmental factors, the consistency means that much of these changes get averaged out during each 13-year period. There are, however, some distinct differences in the distribution of the sea ice cover between the two periods, the most obvious of which are located near the Maud Rise in the Weddell Sea and the continental coastlines near the Ross Sea. The difference map, in Fig. 10.8c, shows more quantitatively the difference in the ice cover over the two periods. In addition to positive changes in the Maud Rise and Ross Sea regions, negative changes in the Bellingshausen/Amundsen Seas are apparent. This is consistent with the negative trends in the latter as indicated in Chap. 7. It is interesting to note that the relatively high negative changes in the Bellingshausen/Amundsen Seas occur in the inner zone, while the high positive changes in the Ross Sea occur in the outer zone. This is basically consistent with the declining sea ice cover in the Bellingshausen/Amundsen Seas in the summer and an increasing ice cover in the Ross Sea in winter. A trend map using the same data set is presented in Fig. 10.8d and it is apparent that the information about the rate of change in ice cover is very similar to that provided by the difference map. Some of the subtle differences may be associated to the averaging and the change in sensor in July 1987 from SMMR to SSM/I. It is, however, notable that the aforementioned difference between the difference map and the trend map for the surface temperature is reflected in the sea ice data that is derived from a totally different type of sensor. The higher degree of warming reflected in Fig. 10.7d compared to that of Fig. 10.7c at around 70°S and 130°E is manifested as a larger decline in the ice cover in the trend map in Fig. 10.8d compared to the difference map in Fig. 10.8c.

The corresponding sets of 13-year average maps for the albedo are presented in Fig. 10.9a and b, respectively. These maps provide information similar to those of the ice concentration data, but there are important differences. First, the albedo data is collected at a much higher resolution than the passive microwave data which are used to generate the ice concentration maps. Secondly, the sensitivity of the albedo to surface effects, such as snow grain size and surface roughness are much higher and can vary significantly in areas of near 100% ice concentration. The albedo images also show a more distinct representation (i.e., low values) in coastal polynyas around Antarctica, indicating improved ability to identify new ice and the persistence of these polynyas. Overall, the average albedo climatologies for the two periods look almost identical with the location and sizes of the coastal polynyas being very similar. Distinct features in the continent, including that of the Transantarctic Mountains, are represented in much the same way in the two images.

The difference map, presented in Fig. 10.9c, shows quantitatively how the albedo in one period differed from that of the other period. It is apparent that the albedo declined in generally the same place as where the sea ice cover declined, and vice-versa. The distribution of the regions of negative decline are, however, significantly different from those of the sea ice data. This is in part because of the higher resolution of the albedo data and differences in sensitivity to different surface features as indicated earlier. Again, the trend map presented in Fig. 10.6d, show basically the same information as the difference map. Negative trends are apparent in the Bellingshausen/Amundsen Seas region and part of the West Pacific region. This is consistent with warming and the loss of sea ice in these regions as shown in Figs. 10.7 and 10.8. The rest of the sea ice covered regions show enhanced albedo, which is consistent with the observed increase in the sea ice cover during the period.

Similar averages but for sea level pressures are presented in Fig. 10.10. The averages for the first thirteen years are again generally similar to those of the second thirteen years, but subtle and significant changes are apparent. In the Antarctic continent, the highs are a little enhanced in the second period. At the latitudinal band between 60 and 70°S, the lows are deeper and more defined in the second 13-year period. The pressure patterns in both periods are mainly mode-3 with the mode being apparent in a more distinct manner in the second period. The difference and trend maps presented in Fig. 10.10c and 10.10d show basically similar patterns with some discrepancies primarily in the Ross Sea region. Both of these maps reflect a deepening of the lows in most places except in the West Atlantic Ocean, western part of the Antarctic plateau, and the eastern tip of the Antarctic plateau. The data shows a deepening of the lows in the Ross Sea, which would enhance the drift of sea ice and increase ice production at the Ross Ice Shelf coastal areas. This is consistent with the observed positive trend of sea ice cover in the Ross Sea.

Plots of multi-year monthly averages of SST for the first and second study period are presented in Fig. 10.11. Again, two regions are considered, but this time, the ice free region at latitudes >55°S (Fig. 10.11a) and the region at latitudes between 35°S and 45°S (Fig. 10.11c). It is apparent that the average monthlies of SST at latitudes >55°S for the two periods do not provide a consistent pattern. The data show that first period was warmer than the second period in January and February; the opposite is true in March and April, and from May through August, the first period was again warmer than the second period. From September to December, the first period was colder than the second period. It is also apparent that the SSTs in 2008 were generally lower than the multiyear averages and also the 2007 values. Towards the north between 35°S and 45°S, the pattern is much more consistent. In this region, the first period was consistently colder than the second period throughout the year with the largest differences occurring in January and February. Otherwise, the difference are relatively minor. For the individual years in 2007 and 2008, the values are also generally lower than those of the multi-year periods for the region >55°S for much of the year. From July to December, 2008 the SSTs were lower than those of the multi-year periods and 2007, while the values in 2007 were consistent with those of the multi-year periods. In the study region between 35°S and 45°S, the SSTs in 2007 and 2008 were actually slightly higher than the climatologies.

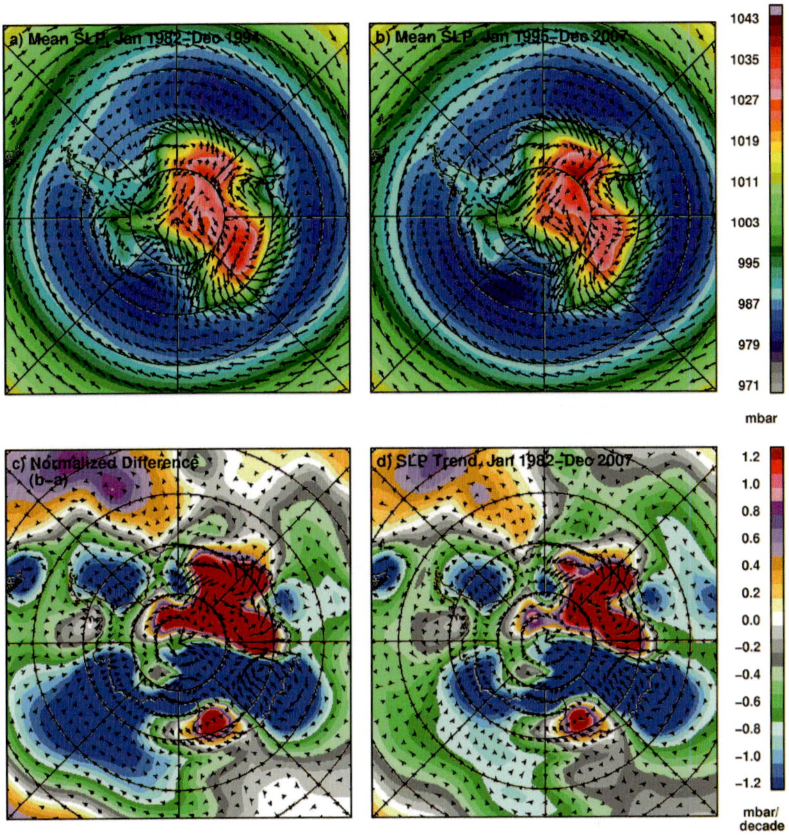

Fig. 10.10 Thirteen-year average of sea level pressures (**a**) from 1982 to 1994 and (**b**) 1995 to 2007 and (**c**) normalized difference map of the two periods while (**d**) represents trends in sea level pressure from 1982 to 2007

The multi-year monthly averages for the surface ice temperature (SIT) at latitudes >66.5°S and the entire ice cover are presented in Fig. 10.11b and 10.11d, respectively. The SITs for the two periods are shown to be consistent, with the second period generally warmer than the first period. The biggest discrepancies occur during the winter period (between June and October). The SITs in 2007 and 2008 are also generally consistent, with 2008 showing the most deviations from the climatologies.

Plots of the seasonal distribution of the multi-year daily averages of the sea ice cover for the first and second periods are presented in Fig. 10.12. The first and second periods are shown to have very consistent ice extents and ice areas with those of the second period usually higher than those of the first period. This is consistent with the observed positive trend in the ice cover in the Southern Hemisphere. Relatively higher values are shown for 2007 and 2008, reflecting the cooling in the region for these years, as described earlier.

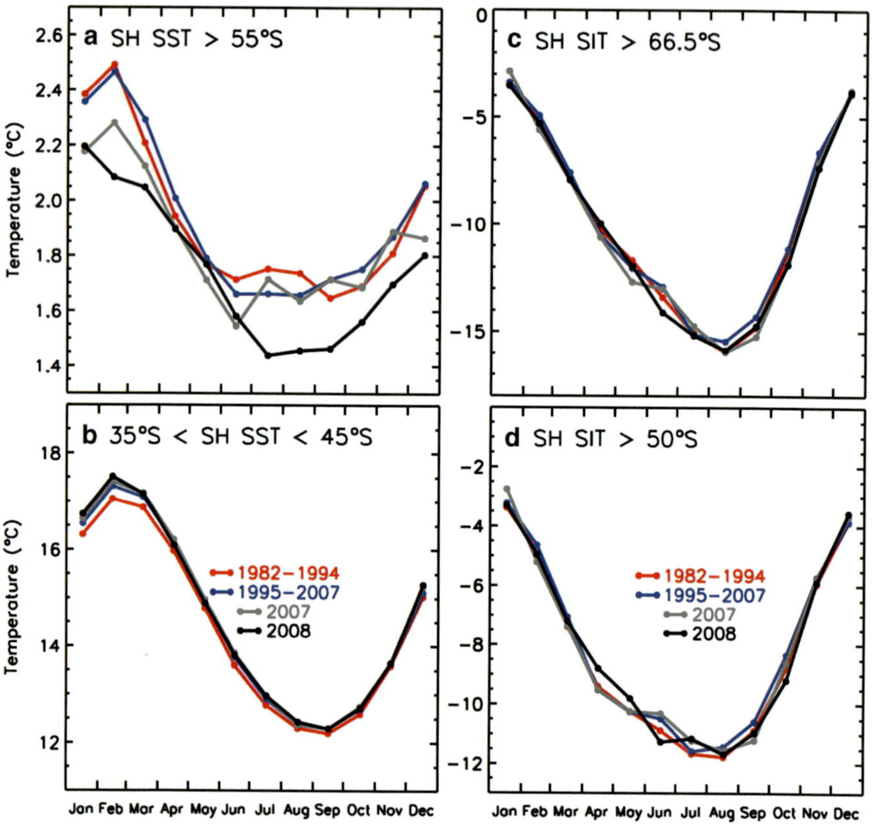

Fig. 10.11 Monthly averages of SST in the Southern Hemisphere over 13-year periods in (**a**) >55°S and (**b**) between 35°S and 45°S. Monthly averages of SITs during two periods and at (**c**) >55°S and (**d**) between 35°S and 45°S

10.3 Seasonal Trends in Surface Temperature and the Sea Ice Cover

To gain additional insights into the changing Arctic climate, we evaluate how the parameters have been changing during the different seasons. This is especially important in ice covered regions where a slight shifts in the onset of freeze-up or melt could significantly change the ice cover and the albedo of the region. The trends in surface temperature and sea ice concentration for the different seasons are generally different but have some similarities as presented in Figs. 10.13 and 10.14, respectively. The trends in surface temperature are shown to be dominantly positive for all seasons, but the magnitude of the trends vary with region. It is interesting that in winter, a large part of Russia is shown to have negative values. This is caused in part by relatively warm temperatures in Russia in the 1980s, as indicated by the February anomaly maps in Chap. 6. Large positive trends in the Baffin Bay, Barents Sea and Greenland Sea are associated with the retreat of the sea ice cover. In the deeper part of the Arctic Basin,

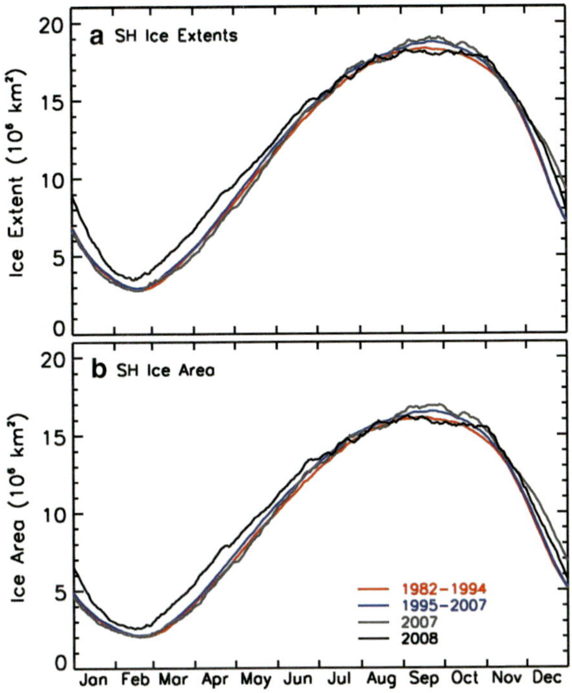

Fig. 10.12 Thirteen year averages of daily (**a**) ice extents and (**b**) ice area in the Southern Hemisphere from January to December, depicting the seasonal variation in the sea ice cover. The 13-year periods are 1982 to 1994 and 1995 to 2007. For comparison, daily values for single years in 2007 and 2008 are presented showing significantly higher ice cover in the summer of 2008 than the 13-year averages

the most positive trends are located in the predominantly multiyear ice regions, north of Canada, and Greenland. The warming trend in the region, especially in winter, makes the thick ice floes, which are the mainstay of the Arctic ice cover, very vulnerable. Such trend shortens the growth season and keeps the ice from growing to its normal thickness. The trend map in spring shows an even more significant warming in the Arctic Basin than in the winter. The onset of melt occurs during the spring and a warming during the period would lead to an acceleration of melt. The areas with highly positive trends during the period in the basin are the Chukchi/Beaufort Seas region, the North Pole and vicinity, and the Lincoln Sea. It is during this season when the fate of the summer ice is, to a large degree, determined. An early spring season would mean early melt of a generally thinner ice cover, which would make them more vulnerable in the summer. In the summer, positive trends are apparent over land, but trends in the sea ice covered regions are usually smaller because the surface temperature of sea ice during this season is generally stable and close to melt temperatures. In autumn, the anomalies are relatively high and positive in the Chukchi/Beaufort Sea regions. By definition, autumn includes September, which is the month of summer ice minimum. As the sea ice cover in the Arctic basin started to decline during the last few years, its surface temperature has also been increasing, and hence the large trends. The cooling

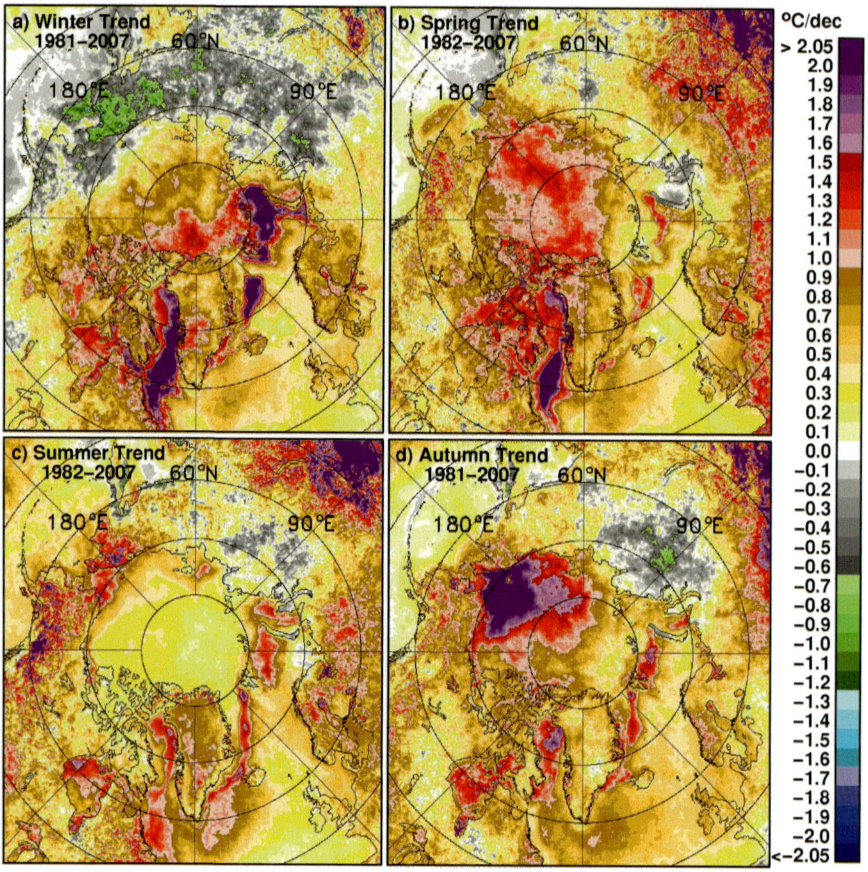

Fig. 10.13 Trends in surface temperatures in the Northern Hemisphere as inferred from AVHRR monthly anomaly data from 1981 to 2007 for (**a**) winter; (**b**) Spring; (**c**) summer and (**d**) Autumn

trend in much of Russia in winter is significantly reduced in spring and summer. In autumn, cooling in a relatively large fraction of Russia is again apparent but the fraction is not as large as in winter. The large temperature anomalies in Baffin Bay in winter and spring are caused primarily by the declining sea ice cover in these regions. The large trends are mainly due to the change in surface types (e.g., ice to liquid ocean) as discussed earlier. Except for the negative trends in Russia, the trends in other areas are dominantly positive. Over land, the changes can be in part due to changes in the vegetation cover and the impact of natural disasters such as flooding and fire.

The seasonal trends in the sea ice cover (Fig. 10.14) show mainly negative trends, reflecting the declining sea ice cover in the Northern Hemisphere shown earlier in Chapter 7. The figure is similar to that presented in Fig. 7.17 but for the period 1981 to 2007 to match the temperature data used in Fig. 10.13. The only region where positive trends are apparent is the Bering Sea in winter and to a lesser degree in spring. This is consistent with the observed cooling in the region in winter, as shown in

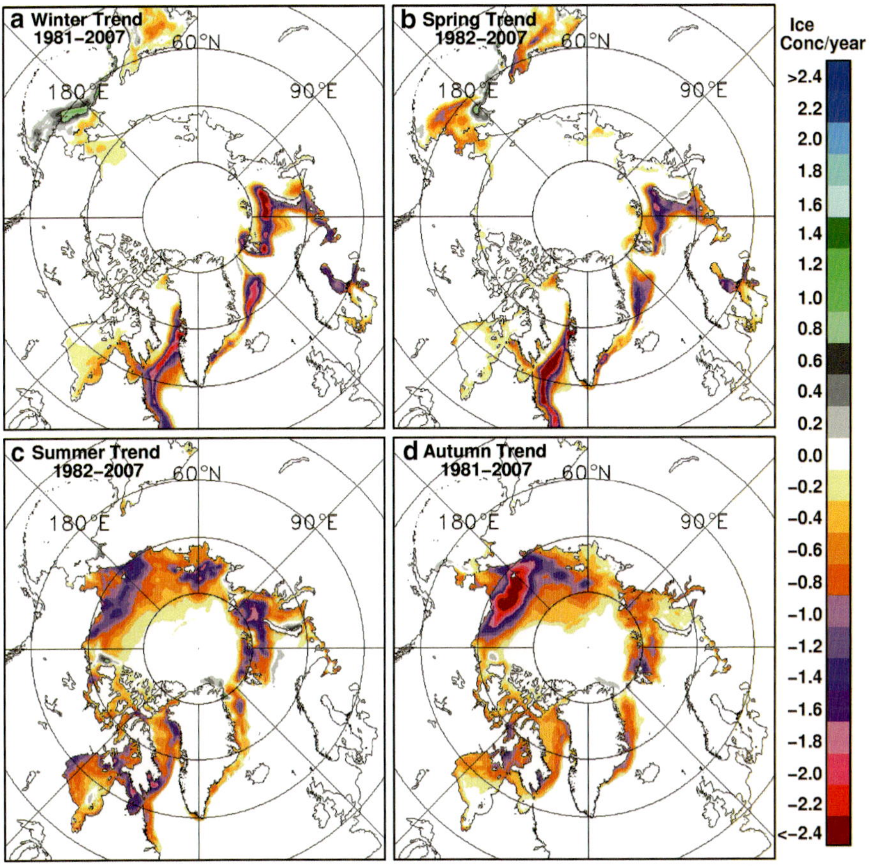

Fig. 10.14 Trends in sea ice concentration in the Northern Hemisphere as inferred from SMMR and SSM/I data from 1981 to 2007 for (**a**) winter; (**b**) Spring; (**c**) summer and (**d**) Autumn

Fig. 10.13a. The large positive trends in ice concentration in the Okhotsk Sea in winter are caused by less extensive ice cover during the latter part of the study period while that for spring are caused by earlier onset of melt in the more recent years. Large declines in sea ice concentrations are also apparent in the Barents Sea, Greenland Sea, and Baffin Bay. In the Barents Sea, the declines are especially large in winter, spring, and summer and are more moderate in autumn. The decline in this region is in part due to the decline in the perennial ice cover in recent years. In autumn, the major trend feature is again the Chukchi/Beaufort Sea region where the biggest declines in the summer ice cover have been observed.

Comparing the trends in Figs. 10.13 and 10.14, it is evident that surface temperature and the sea ice cover are closely linked to each other. The two variables are highly correlated mainly in the seasonal ice region and near the marginal ice zones where most of the interannual changes in the sea ice cover are occurring. The high correlation in these regions is not unexpected since an increase in surface temperature

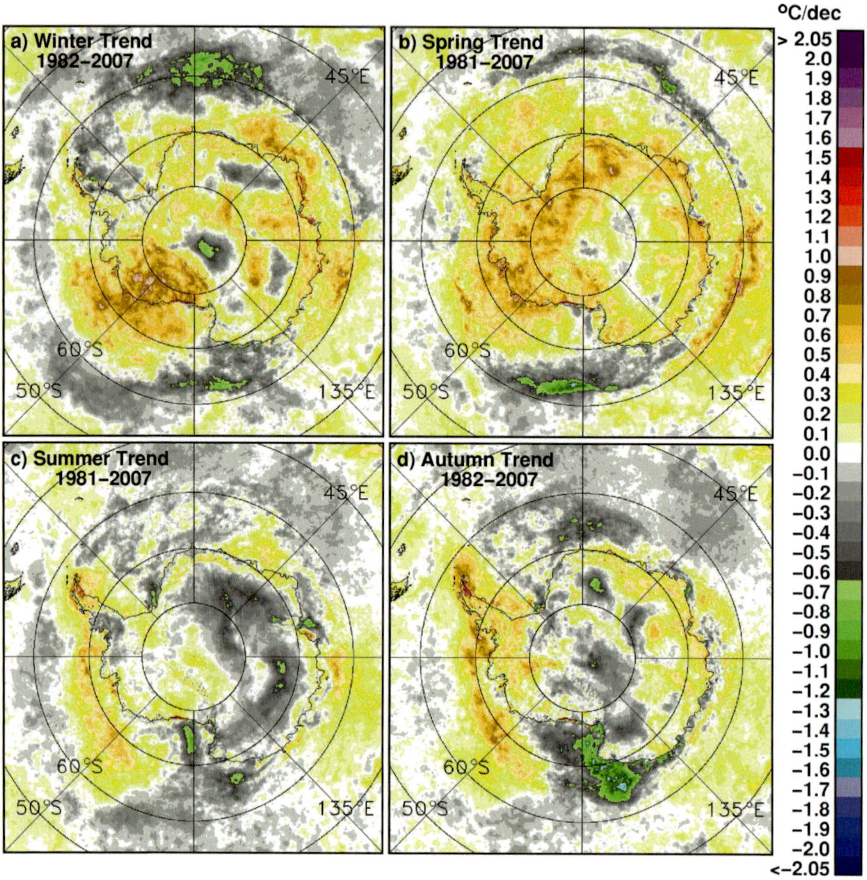

Fig. 10.15 Trends in surface temperature in the Southern Hemisphere as inferred from AVHRR monthly anomaly data from 1981 to 2007 for (**a**) winter; (**b**) spring; (**c**) summer; and (**d**) autumn

would cause a shorter ice season, thinner ice cover and less summer ice cover. Also, less ice concentration, as may be caused by divergence and polynyas would cause higher heat fluxes from ocean to the atmosphere and warmer surface temperatures. Greater fraction of thin ice and open water within the field of view of the sensor also means higher temperature for the surface.

The seasonal trends in surface temperature in the Southern Hemisphere are presented in Fig. 10.15, and it is apparent that the trends are spatially variable and they change substantially from one season to another. The winter trend data show significant warming in the western part of West Antarctica, including the Ross Ice Shelf and the Indian Ocean, and at the same time some cooling around the South Pole, parts of the Antarctic Plateau, and the northern part of the Antarctic Peninsula. In spring, the trends are positive in practically all of Antarctica, the Bellingshausen/Amundsen Seas, the Weddell Sea, and the Indian Ocean, while negative trends are confined mainly to the northern part of the Southern Ocean. In the summer, cooling is evident in much of the Antarctic Plateau, the Ross Sea, and the southern part of the Antarctic Peninsula, while warming is observed over the rest of Antarctica and the Bellingshausen/Amundsen Sea area. In the autumn, cooling is depicted in much of the Ross Sea and the Weddell Sea and also in the Antarctic Plateau and the Ross Ice Shelf, while strong positive trends are apparent in the Antarctic Peninsula, the Bellingshausen and Amundsen Seas, and the northern portion of the Antarctic Plateau.

The seasonal trends in ice concentration in the Souther Hemisphere as presented in Fig. 10.16 are again similar to those presented in Fig. 7.33 but for the period 1981 to 2007 to better match the surface temperature data. As in the Arctic region, the trends in ice concentration are shown to be coherent with those of the surface temperature in that where there was cooling, the average sea ice concentration generally increased and where there was warming, the average sea ice concentration decreased. Some exceptions include the areas of decline in the ice cover north of the Antarctic Peninsula in winter where a significant warming for the same season was not so apparent. The advantage of matching the dates is that the stronger negative trends in the ice concentration in the Western Pacific region in winter and spring correlates better with significant warming in the same region. Improved correlation for the two variables is also apparent with the trends in summer and winter. Although the general characteristics of the trends are relatively similar, these results indicate the importance of using the same periods when comparing trends of different parameters.

10.4 Relationships of Surface Temperature with Sea Ice and Albedo

Surface temperature is a controlling factor in the polar regions because it is the parameter that dictates the timing for the onset and duration of the growth and decay of the sea ice and snow cover and the associated changes in the surface albedo. However, the correlation of changes in the sea ice cover and albedo with changes in surface temperature is not always strong since there are other factors that can be involved. For example, winds can change the ice concentration and albedo of an ice covered area without associated changes in surface temperature. Also, after the formation of consolidated ice cover, further reductions in surface temperature may not lead to an increase in ice concentration or a change in albedo. Furthermore, the sea ice cover is a dynamic entity and the spatial distribution of ice in a certain area may not always reflect that of the surface temperature because of the time lag in the effect of surface temperature on sea ice. It is thus important to consider these factors when making quantitative evaluation about the strength of the relationship between these variables.

The direct relationship between sea ice concentration and surface temperature are studied using regression and correlation analysis using these two variables on a pixel by pixel basis, and the results are expressed in terms of correlation coefficients as presented in Fig. 10.17a. It is apparent that the correlation is very high (−0.8 to −0.9) in the seasonal sea ice regions where sea ice concentration is readily and directly affected by changes in the surface temperature and atmospheric conditions. The correlation coefficients are usually highest in the middle of the seasonal

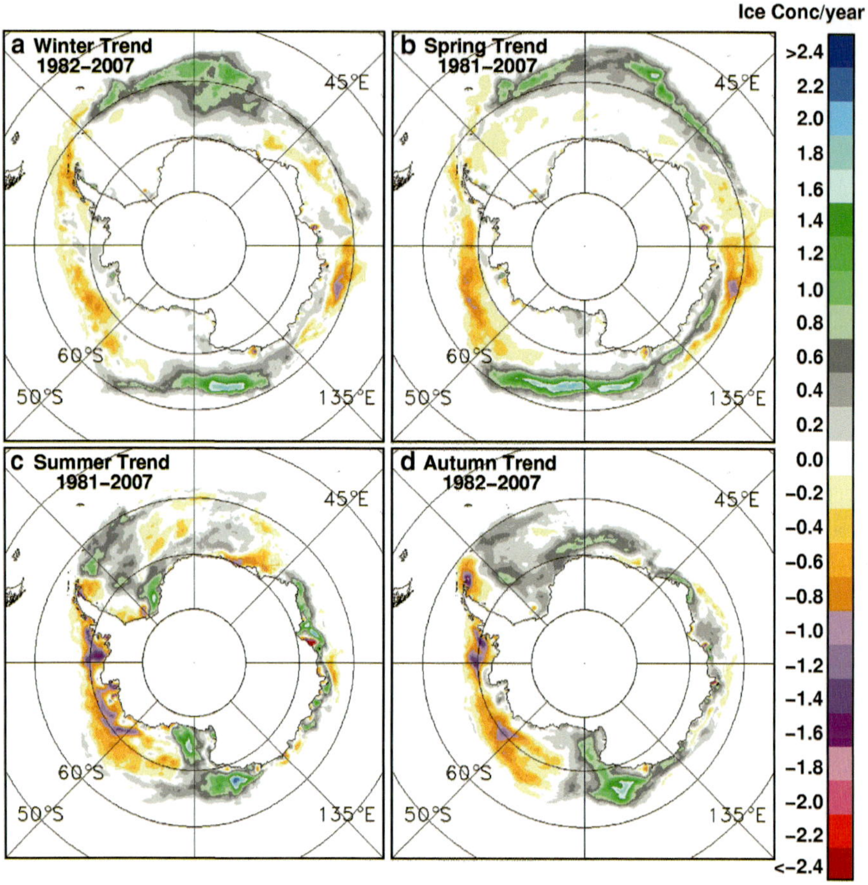

Fig. 10.16 Seasonal trends in ice concentration in the Southern Hemisphere as inferred from SMMR and SMM/I monthly anomaly data from 1981 to 2007 for (**a**) winter; (**b**) spring; (**c**) summer; and (**d**) spring

ice pack because these are the areas where the ice cover fluctuates the most during the season. The correlation is not any higher than shown because there are periods when the region is ice free and although the surface temperature changes during the period, there is no corresponding change in the concentration of the sea ice cover. In the Central Arctic region, the correlation is not so good, and in some places, it even becomes positive. The correlation is especially poor in the perennial ice regions because the ice concentrations are relatively high and do not change much during the year while the surface temperature fluctuates quite substantially. The regions of positive values are the regions that are relatively cold during the summer, and the correlation becomes positive when the surface temperature goes up to melt temperature and the emissivity of the snow surface goes up to almost the value of a blackbody. These results are consistent with those of the comparative analysis of trends presented in Sect. 10.3.

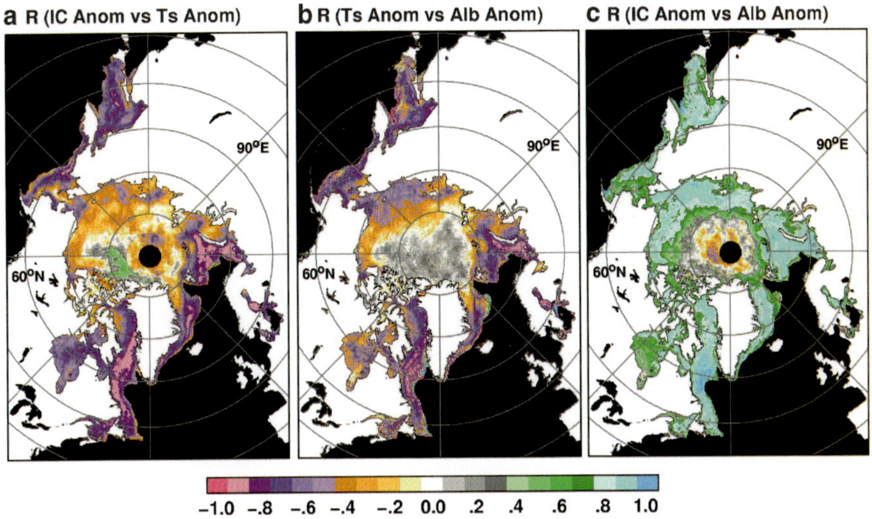

Fig. 10.17 Correlation coefficients derived from regression analysis of (**a**) ice concentration versus surface temperature; (**b**) surface temperature versus albedo; and (**c**) ice concentration versus albedo in the Northern Hemisphere for the period 1998 to 2008

The relationship between albedo and surface temperature is expected to be similar to those between sea ice and surface temperature. A similar analysis is presented in Fig. 10.17b and indeed, this is generally the case. The albedo of the surface varies in much the same way with surface temperature as sea ice. There are discrepancies, in part because the albedo data set is not a year-round data set; hence, there are data gaps during the winter period. The exclusion of the winter data actually made the correlation higher in many places in the albedo vs. temperature map when compared with the ice vs. temperature map. Also, in the Central Arctic, the correlation coefficients are more uniform and are either slightly positive or almost zero.

Albedo and sea ice concentration provide very similar information in sea ice covered regions. In particular, when the ice concentration is low, the albedo is usually low, and when the ice concentration is high, the albedo is usually high. The microwave emissivity of sea ice during ice growth also changes in much the same way as the albedo. Fig. 10.11c shows the correlation of albedo and sea ice on a pixel-by-pixel basis in the Northern Hemisphere, and it is apparent that in the seasonal ice region, the correlation coefficients are relatively high and positive. It is in the seasonal region where the ice concentration and albedo fluctuate considerably, and it is apparent that the correlation coefficients are consistently high in these region. The highest correlations (light blue) usually occur near the marginal ice zone where the ice cover changes constantly in response to changes in wind and surface temperatures. In the central Arctic region, the correlations are close to zero and at times positive. The positives is caused by the narrow range in values of both

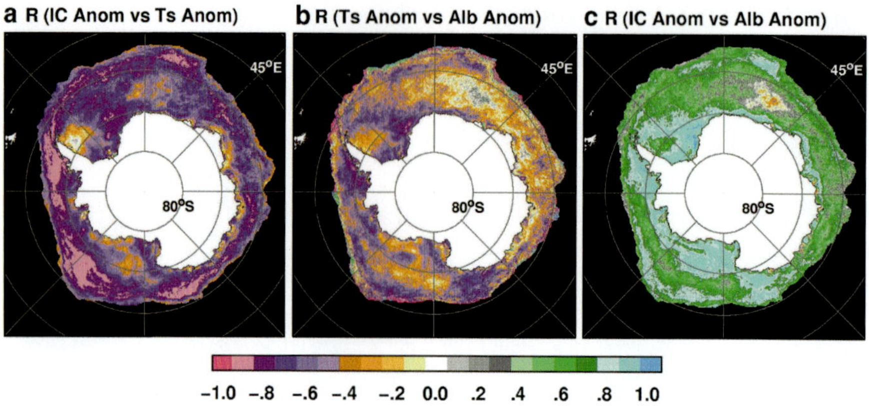

Fig. 10.18 Correlation coefficients of (**a**) ice concentration versus surface temperature; (**b**) surface temperature versus albedo; and (**c**) ice concentration versus albedo in the Southern Hemisphere

albedo and sea ice concentration in the region with the sea ice cover almost always consolidated throughout the year. The pattern of the area where the variables are weakly correlated or negatively correlated in the central Arctic is different from those of the other maps. This is in part because the surface temperature patterns are not necessarily similar to those of sea ice and albedo.

The corresponding correlation maps in the Southern Hemisphere are presented in Fig. 10.18. It is apparent that ice concentration is highly correlated with surface temperature in much of the sea ice covered region (Fig. 10.18a). Low correlations are found usually in the perennial ice regions in the Western Weddell Sea and in the Bellingshausen/Amundsen/Ross Seas regions. Since these are areas where the ice concentration is near 100% most of the time, the correlation is expected to be weak as in the Central Arctic. The surface albedo is shown to be highly correlated with surface temperature in most places, but there are large areas where the correlation is relatively weak or even positive (Fig. 10.18b). These are likely areas where the albedo do not change much because it is consistently a 100% ice covered area or an ice free ocean area during the year. The results of the correlation analysis of sea ice concentration and albedo as presented in Fig. 10.18c are similar to those in Fig. 10.18b but with opposite sign. The regions where the correlation is weak or negative are again likely in where the small changes in ice concentration do not affect the albedo of the same area.

The results of our correlation analysis of surface temperature, ice concentration and albedo indicate good correlation between each pair of variables in most places but not in all places. The results provide a quantitative assessment of where the correlations are expected to be strong and where they are expected to be weak. Such results are useful in the interpretation of the differences in the trends in surface temperature, ice concentration and albedo as presented in Figs. 10.1–10.3 in the Northern Hemisphere and Figs. 10.7–10.9 in the Southern Hemisphere. The results also provide the means to interconnect more effectively and interpret accurately the trend results obtained independently for the different parameters.

10.5 Trends in Chlorophyll Concentration and Primary Productivity and Relations with Sea Surface Temperature and Sea Ice

With a warming Arctic and a retreating sea ice cover, it would be important to know if the observed changes affect the distribution of phytoplankton blooms and the primary productivity of the region. It would also be interesting to know if the unpredictable nature of the Antarctic surface temperature and the sea ice cover is reflected in the productivity of the region. Chlorophyll pigment concentrations in the polar regions are expected to be affected by many variables including the presence of meltwater from sea ice, sea surface temperature, clouds, wind, nutrients, and iron. Some correlation studies have been done between ocean color and these variables using temporally and spatially sparse Nimbus-7/CZCS data in both Northern (Muller et al. 1990; Mitchell et al. 1991) and Southern (Smith et al. 1988; Comiso et al. 1993; Sullivan et al. 1993) Hemispheres. The correlations between variables were found to be generally weak, but results were inconclusive because of too many gaps in the CZCS data. The SeaWiFS data record provide excellent and continuous information but unfortunately is not long enough to be effectively used for evaluating decadal changes. However, it is by far the longest of such global data and is available during a period when rapid changes are occurring in the polar regions. To provide a sense of direction in light of the longer term changes in other parameters, we will present some results from trend and correlation studies using these much more complete and comprehensive data.

In Fig. 10.19, we present trends in the chlorophyll pigment concentration, productivity, temperature and ice concentration for the period from January 1998 to December 2008. We choose this time interval to be able to capture the peaks of plankton blooms during the spring and summer periods of the same year. As in the previous sections, the same time periods are used for all 4 variables to facilitate interpretation of the data. Except for a few regions such as the Bering Sea and North Atlantic Ocean, the trends in phytoplankton concentration and primary productivity, as indicated in Fig. 10.19a and b, are mainly positive showing significant increases in the Central Arctic, Barents Sea and Hudson Bay. The trends in productivity in the Okhotsk Sea, North Pacific Ocean, Baffin Bay, and Greenland Sea are also relatively high as depicted in Fig. 10.19b. Except for Hudson Bay, these areas of increasing plankton concentrations and primary productivity are also where the trends in surface temperatures are relatively high, as indicated in Fig. 10.19c. Slight cooling is also apparent in the Bering Sea and North Atlantic where negative trends in plankton concentration and productivity is observed. During the same period, the trends in ice concentration, as indicated in Fig. 10.19d) are negative in areas where the trends in plankton concentration and primary productivity are positive, except at the Hudson Bay. The Hudson Bay is an exception likely because it is located inland with many other factors contributing such as increases in nutrient and iron availability from river runoff. It is apparent from the set of images that some relationships between the variables exists and that the trends in ocean color and productivity may be related to the trends in surface temperature and ice concentration.

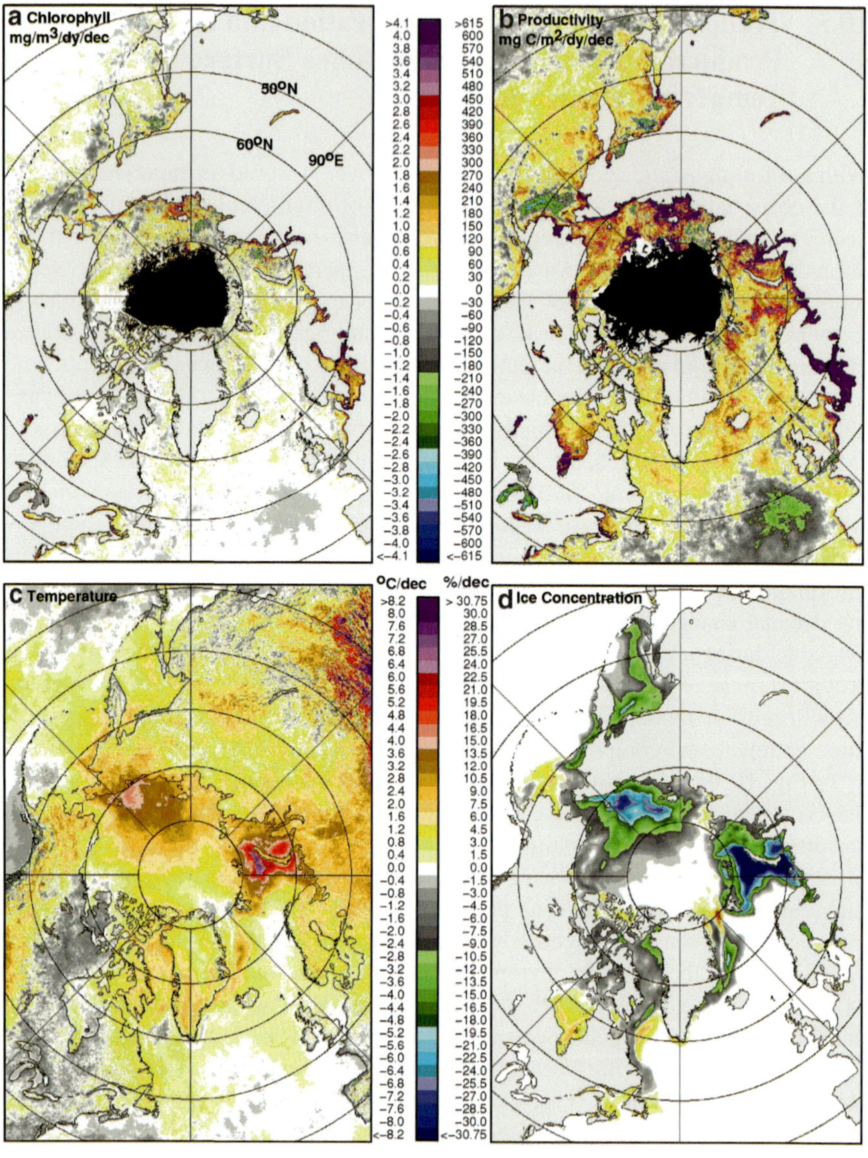

Fig. 10.19 Trends in (**a**) Chlorophyll concentration; (**b**) primary productivity; (**c**) surface temperature; and (**d**) sea ice concentration in the Northern Hemisphere for the period 1998 to 2008

Correlation analysis was done on a pixel-by-pixel basis for surface temperature vs. chlorophyll concentration and the results are shown in Fig. 10.20. The analysis was done for all months of the year (Fig. 10.20a), for spring months only (Fig. 10.20b) and for summer months only (Fig. 10.20c). The seasons were slightly modified to include June in spring and September in summer. The correlation map

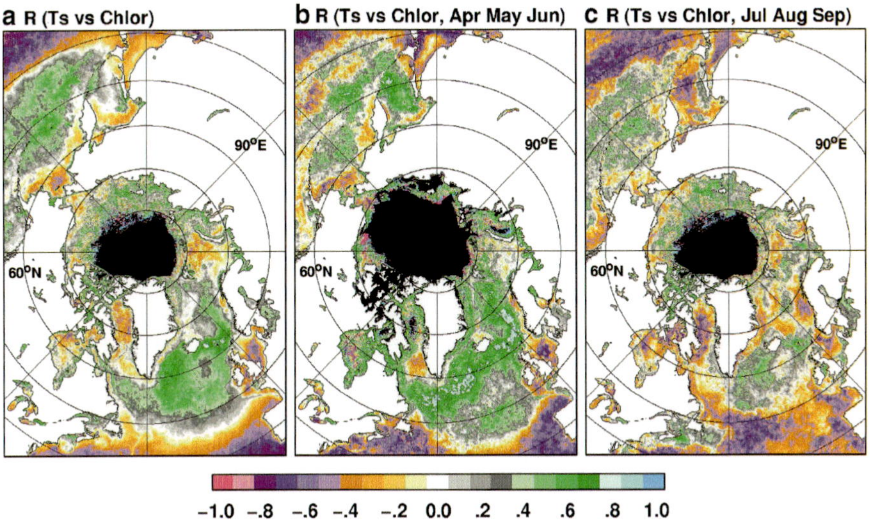

Fig. 10.20 Correlation coefficients from regression analysis between (**a**) surface temperature versus chlorophyll during the entire year; (**b**) surface temperature versus chlorophyll concentration during spring (April, May and June); and surface temperature versus chlorophyll concentration during summer (July, August and September)

in Fig. 10.20a shows large areas of high and positive correlation coefficients (greens and blues) north of 50°N in both Pacific and Atlantic Oceans and areas of high, but negative, correlation coefficients to the south of 50°N. The high and positive correlation at subpolar regions, where surface temperatures are relatively cold, indicates that planktons have preferences for such types of temperatures. The high and negative correlations in the southern regions, where temperatures are relatively warm, suggests that planktons are not acclimated to such temperatures. In the seasonal regions including Bering Sea, part of Okhotsk Sea, Hudson Bay, Baffin Bay and Barents Sea, the correlations are also generally negative but generally weak. To gain insight into this, we present in Fig. 10.20b a similar analysis but for the spring months only. It is apparent in this case that a lot more areas in or near the seasonal ice region have high and positive correlation coefficients. Spring is the time of the most rapid melt of sea ice in the region and the positive influence of meltwater on phytoplankton growth, as discussed in Chap. 9, is likely a factor. The exceptions are some negative correlations at the Bering Sea and southern Baffin Bay, which may be associated with increasing sea ice cover in the region as indicated in Fig. 10.19d. In the summer, the results of similar correlation analysis, as presented in Fig. 10.20c, show patterns similar to those of Fig. 10.20a. During this period, the values are positive in approximately the same regions, but the magnitudes are not as high. The areas of negative correlations have advanced to the north during this period affirming the negative influence of warmer temperatures.

In Fig. 10.21, we present trends in the chlorophyll pigment concentration, productivity, temperature and ice concentration in the Southern Ocean for the period

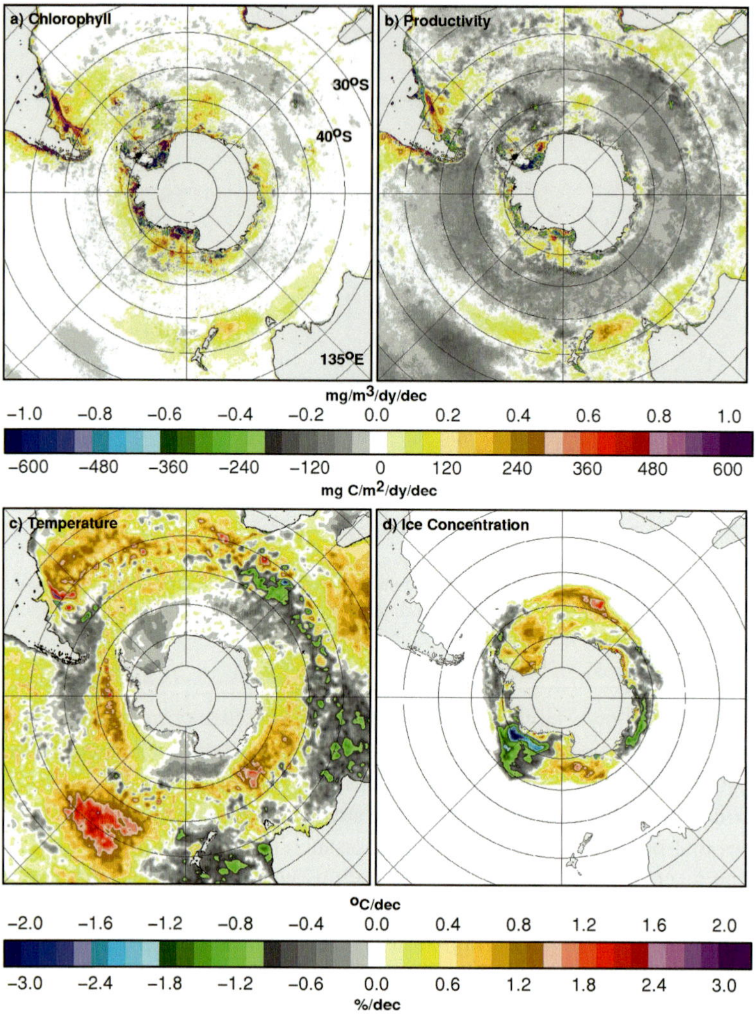

Fig. 10.21 Trends in (**a**) chlorophyll pigment concentration; (**b**) primary productivity; (**c**) surface temperature and (**d**) sea ice concentration for the period from August 1998 to July 2008

from August 1997 to July 2008. As in the previous sections, the same time periods are used for all 4 variables to facilitate interpretation of the data. The time interval is also adjusted such that the peaks of plankton blooms which normally occur between November of one year to February the following year are captured. The trends in pigment concentrations in the Southern Ocean (Fig. 10.21a) are not as high as those in the Arctic, and hence the change in scale, to better illustrate the distribution. The trends are highest near the continental shelf of Argentina and some Antarctic coastal areas and moderately high around Antarctica, New Zealand and off the coast of Chile. Changes along the ring of high plankton concentrations at around

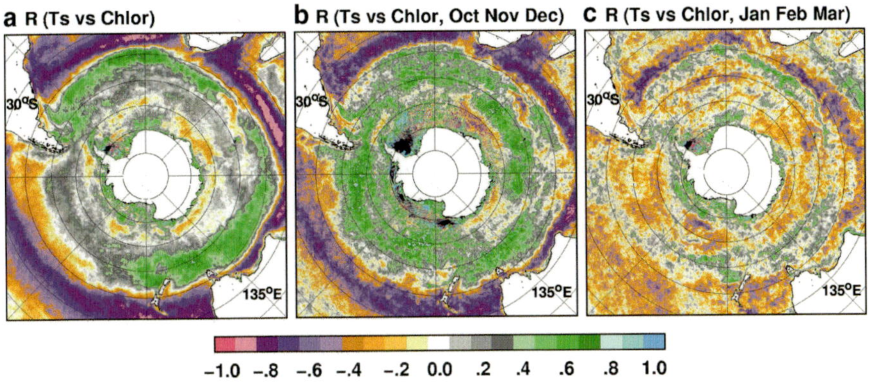

Fig. 10.22 Correlation coefficients as derived from linear regression of (**a**) surface temperature versus chlorophyll; (**b**) surface temperature versus chlorophyll in spring; and (**c**) surface temperature versus chlorophyll in the summer

40°S, as described in Chap. 9, are not apparent in this map. The trends in productivity (Fig. 10.21b) show patterns similar to those in Fig. 10.21a, but between 40°S and 60°S, a halo of slightly negative trends is apparent. Positive trends in primary production are shown to occur only in limited areas primarily north of 40°S including the coastal areas of Argentina and Chile, and around the Antarctic coastal regions. The trends in surface temperature and ice concentration are shown in Fig. 10.21c and d, respectively. These two images show that ice concentration increased in areas where there is cooling as in the Ross Sea and the Weddell Sea while reductions in ice concentration is apparent in the Bellingshausen/Amundsen Seas and Indian and West Pacific Oceans where there is warming. The relationships of these variables with the trends in plankton concentration and primary productivity is, however, not so apparent.

The relationship of surface temperature with plankton concentration is better quantified in the correlation maps presented in Fig. 10.22. Using monthly data throughout the year, a strong positive relationship between surface temperature and plankton concentration is apparent along the ring of high plankton concentration data near 40°S. The high positive correlation area is broader than other areas in the West Pacific and Ross Sea region, while the correlations in the region north of the Bellingshausen and Amundsen Seas are not very strong. It is notable that a similar ring of relatively high correlation values, but with opposite sign, is depicted immediately to the north of the region. This ring is located north of the polar front and it is possible that nutrients (and iron) are not as readily available in the region as at 40°S. Closer to the continent, near 60°S, there is an area of positive but relatively weak correlation basically along the winter ice edge. This is an effect similar that that observed in the Bering Sea and Southern Baffin Bay in the Northern Hemisphere. The reason for such effects is not obvious and requires further investigation. Correlation results in the region can be difficult to interpret because the ocean color algorithm is not valid where there is detectable (10% or higher) ice cover. These may be areas of high productivity because of the abundance of meltwater in the region and yet they are rejected and are not included in the analysis because of the presence of sea ice. In Fig. 10.22b, similar correlation analysis was done during the October to December period when the ice is retreating rapidly. During this period, the high positive correlation data extends all the way to the ice edge, suggesting that the melt of sea ice during the period may indeed make a difference. Also, the correlation coefficients are the highest (close to 1) along the coastal areas where spring polynyas usually form again showing the impact of meltwater on phytoplankton growth. We also show in Fig. 10.22c the results of the correlation analysis for the January to March period. During this time period, negative correlations are more dominant for the entire region and the correlations are not so strong. Positive correlations are basically confined at this time to the area around the Antarctic continent and near the halo at 40°S.

10.6 The Rapidly Shrinking Arctic Multiyear Ice Cover

Among the most visible manifestation of a changing Arctic has been the decline in the perennial ice cover. The rapid decline of about 9% per decade first reported in 2002 (Comiso 2002) was followed by record low values in 2002, 2005, and 2007, with very little recovery in between, and the rate of decline for ice area is now currently at about 12.5% per decade, including 2008 data. The yearly values for both maximum and minimum for ice extent and ice area are presented in Fig. 7.13. The maximum ice cover data provide the means to assess how extensive the ice cover is in winter and how far south the ice edge gets. The minimum ice cover data, on the other hand, reveals how much ice is left in the Arctic before an ice season begins. Both parameters are important climate variables, but it is apparent that one is changing a lot more than the other. The maximum ice extent and ice area are declining at the rate of -2.2 ± 0.4%/decade and -2.5 ± 0.4%/decade, respectively, and in the past 5 years, successive drops for 3 years were followed by 2 years of recovery. The minimum extent and area, on the other hand, show declines that are about 5 times faster at -11.3 ± 1.7 and 12.5 ± 1.7%/decade, respectively. The decline in 2007 was quite drastic and was about 37% lower than average value, and following a recovery in the winter ice cover in 2007 and 2008, there was a recovery in the minimum ice in 2008 as well. These results have been widely reported in the literature (e.g., Comiso et al. 2008; Johannessen 2008; Walsh 2008) and in newspapers and televisions around the world, but for completeness, we present updates and additional insights into the phenomenon.

The minimum ice cover has been used to represent the perennial ice cover but the two are not exactly identical. The perennial ice is usually defined as ice that survives the summer. The Arctic ice cover gradually declines to its minimum value during the summer, but at the time the minimum value is reached, some freeze-up at some high latitude areas has already occurred. The quantified ice minimum extent and area thus actually include a small fraction of new ice. To gain insight into the how the actual perennial ice cover compares with the ice cover during minimum extent, the entire Arctic ice cover was divided into small sectors and the minimum ice extent was estimated in each sector. The minimum extents combined together provides a better estimate of the perennial ice cover and studies show that the actual perennial ice is slightly lower by about 3% (Comiso 2002). However, trend analysis indicates that the trends for minimum and perennial ice cover are very similar (Comiso 2002; Zwally and Gloersen 2008).

Ten year averages of sea ice concentrations during minimum extent from 1979 to 2008 are presented in Figs. 10.23a, 10.23c, and 10.23e while the minimum ice cover in 2006, 2007, and 2008 are presented in Fig. 10.23b, 10.23d, and 10.23f, with a view of summarizing the decline in the perennial ice cover. It is apparent that from one decade to the next decade there is substantial decline, especially from the second to the third decade of data. The black hole in the middle is again part of the limitation of the passive microwave system, which does not cover the high latitude area completely because of relatively narrow swath width. The data for the individual years in 2006, 2007 and 2008 show an even less extensive ice cover than the 10-year averages. The drastic decline from 2006 to 2007 is apparent, and the extent of the ice cover in 2008 is also shown to be comparable to that of 2007. What is remarkable is the massive difference in the area of exposed open water in the last decade and especially in 2007 and 2008 compared to that of the first decade. The Arctic ocean is certainly exposed to a lot more solar heating in the last decade than in previous decade.

The satellite perennial ice data provide only a 2-D characterization of the sea ice cover. Thickness measurements of the sea ice cover have been limited but some progress has been made as indicated in Chapter 7. The change in thickness from one year to another in the last 5 years using ICESat freeboard data has been shown to be large (Kwok et al. 2009). To gain a more long term assessment of the trend in thickness, we use the same passive microwave data to assess how the thicker portions of the perennial ice cover have been changing. A significant percentage of the perennial ice cover as quantified at the end of the summer, consists of second year ice, which has been shown to have signatures closer to that of first year ice. It has been observed that the signature of multiyear ice or ice that has survived at least two summers is different from that of first year ice (Vant et al. 1974), as discussed in Chap. 4. We take advantage of such difference to quantify the distribution of the multiyear ice cover for each year from 1979 to 2008. The multiyear ice fraction is most accurately estimated in the middle of winter when the surface is dry and cold. We choose February of every year and make use of a mixing algorithm to obtain an estimate of monthly multiyear ice concentration for each year. The results are summarized in the color-coded maps shown in Fig. 10.24. It is apparent from the 10-year averages that the multiyear ice fraction has been declining quite rapidly. As expected, the multiyear ice fractions are less extensive than the perennial ice since they represent the thicker ice types and exclude primarily the thinner second year ice types. The result indicate an even greater decline of these generally thicker ice types than that of the perennial ice. The multiyear ice maps for the individual years in 2007, 2008 and 2009 indicate that the extent of the multiyear ice cover has been reduced substantially compared to the 10-year averages. This means an acceleration in the decline in the multiyear ice even in the last ten years alone and a generally thinner ice cover.

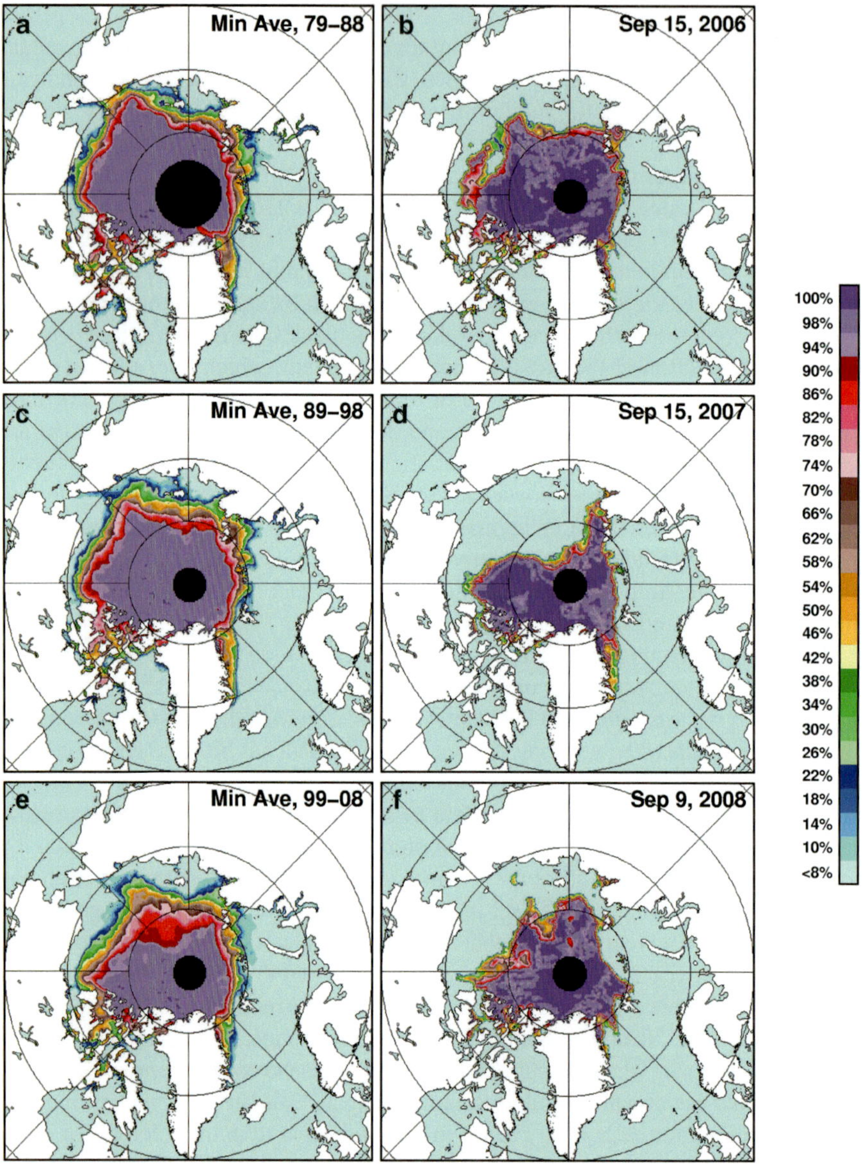

Fig. 10.23 Ten year averages of minimum ice cover for (**a**) 1979 to 1988; (**b**) 1989 to 1998; and (**c**) 1999 to 2008 in the Northern Hemisphere. Daily ice concentration maps during ice minimum ice extent in (**d**) 15 September 2007; (**e**) 15 September 2007; and (**f**) 9 September 2008. The ice concentration gradient is smeared at the ice edges with the ten-year averages and the daily ice maps provide better representation of the distribution of sea ice in the region

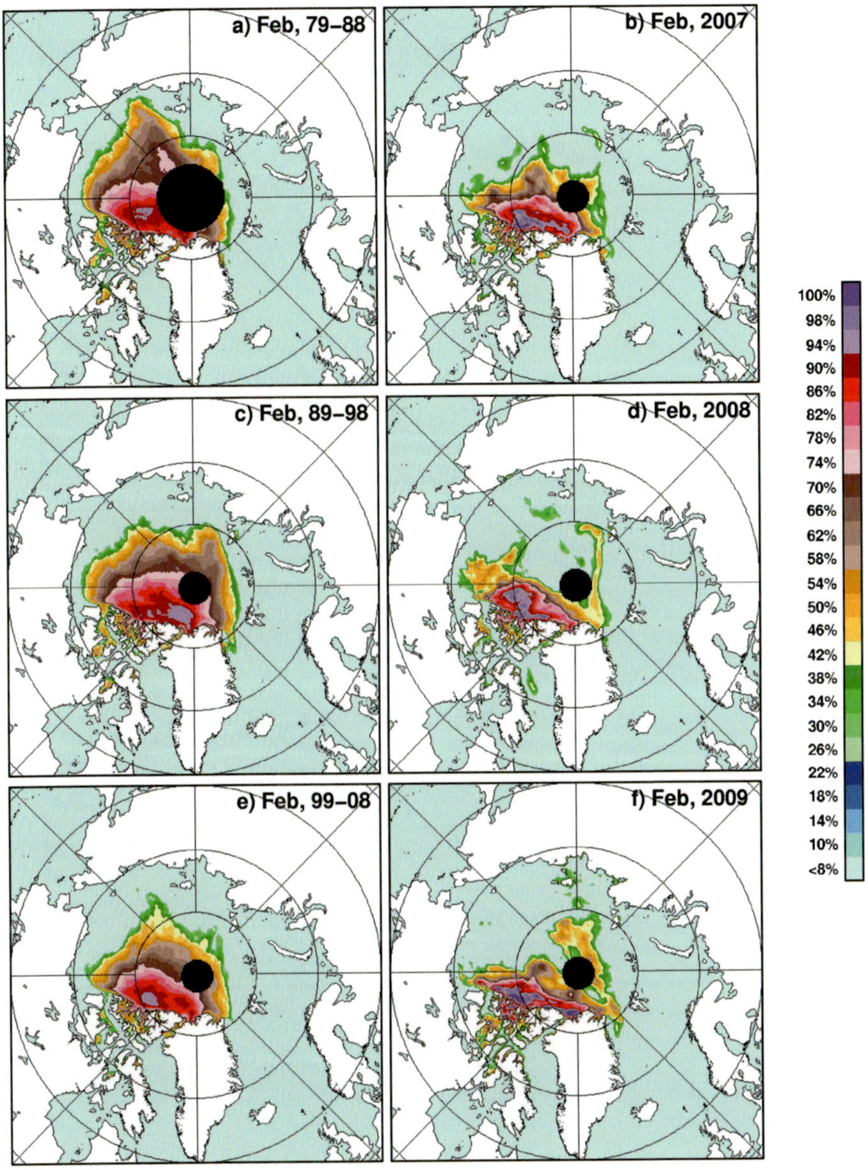

Fig. 10.24 Ten-year averages of multiyear ice concentration in (**a**) Feb 1979-1988; (**b**) Feb 1989 to 1998; and (**c**) Feb 1999 to 2008. Yearly averages in February for (**d**) 2007; (**e**) 2008; and (**f**) 2009 are also shown

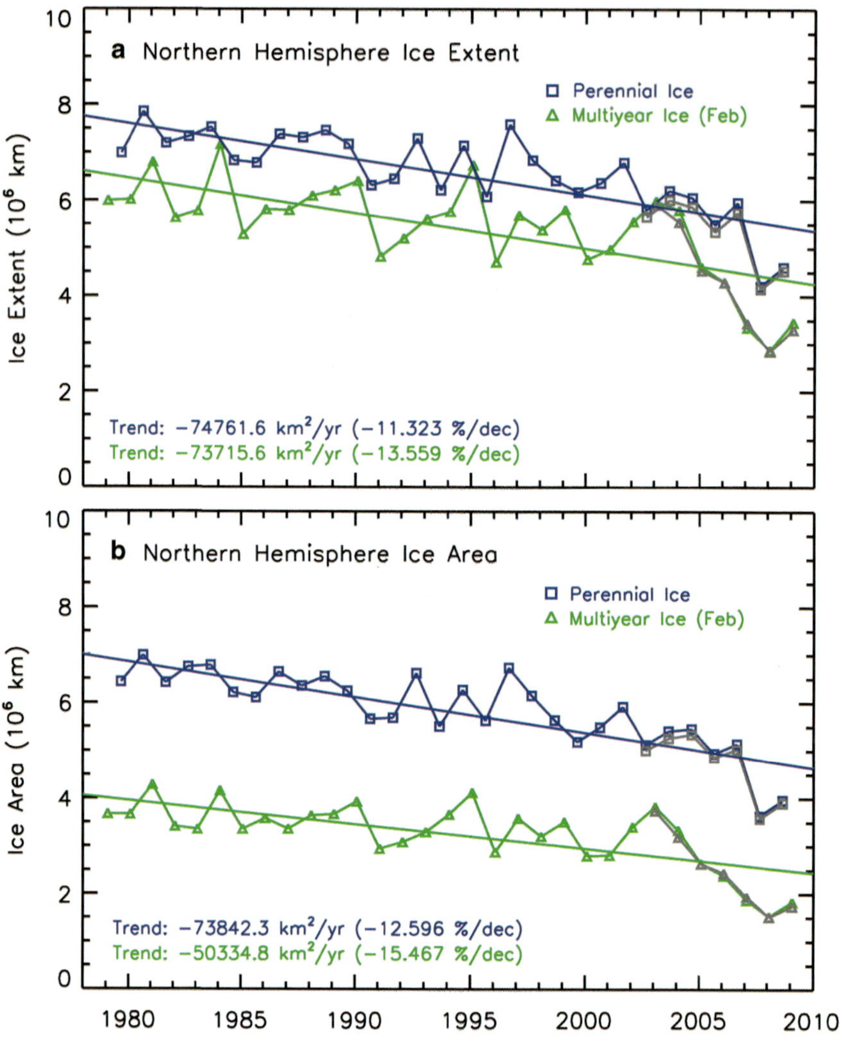

Fig. 10.25 Comparison of the yearly perennial ice cover and the yearly multiyear ice cover as derived during the month of February using SMMR and SSM/I data. Results using AMSR-E data are also shown (in gray)

More quantitative evaluation and comparison of the changes in the perennial and multiyear cover are presented in the plots of ice extent and ice area shown in Fig. 10.25a and 10.25b, respectively. Although both show a recovery in 2008 for the perennial ice and 2009 for the multiyear ice, the multiyear ice area is shown to be declining consistently during the last 7 years. The results of regression analyzes indicate that the decline rate for the perennial ice cover is 12.6% per decade as indicated earlier while that for the multiyear ice is 15.5% per decade. The significantly higher decline of the multiyear ice cover compared to that of the perennial ice suggests that the thicker ice types are declining more rapidly than the perennial ice. This also means that the volume of the perennial ice cover is declining even more rapidly than what would be expected if the average thickness is assumed to be constant. This result also suggests that the mixed layer of the Arctic Ocean may be warming up. A manifestation of such warming is the relatively low extent of the 2008 perennial ice and the 2009 multiyear ice areas, which is the second lowest during the satellite era despite significant cooling in the winter periods of 2008 and 2009.

10.7 Increasing Antarctic Sea Ice Cover and Connections with the Ozone Hole

The trends in the sea ice cover in the Southern Hemisphere are quite different from those of the Northern Hemisphere as discussed in Chap. 7. They are, in fact, going in the opposite direction. In light of the observed global warming, this phenomenon appears counter intuitive. The yearly maximum and minimum extents and areas presented in Fig. 7.29 provide a summary of variability and trend in the ice cover in the Southern Hemisphere. Large interannual variability is apparent with the minimum ice cover more variable than the maximum ice cover. Regression analysis of the data reveals that the ice extents and ice areas during maxima are increasing at the rate of 0.7 ± 0.4%/decade and 1.2 ± 0.4%/decade, respectively. The corresponding values during minima are 2.1 ± 2.6%/decade and 3.1 ± 2.9%/decade. During ice minima, the percentage rate of increase is higher than that of the rate during ice maxima, but the yearly values are small, and because of the large interannual variability, the statistical errors are bigger. It is interesting to note that while the maximum extents and ice area declined in the last 2 years, the minimum ice extents and ice area increased in the same period. Also notable is the large fluctuation of the minimum ice extents and ice areas during the last 5 years.

New insights into the increasing ice cover in the Antarctic has emerged in recent years. A modeling study by Turner et al. (2009) reveals that the ozone hole may have a significant influence on the trend in the ice cover. Numerical analysis using a global circulation model indicates that the ozone hole in 2000 caused a deepening of the lows at 500 hPa geophotential height when compared with pre-ozone hole conditions (Fig. 10.26). The net result is stronger winds, as shown in Fig. 10.27, off the coast at the Ross ice shelf, which has been identified as an ice factory because of frequent occurrences of coastal polynyas in the region as described in Chapter 8. The stronger winds serve to facilitate the formation of such latent heat polynyas and therefore the formation of more ice. This is consistent with the relatively high rate of increase of ice in the Ross Sea sector as discussed in Chap. 7.

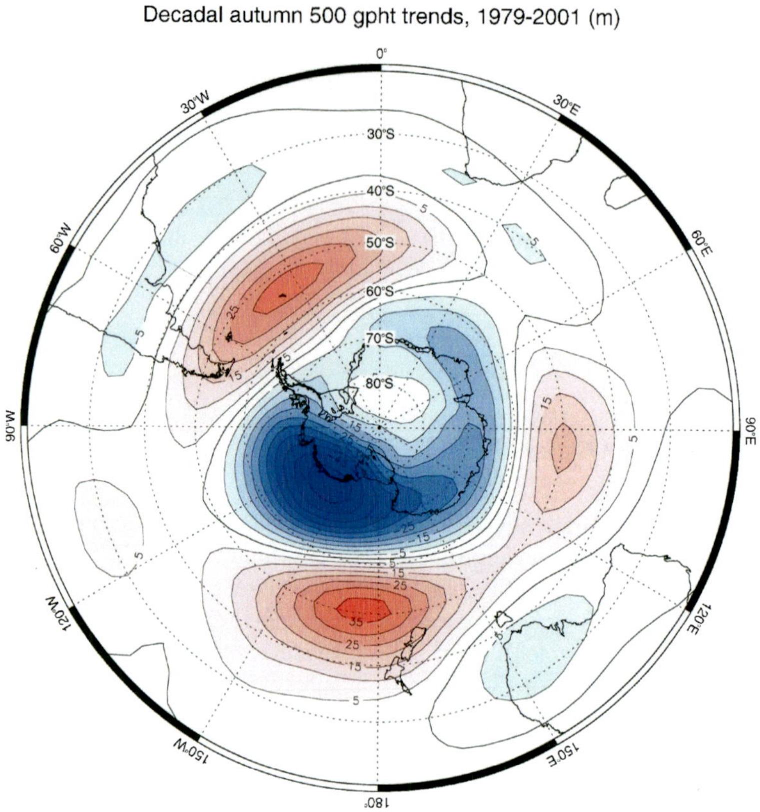

Fig. 10.26 The difference in 500 hPa height (m) between the pre-industrial levels of stratospheric ozone and ozone levels in 2000 as derived from 50 year runs of the HadAM3 model. From Turner et al. (2009) with permission from the American Geophysical Union

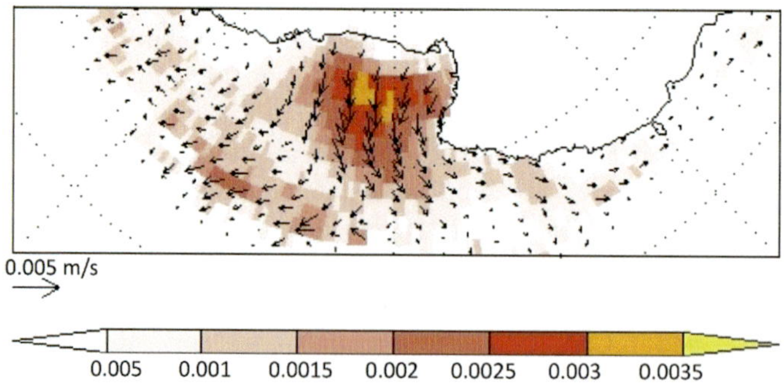

Fig. 10.27 Stronger winds due to deepening of the lows in the West Antarctic region as shown in Fig. 10.26. Courtesy of J. Turner of the British Antarctic Survey, 2009

10.8 Record Length and Trend Analysis Issues

An important issue which needs to be considered when using the satellite data set is the length of the data record and how long it has to be before meaningful trend analysis of the data can be made. If the data record consists of only a few years of data it would not be so useful for this type of analysis because the trend would fluctuate considerably with yearly increases in record length. One way to gain insight into the problem is to evaluate the fluctuation characteristics of the trend as the record length of the data changes, following the study made by Comiso (2000). In particular, we use the relatively long global averages of in situ surface temperature for the period 1900 to 2008 for this analysis. A plot of monthly anomalies of global surface temperature data from 1900 to 2008 from Jones et al. (1999) is presented in Fig. 10.28a and as with the satellite data, the interannual fluctuation is very large. A regression fit to the data yielded a trend for global surface temperature of 0.073 ± 0.001 °C/dec for the 109-year period which is consistent with previous reports (IPCC 2007). This data set is especially interesting because it shows a general increase in surface temperature from about 1920 to 1945, a slight decrease in temperature from 1945 to 1950 followed by a stable period from 1950 to 1975 and a subsequent general warming up to 1998. The temperature for period from 1998 to 2008 has been relatively uniform and the plot shows indications of a downward trend. Some of the periodic interannual fluctuations are more apparent in the 3 year running averages which are the data points connected by a gray line. If different segments of this record are separately taken for trend analysis, different trends would come out. A plot of how the trend changes as the record length is decreased from 108 years to 3 years is presented in Figure 10.28b. In particular, the first point in the plot is the trend from 1900 to 2008, the second point is from 1901 to 2008 and so on. It is apparent that the trend is basically constant up to 1940, increasing gradually from 1940 up to 1993 and then decreased considerably and reaching a negative value in 1998. After that, the trend increased significantly for a year and then declined considerably down to a very negative value in 2005 followed by a small recovery at the last data point which represents the trend from 2006 to 2008. The large fluctuation in the trend from those that starts from 1993 to those that starts in 2006 is obviously unacceptable especially when the trend changed from positive to negative and then to positive in just a period of a few years. The analysis also indicates that the trend is relatively stable from 1960 to 1992 suggesting that maybe 17 to 20 years of data would be needed to have trend results that do not change much with the addition of another year to the data record. Note that the observed decline in surface temperature from 1945 to 1950 is not reflected in the analysis because the trend is calculated with the data in 2008 as the anchor. If the analysis was done with 1900 as the starting point, a similar plot would have shown a slight decline in the trend after 1945. Nevertheless, the trend would have been relatively stable as well after the record length of the data is about 20 years.

The nearly constant value for the trend that starts in 1900 to that which starts in 1940 suggests that at least 70 years of data may be required to obtain a trend that is related to climate change. One can argue that the longer the data record the higher

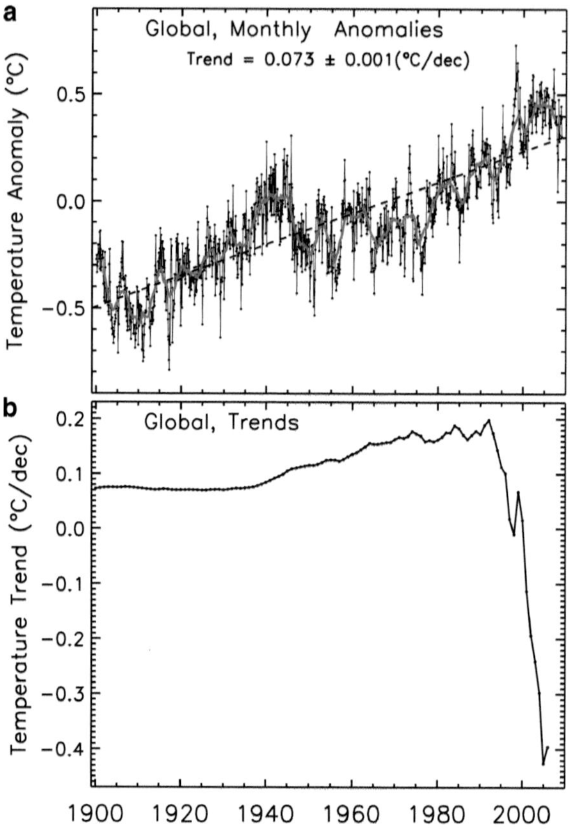

Fig. 10.28 Plots of (**a**) global monthly surface temperature anomalies using Jones et al. (1999) data set from 1900 to 2008; and (**b**) trends in global surface temperature from varying record lengths from 109 years to 3 years

the confidence that the trend is associated with long term changes. However, it is not known if the trend would remain constant if the record length was much longer. The climate system goes through several phases and one can also argue that there may have been a change in phase from the 1940s to the 1970s especially since the trends were also relatively uniform for 20 years starting with the 40-year trend from 1969 to 2008 to the 19-year trend from 1990 to 2008. A basis for the change in phase would be the continued increase in greenhouse gases that would accelerate the warming. The record length of the satellite data which currently span about 30 years falls within the 19 to 40 year data record length and should provide some meaningful results. The more significant the trend is, the more meaningful it becomes. For example, the 30 years of satellite record is probably good enough to establish that the perennial ice cover is undergoing a significant decline. The data record will of course increase in length with time, assuming that the acquisition of satellite data continues.

10.9 Summary and Conclusions

With almost three decades of continuous and synoptic coverage of the polar regions from satellite data, we are now able to study in great detail the large scale changes in the characteristics of several climate parameters in the Polar Oceans. The key instruments are the visible, infrared, and microwave sensors which provide monthly averages of cloud statistics, albedo, surface temperature, plankton concentration, and sea ice concentration. The spatial resolution of the different parameters are not the same but they were generated consistently and stored in a similar gridding format. The scheme is to enable the higher resolution data to be easily degraded to match those of the coarser resolution data for comparative analysis and large scale process studies that requires the use of multi-parameter data. With the unique global coverage that is provided by satellite sensors we are able to examine and identify the important physical changes in the polar environment that have occurred during the last 3 decades. The relatively long data record also allowed for the evaluation of the state of the Arctic and Antarctic climate systems. It is also enableed the evaluation of the possible impact of some phenomena such as the ENSO, SAM, AO and NAO on the various parameters in the polar oceans. The satellite coverage is timely because it coincided with very important changes that have been observed in the polar regions during last three decades.

The Arctic region is now regarded as a region in transformation because of results inferred from satellite data. Among the most remarkable change that has been observed is the rapidly shrinking perennial ice cover in the Arctic Basin (Comiso et al. 2008). Since the beginning of continuous satellite observation of the polar regions in the late 1970s, the ice cover at the end of the summer, called perennial ice, has declined by about 40% of its original area. The expected impact of ice-albedo feedback is currently being observed with the surface temperature in the Arctic region increasing at the rate of 0.7°C per decade which is about one-third the rate of increase globally for the same period. Negative changes in the albedo of the region are being directly observed as well following the decline of summer sea ice and snow cover. Large changes in the thickness and hence the volume have also been observed (Rothrock et al. 1999; Wadhams and Davis 2000; Tucker et al. 2001). Such a decline in thickness makes the perennial ice cover which has been declining at 12.5% per decade and the multiyear ice which is declining at 15% per decade, even more vulnerable. Multiyear ice is ice that has survived at least two summer melt seasons, and they are the thick component of the perennial ice and the mainstay of the Arctic sea ice cover. Buoy studies have shown that these ice types could stay in the Arctic for longer than 7 years. There is already some evidence that more of this type of ice is being advected out of the Arctic through Fram Strait causing a decline in the average age of sea ice in the region. A serious issue is whether the Arctic perennial ice cover has reached a tipping point. A sustained cooling would be needed to enable the perennial ice and some seasonal ice to gain enough thickness in the winter time so they can survive the summer melt. With ice-albedo feedback seemingly in operation and with the current warming trend, which is caused in part by the increasing anthropogenic greenhouse gases in the atmosphere, a reversal of the trend appears unlikely, at least in the immediate future. However, there are still many unknowns about the climate system and nature is full of surprises.

One of these surprises is going on in the Antarctic region. It is intriguing that in the Southern Hemisphere the extent of the sea ice cover is going in the opposite direction. The trend is about 1% per decade which is not large but statistically significant and cooling has been observed in many parts of the region. If there is a global warming, it is counter intuitive that one polar region is showing so much warming while the other polar region does not show it at all. This has been the subject of many research efforts and some results have provided new insights. There is reason to believe that the increases in ice extent may be caused in part by the ozone hole, which started only a few decades ago. Modeling studies indicate that the lows in the Western Antarctic region are amplified due to the ozone hole causing stronger winds to come out of the Ross Ice Shelf and the creation of more ice (Thompson and Solomon 2002; Turner et al. 2009). The coastal region of the Ross Ice Shelf is an ice factory and stronger winds would allow for the formation of more ice. The process leads to the creation of more bottom water that becomes part of the World's thermohaline circulation. While the adjacent seas, called the Bellingshausen/Amundsen Seas, have been losing their ice cover, the Ross Sea ice cover is increasing at a rate that is high enough to more than make up for the loss of sea ice in the former. There are other complications such as the presence of the Antarctic Circumpolar Wave, which appears to exhibit itself as either a Mode-2 or Mode-3 system. It was postulated that this wave circumnavigate the continent in about 8–9 years and for a Mode-2 system, it would revisit a certain area twice during that period. Analysis of 30 years of satellite data, however, reveals that the wave is quite unpredictable and sometimes it is a stationary wave as opposed to a propagating wave. There is also the complicating effects of the Southern Annular Mode that controls the atmospheric circulation pattern in the region and ENSO that can modify the circulation pattern.

The polar oceans are also the site of enhanced phytoplankton blooms in part because of the presence of sea ice that releases melt water in the spring and summer. The melt water becomes a stable platform in the upper ocean layer that is also exposed to abundant sunlight and hence an ideal site for photosynthesis and phytoplankton growth. The seas around the Arctic Basin are highly productive and have been among the richest fishing grounds. Good correlation of sea ice area with plankton concentration in spring is observed. With the study period being relatively short and the interannual variability being large, trend analysis results would have high uncertainties but the primary productivity appears to be increasing in the Arctic region and the trend is more positive in the Atlantic Ocean than in the Pacific Ocean. The pigment concentration and primary productivity also appear to have a good correlation with surface temperature in the region.

In the Southern Ocean, the spring polynyas around the continent appear to have the highest plankton concentrations in part because of the presence of meltwater. At around 60°S, the plankton concentrations are relatively low, likely caused by the lack of nutrients and iron. It is surprising, however, that toward the north at about 35°S, enhanced phytoplankton blooms are all over, forming a ring of highly productive area north of the Antarctic circumpolar current. The location is near the northern fronts that have been the sites of upwelling of relatively warm water and nutrients. Its proximity to land areas also makes the region iron-rich.

In this book, we have provided a detailed history of the changing climate of the polar oceans as observed by satellite sensors in the last three decades. The data show in good detail the spatial distributions of surface temperature, sea ice, albedo, chlorophyll concentration, clouds and sea level pressure. One needs to examine carefully some of the images to be able to appreciate how uniquely the surface parameters are being revealed by satellite sensors. A look at the time sequence of these images not just from month to month but also from 1 year to another also enables the realization that our planet Earth is changing at many times scales and that some of the trends observed need special attention. The satellite data provide us with a powerful tool for studying the polar regions and with the advent of new and even more advanced systems, they will undoubtedly provide us with additional insights into the current state of the climate system and how it is being impacted by human activities.

References

Abdalati W (2006) Recent changes in high-latitude glaciers, ice caps and ice sheets. Weather 61(4):95–101

Arblaster JM, MKeehl GA (2006) Contributions of external forcings to Southern Annular Mode. J Climate 19:2896–2905

Armstrong RL, Brodzik MJ (2001) Recent northern hemisphere snow extent: a comparison of data derived from visible and microwave satellite sensors. Geophys Res Lett 28(19): 3673–3676

Arendt AA, Echelmeyer KA, Harrison WD, Lingle CS, Valentine VB (2002) Rapid wastage of Alaska glaciers and their contributionb to rising sea level. Science 297:382–386

Broecker WS (1992) Global warming on trial. Natural History 4:6–14

Cavalieri DJ, Gloersen P, Parkinson C, Comiso J, Zwally HJ (1997) Observed hemispheric asymmetry in global sea ice changes. Science 278(7):1104–1106

Comiso JC (2000) Variability and trends in Antarctic surface temperatures from in situ and satellite infrared measurements. J Climate 13:1674–1696

Comiso JC, McClain C, Sullivan C, Ryan J, Leonard CL (1993) CZCS pigment concentrations in the Southern Ocean and their relationships to some geophysical parameters. J Geophys Res 98(C2):2419–2451

Comiso JC (2002) A rapidly declining Arctic perennial ice cover. Geophys Res Lett 29(20):1956, doi:10.1029/2002GL015650

Comiso JC, Parkinson CL, Gersten R, Stock LV (2008) Accelerated decline in the Arctic sea ice cover. Geophy Res Lett 35:L01703, doi:10.1029/2007GL031972

Comiso JC (2006) Arctic warming signals from satellite observations. Weather 61(3):70–76

Deser C, Walsh JE, Timlin MS (2000) Arctic sea ice variability in the context of recent atmospheric circulation trends. J Clim 13:617–633

Doran PT, Priscu JC, Lyons WB, Walsh JE, Fountain AG, McKnight DM, Moorhead DL, Virginia, RA, Wall DH, Clow GD, Fritsen CH, McKay CP, Parsons AN (2002) Antarctic climate cooling and terrestrial ecosystem response. Nature doi:10.1038/nature710

Frohlich C (2000) Observations of irradiance variations. Space Sci Rev 94:15-24

Hanna E (1996) The role of Antarctic sea ice in global climate change. Prog Phys Geogr 20(4):371–401

Hansen J, Ruedy R, Sato M, Imhoff M, Lawrence W, Easterling D, Peterson T, Karl T (2001) A closer look at United States and global surface temperature change. J Geophys Res 106: 23947–23963

Holland MM, Bitz CM (2003) Polar amplification of climate change in coupled models. Clim Dyn 23:221–232

Jacobs S (2006) Observations of change in the Southern Ocean. Phil Trans R Soc A, doi:10.1098/rsta.2006.1794

Johannessen OM (2008) Decreasing Arctic sea ice mirrors increasing CO2 on decadal time scale. Atmospheric and Oceanic Science Letters 1:51–56

Jones PD, New M, Parker DE, Marin S, Rigor IG (1999) Surface air temperature and its changes over the past 150 years. Rev Geophys 37:173–199

Karl TR, Trenberth KE (2003) Modern climate change. Science 302(5):1719–1723

Kwok R, Cunningham GF, Wensnahan M, Rigor I, Zwally HJ, Yi D (2009) Thinning and volume loss of the Arctic Ocean sea ice cover:2003–20089, J Geophys Res 114, C07005, doi:10.1029/2009JC005312

Miller RL, Schmidt GA, Shindell DT (2006) Forced annular variations in the 20th century Intergovernmental panel on climate change fourth assessment report models. J Geophys Res 111, D18101, doi:10.1029/2005JD006323

Mitchell BG, Brody E, Yeh EN, McClain C, Comiso JC, Maynard NC (1991) Meridional zonation of the Barents Sea ecosystem inferred from satellite remote sensing and in-situ bio-optical observations, Pro Mare symposium. Polar Res 10(1):147–162

Monaghan AJ, Bromwich DH, Chapman W, Comiso JC (2008) Recent variability and trends of Antarctic near-surface temperature. J Geophys Res 113, D04105, doi:10.1029/2007JD009094

Muller-Karger FE, McClain CR, Sambrotto RN, Ray GC (1990) A comparison of ship and coastal zone color scanner mapped distribution of phytoplankton the southereastern Bering Sea. J Geophys Res 95()97):11,483–11,500

Overland JE, Wang M (2005) The Arctic climate paradox: the recent decrease of the Arctic oscillation. Geophys Res Lett 32:L06701, doi:10.1029/2004GL021752

Platt T, Sathyendranath S, Longhurst A (1995) Remote sensing of primary production in the ocean: promise and fulfillment. Phil Trans R Soc Lond Series B 348:191–202

Romanosky VE, Osterkamp TE (2006) Interannual variations of the thermal regime of the active layer and near-surface permafrost in northern Alaska, Permafrost and Periglacial Processes, 6(4):313–335

Rothrock DA, Yu Y, Maykut GA (1999) Thinning of the Arctic sea-ice cover. Geophys Res Lett 26(23):3469–3472

Serreze MC, Holland MM, Stroeve J (2007) Perspective on the Arctic's shrinking sea-ice cover. Science 315:1533–1536

Shibata A, Murakami H, Comiso JC (2010) Anomalous Warming in the Arctic Ocean in the Summer of 2007, Japan J. Remote Sensing (in press)

Smith Jr WO, Keene NK, Comiso JC (1988) Potential interannual variability in primary productivity of the Antarctic Marginal Ice Zone. In: D. Sahrhage (ed) Antarctic Ocean and resources variability. Springer, New York, pp 131–139

Spencer RW (2008) Climate confusion, Encounter Books, New York, 191 pp

Stamerjohn, Smith (1997) – 10.2.2

Steig EJ, Schneider DP, Rutherford SD, Mann ME, Comiso JC, Shindell DT (2009) Warming of the Antarctic ice sheet surface since the 1957 International Geophysical Year. Nature 457:459–463

Sullivan CW, Arrigo KR, McClain CR, Comiso JC, Firestone J (1993) Distributions of phytoplankton blooms in the Southern Ocean. Science 262:1832–1837

Thompson DWJ, Solomon S (2002) Interpretation of recent southern hemisphere climate change. Science 296:895–899

Tucker III WB, Weatherly JW, Eppler DT, Farmer LD, BEntley DL (2001) Evidence for rapid thinning of sea ice in the western Arctic Ocean at the end of the 1980s. Geophys Res Letts 28:2851–2854

Turner J, Comiso JC, Marshall GJ, Connolley WM, Lachlan-Cope TA, Bracegirdle T, Wang Z, Meredith M, Maksym T (2009) Antarctic sea ice extent increases as a result of anthropogenic activity. Geophy Res Lett 36:L08502, doi:10.1029/2009GL037524

Vant MR, Gray RB, Ramseier RO, Makios V (1974) Dielectric properties of fresh and sea ice at 10 and 35 GHz. J Applied Physics 45(11):4712–4717

Walsh JE (2008) Climate of the Arctic Marine Environment. Ecological Applications 18(2):S3–S22

Yang J, Comiso JC, Krishfield R, Honjo S (2001) The role of synoptic storms in the development of the 1997 warming and freshening event in the Beaufort Sea. Geophys Res Lett 28(5): 799–802

Zwally HJ, Gloersen P (2008) Arctic sea ice surviving the summer melt interannual variability and decreasing trend. J Glaciol 54(185):279–296

Index